Enabling Tools and Techniques for Organic Synthesis

Enabling Tools and Techniques for Organic Synthesis

A Practical Guide to Experimentation, Automation, and Computation

Edited by

Stephen G. Newman
Department of Chemistry & Biomolecular Sciences
University of Ottawa
Ottawa, Canada

WILEY

Published by John Wiley & Sons, Inc., Hoboken, New Jersey.
Published simultaneously in Canada.

Library of Congress Cataloging-in-Publication Data Applied for

ISBN: 9781119855637, ePDF: 9781119855651, epub:9781119855644, oBook:9781119855668

Cover Design: Wiley
Cover Image: © Sergey Tarasov/Adobe Stock Photos

Set in 9.5/12.5pt STIXTwoText by Straive, Pondicherry, India

Contents

List of Contributors

Yosuke Ashikari
Department of Chemistry
Faculty of Science
Hokkaido University
Sapporo, Japan

Martin Breugst
Institut für Chemie
Technische Universität Chemnitz
Chemnitz, Germany

Sylvain Charvet
Institut de Chimie et Biochimie
Moléculaires et Supramoléculaires
(ICMBS, UMR 5246 du CNRS)
Université Lyon
Villeurbanne, France

Connor W. Coley
Department of Chemical Engineering
Massachusetts Institute of Technology
Cambridge, MA, USA

and

Department of Electrical
Engineering and Computer Science
Massachusetts Institute of Technology
Cambridge, MA, USA

Pablo Díaz-Kruik
Department of Chemistry
Biochemistry and Pharmaceutical
Sciences
Bern, Switzerland

Marion H. Emmert
Process Research & Development
Merck & Co., Inc.
Rahway, NJ, USA

Maurizio Fagnoni
PhotoGreen Laboratory
Department of Chemistry
University of Pavia
Pavia, Italy

Stephanie Felten
Process Research & Development
Merck & Co., Inc.
Rahway, NJ, USA

A. Filipa de Almeida
Instituto de Investigação do
Medicamento (iMed)
Faculdade de Farmácia
Universidade de Lisboa
Lisbon, Portugal

Laura Forfar
Paul Murray Catalysis Consulting Ltd
Yate, Bristol, UK

Wentao Guo
Department of Chemistry
University of California
Davis, CA, USA

Taline Kerackian
Institut de Chimie et Biochimie
Moléculaires et Supramoléculaires
(ICMBS, UMR 5246 du CNRS)
Université Lyon
Villeurbanne, France

Itai Levin
Synthetic Biology Center
Department of Biological Engineering
Massachusetts Institute of Technology
Cambridge, MA, USA

and

Department of Chemical Engineering
Massachusetts Institute of Technology
Cambridge, MA, USA

David Lim
Department of Chemistry
Biochemistry and Pharmaceutical
Sciences
Bern, Switzerland

Amy T. Merrill
Department of Chemistry
University of California
Davis, CA, USA

Paul Murray
Paul Murray Catalysis Consulting Ltd
Yate, Bristol, UK

Aiichiro Nagaki
Department of Chemistry
Faculty of Science
Hokkaido University
Sapporo, Japan

Francesca Paradisi
Department of Chemistry
Biochemistry and Pharmaceutical
Sciences
Bern, Switzerland

Stefano Protti
PhotoGreen Laboratory
Department of Chemistry
University of Pavia
Pavia, Italy

Davide Ravelli
PhotoGreen Laboratory
Department of Chemistry
University of Pavia
Pavia, Italy

Tiago Rodrigues
Instituto de Investigação do
Medicamento (iMed)
Faculdade de Farmácia
Universidade de Lisboa
Lisbon, Portugal

Camille Z. Rubel
Institut de Chimie et Biochimie
Moléculaires et Supramoléculaires
(ICMBS, UMR 5246 du CNRS)
Université Lyon
Villeurbanne, France

and

Department of Chemistry
The Scripps Research Institute
La Jolla, CA, USA

Michael Shevlin
Process Research & Development
Merck & Co., Inc.
Rahway, NJ, USA

Jason M. Stevens
Bristol Myers Squibb
Summit, NJ, USA

Dean J. Tantillo
Department of Chemistry
University of California
Davis, CA, USA

Zhengkai Tu
Computational Science and
Engineering
Massachusetts Institute of Technology
Cambridge, MA, USA

Julien C. Vantourout
Institut de Chimie et Biochimie
Moléculaires et Supramoléculaires
(ICMBS, UMR 5246 du CNRS)
Université Lyon
Villeurbanne, France

Preface

At the undergraduate level, the organic chemistry curriculum at most universities is similar. Professors emphasize the fundamental concepts necessary to understand how, when, and why organic molecules interact, while lab instructors familiarize students with important hands-on aspects of carrying out experiments. Bachelor's students can expect to finish their studies with an idea of how molecules behave, how they are made, and how technologies such as NMR and IR can be used for their characterization. Those that enter graduate school are often surprised at the breadth of powerful technologies that make advanced organic chemistry the discipline it is today. Instead of a typical stirred round-bottomed flask, many reactions are better done using photochemical, electrochemical, or flow reactors. Computational chemistry, once reserved for dedicated experts, can now be used by organic chemists to help predict outcomes, understand selectivity, and decipher reaction mechanisms. Automation technology can be used to generate large amounts of data with limited amounts of material, and data processing software can be used to extract subtle trends.

Due to their prominence in the recent literature, trainees and established chemists alike would benefit from gaining expertise with these technologies to be best prepared for solving the diverse synthetic challenges that come their way. However, the barrier to learning techniques without formal instruction can be high. Even if one is fortunate enough to have access to advanced training, the expert instructor may not necessarily curate the course to the needs and the background of a synthetic chemist. The primary literature and recent textbooks have similar limitations – while there is no shortage of resources, experts generally write to other experts, and the interested organic chemist may have critical gaps in their understanding and struggle with subdiscipline-specific jargon.

The goal of this text is to help fill this gap by providing synthetic chemists with a user-friendly starting point to initiate their journey in developing new skills and knowledge. In each of the 11 chapters, experts communicate basic information about an impactful technology in a manner accessible to a classically trained synthetic chemist. Chapters also includes a glossary of common terminology, a general introduction to the technology of interest, case-study examples of how it may useful to synthetic chemists, a practical discussion about steps one may take to put knowledge into practice, and references to recommended further reading. The book seeks to be a go-to resource for organic chemists at or above the graduate level that wish to expand the breadth of tools they can use to perform, analyze, and interpret chemistry experiments. After completion, the reader will be armed with the practical knowledge needed to comprehend the literature, to assess the strengths and limitations of each technique, and to begin applying modern tools to solve synthetic challenges. This will make it useful as a general resource for graduate students looking to expand their expertise, for instructors of graduate-level courses on advanced techniques for organic synthesis, and for industrial scientists seeking a beginner-friendly way to expand their knowledge.

The book is organized into four subsections. Chapters 1–4 describe different enabling technologies for performing chemical experiments – biocatalysis, photochemistry, electrochemistry, and flow chemistry. While none of these topics are fundamentally new, their power as a tool for organic synthesis is becoming increasingly evident. These chapters will help the reader overcome the technical barrier hindering them from comfortably replicating experiments and designing their own. Chapters 5 and 6 focus on improved approaches to select, carry out, and analyze experiments. Specifically, Chapter 5 describes a statistical approach to experimentation that can be used to understand and optimize chemical reactions. This Design of Experiments (DoE) technique is commonly employed by practicing scientists in many fields but is seldom taught to chemists. Chapter 6 describes techniques that researchers can use to get more data using less time and fewer resources. This high-throughput experimentation (HTE) approach shows the reader how to carry out reactions in parallel and how the collected data can be interpreted to gain insights that might otherwise be missed. Chapters 7 and 8 introduce the reader to computational chemistry tools that enable molecules and reactions to be modeled *in silico*, providing predictions and mechanistic insight to supplement experimentation. Chapter 7 provides a general overview of the most common computational tasks that an organic chemist may want to carry out and walks the reader through a beginner-friendly case study wherein the reactants, transition states, and products of a Diels–Alder reaction are calculated. Chapter 8 builds upon the general knowledge given in the previous chapter and describes how computational chemistry can be used to predict the NMR spectrum of organic molecules. The goal of this chapter is to put this powerful technique into the hands of experimental chemists, which should be achievable after familiarizing themselves with the simplified approach detailed throughout. Chapters 9–11 provide the reader with an introduction to programming and machine learning. Computers already play a critical role in the daily life of a synthetic chemist, and a little bit of familiarity with modern techniques can go a long way. Chapter 9 provides a blueprint for understanding how and why a chemist may go about familiarizing themselves with programming. Chapter 10 describes a deep dive case study for using machine learning to facilitate reaction optimization, providing a step-by-step guide that a beginner may follow to use the tool and to gain confidence in harnessing other published algorithms. Chapter 11 explains how computers can facilitate the planning of multistep synthesis by suggesting synthetic routes and reaction conditions. Helpful discussions on the current tools available, how they work, and their associated strengths and weaknesses are also described.

This project was only possible due to an immense amount of work by the authors who generously agreed to share their knowledge and meet the formidable task of communicating with a general audience. I am also indebted to the many students and postdoctoral fellows at the University of Ottawa that served as reviewers to help ensure that the content serves as a welcoming and beginner-friendly introduction to these topics that are becoming increasingly important to the modern synthetic chemists. I hope the readers agree that this goal has been met and that this marks the beginning of their journey to being a more well-rounded scientist capable of tackling diverse problems that come their way.

Stephen G. Newman
Ottawa

July 2023

1

Biocatalysis 101 – A Chemist's Guide to Starting Biocatalysis

Pablo Díaz-Kruik, David Lim, and Francesca Paradisi

Department of Chemistry, Biochemistry and Pharmaceutical Sciences, Bern, Switzerland

Glossary

API Active pharmaceutical ingredient
BRENDA A comprehensive enzyme information system
CALB Lipase B from *Candida antarctica*
Cofactor A non-protein chemical compound or metallic ion that is required for an enzyme's role as a catalyst
DKR Dynamic kinetic resolution
IRED Imine reductase
NAD^+/NADH Nicotinamide adenine dinucleotide
Protein expression Biological process where the protein is synthesized inside a cell
Recombinant DNA DNA scaffold that contains the protein sequence of interest
TRIS Tris(hydroxymethyl)aminomethane

1.1 Introduction

1.1.1 Enzymes – the Green and Sustainable Way of the Future

Recent efforts by chemists to actively reduce toxic waste production and minimize costs have led to the discovery of many green and sustainable technologies. Not surprisingly, the use of enzymes, Nature's catalysts, has seen a major resurgence in academic and industrial interest over the past decade – not only for their sustainability and natural activities but for engineering them to perform novel transformations beyond capabilities observed in a synthetic organic lab [1, 2].

Enabling Tools and Techniques for Organic Synthesis: A Practical Guide to Experimentation, Automation, and Computation, First Edition. Edited by Stephen G. Newman.

The attractiveness of using enzymes for transformations stems from their exquisite regio- and stereoselectivities – something that traditional chemists still struggle to achieve in the lab – that enzymes often execute effortlessly. Moreover, we have seen the emergence of multienzyme cascades for the synthesis of active pharmaceutical ingredients (APIs). A recent landmark example involves the synthesis of molnupiravir (MK-4482), an orally dosed ribonucleoside analogue and inhibitor of influenza viruses, which has demonstrated activity against COVID-19 when administered in animal models [3, 4]. In this work, McIntosh et al. developed a scalable three-step route toward MK-4482 [5]. Using a cascade of five enzymes, MK-4482 could be accessed from 5-isobutyrylribose (Figure 1.1).

To the uninitiated, entering the world of enzyme-catalyzed chemical transformations can be incredibly daunting, especially when one is not equipped with a foundational understanding of what an enzyme is and how these macromolecules work. However, you may be surprised to hear that enzymology and chemistry are not too different from each other at all! With an undergraduate chemistry background, a chemist can easily harness the power of enzymes to perform desired transformations – a *fact* that we aim to convince you of over the next few pages.

However, while this chapter aims to illustrate the power of enzymes for novel and sustainable transformations, we do not want to inadvertently imply the use of these macromolecules is the be-all-end-all solution – sometimes the use of traditional organic synthesis to access target molecules is the more logical solution. Therefore, when an enzyme *might* be used is a weighted question often involving the combination of various intricate factors, including efficiency and cost.

Over the following sections, we will do our best to educate you on these factors so that you can begin making an informed decision on this matter. We also aim to

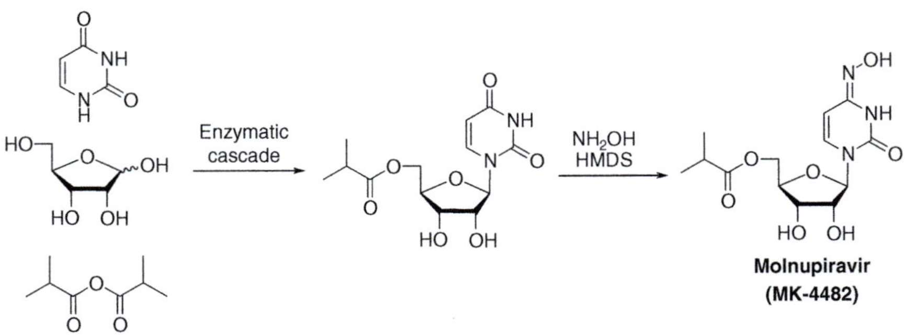

Figure 1.1 A combined enzymatic cascade/hydroxylamination for the synthesis of molnupiravir (MK-4482).

convince the reader that the use of enzymes is not limited to biologists and biochemists but also readily available for use by synthetic chemists. With the following breakdown of important considerations to make when using an enzyme, we hope to instill confidence in the reader that a biological catalyst is not too dissimilar to a chemical catalyst and can be readily obtainable from common suppliers.

We will also dispel common misconceptions and myths surrounding the use of enzymes and then give an overview of several classes of reactions that can be performed with enzymes, including recent developments into more exotic transformations such as photobiocatalysis.

This chapter will then conclude with a snippet into recent trends and technologies that have harnessed the use of enzymes in novel ways. We hope that the information gained from reading this chapter will provide a strong foundation for the reader to develop confidence in the use of enzymes and begin their venture into the world of biocatalysis.

1.1.2 Enzymatic and Organic Catalysis Are Not too Different from Each Other

A seasoned chemist may be quite familiar with several stereoselective reactions whereby stereocontrol is dictated by the chiral environment of the reaction. For example, one model for the Corey–Bakshi–Shibata (CBS) reduction involves coordination of the respective carbonyl to the CBS catalyst in a specific spatial orientation, leading to stereoselective reduction of the carbonyl to the corresponding alcohol (Figure 1.2) [6]. Enzymes utilize a very similar concept to this reaction – the catalyst (enzyme) places a reactant (substrate) in a chiral environment (the active site), whereby stereoselectivity is dictated by the local reactive environment, leading to a selective reaction outcome. In the next subsection, we will look at how an enzyme achieves these feats.

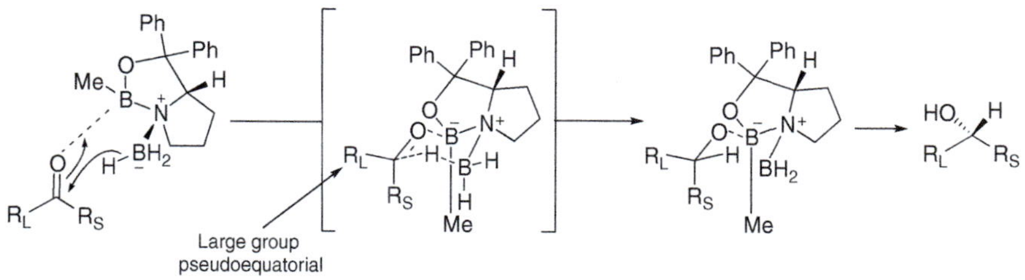

Figure 1.2 The CBS reaction has been used in undergraduate texts as a classic example of where an achiral reactant is stereoselectively transformed in a chiral environment to the corresponding product in high enantiomeric excesses.

1.1.3 Enzymes 101

Enzymes are known to accelerate reactions by more than 10^{17}-fold [7]. How an enzyme achieves these colossal rate increases under aqueous conditions requires an understanding of the active site architecture in great molecular detail.

There are 20 essential amino acids found in nature. Enzymes are formed by cellular machinery, which stitch together combinations of these amino acids in a genetically pre-defined sequence, making one very long polymer. This polymer is folded to give a precise three-dimensional structure (Figure 1.3). The active site is defined as the region of the enzyme where substrates bind and undergo catalysis. The catalytic cycle begins with the binding of the substrate in the active site. This process precisely positions all molecules involved in the catalysis (metals, solvents, cofactors, etc.) in their respective orientations ready to achieve regio- and stereoselectivity. Subsequent activation of the substrate initiates the reaction, generating a transition state, which is stabilized by interactions with the active site residues of the enzyme. Following effective conversion of the substrate, the product is then released from the active site of the enzyme, completing one turnover and returning the catalyst back to its original state.

1.2 When Should I Choose an Enzyme over a Chemical Catalyst?

The choice of using a chemical catalyst over a biochemical solution needs to be assessed on a case-by-case basis, often involving a detailed cost–benefit analysis. For example, chemical asymmetric imine reduction often requires the use of expensive precious metals, such as Ir, Rh, Ru, and Pd (Figure 1.4) [8]. While recent methods have moved toward Earth-abundant solutions, such as employing iron or nickel, all these still require decoration with expensive chiral ligands that cannot be recycled [8], making the overall synthesis very environmentally and economically demanding.

In contrast, imine reductases (IREDs) can perform stereoselective reductions without the use of expensive metals and can be performed under aqueous conditions mitigating the need for organic solvents. Since the initial report of IREDs in 2010 [9], many advancements have been made to use these enzymes for novel synthetic transformations [9, 10]. In fact, Matzel et al. published an elegant procedure for performing biocatalytic dynamic kinetic resolutions (DKRs) of aldehydes using IREDs (Figure 1.5). This method exploits the stereo-preference of the enzyme for either the *R*- or *S*-chiral center [11].

The use of enzymes in this case showcases the re-opening of the chemical window, enabling unprecedented reaction conditions, merging asymmetric reduction

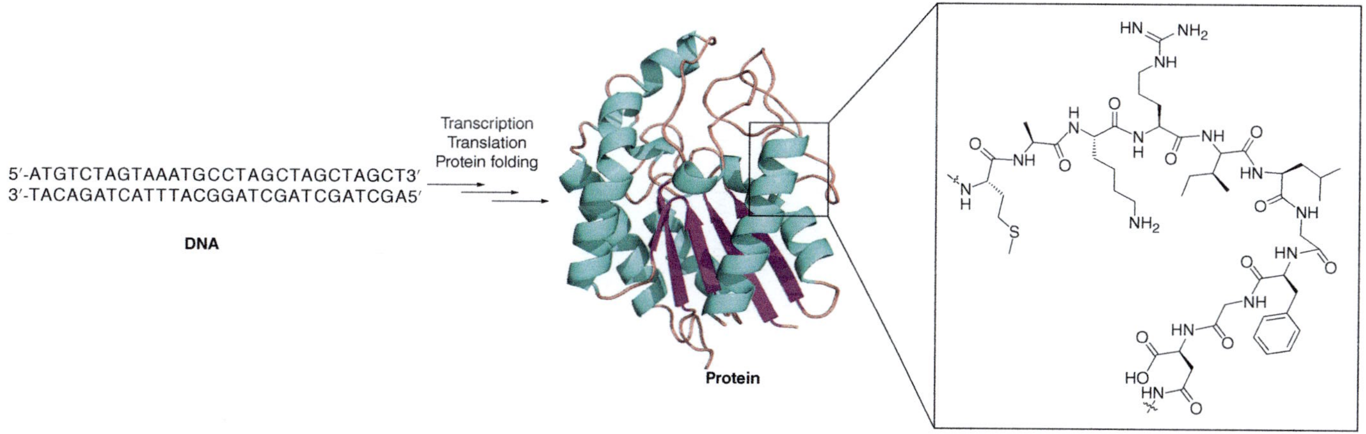

Figure 1.3 Proteins are formed by intracellular machinery that uses genetic information (DNA) to form a polymeric amino acid chain, which is then precisely folded to give a protein.

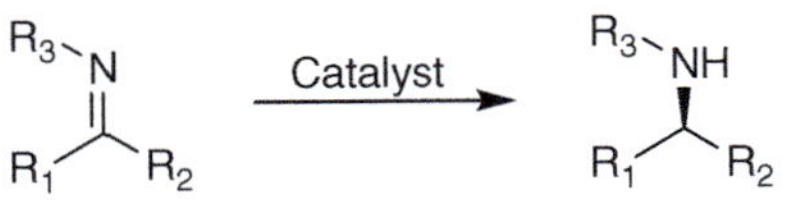

Figure 1.4 Asymmetric reduction of imines to amines in the presence of a chiral catalyst.

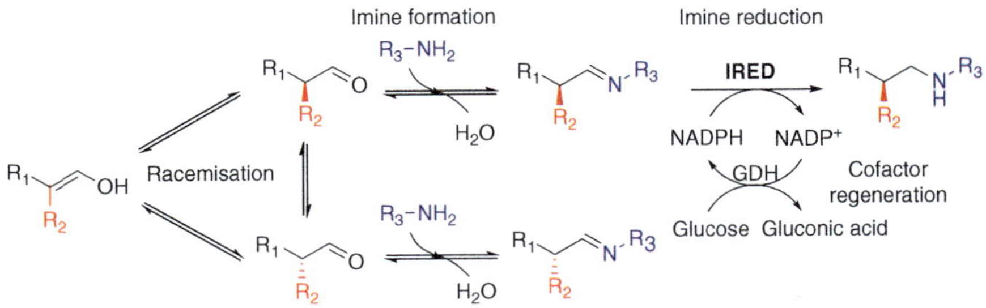

Figure 1.5 Biocatalytic dynamic kinetic resolutions of aldehydes using imine reductases. GDH = glucose dehydrogenase.

and water media. This would be near impossible to achieve with classical reducing agents, such as $NaBH_4$ or $Na(CH_3COO)_3BH$.

Next, we move the reader on to build an understanding of the variables associated with using a biocatalyst and here, they will gain foundational knowledge on how to gauge whether a chemical or biochemical solution is appropriate for solving a target problem.

1.3 Key Considerations for Running Biocatalytic Reactions

With a primary level of appreciation of the power of enzymes, we can now continue our journey by addressing the variables that define a biocatalytic reaction. We will also illustrate how these variables change depending on the system that is being applied for the reaction. Before we advance in that direction, we will first begin by dissipating common myths surrounding the use of enzymes.

1.3.1 Dispelling Myths

The uptake of biocatalysis in academic and industrial applications has increased significantly in recent years [12, 13]. Despite the positive perception of the technology, the breadth of applications remains rather modest. A factor that contributes to this lack of progress may be associated with the perceived notion of the limitations of biocatalysts – their availability, cost, ease of use, substrate scope, and operational stability. We aim to address these factors in the following subsections.

1.3.1.1 Enzymes Are Not Safe to Use

Enzymes are very safe to use. In fact, proteins (as well as their essential metals and cofactors) are biodegradable and therefore considered to be environmentally friendly and can eliminate the need to use of noble metals (e.g. Pt, Pd, or Rh), which can complicate purification processes and are often subject to unforeseeable price variations. However, one must be careful not to inhale enzymes as they can illicit an immune response from susceptible individuals. Incidents such as these can be avoided by performing manipulations of reaction components under a fume hood wearing the appropriate protective wear.

1.3.1.2 Enzymes Are Not as Readily Available as Chemical Catalysts

One concern revolves around the availability of active biocatalysts. The acquisition of viable enzymes has come a long way. It used to be common for chemists to perform exhaustive screening of a broad variety of microorganisms, ranging from bacterial to fungal systems, to acquire a desired enzyme. Significant advancements to the methodology in which enzymes are manufactured has allowed them to be accessed in large quantities that may be readily available [14]. Global efforts have allowed the modification and enhancement of enzymatic activities with site-directed mutagenesis. This progress has been coupled with significant improvements in genomics and bioinformatics data, further allowing the rational and directional engineering of enzymes for unique activities.

1.3.1.3 Enzymes Are Seldom Useful Due to Their Limited Substrate Scope

Another misconception, which discourages the use of enzymes, is the myth that any given enzyme is only capable of performing a reaction with one specific substrate, whereby a different enzyme is required for each different substrate. While it is true that most enzymes have a strict substrate specificity, many can accept very different substrates and are not limited to their natural starting materials. Furthermore, significant advances in protein engineering techniques, such as directed evolution, have allowed biochemists to tailor the active site of an enzyme to accommodate non-native substrates, allowing a broader range of novel biotransformations [12]. For example, Contente et al. demonstrated that a single point mutation in the active site of an acyl transferase in which a serine was substituted with a cysteine allowed the formation of thioesters, an observation that had not been previously reported with an acyltransferase [15]. Further examples include the work reported by Xu et al., where the conversion of an FAD-dependent Baeyer–Villiger monoxygenase into a ketoreductase was achieved [16]. Moreover, Crawshaw et al. recently reported the engineering of a Morita–Baylis–Hillman-ase to accept unnatural molecules as substrates, converting 2-cyclohexen-1-one and 4-nitrobenzaldehyde into 2-(hydroxy(4-nitrophenyl)methyl) cyclohex-2-en-1-one [17].

1.3.1.4 The Cost of Enzyme Production Is Very High

The use of biocatalysts in industry is often curbed by the myth of high production costs potentially associated with enzyme production. However, enzymes can often be purchased in different forms (see Section 1.3.2.7) each varying in price range. As depicted in Figure 1.6, producing 1 kg of unpurified enzyme ranges between €100 and €250, whereas the production of whole cells containing high levels of the desired enzyme is significantly cheaper at €35–€100 kg^{-1} [18]. It is important to also note that the price trend of purified enzyme formulations (e.g. FDH-101 from Codexis or the immobilized lipase B) starts to skyrocket rapidly, approaching close to the prices of precious metals. However, this trend can also be seen in the case of metal catalysts.

Moreover, even if the price of the metal used is insignificant, these often need to be further decorated with organic ligands which significantly increases the overall price. Case-in-point, Crabtree's catalyst, which contains an iridium center, costs US$ 1 091 800 kg^{-1} of catalyst – the price of the adorned catalyst being *six times* higher than the metal alone. In addition, the prices of the metal cores are subjected to fluctuations depending on the stock market and their abundance in the earth's crust.

While the decision to choose a chemical or a biological solution may sound confusing, we can advise the reader to follow a simple checklist to assist in their choice of catalyst below.

1) Compare the performance of both chemical and biocatalyst reported in literature and choose the best option for the desired reaction.

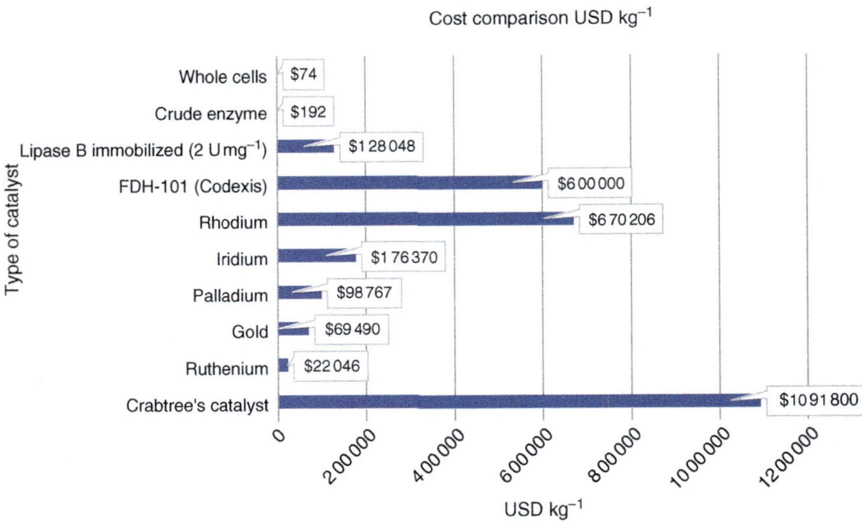

Figure 1.6 Cost comparison of chemical and biological catalysts in USD kg^{-1} of catalyst. Spot market @ 11 March 2022.

2) Evaluate if either catalyst offers a significant benefit in, for example, selectivity, price, waste, pollution, etc.
3) If the chosen option is the biocatalyst, carefully read the literature to understand the physical form of the biocatalyst (see Section 1.3.2.7). In some cases, use of the enzyme in one or another form is preferred, for example, often when using whole cells the product accumulates inside the cell, or when using crude extracts one can have cross reactivity with the cell metabolites.
4) The most common choice for preferred enzyme form are purified enzymes. However, in some cases, the use of this enzyme form can be excessive since the reaction might also be efficiently performed with crude extracts or whole cells, thus reducing significantly the cost of the catalyst.

1.3.1.5 Enzymes Are Functionally Unstable Under Organic Conditions

The functional stability of biocatalysts is one of the principal parameters to address in the design of a viable enzyme and has steadily improved over the past few years. Such developments are crucial, especially if the reaction conditions stray from aqueous conditions, such as in the use of high ratios of organic co-solvents. Genetic engineering has allowed the design of highly active enzymes which operate under conditions of high temperatures, non-buffer salts, solid-state buffers, crown ethers, and cyclodextrins [19].

Furthermore, several remedies can be taken to boost the performance and stability of the biocatalyst, such as entrapment or immobilization onto a solid support (see Section 1.5.2).

1.3.1.6 Sustainability

Biocatalysis has been described many times as the Holy grail and perfect solution for reducing the waste and contaminants produced in industry – but is this always the case? The answer is definitely no. Even if most of the biocatalytic transformations are more environmentally friendly than classical chemical synthesis, sometimes it is better to stick to the chemical process. For example, we would not recommend that an enzymatic process is used to access multi-kilo quantities of product since some biocatalytic processes are limited to very low substrate concentrations (μM–mM). However, while making this point, we truly believe that the future of chemical manufacturing will make chemoenzymatic approaches feasible, where the advantages of organic synthesis and biocatalysis can be merged [20].

1.3.2 Challenges of Using Enzymes: the Need for Strict Reaction Conditions

Now that we have addressed some common myths, let us look at important considerations needed when addressing an enzymatic reaction. While procedures for biocatalytic reactions are well documented, the identical replication of a published procedure gives no guarantee of a similar result. This observation is often due to the multitude of factors that contribute to the biocatalytic transformation in question, including that of pH, ionic strength, and temperature of the reaction media, to name a few. Moreover, we should not lose sight that enzymes are naturally evolved biomacromolecules coming from, and performing reactions inside, living cells and we, as chemists, are essentially "hijacking" this natural process to carry out novel synthetic transformations. Therefore, we need to make careful considerations to ensure the enzyme is fully functional to perform catalysis in the chemistry lab. Here, we outline a few key conditions to be aware of when designing your biocatalytic assay.

1.3.2.1 Enzymes from Extremophiles

An understanding on the bacterial origins of any enzyme is key in exploiting the full potential of its reactivity. For example, enzymes from thermophilic organisms are endowed with structure–function properties of high thermostability and optimal activity at these high temperature ranges while having poor functionality at moderate temperatures below 40 °C [21]. Psychrophiles, on the other hand, produce enzymes that may only function at cold temperatures (15 °C or lower). Halophiles thrive in environments of high salt concentrations and therefore their enzymes may require such conditions when used in biocatalytic applications. Finally, acidophiles grow in environments of high acidity, at pH 2.0 or below.

1.3.2.2 Solvents (and Co-solvents)

A hallmark feature on the use of enzymes is the ability to perform reactions in water. However, in some cases, the aid of a co-solvent, such as ethanol, acetone, DMSO, or acetonitrile, is necessary. Not only does the addition of a co-solvent assist in the solubility of nonpolar substrates under aqueous conditions, it can also enhance the activity and stability of enzymes, aid in product recovery, or even shift the equilibrium toward new reactions [22–24].

With significant advances made in the understanding of protein structure and dynamics in the past few decades, there are many reports on the engineering of proteins for nonaqueous biocatalysis [25]. Furthermore, with the significant leaps made in computational methodologies, more options are becoming available for the prediction and design of enzymes, which work better under a set of reaction conditions. For example, Cui et al. recently published a guide on using molecular dynamics (MD) simulations to assist in the rational design of proteins under organic co-solvent conditions [26].

1.3.2.3 Concentration and Ionic Strength of the Buffer

Enzymes typically only function in aqueous environments at a specific pH, the latter often being dictated by the natural environment of the organism from which the enzyme originated. Consequently, buffers are used to adjust and stabilize the pH of the solution during an enzyme reaction. While the design of buffer systems is beyond the scope of this chapter, the ionic strength and concentration, as well as the choice of buffer components will be addressed here.

Salt concentration can have dramatic effects on the binding and turnover of a substrate by an enzyme. Moreover, the choice of ion (i.e. Na^+ vs. K^+, Cl^- vs. Br) can have a significant effect on the ionic strength of the buffer, as well as enzymatic activity. Case-in-point are enzymes that require either a mono- or divalent cation for catalysis [27]. Supplementation of the buffer with the wrong ion can lead to a slow or inactive enzyme. Some buffers, for example those containing TRIS (tris(hydroxymethyl)aminomethane) or even the commonly used phosphate buffer, can also have destabilizing effects on protein structure [28]. Therefore, it may be useful to dedicate a small amount of time investigating previously reported conditions for a particular enzyme in the literature to determine what salts are compatible.

The buffer concentration dictates the capacity of a buffer to stabilize the pH, with higher concentrations having higher buffering capacity. This is typically 0.05–0.2 M for most enzymes; however, halophilic and thermophilic enzymes are known to prefer higher concentrations of up to 1 M. Effects of substrates and additives (whether these are sold in their salt forms) on the ionic strength of the buffer must also be considered.

1.3.2.4 pH Dependence

The activity of an enzyme is strictly dependent on the pH of the reaction, which can be explained by a couple of factors; first, the state of protonation of the functional groups of amino acids and cofactors involved in the reaction will invariably change with change in acidity. This effect could have drastic consequences to catalytic activity, especially if the residues directly involved in catalysis are dependent on specific protonation states. While the pH optima of many enzymes are found near physiological pH, there are some that deviate from this rule – for example, the stomach protease, pepsin, works with a pH optimum of 2, while alkaline phosphatase has a pH optimum of 10.5. Second, variations in pH can lead to irreversible damage to the three-dimensional structure of the protein. Therefore, assays of an enzyme to find the pH optima may be necessary and significant deviations from the pH optimum should be avoided.

1.3.2.5 Concentration of Reactants

Reactant concentration is a key parameter that needs to be carefully analyzed while planning a biocatalytic reaction. The first obstacle that one can encounter is the solubility of the substrate; many organic compounds are not soluble at high concentrations in the reaction media (generally water). To tackle this issue, co-solvents can be added.

One also needs to keep in mind that the inherent nature of enzymes is to work inside cells where the substrate concentrations are generally very low compared to the protein [29]. Therefore, increasing the substrate concentration may lead to initial saturation of the enzyme activity, followed by subsequent inhibition.

1.3.2.6 Enzyme Concentration

In a manner similar to using a chemical catalyst, the amount of enzyme to add to a reaction is a balance between achieving the highest reaction conversion in an appropriate amount of time – this differs from enzyme to enzyme. However, if an enzyme is commercially available, then it is likely that a procedure is already available. In such a case, one just needs to emulate the published reaction conditions using the reported values as a starting point for determining the amount of enzyme to add.

1.3.2.7 Enzyme Forms

Enzymes can be stored/sold in purified form (only the desired enzyme is present) or as a crude extract (where the desired enzyme is present along with other proteins and cell metabolites released when the enzymes are extracted out of the cell). More commonly, enzymes are stored in solution since they come from microorganisms and chemistry in nature is mostly performed in solution. However, in the lab they can be prepared/formulated in different ways depending on the desired purpose (Figure 1.7).

Whole cells – most enzymes are first synthesized within a host cell using the RNA/DNA machinery and then lyophilized to give lyophilized cells containing the desired enzyme. These cells can also be engineered to produce particularly large amounts of a given enzyme and may be harvested and used directly for catalysis. This method is useful when the biocatalyst is unstable under extracellular conditions. From an economical point of view, this form is very attractive considering that isolation and purification of the biocatalyst is avoided.

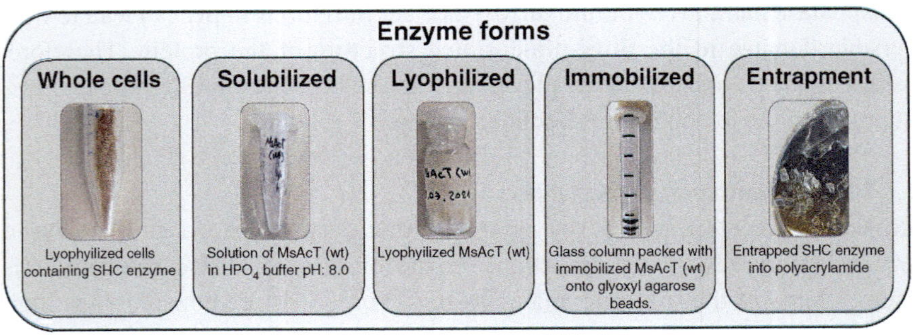

Figure 1.7 Forms that enzymes can be stored/sold in.

Solubilized – the most common way of using enzymes *in vitro* involves storing them in an appropriate buffer. In this case, no additional treatment of the enzyme is required after purification (cf. whole cells or lyophilized enzyme, which need rehydration).

Lyophilized – this technique is often employed when long-term storage of the enzyme is required. However, it is important to note that some enzymes may suffer from deactivation since water may be required to stabilize their quaternary structures. This method can be applied to either purified or crude extracts.

Immobilized – immobilization of an enzyme onto a solid support is particularly advantageous when enzyme recovery and re-use is desired, i.e. flow biocatalysis. This method is applicable to purified enzymes and whole cells.

Entrapment – entrapment of an enzyme into a polymeric matrix is a milder immobilization technique, since in this case, the enzyme does not suffer conformational changes due to chemical bonding to the support.

1.3.2.8 Toxicity

When addressing the toxicity of enzymes, one must distinguish between pure enzymes and whole cell biocatalysts. While, to the best of our knowledge, pure enzymes are not toxic (aside from some reported hypersensitivity cases in enzyme replacement therapies [30]), using whole cells can pose risks due to the potential toxicity of cell metabolites. Even the cell wall fragments can present pathogenic roles [31]. In addition, the recombinant DNA of these type of cells encodes an antibiotic resistance gene that can be transformed or eventually conjugated between bacteria, leading to populations of antibiotic-resistant bacteria [32].

No special considerations need to be taken into account when working with regular *Escherichia coli* cells (most of the enzymes are synthesized in *E. coli*), apart from the standard personal protective equipment used in a regular chemistry lab (lab coat, safety goggles, and gloves). Even if some variants of *E. coli* are completely innocuous, as are the species commonly found in the large intestine of humans [33], others may contain virulence factors. Therefore, one needs to treat this situation similar to organic synthesis, whereby a newly synthesized compound is treated with utmost care due to no available safety data for that molecule.

1.3.3 What Do I Need to Start Biocatalytic Experiments in My Lab?

The running of a biocatalytic experiment in a chemistry lab does not require many unfamiliar pieces of equipment (Figure 1.8), with perhaps the most exotic piece of equipment needed being a spectrophotometer (not compulsory but highly recommended, see Section 1.6). In a similar manner to an organic wet lab, a protocol needs to be written down, conditions thoroughly considered, and reagents and

Equipment list
1. Set of micro-pipettes (1–20, 20–200, 100–1000) μL
2. Round-bottomed flasks
3. Bench-top spectrophotometer
4. Set of centrifuge tubes (0.5, 1.5 and 2.0) mL
5. TLC plates
6. Hot plate magnetic stirrer

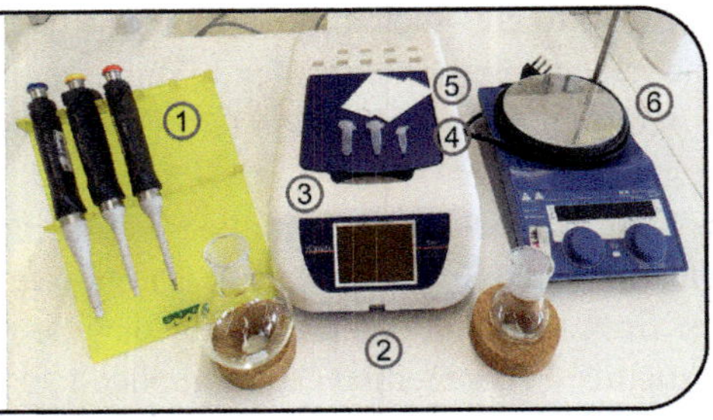

Figure 1.8 Standard pieces of equipment needed for running a biocatalytic reaction can be found in most organic chemistry wet lab setups.

equipment pre-prepared before mixing. We point the reader toward a detailed outline by Bisswanger [34], where practical considerations when starting a biocatalytic reaction are neatly summarized.

1.3.4 Additional Considerations

Upon storage, enzymes can suffer from aggregation, precipitation, and other structural changes that may affect their activity. This parameter is key in every enzymatic reaction; it not only tells us about the performance of the enzyme towards a specific reaction, but also sets a baseline for the maximal amount of activity expected from that specific batch of the enzyme over time. Let us observe the following example: a chemist is attempting a transesterification with the commercially available lipase, CalB, that comes lyophilized. Prior to setting up the reaction, the chemist makes a stock solution of 1 mg mL^{-1} enzyme in phosphate buffer at pH 8.0. After running the first set of reactions, they obtain an average of 90% conversion. Repetition of same transformations by the same chemist after two weeks with the same stock solution still furnishes the same product, but surprisingly, no greater than 40% conversion can be obtained.

How is it possible to obtain such a dissimilar result? Did the chemist do something wrong? The answer is: probably not. Many enzymes are not stable over time, meaning that they gradually lose activity. Therefore, as a rule in biocatalysis, prior to each reaction, one must check the initial activity of the enzyme toward a known substrate. In this way, a baseline level of activity is set and, if now after a certain time the chemist wants to retry the reaction with the same enzyme, can have an idea of how much activity was lost during storage. The specific activity of an enzyme can be garnered by the enzyme unit (U), where 1 U is defined as the amount of enzyme required to convert 1 μmol of substrate per minute under the specified conditions [35]. If purchased commercially, the amount of enzymes which constitutes to 1 U is provided on the commercial bottle in units of U mg^{-1};

however, if you have a non-commercially available or non-reported enzyme, you may need to create an assay that allows you to define 1 U for that particular enzyme.

The use of the unit allows a quantification of the amount of an enzyme with respect to its function rather than its mass. While the use of the enzyme unit is not presently valid according to the SI system (the unit was replaced with the *katal* in 1978 [36]), most commercial suppliers of enzymes continue to market enzyme preparations in enzyme units.

As in any other chemical transformation, a reaction work-up is usually necessary in order to separate the products from the biocatalyst. To do so, liquid–liquid extractions in the case of soluble enzymes are generally effective. In the case of supported/immobilized enzymes, an additional filtration may be required.

1.4 Transformations Catalyzed by Enzymes

In this section, we will glance over how enzymes are categorized into various classes and provide some examples of commercially available enzymes that have been used to perform biotransformations in standard chemistry laboratories. This section aims to provide the reader of a scope of possible reaction types they may want to conduct.

1.4.1 EC – The Enzyme Commission Number

The classification of enzymes follows the Enzyme Commission (EC) number, which categorizes each protein according to the type of reaction the enzyme catalyzes. An EC number is composed of four digits in the form, EC x.x.x.x. The first number denotes the class of the enzyme. For example, one class of enzymes is called "oxoreductases." All oxoreductases will have an EC classification of the format, EC 1.x.x.x. The second number gives the subclass, with the third number further dividing the subclass into the sub-subclass. Finally, the fourth number represents the serial number of the enzyme in its sub-subclass. As of 2022, there are seven EC levels: the oxoreductases, transferases, hydrolases, lyases, isomerases, ligases, and translocases (Figure 1.9).

1.4.1.1 EC 1 – Oxoreductases

Oxoreductases (also known as oxireductases) catalyze the transfer of electrons from an electron donor (reductant) to an electron acceptor (oxidant), often using cofactors (flavin, heme, and other metal ions). One important class of oxoreductases – the dehydrogenases – catalyze the oxidation of primary or secondary alcohols to the respective carbonyl product (Figure 1.10) [37]. Oxoreductases make up one-third of all proteins classified in the BRENDA database [38].

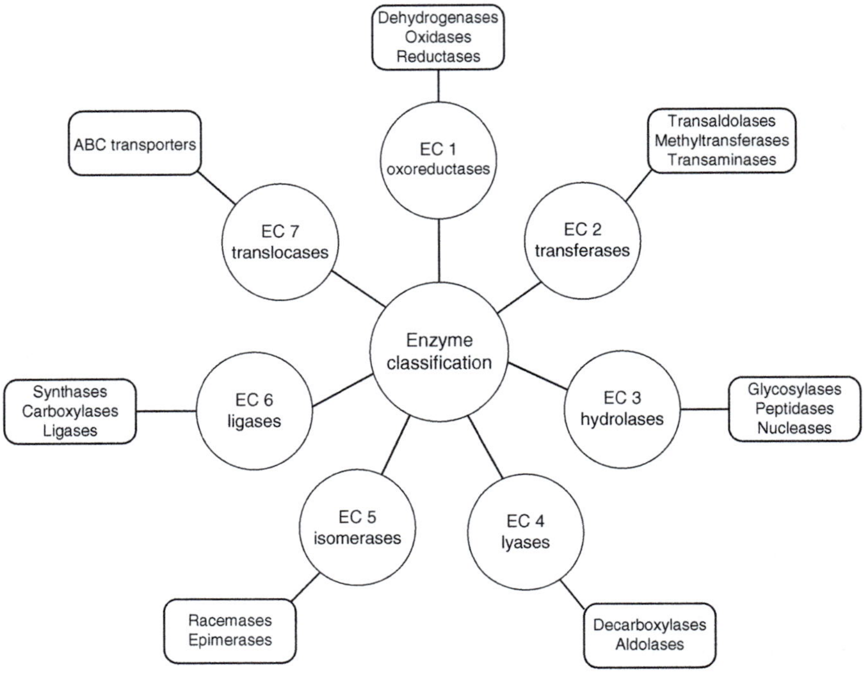

Figure 1.9 Classification of enzymes based on reaction type with several examples from each class.

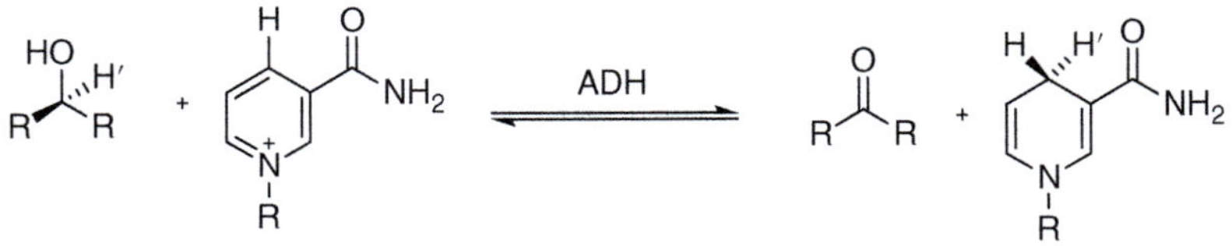

Figure 1.10 Generic reaction catalyzed by alcohol dehydrogenase.

1.4.1.2 EC 2 – Transferases

This class of enzymes catalyze the transfer of a specific group from one molecule to another. Groups that can be transferred include, but are not limited to, methyl, acyl, amino, glycosyl, and phospho-groups. For example, aspartate aminotransferase (AspAT) catalyzes the transfer of an amino group from L-aspartate to 2-oxoglutarate to give L-glutamate and oxalacetate (Figure 1.11) [39].

1.4.1.3 EC 3 – Hydrolases

Hydrolases have the capacity to catalyze the hydrolysis of various substrates. Depending on their substrate specificity, the hydrolases are further divided into

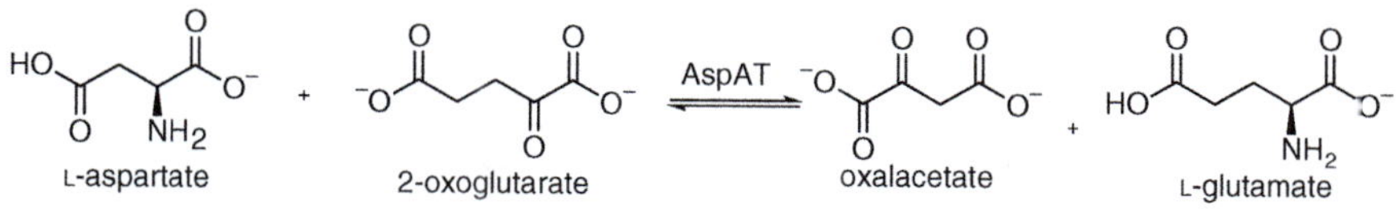

Figure 1.11 Interconversion mediated by aspartate aminotransferase.

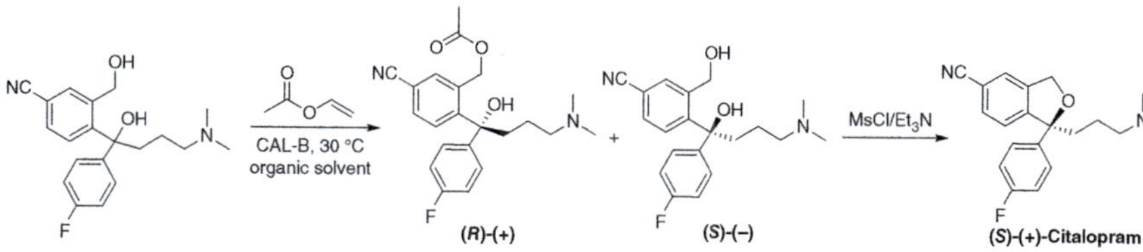

Figure 1.12 Chemoenzymatic synthesis of (*S*)-(+)-Citalopram.

their respective subclasses. One class of hydrolases, the lipases, have been particularly useful in the synthesis of novel pharmaceuticals. For example, an enzymatic acylation of racemic 4-[4-dimethylamino]-1-(4′-fluorophenyl)-1-hydrox-1-butyl]-3-(hydroxymethyl)-benzonitrile using CAL-B and vinyl acetate as the acyl donor allowing the synthesis of (*S*)-(+)-Citalopram (Figure 1.12) [40]. The hydrolases are further subclassified into – EC 3.1 ester hydrolases; EC 3.2 glycosylases; EC 3.3 enzymes hydrolyze ether linkages; EC 3.4 peptidase/protease, etc.

1.4.1.4 EC 4 – Lyases

Lyases catalyze non-hydrolytic bond cleavage reactions, ranging from C–C, C–O, C–N, and C–S, leading to the formation of a double bond, a new ring, or addition of groups to double bonds. The lyases are consequently subdivided into the bond type involved in the reaction: EC 4.1 carbon–carbon lyases, EC 4.2 carbon–oxygen lyases, EC 4.3 carbon–nitrogen lyases, EC 4.4 carbon–sulfur lyases, EC 4.5 carbon–halide lyases, EC 4.6 phosphorus–oxygen lyases, EC 4.7 carbon–phosphorus lyases, and EC 4.99 other lyases. Lyases are important in many biological processes – for example, as the first committed step towards the phenyl propanoid pathway, phenylalanine ammonia lyase (PAL, EC 4.3.1.24) catalyzes the conversion of L-phenylalanine to ammonia and *trans*-cinnamic acid (Figure 1.13) [41].

Figure 1.13 Reaction catalyzed by phenylalanine ammonia lyase (PAL).

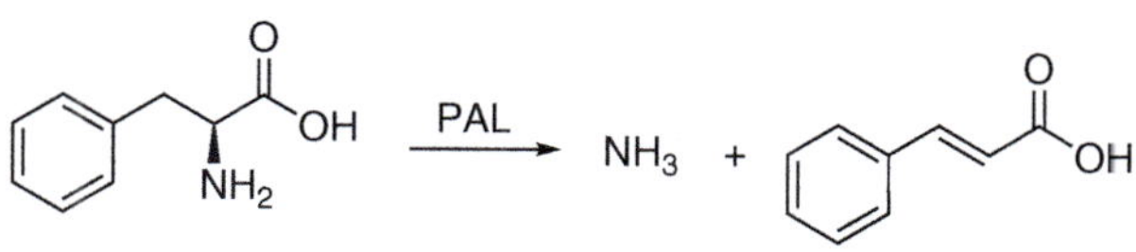

1.4.1.5 EC 5 – Isomerases

Isomerases catalyze intramolecular rearrangement or isomerization reactions. This class of enzymes is subdivided into seven subclasses based on the type of catalyzed reaction. EC 5.1 racemases and epimerases, EC 5.2 *cis–trans* isomerases, EC 5.3 intramolecular oxidoreductases, EC 5.4 intramolecular transferases, EC 5.5 intramolecular lyases, EC 5.6 isomerases altering macromolecular conformation, and EC 5.99 other isomerases. One classic example from undergraduate biochemistry classes is glucose-6-phosphate isomerase [42], which functions in glycolysis by interconverting glucose-6-phosphate and fructose-6-phosphate, the latter which proceeds through the glycolytic pathway to generate ATP (Figure 1.14).

1.4.1.6 EC 6 – Ligases

Ligases, as implied by their name, catalyze the attachment of two molecules. For example, the enzyme pyruvate carboxylase (EC 6.4.1.1) catalyzes the carboxylation of pyruvate to form oxaloacetate (Figure 1.15) [43]. Reactions catalyzed by ligases require energy from nucleoside triphosphates [44].

1.4.1.7 EC 7 – Translocases

Translocases assist in the movement of ions or molecules across the cell, often involving the hydrolysis of ATP. The translocases can be further subdivided into six subclasses, depending on the type of ion or molecule being translocated. EC 7.1 enzymes catalyze the translocation of hydrons, EC 7.2. inorganic cations, EC 7.3. inorganic anions, EC.7.4. amino acids and peptides, EC 7.5. carbohydrates, and EC 7.6. other compounds.

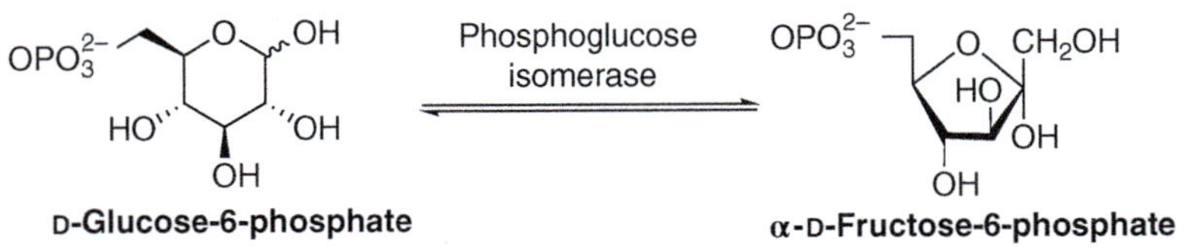

Figure 1.14 Reaction catalyzed by glucose-6-phosphate isomerase.

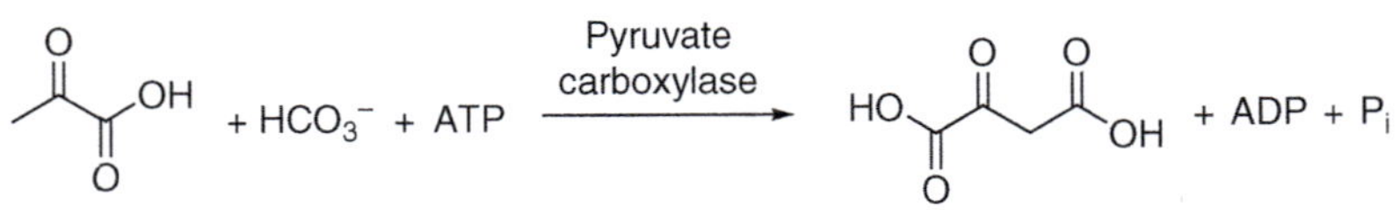

Figure 1.15 Reaction catalyzed by pyruvate carboxylase.

1.4.2 Some Applications of Selected Commercially Available Enzymes

1.4.2.1 Horseradish Peroxidase

Horseradish peroxidase (HRP) is an enzyme that is found in the roots of horseradish and is commonly used to catalyze the oxidation of various organic substrates using hydrogen peroxide. In biochemistry, HRP is used to amplify signals in photometric assays, whereby chromogenic or chemiluminescent substrates are oxidized for the detection of targets, such as proteins and nucleic acids [45]. For example, in chemiluminescent Western blotting, a technique that enables indirect detection of protein samples immobilized on membranes, a primary antibody is applied against the protein of interest. After subsequent washing of excess primary antibody, a secondary antibody, which targets the primary antibody and is conjugated with HRP, is added, before another washing procedure is applied. Following this, a buffer containing peroxide and luminol is added, in which the latter oxidizes and forms an excited state product that emits light as it decays to the ground state (Figure 1.16).

1.4.2.2 Lysozyme

Lysozyme, also known as muramidase, is a glycosyl hydrolase that catalyzes the hydrolysis of 1,4-β-linkages between *N*-acetylmuramic acid and *N*-acetyl-D-glucosamine residues in peptidoglycan, a major component of gram-positive bacterial cell walls (Figure 1.17). This reactivity consequently disrupts the integrity of the cell wall, causing osmotic shock and leading to lysis of the bacteria. Commercially available hen egg white lysozyme is commonly used for the aforementioned purpose; however, this enzyme has also enjoyed the spotlight in several biotechnological and catalytic applications [46, 47].

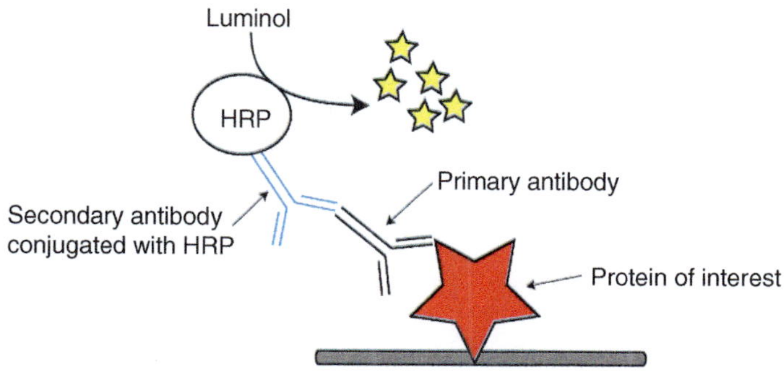

Figure 1.16 Use of horseradish peroxidase in Western blotting.

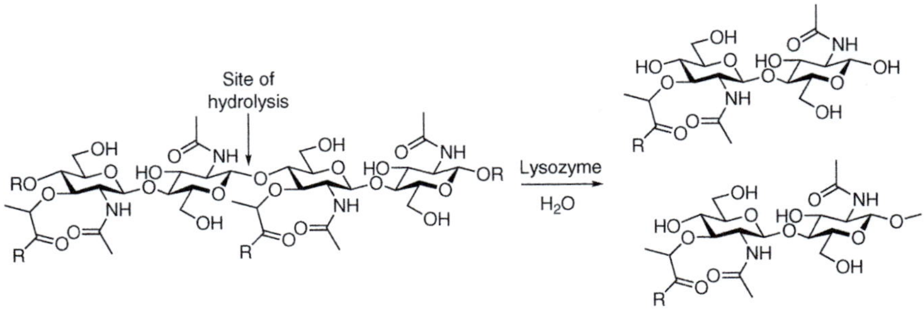

Figure 1.17 Reaction catalyzed by lysozyme.

1.4.2.3 Trypsin

Trypsin is a serine protease found in the digestive tract of many vertebrates. This enzyme catalyzes the hydrolysis of proteins, cleaving the carboxyl end of lysine or arginine residues. As such, trypsin has been very useful in proteomics where it has been used to digest proteins for mass spectrometric analysis [48].

1.4.2.4 Candida Lipase B

Candida antarctica: lipase B (CALB) is a hydrolase that has been used in a broad range of organic reactions, including kinetic resolutions [49], aminolysis [50], and even reactions in organic solvents [51]. Having been extensively studied, CALB has undergone several modifications via mutagenesis, for example, to improve its thermal stability [52] and activity under immobilized conditions [53, 54]. A commercially available immobilized version of CALB, Novozyme® 435, has been used for the stereoselective hydrolysis of esters, transesterification, and DKR of alcohols [55, 56].

1.4.2.5 Amino Acid Dehydrogenase

Amino acid dehydrogenase enzymes employ NAD^+ and H_2O to catalyze the deamination of amino acids (Figure 1.18) [57]. Depending on the specificity of the amino acid dehydrogenase, either L- or D-amino acids can be converted to their corresponding oxo-acid. D-Amino acid dehydrogenase has been particularly useful in the synthesis of D-amino acids in both high enantioselectivity and high yields [58].

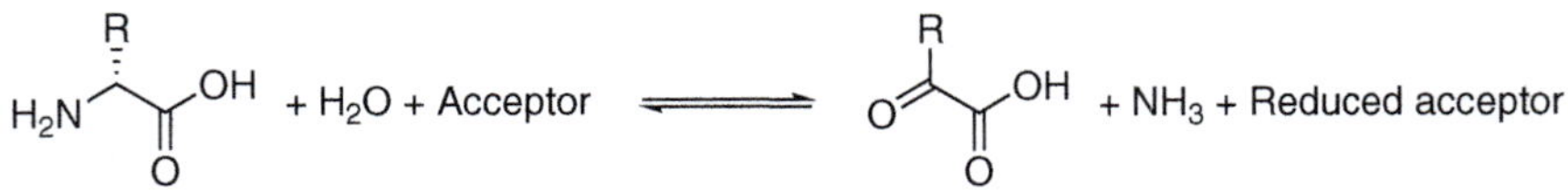

Figure 1.18 Transformation catalyzed by D-amino acid dehydrogenase. Acceptor can be various electron acceptors, but commonly Coenzyme Q is used.

1.4.2.6 Glycosidases

The inventory of commercially available enzymes which catalyze the hydrolysis of sugars is vast, largely due to the strict specificity of each enzyme for their respective glycosidic linkage. Glycosidases have been particularly useful in biotechnology where they have been used for various purposes, including improving the quality of bakery products, brewing beer, and biomodification of materials [59].

1.4.3 Engineered (Unnatural) Reactions

As our understanding of enzymes at the atomic level improves, the development of powerful bioinformatic tools and machine learning algorithms has also flourished. Consequently, these advancements have allowed the development of "designer" enzymes that catalyze novel reactions [16]. Tools such as AlphaFold, an AI system developed by DeepMind, has the power to predict protein structure based on genomic datasets [60]. RetroBioCat, a tool for computer-aided design of biocatalytic cascades, also allows for easy searching of a set of enzymes that might perform a desired transformation [61]. We refer to recent advancements in these fields in Section 1.5.

1.5 New Trends and Technologies in Biocatalysis

1.5.1 Flow Biocatalysis and New Technologies

1.5.1.1 What Is Flow Biocatalysis?

In the last few decades, flow chemistry has risen as a promising alternative to perform chemical synthesis in a more efficient, selective, sustainable, and safer manner (see Chapter 4) [62–64]. Flow *biocatalysis* has allowed a merger between continuous processing and enzyme-mediated reactions, forming the perfect marriage for sustainable and efficient biocatalytic transformations [65–67].

As with any continuous processing setup, the reagents are pumped through a system by means of pumps through modules. Implementation of flow biocatalysis begs several questions – how does this process work? How does one decide when a flow compared to a batch process is beneficial for a specific transformation? Can every enzymatic reaction be implemented in flow?

In the following paragraphs, we aim to guide you on when to consider a flow setup.

1.5.1.2 How Does Flow Biocatalysis Work?

The equipment required for flow biocatalysis does not far exceed a generic flow setup one may already have set up in a chemistry lab so can be easily implemented into an existing framework. Due to its modular nature, flow systems can be viewed

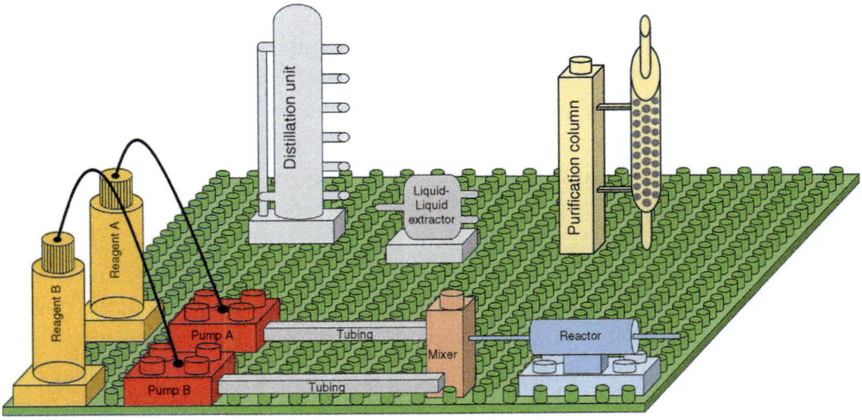

Figure 1.19 Representation of the typical units and general setup for flow chemistry experiments.

in a similar manner as "Lego" – the user can change the bricks (modules) depending on their specific needs (Figure 1.19). The different "Lego" bricks can represent the analytical, purification, solvent switchers, reactors, and an infinite number of other units.

However, where does the enzyme take part in this scientific Lego? For optimal sustainability, the enzyme needs to be recycled and reused after each synthesis. Recently, a plethora of techniques to immobilize the enzyme inside the reactor have been reported [68–70], enabling a continuous flow of reagents without losing the precious catalyst [71]. Different methods of immobilization can be found throughout the literature and generally classified into chemical (covalent) and physical methods (embedding) (see Figure 1.20).

1) *Chemical immobilization* of enzymes refers to the covalent binding of the enzyme onto a solid support. This technique offers several advantages, such as stabilization of the biocatalyst due to rigidification [72] and low leaching after multiple catalytic cycles.
2) *Physical methods* involve the embedding of an enzyme into an inorganic or organic polymer. Physical entrapment of the enzyme is a milder procedure compared to chemical immobilization since the enzyme is not stressed to undergo conformational changes due to the chemical linkage to the solid supports [73].

An innovative example of physical immobilization is 3D bioprinting, where an enzyme is extruded together with a polymer into a desired shape, opening the door to an infinite type of micro- and meso-reactors [74]. This method can prove advantageous when dealing with enzymes that suffer inactivation upon

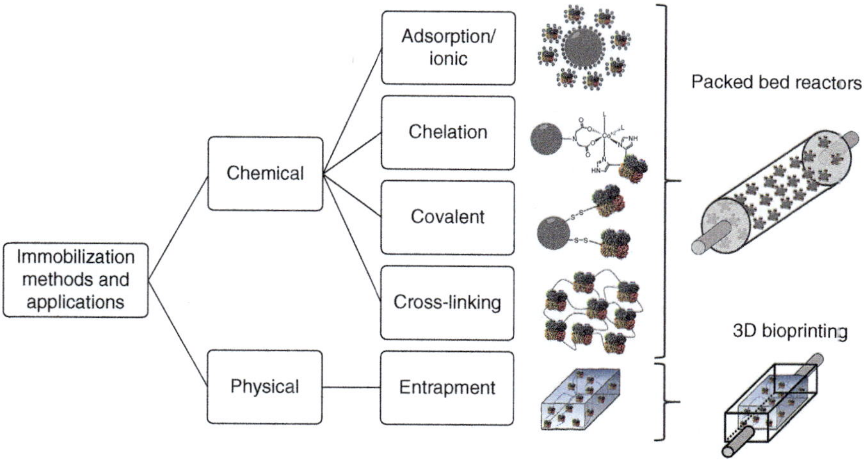

Figure 1.20 General enzyme immobilization techniques and their potential applications.

covalent or ionic immobilization due to conformational changes in their quaternary structure [75, 76].

1.5.1.3 When Is a Flow Process More Beneficial for a Specific Transformation?

Several considerations need to be made when deciding whether to switch from a classical batch process to a flow setup (applies in both classical synthesis and biocatalysis). One of these is process safety – flow setups offer safety levels that are astonishingly higher than batch for handling dangerous or very reactive intermediates due to enhanced heat transfer. Even though the reactions are usually performed under milder conditions in biocatalysis, more harsher conditions may be introduced downstream if running a chemo-enzymatic synthesis. Another common issue of batch reactions involves product build-up, leading to the inhibition of the enzyme, and therefore decreasing the productivity of the reaction. In the case of a flow setup, the products are continuously evacuated from the reactor where the enzyme is confined, mitigating this problem [67]. Moreover, when a reaction scale-up is necessary, a flow setup can prove very beneficial since the process follows a more linear trend, and reaction parameters can be screened much faster than in batch mode.

1.5.1.4 Should One Implement Every Enzymatic Reaction in Flow?

Flow setups can have weak points (long reaction times, insoluble compounds, etc.), and in some cases, a batch reaction is still the best option. The typical approach for deciding if a flow setup could be beneficial for a specific reaction would be first running the reaction initially in batch to identify the critical parameters (such as reaction times, solubilities, temperatures, etc.) and then moving the reaction to flow to determine if any improvement is observed.

1.5.2 Enzyme Engineering

Generally, enzymes exhibit different activities toward different substrates. To improve the enzymatic activity toward a specific substrate or class of molecules, the structure of the enzyme can be finely tuned using enzyme engineering, which, in classical organic chemistry would be akin to changing the ligands around the metal center in order to tune the selectivity of the catalyst. This can lead to the improvement of different parameters, such as temperature, pH, solvent tolerance, substrate, and reagent specificity, among others.

There are several methods to engineer new enzyme variants (mutants), which are summarized in Figure 1.21. The model-driven approach is the most classical one in which a mutation (changing a ligand by another one in organometallic chemistry) (or a set of mutations) can be planned by taking into consideration the physicochemical properties and interactions of the different residues of the protein. Complementary to the model-driven approaches in the last decades, data-driven approaches are proving to be very fruitful [77]. The main difference between both methodologies is that the latter is based on finding patterns and trends in pre-existing data to predict properties (i.e. solubility,

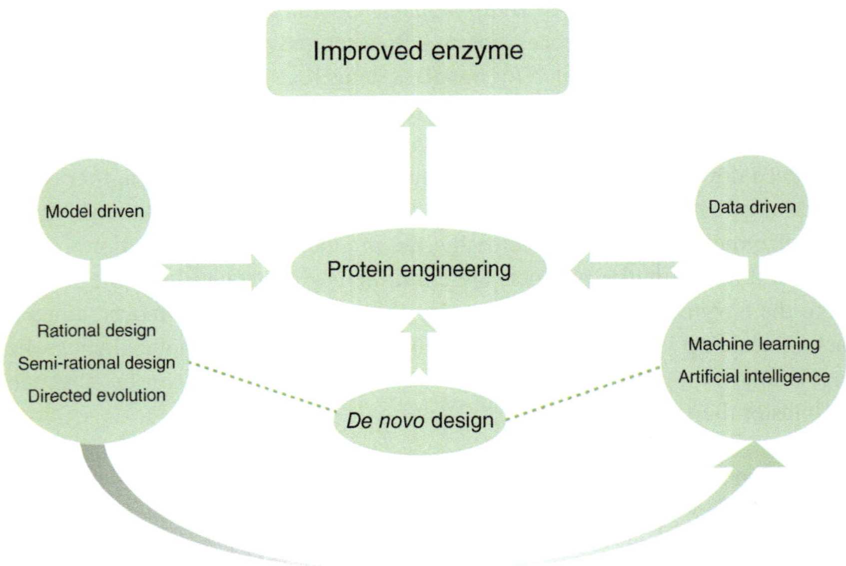

Figure 1.21 Two approaches of enzyme engineering.

thermal stability, activity). Machine learning and artificial intelligence have been demonstrated as excellent tools for analyzing multidimensional relations that the human brain is not able to identify [78]. In between model- and data-driven approaches, *de novo* design is emerging as a promising alternative to fabricate new proteins from scratch, based on experimental data and computer-aided design [79, 80].

1.5.3 Photobiocatalysis

Photobiocatalysis has recently gained a significant amount of interest by the scientific community [81]. The biocatalyst can act either as a chemical environment to favor the transformation of the photochemically activated substrate [82] or as mediator in which the enzyme converts light energy into redox equivalents [83]. Not surprisingly, in the last few years, photobiocatalysis has been merged with continuous flow technologies [84, 85].

1.6 Flow Chart to Biocatalysis

To simplify the task of choosing the right enzyme, we have developed the following flow chart in which we guide the reader through the crucial steps when dealing with a biocatalytic reaction (Figure 1.22).

1) *Identify the target reaction*
   First, we need to determine the transformation we want to achieve so that we may ascertain the enzyme required. These transformations may be oxidation, reduction, C–C bond formation, C–C bond cleavage, transamination, reductive amination, etc.
   Hint: Many enzymes are named after the reaction they catalyze.
2) *Is there a biocatalytic route available?*
   The chemist now needs to explore the literature to identify an enzyme or a class of enzymes that can perform the desired transformation (SciFinder, Web of Science, Reaxys, etc.). If one is available, you can go on to step 3. However, if the desired enzymatic transformation has not been reported, it is unlikely to be easily achieved and beyond the scope of the expertise of an average organic chemist.
3) *Is there a commercial enzyme available?*
   When looking for commercial enzymes, we suggest browsing through: Sigma Aldrich, Gecco, Codexis, Creative-enzymes, Eucodis, Purolite, Resindion, and Novozymes. These sources have a large repertoire of enzymes neatly catalogued for perusal. If the enzyme is commercially available, the chances of success are much higher, since it is very likely the protein has been thoroughly

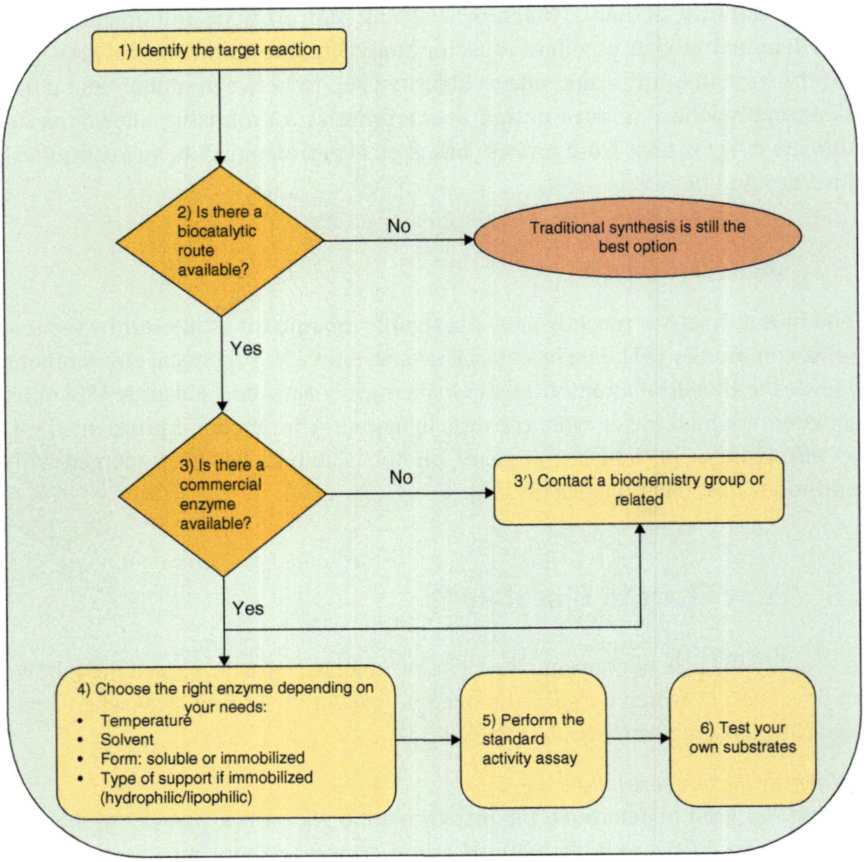

Figure 1.22 Flow chart to biocatalysis.

characterized. Often, many of those suppliers will offer screening kits to rapidly find a hit.

However, if the biocatalyst is not commercially available, it is still possible to engineer an enzyme for the target transformation, but this process is very complex and will likely require a collaboration with a specialized biochemistry group.

4) *Choosing an enzyme based on your needs*
Once we know that a commercial supplier or a biochemistry group can provide us with the biocatalyst, we need to consider the compatibility of the enzyme with the reaction conditions. For example, if the substrate is only soluble in a specific organic solvent, the enzyme must be compatible/stable in it. Moreover, one needs to decide whether enzyme recovery is required at the end of the reaction. For example, if your enzyme is difficult to make in large quantities,

requiring multiple rounds of expression and purification, then in such a case, an immobilized enzyme may be the best option, allowing you to retain the enzyme for multiple rounds of catalysis before disposal (see Section 1.5.2). Depending on the polarity of the substrate, a lipo- or hydrophilic support may prove advantageous. Contrastingly, the use of very apolar substrates would not be suitable for such a support. Vendors such as Sigma Aldrich, Purolite, and the other previously reported suppliers have different supports and immobilization chemistries available. We invite the reader to consult the respective catalogues to be informed of what options are available.

5) *Perform a standard activity assay*
   Once we have the enzyme at hand, a standard activity assay must be performed to assure the healthiness of the enzyme, as discussed in Section 1.3.3. Ideally but not compulsory (NMR, HPLC, GC can also be used), this assay will be carried out in a classical spectrophotometer since most of the reported activity assays are a measure of the change in absorbance of a certain molecule therefore, they are usually very fast. When using commercial or published enzymes, the activity assays should be thoroughly described in the Supplementary Information section of the respective publication. If the recovered activity fits the reported values, one can move on to the last step of the process.
6) *Test your own substrates*
   Finally, screening of the desired substrates can be performed. As a rule of thumb first, the published reaction conditions will be tested (enzyme concentrations, solvents or co-solvents, temperature, reagents concentrations, etc.) and then adjusted to the specific starting materials if needed. Regarding the workup of the reactions, usually a regular reaction workup should enable product recovery, nevertheless when using immobilized enzymes an additional filtering step should be performed. The reactions can be analyzed with the usual techniques such as NMR, HPLC, or GC. No special techniques or skills should be needed for an organic chemist to perform biocatalytic reactions.

1.7 Case Study: Setting up a Biotransformation

In this section, we aim to guide the reader through a case study of the process of setting up a biocatalytic reaction, from identifying the transformation in the literature until the final reaction setup, using the flow chart depicted in Figure 1.22. We have chosen the "Enzymatic Kinetic Resolution of (*R*/*S*)-1-phenylethanol" [86] as an elementary example for the reader to follow.

1) *Identify the target reaction*
   The reaction we are interested in here is the kinetic resolution of a racemic mixture of 1-phenylethanol (Figure 1.23).

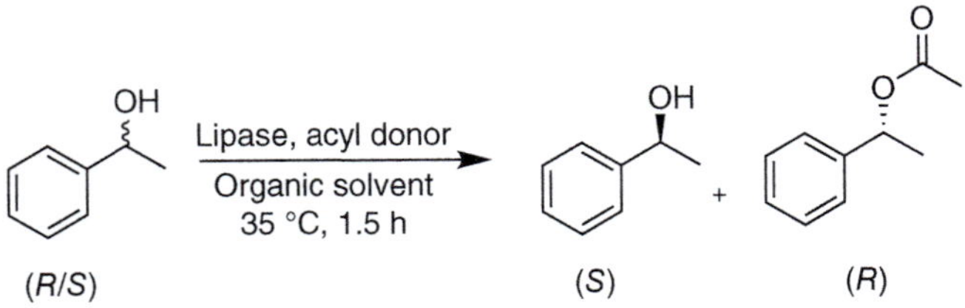

Figure 1.23 Enzymatic kinetic resolution of (*R*/*S*)-1-phenylethanol.

2) *Is there a biocatalytic route available?*
   Once we identify a publication or set of publications for our desired transformation, the workflow is very similar to a classical organic synthesis approach. What reagents do we need? What type of equipment/instruments do we need?
3) *Is there a commercial enzyme available?*
   This is perhaps the less intuitive step for a classical synthetic chemist because often when browsing for commercial enzymes, one will find multiple formulations of the same enzyme (see example below).

   In this case study, the authors screened three different enzymes: Porcine pancreas lipase (PPL), lipase from *Thermomyces lanuginous* (TL-IM), and lipase-B from *Candida antarctica* (CAL-B). Looking into Sigma Aldrich for Candida antartica lipase B, one will find multiple entries. We can simplify the search by filtering the results to only "enzymes" and now only four different CAL-B formulations will appear (see Figure 1.24).

   In order to decide which of the multiple enzymes are the most appropriate for the desired transformation, one needs to evaluate several parameters, for instance, is enzyme re-use desired? If yes, then the enzyme will need to be immobilized on a support (entries 2 and 3 from Figure 1.24). If not, one could go for the lyophilized powder (entry 1 from Figure 1.24). Analyzing the procedure from the original work of the case study, one can easily deduce that the enzymes used are immobilized since after the defined reaction time, and the enzymes are subsequently filtered.

   The options are now narrowed down to two; in this case the only difference between both formulations is the recombinant micro-organism that was used as host for the enzyme synthesis (i.e. both enzymes are *identical*). If nothing is stated in the procedure (as in our case study), the simplest solution is to go for the formulation that contains the most units per enzyme (see explanation below). Thus, in our case and considering all the parameters analyzed above, the enzyme of choice would be "Lipase B *Candida antarctica* immobilized on Immobead 150, recombinant from yeast."

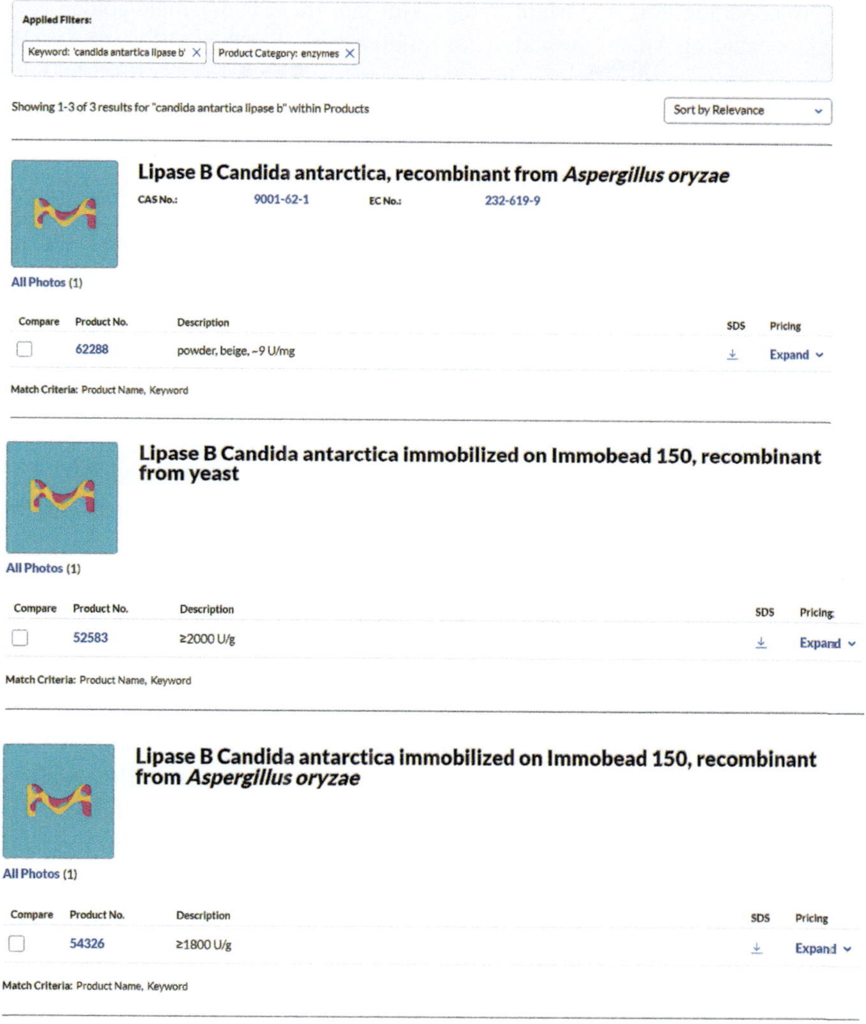

Figure 1.24 Sigma Aldrich search results for "Candida Antartica lipase B" (accessed 11 April 2023).

The same reasoning must be applied for the rest of the lipases. In any case, we should expect that, as with any published synthetic organic procedure, the commercially available reagents are reported in the materials and methods section or in the supplementary information (this case study).

To conclude, it is important to highlight that there is no single golden rule when deciding which biocatalyst/formulation/supplier that will *a priori* work best. Therefore, sometimes a broader screening is necessary to find the best candidate for a specific reaction. This task does not differ from the screening of, for example, a Pd/C catalyst within different suppliers with different loadings and activations in organic synthesis.

4) *Choosing an enzyme based on your needs*
In our published case study, the rationale for their choice of enzyme is due to the fact that they are known to be more stable when exposed to organic solvents [87]. Additionally, having the enzymes immobilized enables future reusage of the catalyst.

5) *Perform a standard activity assay*
Let us take the example of "Lipase B *Candida antarctica* immobilized on Immobead 150, recombinant from yeast." The product description contains a section where it states the unit definition. In this case, 1 U corresponds to the amount of enzyme (mg) needed to produce 1 μmol of butyric acid per minute at pH 7.5 and 40 °C, using glyceryl tributyrate as starting material. In this case, the supplier does not have a step-by-step protocol for the activity assay. However, one can find several suitable activity assay protocols for lipase to perform, depending on the equipment available in the lab, in a review published by Hasan et al. [88].

One can readily notice that this reaction is opposite to the one that we want to carry out (acylation). For an organic chemist, this may be counterintuitive since the natural way of testing the activity of a catalyst is in the reaction direction that one wants to perform. This does not mean that in the biocatalytic synthesis, the "real" reaction is not tested, but that a standard reaction needs to be performed prior to the desired transformation, to ensure that the biocatalyst has not lost activity over time due to, for instance, exposure to temperature, organic solvents, or high salt concentrations.

Generally, before first use of the enzyme for experimentation, a standard activity assay using the conditions described by the supplier should be performed first to determine the baseline activity. This ensures that in the future, one can easily detect if the enzyme has lost activity over time and, second, to ensure that the purchased enzyme is complying with the specifications provided by the supplier.

6) *Test your own substrates*
This part is probably the most important and interesting section of the whole process. Here, standard organic chemistry techniques are used. One simply needs to design reaction conditions based on a reported procedure and adjusting accordingly. Reactions are run exactly in the same manner as a synthesis

procedure (weighing starting materials, calculating stoichiometries, monitoring reaction progress using chromatographies). Once initial results are obtained, one can then screen various conditions to achieve optimal parameters for high yields/conversions, enantioselectivity, reaction time, etc.

1.8 Concluding Remarks

Over the past few sections, we gave insight into the feasibility of introducing biocatalysis in a standard chemistry lab. We hope we have instilled the reader with enough foundational knowledge to take their first steps into establishing biocatalytic reactions in their laboratory. Taking that first step toward discovering the power of enzymes for chemical synthesis will allow the reader to not only see the ease of incorporating enzymes into the workflow but also give one confidence in trying to expand the toolbox of available enzymes for alternative transformations. Furthermore, we provide several references below for more advanced reading.

Additional Resources

- Enzymatic parameters and reporting results [34, 89]
- Advanced biocatalysis book with an overview of typical reactions and industrial biotransformations [90]
- "The Hitchhiker's guide to biocatalysis: recent advances in the use of enzymes in organic synthesis" [91]
- *Biotransformations in Organic Chemistry* [92]
- *The Transformative Power of Biocatalysis in Convergent Synthesis* [93]

References

1 Alcántara, A.R., Domínguez de María, P., Littlechild, J.A. et al. (2022). *ChemSusChem* 15: e202102709.

2 Winkler, C.K., Schrittwieser, J.H., and Kroutil, W. (2021). *ACS Cent. Sci.* 7 (1): 55–71.

3 Cox, R.M., Wolf, J.D., and Plemper, R.K. (2021). *Nat. Microbiol.* 6 (1): 11–18.

4 Sheahan, T.P., Sims, A.C., Zhou, S. et al. (2020). *Sci. Transl. Med.* 12 (541): eabb5883.

5 McIntosh, J.A., Benkovics, T., Silverman, S.M. et al. (2021). *ACS Cent. Sci.* 7 (12): 1980–1985.

6 Corey, E.J., Bakshi, R.K., and Shibata, S. (1987). *J. Am. Chem. Soc.* 109 (18): 5551–5553.
7 Richard, J.P. (2013). *Biochemistry* 52 (12): 2009–2011.
8 Barrios-Rivera, J., Xu, Y., Wills, M., and Vyas, V.K. (2020). *Org. Chem. Front.* 7 (20): 3312–3342.
9 Mitsukura, K., Suzuki, M., Tada, K. et al. (2010). *Org. Biomol. Chem.* 8 (20): 4533–4535.
10 Roura Padrosa, D., Benítez-Mateos, A.I., Calvey, L., and Paradisi, F. (2020). *Green Chem.* 22 (16): 5310–5316.
11 Matzel, P., Wenske, S., Merdivan, S. et al. (2019). *ChemCatChem* 11 (17): 4281–4285.
12 Schmid, A., Dordick, J.S., Hauer, B. et al. (2001). *Nature* 409 (6817): 258–268.
13 Wu, S., Snajdrova, R., Moore, J.C. et al. (2021). *Angew. Chem Int. Ed.* 60 (1): 88–119.
14 Stewart, J.D. (2001). *Curr. Opin. Chem. Biol.* 5 (2): 120–129.
15 Contente, M.L., Roura Padrosa, D., Molinari, F., and Paradisi, F. (2020). *Nat. Catal.* 3 (12): 1020–1026.
16 Xu, J., Peng, Y., Wang, Z. et al. (2019). *Angew. Chem Int. Ed.* 58 (41): 14499–14503.
17 Crawshaw, R., Crossley, A.E., Johannissen, L. et al. (2021). *Nat. Chem.* 14 (3): 313–320.
18 Hollmann, F., Opperman, D.J., and Paul, C.E. (2021). *Angew. Chem Int. Ed.* 60 (11): 5644–5665.
19 Lee, M.-Y. and Dordick, J.S. (2002). *Curr. Opin. Biotechnol.* 13 (4): 376–384.
20 Ni, Y., Holtmann, D., and Hollmann, F. (2014). *ChemCatChem* 6 (4): 930–943.
21 Vieille, C. and Zeikus, G.J. (2001). *Microbiol. Mol. Biol. Rev.* 65 (1): 1–43.
22 Carrea, G. and Riva, S. (2000). *Angew. Chem Int. Ed.* 39 (13): 2226–2254.
23 Klibanov, A.M. (2001). *Nature* 409 (6817): 241–246.
24 Zaks, A. and Klibanov, A.M. (1986). *J. Am. Chem. Soc.* 108 (10): 2767–2768.
25 Arnold, F.H. (1990). *Trends Biotechnol.* 8 (C): 244–249.
26 Cui, H., Vedder, M., Schwaneberg, U., and Davari, M.D. (2002). Using Molecular Simulation to Guide Protein Engineering for Biocatalysis in Organic Solvents. In: *Enzyme Engineering. Methods in Molecular Biology*, vol. 2397 (ed. F. Magnani, C. Marabelli, and F. Paradisi). New York, NY: Humana.
27 Di Cera, E. (2006). *J. Biol. Chem.* 281 (3): 1305–1308.
28 Schmidt, J., Wei, R., Oeser, T. et al. (2016). *FEBS Open Bio* 6 (9): 919–927.
29 Albe, K.R., Butler, M.H., and Wright, B.E. (1990). *J. Theor. Biol.* 143 (2): 163–195.
30 Turgay Yagmur, I., Unal Uzun, O., Kucukcongar Yavas, A. et al. (2020). *Ann. Allergy Asthma Immunol.* 125: 460–467.
31 Dörr, T., Moynihan, P.J., and Mayer, C. (2019). *Front. Microbiol.* 10: 2051.
32 Amarasiri, M., Sano, D., and Suzuki, S. (2020). *Crit. Rev. Environ. Sci. Technol.* 50 (19): 2016–2059.

33 Katouli, M. (2010). *Iran. J. Microbiol.* 2 (2): 59–72.
34 Bisswanger, H. (2014). *Perspect. Sci.* 1 (1–6): 41–55.
35 Labuda, J., Bowater, R.P., Fojta, M. et al. (2018). *Pure Appl. Chem.* 90 (7): 1121–1198.
36 Units of Enzyme Activity (1979). *Eur. J. Biochem.* 97 (2): 319–320.
37 Sofer, W. and Martin, P.F. (1987). *Annu. Rev. Genet.* 21 (1): 203–225.
38 Sellés Vidal, L., Kelly, C.L., Mordaka, P.M., and Heap, J.T. (2018). *Biochim, Biophys. Acta – Proteins Proteomics* 1866 (2): 327–347.
39 Rej, R. and Shaw, L.M. (1984). *CRC Crit. Rev. Clin. Lab. Sci.* 21 (2): 99–186.
40 Solares, L.F., Brieva, R., Quirós, M. et al. (2004). *Tetrahedron Asymmetry* 15 (2): 341–345.
41 Camm, E.L. and Towers, G.H.N. (1973). *Phytochemistry* 12 (5): 961–973.
42 Read, J., Pearce, J., Li, X. et al. (2001). *J. Mol. Biol.* 309 (2): 447–463.
43 Utter, M.F. and Keech, D.B.J. (1960). *Biol. Chem.* 235 (5): PC17–PC18.
44 Holliday, G.L., Rahman, S.A., Furnham, N., and Thornton, J.M. (2014). *J. Mol. Biol.* 426 (10): 2098–2111.
45 Alegria-Schaffer, A. (2014). *Methods Enzymol.* 541: 251–259.
46 Wei, H., Wang, Z., Zhang, J. et al. (2011). *Nat. Nanotechnol.* 6 (2): 93–97.
47 Sanghamitra, N.J.M. and Ueno, T. (2013). *Chem. Commun.* 49 (39): 4114–4126.
48 Fang, P., Liu, M., Xue, Y. et al. (2015). *Analyst* 140 (22): 7613–7621.
49 Lund, I.T., Bøckmann, P.L., and Jacobsen, E.E. (2016). *Tetrahedron* 72 (46): 7288–7292.
50 Mouad, A.M., Taupin, D., Lehr, L. et al. (2016). *Mol. Catal. B Enzyme* 126: 64–68.
51 Hasegawa, S., Azuma, M., and Takahashi, K. (2008). *J. Chem. Technol. Biotechnol.* 83 (11): 1503–1510.
52 Mei, Y., Kumar, A., and Gross, R.A. (2002). *Macromolecules* 35 (14): 5444–5448.
53 Mateo, C., Palomo, J.M., Fernandez-Lorente, G. et al. (2007). *Enzyme Microb. Technol.* 40 (6): 1451–1463.
54 Idris, A. and Bukhari, A. (2012). *Biotechnol. Adv.* 30 (3): 550–563.
55 Ortiz, C., Ferreira, M.L., Barbosa, O. et al. (2019). *Catal. Sci. Technol.* 9 (10): 2380–2420.
56 Hvidsten, I.B. and Marchetti, J.M. (2021). *Energy Convers. Manag. X* 10: 100061.
57 Jones, H. and Venables, W.A. (1983). *FEBS Lett.* 151 (2): 189–192.
58 Akita, H., Suzuki, H., Doi, K., and Ohshima, T. (2014). *Appl. Microbiol. Biotechnol.* 98 (3): 1135–1143.
59 Bhat, M.K. (2000). *Biotechnol. Adv.* 18 (5): 355–383.
60 Senior, A.W., Evans, R., Jumper, J. et al. (2020). *Nature* 577 (7792): 706–710.
61 Finnigan, W., Hepworth, L.J., Flitsch, S.L., and Turner, N.J. (2021). *Nat. Catal.* 4 (2): 98–104.
62 Mándity, I.M., Ötvös, S.B., and Fülöp, F. (2015). *ChemistryOpen* 4 (3): 212–223.
63 Yoshida, J.I., Kim, H., and Nagaki, A. (2011). *ChemSusChem* 4 (3): 331–340.

64 Rahman, M.T. and Wirth, T. (2018). Safe use of hazardous chemicals in flow. In: *Flow Chemistry for the Synthesis of Heterocycles. Topics in Heterocyclic Chemistry*, vol. 56 (ed. U. Sharma and E. Van der Eycken). Cham: Springer.

65 Benítez-Mateos, A.I., Contente, M.L., Roura Padrosa, D., and Paradisi, F. (2021). *React. Chem. Eng.* 6 (4): 599–611.

66 Britton, J., Majumdar, S., and Weiss, G.A. (2018). *Chem. Soc. Rev.* 47 (15): 5891–5918.

67 Santi, M., Sancineto, L., Nascimento, V. et al. (2021). *Int. J. Mol. Sci.* 22 (3): 1–32.

68 Khan, M.R. (2021). *Bull. Natl. Res. Cent.* 45: 1.

69 Thangaraj, B. and Solomon, P.R. (2019). *ChemBioEng Rev.* 6 (5): 157–166.

70 Sheldon, R.A., Basso, A., and Brady, D. (2021). *Chem. Soc. Rev.* 50 (10): 5850–5862.

71 Sheldon, R.A. and Pelt, S.V. (2013). *Chem. Soc. Rev* 42: 6223.

72 Rodrigues, R.C., Berenguer-Murcia, Á., Carballares, D. et al. (2021). *Biotchnol. Adv.* 52: 107821–107858.

73 Romero-Fernández, M. and Paradisi, F. (2020). *Curr. Opin. Chem. Biol.* 55: 1–8.

74 Wenger, L., Radtke, C.P., Göpper, J. et al. (2020). *Front. Bioeng. Biotechnol.* 8: 713.

75 Schmieg, B., Döbber, J., Kirschhöfer, F. et al. (2019). *Front. Bioeng. Biotechnol.* 6 (JAN): 211.

76 Maier, M., Radtke, C.P., Hubbuch, J. et al. (2018). *Angew. Chem Int. Ed.* 57 (19): 5539–5543.

77 Mazurenko, S., Prokop, Z., and Damborsky, J. (2020). *ACS Catal.* 10 (2): 1210–1223.

78 Yi, D., Bayer, T., Badenhorst, C.P.S. et al. (2021). *Chem. Soc. Rev.* 50 (14): 8003–8049.

79 Huang, P.S., Boyken, S.E., and Baker, D. (2016). *Nature* 537 (7620): 320–327.

80 Xiong, W., Liu, B., Shen, Y. et al. (2021). *Biochem. Eng. J.* 174: 108096.

81 Schmermund, L., Jurkaš, V., ss et al. (2019). *ACS Catal.* 9 (5): 4115–4144.

82 Hyster, T.K. (2020). *Synlett* 31 (03): 248–254.

83 Schmermund, L., Jurkaš, V., Özgen, F.F. et al. (2019). *ACS Catal.* 9 (5): 4115–4144.

84 Chanquia, S.N., Valotta, A., Gruber-Woelfler, H., and Kara, S. (2021). *Front. Catal.* 1: 816538.

85 Lee, S.H., Choi, D.S., Kuk, S.K., and Park, C.B. (2018). *Angew. Chem Int. Ed.* 57 (27): 7958–7985.

86 Bandeira, P.T., Thomas, J.C., De Oliveira, A.R.M., and Piovan, L. (2017). *J. Chem. Educ.* 94 (6): 800–805.

87 Stepankova, V., Bidmanova, S., Koudelakova, T. et al. (2013). *ACS Catal.* 3 (12): 2823–2836.

88 Hasan, F., Shah, A.A., and Hameed, A. (2009). *Biotechnol. Adv.* 27 (6): 782–798.

89 Halling, P.J. and Gupta, M.N. (2014). *Perspect. Sci.* 1 (1–6): 98–109.

90 Tibhe, J.D. (2012). *Green Process. Synth.* 1: 6.

91 Sheldon, R.A., Brady, D., and Bode, M.L. (2020). *Chem. Sci.* 11 (10): 2587–2605.

92 Faber, K. (2011). *Biotransformations in Organic Chemistry*, 6ee, 315–324. Heidelberg: Springer.

93 Zetzsche, L.E., Chakrabarty, S., and Narayan, A.R.H. (2022). *Am. Chem. Soc.* 144: 5214–5225.

2

Introduction to Photochemistry for the Synthetic Chemist

Stefano Protti, Davide Ravelli, and Maurizio Fagnoni

PhotoGreen Laboratory, Department of Chemistry, University of Pavia, Pavia, Italy

Glossary

Absorbance (A) Absorbance is defined (and calculated) as the logarithm of the ratio of incident to transmitted radiant power (light) through a sample.

Chromophore The part (atom or group of atoms) of a molecular entity in which the electronic transition responsible for a given spectral band is approximately localized. The term arose in the dyestuff industry, referring originally to the groupings in the molecule that are responsible for the dye's color.

Excited state State of a system with energy higher than that of the ground state. This term is most commonly used to characterize a molecule in one of its electronically excited states but can also refer to vibrational and/or rotational excitation in the electronic ground state.

Ground state The state of lowest Gibbs energy of a system.

HOMO Highest-energy filled (or partially filled) molecular orbital of a molecular entity.

Intersystem Crossing (ISC) A photophysical process. An isoenergetic radiationless transition between two electronic states having different multiplicities. It often results in a vibrationally excited molecular entity in the lower electronic state, which then usually deactivates to its lowest vibrational level.

For further information on photochemical terminology, please refer to https://goldbook.iupac.org/.

Enabling Tools and Techniques for Organic Synthesis: A Practical Guide to Experimentation, Automation, and Computation, First Edition. Edited by Stephen G. Newman.

LUMO Lowest-energy unoccupied (completely or partially vacant) molecular orbital of a molecular entity.

(Molar) Extinction coefficient (ε) The linear absorption coefficient divided by the amount concentration.

Photocatalysis A change in the rate of a chemical reaction, or its initiation due to the action of a radiation in the presence of a chemical entity (dubbed as photocatalyst) that absorbs light and is involved in the chemical transformation of the reaction partners.

Photochemical reaction Generally used to describe a chemical reaction caused by absorption of ultraviolet, visible, or infrared radiation. There are many ground-state reactions, which have photochemical counterparts. Among these are photochemical nitrogen extrusions, photocycloadditions, photodecarbonylations, photodecarboxylations, photoenolizations, photo-Fries rearrangement, photoisomerizations, photooxidations, photorearrangements, photoreductions, photosubstitutions, etc.

Photon Particle of zero charge, zero rest mass, spin quantum number 1, energy $h\nu$ and momentum $h\nu/c$ (h is the Planck constant, $\boldsymbol{\nu}$ the frequency of radiation and c the speed of light), carrier of electromagnetic force.

Photosensitized reaction A photochemical or photophysical alteration that takes place in one molecular entity as a result of light absorption by another molecular entity defined as "photosensitizer."

Singlet state A spin state that shows a total electron spin quantum number equal to 0.

Transmittance (T) The ratio of the transmitted radiant power (P) to that incident on the sample (P_0). $T = P/P_0$.

Triplet state A spin state that shows a total electron spin quantum number of 1.

2.1 Introduction

2.1.1 Light to Make Your Synthesis Greener

In common organic chemistry textbooks, most of the reactions considered useful for practical applications make use of "thermal" conditions, namely taking place through processes that simply require mixing the reagents in the presence of an additive (e.g. an acid, a base, or an (organo)catalyst) at a suitable temperature (usually involving [mild] heating). However, the concerns raised by the pollution connected with chemical processes have recently pushed chemists to (re)consider alternative approaches, until now limited primarily to academic studies. Thus, different strategies have been proposed to surmount this problem, and photochemistry, which studies the use of light to promote a chemical conversion, is widely considered as one of the best choices. In fact, it is reasonable to speculate that green chemistry and photochemistry were born at the same time [1]. A pioneer in this field was Giacomo Ciamician and his work at the beginning of the twentieth century deeply contributed to the foundation of both photochemistry and sustainable chemistry [2]. He recognized that an organic reaction may be carried out under conditions milder than those typical of thermal processes by using a renewable source of energy, such as solar light, to mimic what plants realize in everyday life, namely converting luminous light to chemical bonds. In fact, he stressed that "using aggressive reagents and high temperatures is almost always unavoidable when carrying out an organic synthesis in the laboratory. Deploying energy would, on the other hand, not be so frustrating for modern organic chemistry, were it not that the living world, in particular plants, gives us the marvelous example of great results obtained, at least from what appears, by using minimal means" [2]. During the last decades, it has become clear that the ideal traceless reagent able to promote a chemical reaction leaving no residues behind is the photon [3, 4]. The concept that photochemistry uses the energy of the photon *only* as a simple heating surrogate, however, is a trivial simplification.

2.1.2 A Way to Overcome HOMO/LUMO Interactions

It is commonsense that the nature and the energy distribution of the molecular orbitals (in particular for what concerns frontier orbitals, namely HOMO and LUMO) dictates the chemical behavior of a given compound under thermal conditions. Thus, the electrophilicity of the carbonyl group or of an alkyl halide is due to the π^* (C=O) or the σ^* (C—X) orbitals, respectively, whereas the nucleophilicity of an alkene is set by the π (C=C) orbital. The outcome of a photochemical reaction, on the contrary, depends on the somewhat different electronic distribution of the reacting species (the "excited state") with respect to the common chemistry stemming from the "ground state," as detailed in the following. The fundamental law of photochemistry (the so-called Grotthuss–Draper law) says that light must be absorbed by a chemical substance for a photochemical reaction to take place. The absorption of light by a chemical entity is due to the incorporation of a "chromophore" (e.g. a carbonyl group, a double bond, an aromatic ring, or an azo group). In order to have a chemical modification, the organic compound must absorb light, generally in the UV (wavelength of 100–400 nm) or/and visible (400–800 nm) range. The UV spectrum may be roughly divided into three regions, namely UV-A (320–400 nm), UV-B (280–320 nm), and UV-C (100–280 nm). However, for a practical use the wavelength should be no shorter than 254 nm due to the lack of convenient light sources (*vide infra*).

The absorption of UV and visible light by an organic molecule causes the promotion of an electron from an occupied molecular orbital (σ, π, or a nonbonding orbital, n) to an unoccupied molecular orbital (σ*, π*) forming an "electronically excited state." The most common case is the excitation of an electron from the HOMO to the LUMO orbital leading to a species having two unpaired electrons in two distinct orbitals, mimicking a "diradical species" where a bond is weakened by the presence of an electron in the antibonding orbital.

The two electrons in the excited state may accommodate in two ways, viz. with the spin anti-parallel or parallel to each other, thus forming the corresponding singlet or triplet states, respectively. Singlets are directly accessed by excitation, whereas triplets may be obtained by Intersystem Crossing (ISC) from the singlets or via energy transfer with a triplet excited state of a suitable photosensitizer (see the following sections for further details). Usually both spin states lead to the same product, though exceptions exist, and a spin-selective chemistry may take place [5].

The excited states are short-lived species (in the μs range for triplets and ns for singlets) and may undergo a rapid physical decay, back to the more stable ground state [6]. Accordingly, a photochemical reaction must be by necessity faster than decay to avoid electronic deactivation. An organic compound, when in the excited state, shows a chemical reactivity markedly different from that of ground-state molecules, taking advantage of the extra amount of energy provided by the photon absorbed. This contrasts with thermal reactions that may encounter high barriers to surmount. This was clearly stated by Prof. Noiory who highlighted how photochemistry "enhances the power of chemical synthesis by removing current thermodynamic restrictions" [7], thus representing an appealing tool for organic synthesis. Figure 2.1 shows some paradigmatic cases of excited states. Alkanes may be involved in a σ→σ* transition to give a σσ* state, but this is poorly useful since this transition requires high-energy photons. The n→σ* transition in alkyl halides may have some applications (e.g. in the cleavage of C—X bonds); however, the π→π* transition in alkenes (or polyenes) and aromatics along with the π→π* (or n→π*) transition in carbonyl derivatives account for most photochemical transformations of synthetic interest.

For a given compound, the excited state is an "electronic isomer" of the ground state, and this completely changes the known behavior of the relevant functional group typically studied in organic chemistry courses. The other side of the medal is that photochemistry may appear somewhat "magic" and, as stated by George Büchi, "the course of the transformations could rarely be predicted, thus robbing the investigator of the pleasure derived from designing new reactions" [8]. In fact, the introduction of different substituents in a given compound can interfere with the nature and the lifetime of the involved excited states, thus affecting the reaction course. Nonetheless, these changes can be studied and rationalized by detailed photophysical investigations.

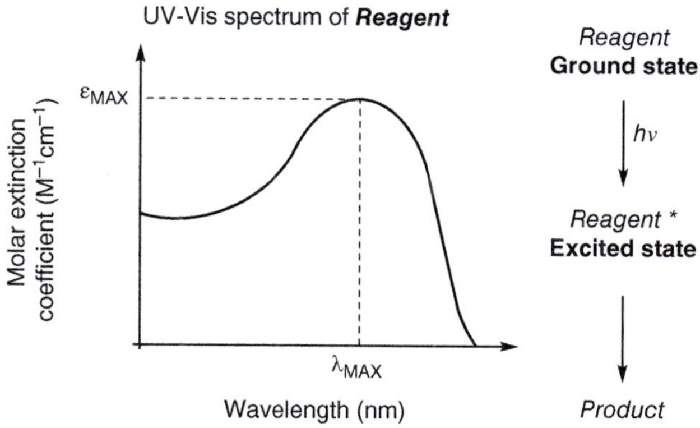

Chromophore	Transition	λ_{MAX} (nm)	ε_{MAX} ($M^{-1}cm^{-1}$)
CH_3–CH_3	$\sigma \rightarrow \sigma^*$	133	10 000
Double C=C bond	$\pi \rightarrow \pi^*$	165	1600
CH_3–Cl	$n \rightarrow \sigma^*$	173	200
CH_3–OH	$n \rightarrow \sigma^*$	177	200
Triple C≡C bond	$\pi \rightarrow \pi^*$	185	2000
CH_3–I	$n \rightarrow \sigma^*$	259	400
Acetone	$n \rightarrow \pi^*$	280	12

Figure 2.1 Transitions of selected chromophores.

How light is absorbed is crucial. The Lambert–Beer law states that the absorbance of a solution is $A = \varepsilon bc$, where c is the molar concentration (M) and b the optical path (cm). The value of ε_{MAX} ($M^{-1}\,cm^{-1}$) (see Figure 2.1 above) is an indication of the feasibility of a transition, being low for a "partially forbidden" band (e.g. *ca.* 10–100 $M^{-1}\,cm^{-1}$ for the nπ^* band of aliphatic ketones at 280 nm) and high (at least $10^4\,M^{-1}\,cm^{-1}$) for an "allowed" transition (e.g. the $\pi\pi^*$ band in polyenes, aromatics, and dyes). Table 2.1 collects the longest wavelength where selected reagents have an absorbance $A = 1$ at a 0.01 M concentration, as a rough guide in the choice of a suitable λ for excitation.

The two most important functional groups that are commonly involved in direct photochemical processes are the C=C bond in alkenes (dienes) and the C=O bond in ketones. Schemes 2.1a–e describe some of the most important photochemical reactions arising from these two chromophores. The knowledge of the peculiar behavior of the functional group when in the excited state is mandatory to design a selective photochemical reaction. The reaction occurring depends on the promoted electronic transition and, accordingly, on the excited state populated. However, in alkenes the promotion of an electron from a bonding orbital (π) to the

Table 2.1 Absorption of selected organic compounds.

Reagent	λ^a (nm)	Reagent	λ^a (nm)
Acetanilide	288	2,5-Dimethyl-2,4-hexadiene	275
Acetophenone	304	Fluorobenzene	270
Acridine	395	Indene	291
Allyl alcohol	205	Iodobenzene	280
Allyl amine	232	Nitrobenzene	350
Allyl bromide	253	Phenanthrene	345
Aniline	308	Phenol	285
Anthracene	378	Pyrrole	238
Benzaldehyde	298	Quinine	370
Benzophenone	360	Retinol (Vitamin A)	395
Bromobenzene	276	Stilbene	333
Carbazole	350	1,1,2,2-Tetrachloroethylene	260
Chlorobenzene	275	Thiophenol	280
2-Cyclohexenone	310	Uracil	285
cis-1,2-Dichloroethylene	230		

[a] The longest wavelength at which the reagent has absorbance $A = 1$ at a 0.01 M concentration (optical path 1 cm).

corresponding antibonding orbital (π*) generates a ππ* state, where formally the double bond is not present anymore and the rigidity of the C=C bond is removed (Figure 2.2, left part). As a result, an easy *E* to *Z* (or vice versa) conversion may take place. The reaction is feasible even when using azoderivatives, having a N=N rather than a C=C bond (Scheme 2.1a). The diradical nature of the excited C=C bond (especially in enones) is likewise useful for the construction of four membered rings by reaction with other alkenes via a [2+2] cycloaddition process (Scheme 2.1b).

Likewise, the chemistry of the carbonyl group is largely affected by excitation (Figure 2.2, right part). In fact, the ground-state chemistry is dominated by the "basicity" of the n orbitals (functioning as the HOMO) and the electrophilicity imparted by the π* orbital (the LUMO). Considering the n→π* transition, the resulting excited state (being diradical in nature) mimics, however, the behavior of alkoxyl radicals [9]. These radicals are known to fragment to restore the carbonyl group [10] or to abstract hydrogen atoms to form a strong O—H bond [11]. This is reflected in the C(=O)—Cα bond fragmentation to form a radical couple (Norrish type I reaction), from which CO may be lost to form a photoextrusion product

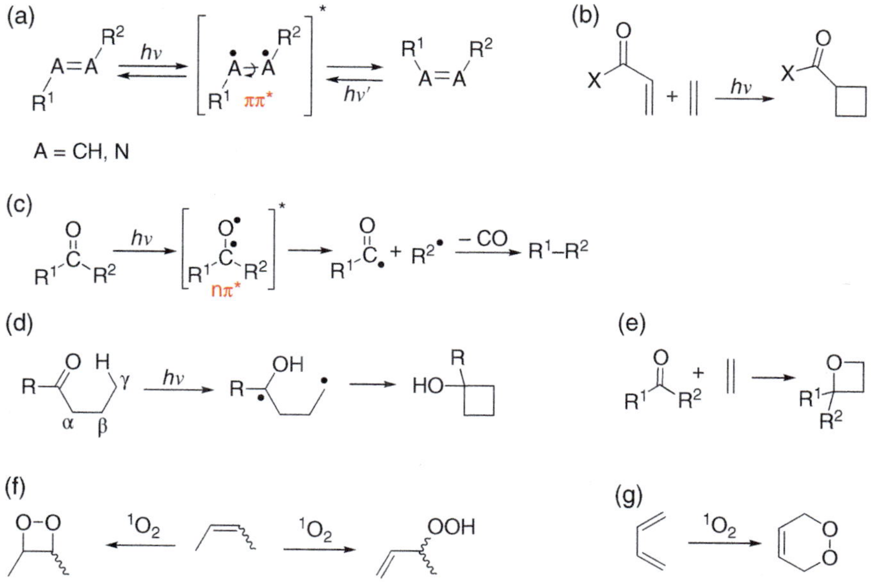

Scheme 2.1 Selected photochemical reactions of double bonds and carbonyls: (a) *E/Z* isomerization, (b) [2+2] photocycloaddition, (c) Norrish type I reaction, (d) Norrish–Yang reaction, (e) Paternò–Büchi reaction; and those involving singlet oxygen: (f) Schenk-ene reaction (right) and synthesis of 1,2-dioxetanes (left); (g) synthesis of 3,6-dihydro-1,2-dioxines via a hetero-Diels-Alder reaction.

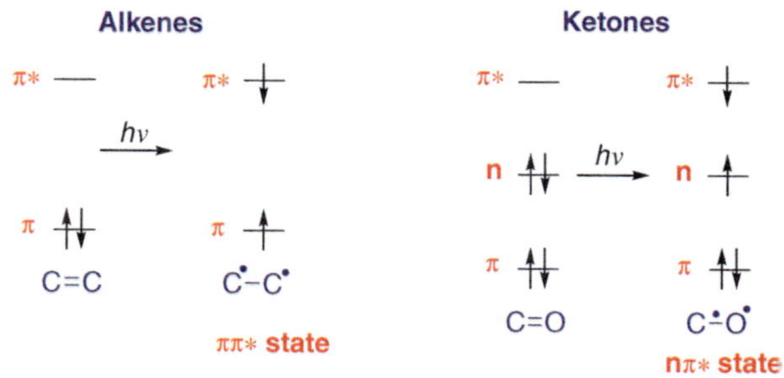

Figure 2.2 Electronic transitions occurring in alkenes (π→π*) and ketones (n→π*).

(Scheme 2.1c). Alternatively, an intramolecular hydrogen abstraction from the γ-position may lead to a biradical intermediate, which upon intramolecular radical coupling forms a cyclobutanol (the Norrish–Yang reaction, Scheme 2.1d). In analogy with the C=C bond, the C=O group could be engaged in a [2+2] cycloaddition reaction to form oxetanes (the Paternò–Büchi reaction, Scheme 2.1e) [12].

In the last decade, photochemistry strongly emerged also as a tool for the photogeneration of reactive ground-state intermediates (carbon- and heteroatom-based radicals, radical ions, diradicals, ions, carbenes, and nitrenes) [13–15]. Most intermediates can be easily detected and characterized by time-resolved spectroscopic techniques (not treated in this chapter), such as Laser Flash Photolysis [16]. This is a precious tool in the understanding and engineering of a photochemical process, since, once the generation of a given intermediate has been confirmed under the employed reaction conditions, practitioners can then investigate how to govern the ensuing chemistry.

These reactive intermediates may be photochemically formed in different ways. Typical examples are the photohomolysis [13–15, 17] or photoheterolysis [18] of chemical bonds, taking place with a negligible activation energy, provided that the energy of the photon is sufficient to achieve the excited state and promote the desired cleavage (Figure 2.3, path b). In most cases, only a given bond is broken, leading to a clean reaction.

Interestingly, a reaction may derive from the excited state of a reactant that did not absorb light. The typical case is that of a photosensitized reaction, wherein a photosensitizer (PS) is responsible of light absorption and then its excited state (PS*) engages an energy transfer with the reactant, ultimately populating its excited state (Figure 2.3, path c) [19].

This is a common strategy for the generation of the triplet state of a reactant while avoiding the intermediacy (and the chemistry) of the singlet excited state that is usually formed upon direct irradiation of the same reactant. A particular

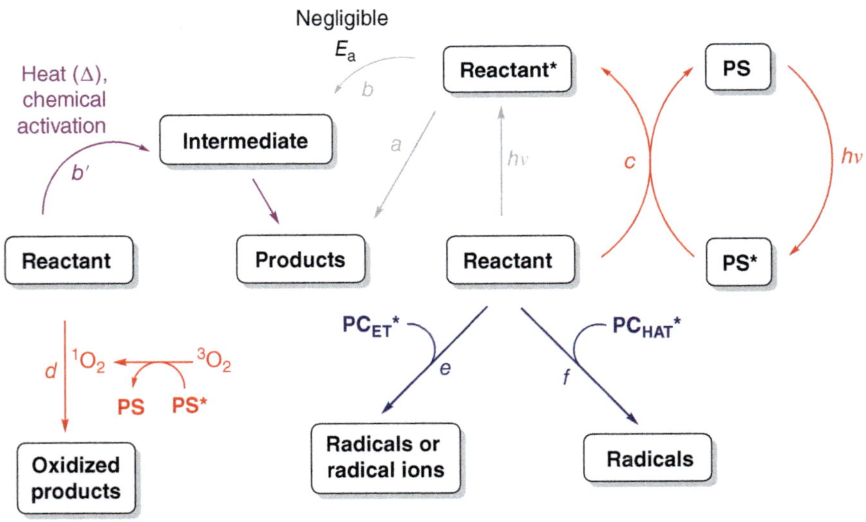

Figure 2.3 General scheme describing thermal (in magenta), photochemical (in gray), photosensitized (in red), and photocatalytic (in blue) processes occurring on a reactant. PC = Photocatalyst. PS = Photosensitizer. HAT = Hydrogen Atom Transfer. ET = Electron Transfer.

case is the generation of the excited state of molecular oxygen. In fact, the ground state of oxygen is a triplet since it contains two unpaired electrons ($^3O_2$). Direct excitation of $^3O_2$, however, is unpractical due to the prohibitive wavelengths required. Accordingly, the formation of singlet oxygen ($^1O_2$), a well-known oxidizing agent, can be triggered through the adoption of a suitable photosensitizer (e.g. a dye; Figure 2.3, path d) [20–22]. Widely used is the reaction of $^1O_2$ with alkenes to form 1,2-dioxetanes or allylic hydroperoxides (the Schenk-ene reaction, Scheme 2.1f) or with dienes to give 3,6-dihydro-1,2-dioxines (Scheme 2.1g) [20–22].

In other instances, the reactive molecule neither absorbs light nor is its excited state involved in the reaction. This is the case of photocatalyzed reactions [23–30], wherein an excited photocatalyst (PC_{ET}* or PC_{HAT}*) activates a light-transparent reagent by a chemical reaction, usually involving an electron transfer (ET, Figure 2.3, path e) [23–30] or a hydrogen atom transfer (HAT, path f) step [31–33]. In the former case, the process takes advantage of the enhanced oxidation or reduction capability of PC_{ET} when in the excited state to generate radical ions (or radicals when charged substrates are used) upon reaction with an organic compound. In the latter case, PC_{HAT}* mimics the alkoxy radical behavior and is thus able to cleave homolytically a C—H bond to form a carbon-centered radical. Figure 2.4 collects representative photocatalysts/photosensitizers currently used in synthesis, some of them being simple organic compounds (PhotoOrganoCatalysts, POC) [34–37].

2.2 How to Plan a Photochemical Synthesis

A crucial issue in the design of a photochemical reaction is assuring an efficient light absorption of the photoreacting species [38]. In this respect, the preliminary measurement of the UV-visible spectrum of the reaction mixture components and the identification of the major absorption bands is mandatory. Later on, other factors must be taken into account as detailed in the following.

2.2.1 The Choice of the Solvent

An important parameter when planning a photochemical process is the choice of the solvent. Apart from the obvious issue of the solubility of reagents/additives, the solvent must not absorb competitively with the photoreacting species. Figure 2.5 gives an indication on the transparency of common solvents, showing that solvents having different polarities can be conveniently chosen. Having a look at the absorption edge for these solvents, it is apparent that acetonitrile and alcohols (MeOH or EtOH) are ideal polar solvents, whereas alkanes (when deprived of other hydrocarbon impurities) and ethers may be suitable as apolar or medium polar media, respectively. All solvents reported are transparent in the visible region.

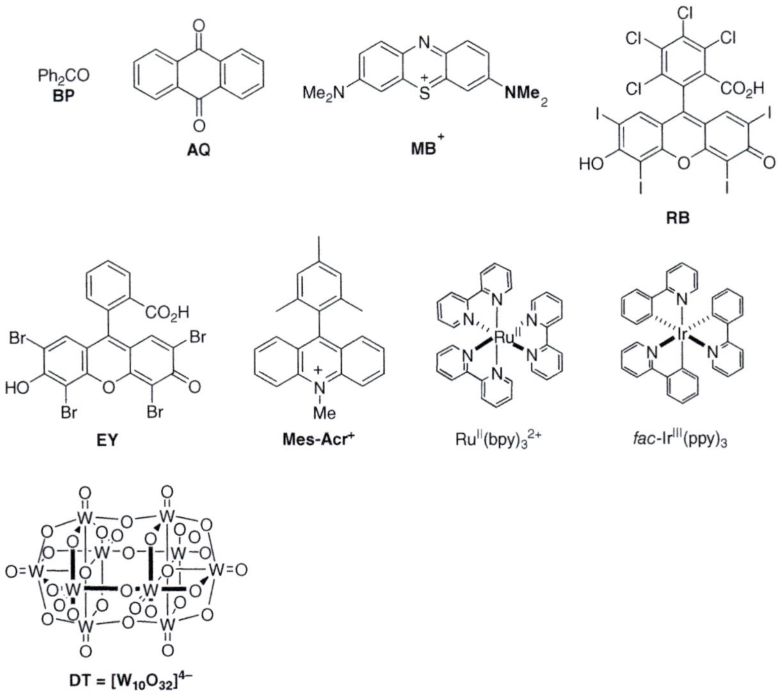

Figure 2.4 Common photocatalysts/photosensitizers used in synthetic applications. **BP** = benzophenone; **AQ** = anthraquinone; **MB^+** = methylene blue; **RB** = rose bengal; **EY** = eosin Y; **Mes-Acr^+** = 9-mesityl-10-methylacridinium salt; **$Ru^{II}(bpy)_3^{2+}$** = tris-(2,2′-bipyridine) ruthenium; **fac-$Ir^{III}(ppy)_3$** = *fac*-(tris-(2,2′-phenylpyridine))iridium; **DT** = decatungstate anion.

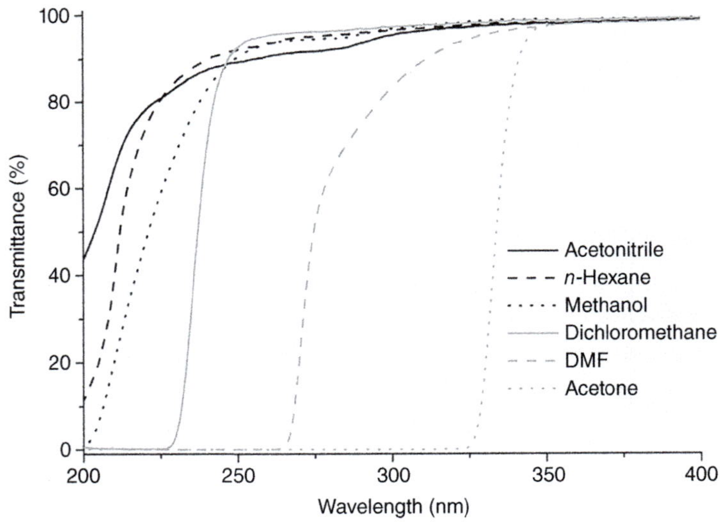

Figure 2.5 Absorption edge of some commonly used solvents (1 cm optical path). *Source:* Albini and Fagnoni [13]/with permission from John Wiley & Sons.

Albeit counterintuitive, a photochemical reaction may take place even when the reagent is in the solid state (e.g. by irradiating crystals) leading even to enantioselective processes [39, 40]. An alternative is the photolysis of a suspension of microcrystals obtained by precipitating the reagent in a proper solvent while stirring [41]. In some instances, photocatalyzed reaction may be carried out under heterogeneous conditions where the PC is not soluble in the reaction mixture [13].

2.2.2 Concentration of the Absorbing Species

Other crucial issues in a photochemical reaction are the concentration and the molar absorption coefficient (which represents a measure of how strongly a compound absorbs light at a given wavelength) of the photoreacting species, which reflect the absorbance of the system. Roughly, adopting a high concentration lengthens the required irradiation time, in turn making the process sluggish. As an example, irradiating a 0.1 M solution of the reagent (vessel having a 1 cm path) on a wavelength maximum with $\varepsilon_{MAX} = 10\,M^{-1}\,cm^{-1}$, an $A = 1$ results. From the Lambert–Beer law, $A = \log(1/T)$, where T = transmittance, a measure of the fraction of light that passes through a material (see Glossary). Accordingly, 90% of the light intensity is absorbed by the solution, but 10% is transmitted (and is thus not useful for the transformation). If $\varepsilon_{MAX} = 10^4\,M^{-1}\,cm^{-1}$, however, 99.9% of the light is absorbed, since $A = 10^3$, and this means that 90% of the photons are absorbed in the first 0.1 mm layer of the solution, while the remainder of the solution cannot interact with light, remaining unreactive. A balance is then required since a too low concentration requires a huge amount of solvent (and of solution, accordingly) for the obtainment of the required product(s) on the desired scale, while a too high concentration may call for a long irradiation time rendering the process unpractical. An effective mixing of the solution, however, can alleviate (at least, in part) such a drawback. A general improvement in this respect is the use of flow conditions [42–45] or microreactors [46–48] (*vide infra*), where the limited optical path of the reacting solution allows for an efficient absorption of light.

A further problem may arise from the competitive absorption of the thus-formed photochemical product(s) that could completely hinder the photoreaction, when photostable (the so-called inner filter effect), or yield further undesired secondary photoproducts, if photoreactive. In some instances, the products (or by-products, such as polymers) may form films on the wall of the vessel, thus stopping the reaction because the remaining reagent in solution receives little, if any, light. As apparent from the latter consideration, there is not a general rule for the design of a given photochemical process and, in each case, a fine tuning of the reaction conditions is required when investigating a new reaction.

2.2.3 The Reaction Vessel

A photochemical reaction may be roughly carried out in two ways: by external or internal irradiation. In the first case, the solution to be irradiated is confined in a vessel (usually a tube or a vial) and the reaction is promoted by an external light source (see Section 2.2.4). This situation is particularly suited for small-scale photoreactions, where small amounts of solution (*ca.* 1 mL) may be used. This setup is indicated in the optimization of the various reaction parameters (concentration of the reagents and/or additives, best wavelength, etc.), avoiding the need to waste (expensive) starting materials. Furthermore, the vessel may be purposely designed, allowing to tune the scale of the reaction.

As apparent from Figure 2.6, with a light source emitting at 254 nm only rather expensive vessels made of fused quartz (that are transparent above 170 nm) can be used. The transparence limit of Pyrex glassware (or borosilicate glass) is around 300–310 nm; albeit less expensive, these vessels (typical examples include vials or round-bottomed flasks commonly available in a chemistry laboratory) are practical only when irradiating with $\lambda > 350$ nm, therefore including the entire range of visible light.

2.2.4 Light Sources

A judicious choice of the light source is fundamental for increasing the efficiency of the reaction, and a good matching between the absorption band of the photoreacting species and the wavelength emitted by the light source is required, also

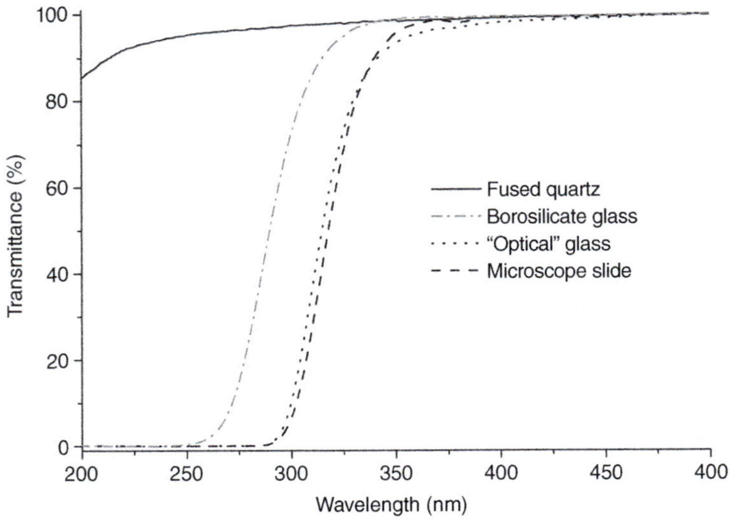

Figure 2.6 Transparency of materials used for the reaction vessel (1 mm thick). *Source:* Albini and Fagnoni [13]/with permission from John Wiley & Sons.

verifying that nothing else interferes (e.g. the wall of the vessel or the cooling jacket of the lamp [*vide infra*]) with the photons in reaching the target molecule. A photochemical system should be engineered considering the dimension and the shape of the lamp along with the geometrical arrangement with respect to the reactor vessel. In this regard, performing a photochemical reaction at the industrial level is still challenging, despite a hard work has been devoted in the setup of photochemical reactors suitable to promote an efficient synthesis on a large scale (kg) [49, 50].

As it has been previously stressed, in some cases the reagent may not be the light absorbing component, as in the case of photosensitized or photocatalyzed reactions. In rare instances, the solvent itself may be the absorbing species, as in the case of acetone when adopted as photosensitizer (see Glossary).

The type of light sources used depends on the range of wavelength required for triggering the desired reaction. It is possible to find focalized sources, where the beam is directed on the object to be irradiated, or non-focalized sources, where light is emitted over the entire sphere described by the light source, thus decreasing the light power that hits the target. In the UV range, the most largely applied light sources are mercury vapor arcs, which can be divided according to the operating pressure. The main drawback of these lamps is the low efficiency in converting the electrical power into light that is dispersed over a range of wavelengths and toward all directions. Nowadays, there are several firms supplying photochemical apparatuses including the light source [51], the appropriate power supply, and in some cases the reaction flask with accessories, e.g. for gas inlet. The main characteristics of lamps used for photochemical synthesis are presented in the following.

2.2.4.1 Low-Pressure Mercury Arcs

Low pressure (*ca.* 10^{-5} atm under operating conditions) Hg arcs (often dubbed as germicidal lamps) are long-lived (>10^4 hours) light sources traditionally used in organic photochemistry. These are supplied as quartz tubes of various lengths, typically from 20 to 60 cm and 6–16 W power (Figure 2.7a), exhibiting an emission at 254 nm. However, when the lamps are internally coated with a phosphor (or a combination of phosphors) that absorbs the Hg radiation, the lamp emission can be shifted to longer wavelengths. These light sources include:

- lamps with emission centered in the UV-B region, at 305–310 nm (a white phosphor is used);
- "black light" (also called "wood light") lamps (see again Figure 2.7a), painted with $BaSi_2O_5 : Pb^+$ phosphor (emitting at 350–355 nm) or a $SrB_4O_7F : Eu^{2+}$ phosphor (emitting at 368–370 nm);
- lamps with the emission centered in the visible range at various wavelengths, e.g. in the blue around 450 nm or over the whole spectrum ("compact fluorescent lamps," CFL).

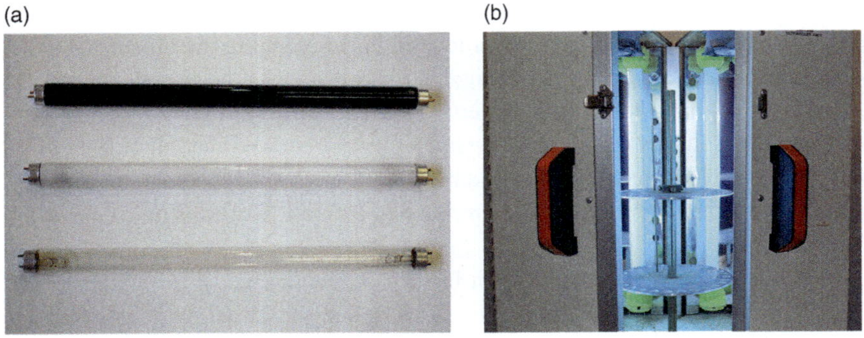

Figure 2.7 (a) Low-pressure mercury arcs used for photochemical reactions. From top to bottom: 350–370 nm phosphor-coated "wood light" lamp, 305–310 nm phosphor-coated lamp, and 254 nm emitting lamp. (b) A typical multilamp apparatus fitted with low pressure mercury lamps.

A multilamp apparatus where 8–12 Hg lamps are circularly disposed is shown in Figure 2.7b; the test tubes containing the solution to be irradiated are placed inside. A rotating "merry-go-round" setup ensures a uniform illumination of the vessels and a cooling fan maintains the temperature <40 °C inside the apparatus to avoid undesired thermal contributions to the reaction. Such multilamp systems are commercially available [51], but everybody can build a home-made version by assembling the lamps in a lamp holder.

2.2.4.2 Medium- and High-Pressure Mercury Arcs

Medium-pressure mercury arcs (sometimes also dubbed as "high," 1–10 atm operating pressure) are available in the 100–1000 wattage range and are characterized by some emission peaks (at 254, 313, 366, 405, 550 nm) over a continuum background. Such light sources develop a considerable amount of heat and cooling is mandatory (e.g. by circulating tap water in a suitable cooling jacket) to maintain the reaction temperature around 20 °C. These lamps are well suited for an internal irradiation of solutions having volumes between 100 and 1000 mL, having recourse to an immersion well apparatus (Figure 2.8). In this case, the reactor vessel and/or the cooling well should be made of a material (quartz or Pyrex) transparent to the wavelength absorbed by the desired species. If the cooling well is made of Pyrex, a filter effect operates and, as a result, the more energetic wavelengths do not reach the reacting mixture.

2.2.4.3 Other Light Sources

Tungsten (incandescent) and tungsten–halogen lamps are based on the emission from a tungsten wire heated by conduction. Such lamps find a limited application (mainly in photoinitiated chain reactions) because they liberate much heat.

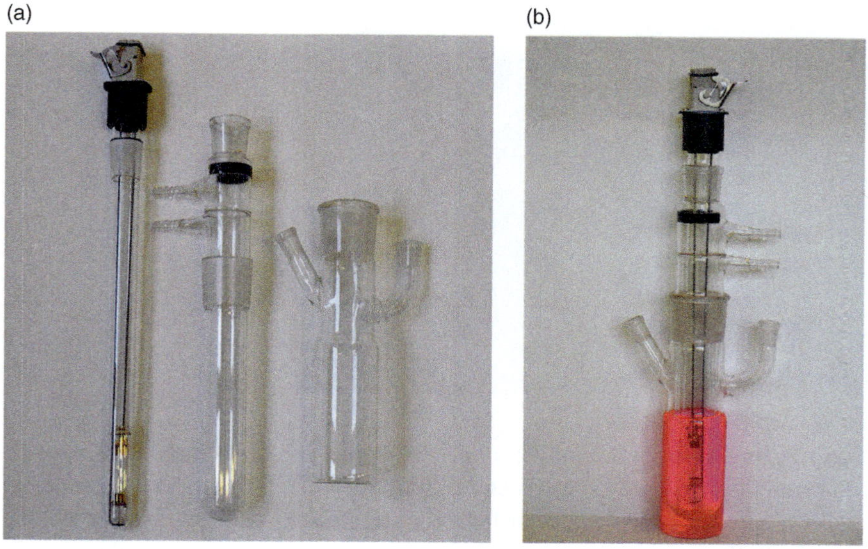

Figure 2.8 (a) Parts of an immersion well irradiation apparatus (from left to right): medium-pressure mercury arc inserted in an internal tube (made of quartz), cooling vessel (made of quartz if emission below 310 nm is required, otherwise of Pyrex), irradiation vessel made of Pyrex. (b) Assembled apparatus.

Incandescent lamps exhibit a continuous emission with no UV component and peaking in the 550–650 nm region, having a <5% conversion of electric power to light, whereas the tungsten-halogen lamps have a center of emission shifted toward the blue. Energy-saving **compact fluorescent lamps (CFLs)** were designed to replace incandescent or halogen lamps and emit "white" light from a mix of phosphors inside the bulb. The lighting efficiency is increased to 7–10% and these lamps have an operation time of 6000–15000 hours. This source has been now replaced with toxic mercury-free sources, such as LEDs.

Light-Emitting Diodes (LEDs) are intrinsically focalized sources made of a semiconductor material that emits an incoherent electroluminescence. LEDs are nowadays widely used and, although the emission covers the whole of the visible and the UV-A spectrum (down to *ca.* 300 nm), in the latter case the emission is much less intense and the LEDs more expensive. The conversion of the input electric power into light depends on the output wavelength, usually having a narrow range of emission (typically a 20 nm range). Visible LEDs are however cheap and long-lived ($>10^4$ hours) allowing the fine-tuning of emission wavelength. Among the different LED-based light sources, the devices manufactured by Kessil are the most used in organic synthesis. Such lamps exhibit a high and stable output, precise wavelengths, and strong penetration in the solution, contributing to the fast development of photoredox catalysis (Figure 2.9) [52].

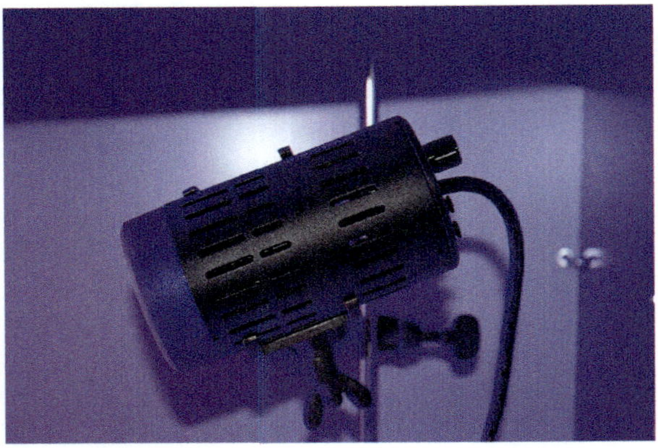

Figure 2.9 Kessil lamp emitting at 427 nm.

Sunlight is a perennial source, is freely available, and can be used with no need of consuming electricity. This is certainly a convenient alternative to the commonly employed lamps, but solar light is characterized by a low density, lack of reproducibility (the geographical location, time of the year, and weather conditions largely affect the flux impinging), and a limited UV component (mainly in the UV-A region) [53, 54]. Nonetheless, putting a solution on a window ledge in a sunny day could be an eco-sustainable way for carrying out a photochemical reaction, provided that the compound absorbs at 330 nm or longer wavelengths [55]. To mimic solar light with high reproducibility, a *high-pressure xenon* arc lamp may be fitted in a photochemical reactor (called "SolarBox"), mimicking the emission spectrum of natural sunlight [56]. In the past, solar collectors were designed to promote photochemical reactions on a large scale (at least in the kg scale) [57], while nowadays these are not extensively used.

2.2.5 From Batch to Flow Conditions

As stated in Section 2.2.2, if the reagent (or the catalyst) has a high absorbance (or is present in a high concentration), practically most of the solution is not affected by light, except for a thin layer adjacent to the surface exposed to the radiation source. This is the main drawback when designing large-scale photoreactors, where an effective mixing may partially solve the so-called mass transfer problem.

These limitations can be overcome by having recourse to flow conditions (see Chapter 4), wherein the irradiated solution, stored in a suitable reservoir, is circulated through the actual reactor with a small optical path, thus exposing a thin layer of solution to the lamp at any time [42–44]. The main advantages of this approach are related to reducing the reaction time and improving productivity, usually with no negative effect on the overall yield. Thus, the solution containing the product is exposed to the light source for a strongly reduced period of time with respect to batch conditions, thus limiting the likelihood of product degradation and/or by-product formation. Flow photochemistry reactors come in a range of designs.

The most common way for irradiating the solution under flow conditions is through a tubing containing the flowing reacting mixture. The output of the reaction strongly depends on the material of the tubing, the thickness of the channels where the reaction mixture is circulated, and the flow rate. It is possible to classify such reactors in different groups, including mesoreactors (optical path of the channel >0.5 mm, flow rate > 1 mL min^{-1}) or microflow reactors (having micro-sized channels usually <0.5 mm in diameter) [42–44]. In both cases, the material of the tubing (or of the assembled photoreactor) must assure a high light transmission and complete inertness to the reactants and solvents used. Several tubings made of inert polymeric materials are used for such application, perfluoroalkoxyalkanes (PFA) and fluorinated ethylene propylene (FEP) being the elective choice for the high transmission (>90%) in the visible and UV light ranges. Moreover, these polymers exhibit high flexibility and resistance to relatively high pressures, and are inert to strong acids and bases, as well as do not suffer from swelling phenomena.

Roughly, the assembling of a flow photoreactor may be designed in two ways. The reactor could be built *around* the lamp assuring an internal irradiation (see the case of the apparatus shown in Figure 2.10, left) [42], where the solution is flown in a transparent tubing wrapped around the lamp. By using this apparatus, considerable volumes (up to several liters) of solution may be irradiated in a limited period of time [42, 58]. External irradiation is well suited for "lab-on-a-chip" microreactors, where the solution flows into microchannels while being irradiated by a lamp placed in front of the reactor (Figure 2.10, right). More recently, the design and preparation of 3D-printed microreactors [46–48] for light-driven organic synthesis was proposed [59]. However, apart from lab-made apparatuses, different continuous flow devices are commercially available [60, 61].

A peculiar case deals with "falling film" photoreactors, wherein a solution from a reservoir is sprayed at the top of the (cooled) lamp by means of a pump [62]. Alternatively, a thin layer of solution may be made to fall in the proximity of the light source or on the wall of the cooling jacket.

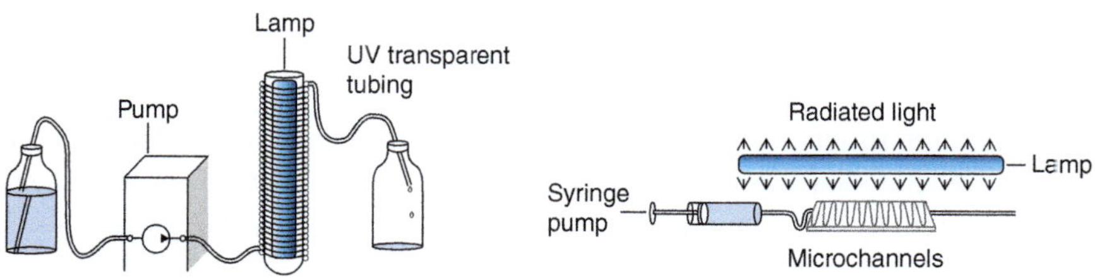

Figure 2.10 Reactors using internal (left) and external (right) irradiation. *Source:* Knowles et al. [42]/Beilstein Institute for the Advancement of Chemical Sciences/CC BY-2.0.

2.2.6 Preparation of the Sample

As hinted above (Section 2.1.2), the lifetime of excited states is rather short (roughly lifetime ca.1 μs for triplets and ca.1 ns for singlets) and a photochemical process must be very fast to favorably compete with physical decay. This has a positive implication in reducing the effect of impurities present in solution (thus the extensive purification of reagents and solvents is not generally required) but a large excess of the reaction partner may be mandatory in case of bimolecular reactions for intercepting the short-lived excited state.

A peculiar case is represented by molecular oxygen, which may quench excited states at diffusion-controlled rate. The oxygen present in many air-equilibrated organic solvents can be estimated in the range $2–3 \times 10^{-3}$ M (5×10^{-4} M in water), thus, enough to quench >90% of long-lived triplets. In some cases, oxygen is purposely added to the reaction mixture to ascertain if a triplet excited state is involved in the process or not. Moreover, O_2 may smoothly react also with radical or radical ion species formed to deliver oxidation products, thus diverting the expected reaction outcome. At any rate, the amount of oxygen dissolved drops by *ca.* three orders of magnitude when flushing the solution with an inert gas (nitrogen or, better, argon) for some minutes to ensure "deoxygenated" conditions. In particular cases, where oxygen exerts a profound effect, freeze-pump-thaw cycles, where the solution is frozen (thanks to the use of liquid nitrogen) and the dissolved gas is then removed under vacuum, are compulsory. This precaution may be skipped, however, thus further simplifying the experimental procedure if it is proven that oxygen has no appreciable effect on the reaction outcome.

A great advantage is the negligible dependence of many photochemical processes on temperature, thus the reactions are routinely carried out at room temperature, unless the nature and/or the lifetime of the thus formed products (or ground state intermediates) call for a specific temperature control (e.g. lowering or increasing).

2.2.7 Safety Equipment

The use of intense light sources (including solar light) may raise some safety issues. In fact, UV-B radiation is dangerous for the skin, causing erythema, and UV-C is genotoxic. Likewise, the exposure to a strong UV-A (up to 400 nm) or visible light radiation may cause damage.

For this aim, irradiation apparatuses must be placed in a protected area. Direct eyes exposure to short-wavelength light must be avoided during the execution of the experiment by wearing purposely designed protective glasses. However, accidental exposure may occur, which might cause a quite annoying *conjuctivitis*.

When using high-pressure mercury arcs, ozone may be formed locally in a high concentration. In this case, it is highly recommended to flush nitrogen in the close vicinity of the arc (to avoid a buildup of ozone concentration) or carrying out the reaction under a fume hood. In addition, a proper disposal of the mercury-containing exhausted lamps is always mandatory.

2.3 Selected Applications of Photochemical/Photocatalyzed Reactions

In the following, selected examples are presented and discussed with no particular emphasis to the "mechanistic" detail of the reactions [63, 64].

This section provides several recent (mostly in the last 10 years) examples of diverse photochemical reactions involving the most common chromophores, including demonstration of a range of representative activation modes described in Figure 2.3 (the absorbing species in each scheme is highlighted in red). In addition to a general description of the chemical reaction, an abbreviated procedure is reported to give the reader insights into the technical aspects of experimental setup. Lastly, key information about the equipment used is given.

2.3.1 Reactions Involving the C=C Double Bond

Rotation around a C=C double bond is energetically disfavored. Therefore, alkenes can exist as non-interconverting *E*/*Z* isomers. These species can have very different shapes and properties, the case of vision being a notable example of the profound impact of this type of isomerism [65]. Very recently, photo-induced *E*/*Z* isomerization of enamines has been exploited to achieve the deracemization of α-branched aldehydes [66]. As shown in Scheme 2.2, this approach capitalized on a chiral primary amine organocatalyst and a readily available photosensitizer.

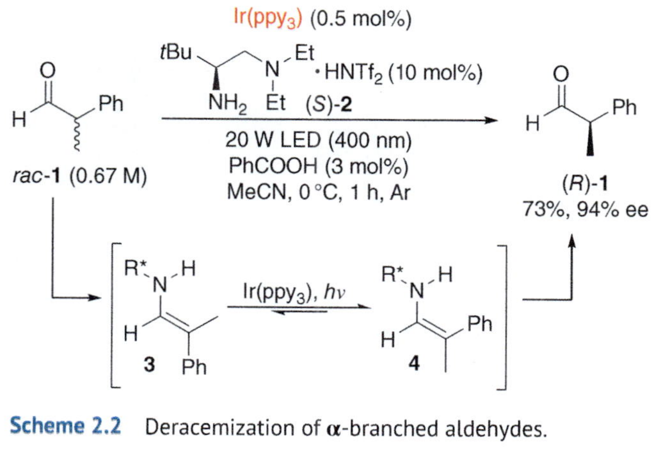

Scheme 2.2 Deracemization of α-branched aldehydes.

In the process, the racemic aldehyde (*rac*-**1**) reacted with the aminocatalyst (*S*)-**2** to preferentially deliver the *E*-enamine **3**. 400 nm LED excitation of the $Ir(ppy)_3$ photosensitizer followed by energy transfer to **3** triggered the conversion to the *Z*-isomer **4** that underwent facially selective protonation leading preferentially to (*R*)-**1**. Noteworthy, this protocol demonstrated a remarkable functional group tolerance, including ketones. However, the presence of an α-aromatic substituent in the starting aldehyde is a prerequisite to obtain a good performance.

*Typical procedure for the synthesis of (R)-**1**:* To an oven-dried 10 mL Schlenk tube equipped with a magnetic stir bar was added (*S*)-**2**/$HNTf_2$ (10 mol%), $Ir(ppy)_3$ (0.5 mol%) and benzoic acid (3 mol%). Freshly prepared *rac*-**1** (0.2 mmol) dissolved in 0.3 mL MeCN was then added into the tube under Ar atmosphere. The mixture was degassed for three times using standard freeze–thaw method and stirred at 0 °C under irradiation of 20 W 400 nm LED for one hour. After the completion of deracemization, the solution was purified by flash column chromatography to give (*R*)-**1** (73%). The ee value of the isolated aldehyde was determined by HPLC analysis.

Experimental setup: A 20 W 400 nm LED fixed on an aluminum block was glued at the bottom of a crystallizing dish by silicone gasket and grease (Figure 2.11). The temperature was controlled by the ethanol bath and coolant circulation pump.

Vitamin D analogues endowed with biological activity are commonly prepared through a photosensitized protocol exploiting an *E*/*Z* isomerization of the C=C bond. As an example, calcipotriol (**8**), routinely used for the treatment of psoriasis can be obtained by deprotection of silyl ether **7** in turn formed from the conjugated triene derivative **5** upon irradiation with UV light in the presence of a catalytic amount of 9-acetyl anthracene (**6**) as the photosensitizer (Scheme 2.3) [67]. The reaction has been performed both under batch and flow conditions. In the

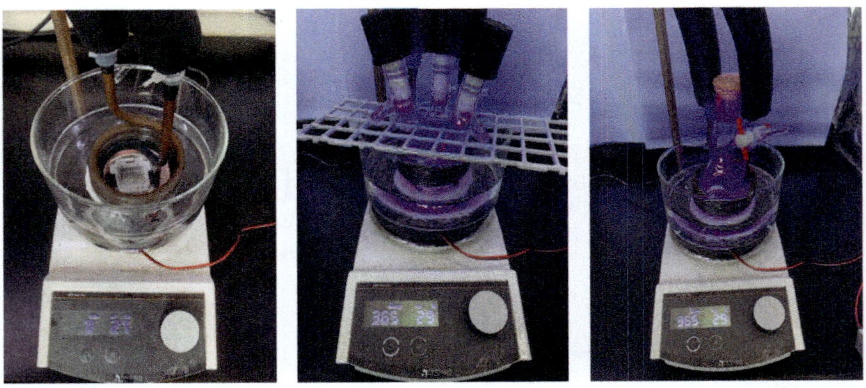

Figure 2.11 (Left) Photoreactor. (Middle) 0.2 mmol scale reaction setup. (Right) 5 mmol scale reaction setup. *Source:* Huang et al. [66], Reproduced with permission from American Association for the Advancement of Science – AAAS.

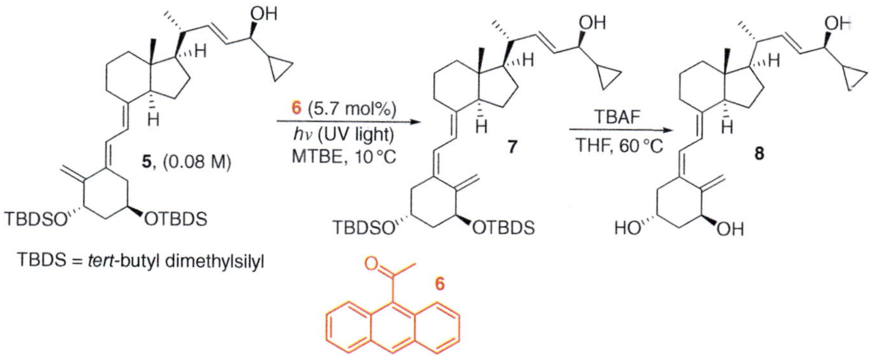

Scheme 2.3 Photosensitized synthesis of Calcipotriol **8**.

former case, the reaction was slower with a lower productivity (1.64 min g^{-1} product vs. 0.72 min g^{-1} in flow) and led to much less pure product (65% vs. 80% yield, under batch and flow conditions, respectively).

Typical procedure for the synthesis of **7**: An oxygen-free solution of **5** in MTBE (1 g/20 mL) containing 450 mg of **6** was photo-isomerized under batch and flow conditions. In both setups, the lamp providing UV-light (medium pressure mercury lamp doped with iron from Heraeus: TQ718 Z4, 800 W power supply), surrounded by a quartz lamp-housing comprising inner and outer jackets for water cooling, was placed in the center of a standard immersion well photoreactor having closable inlet and outlet openings. The lamp was temperature-controlled by cooling the light housing with cooling water (10 °C). The irradiation layer thickness of the photoreactor was 9.7 mm. The reaction was kept under nitrogen at all times. In the flow version, the flow was controlled to be in the range of 3600–4800 mL min^{-1}.

Very recently, the enantioselective preparation of bridged cyclobutanes via a crossed intramolecular [2+2] cycloaddition has been reported (Scheme 2.4) [68]. The reaction occurred between the tethered C=C bond of the allyloxy moiety within 2-(alkenyloxy)cyclohex-2-enones **9a-c** and took place in the presence of a chiral-at-Rh complex chelating Lewis acid (**10**, 2 mol%) upon irradiation with visible light (437 nm). Thus, **9a-c** bound to the catalyst to form a chiral complex and, upon excitation, delivered the desired cyclobutanes **11a-c** containing an otherwise elusive bridged framework with an excellent control over the absolute stereochemistry. Notably, the obtained tricyclic photocycloaddition products could be further elaborated to give different cyclobutane-containing derivatives.

Typical procedure for the synthesis of **11a-c**: An oven-dried Schlenk tube, equipped with a magnetic stir bar, was charged with **10** (2 mol%), the chosen **9a-c** and 10 mL of dry DCE. The reaction mixture was deoxygenated by three cycles of

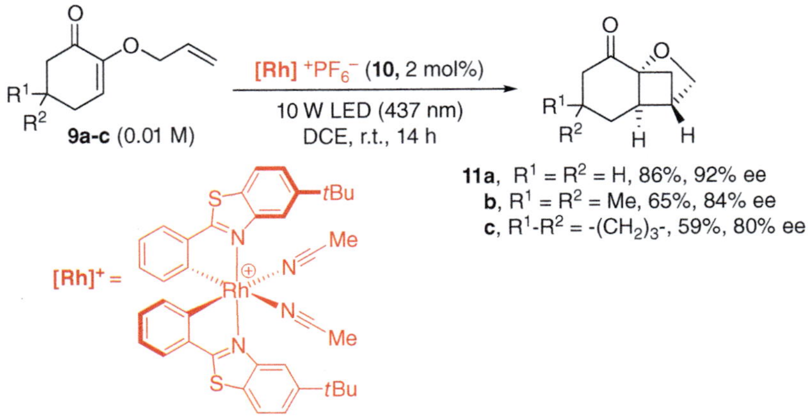

Scheme 2.4 Synthesis of bridged cyclobutanes.

"freeze-pump-thaw" and stirred at room temperature under blue LED irradiation (λ = 437 nm) to afford the corresponding cyclobutanes **11a-c** after purification by flash chromatography.

Experimental setup: Photochemical experiments using a LED (λ = 437 nm) were carried out in a Schlenk tube (diameter = 1.0 cm) with a polished quartz rod as an optical fiber, which was roughened by sandblasting at one end (Figure 2.12). The roughed end must be completely submerged in the solvent during the reaction in order to guarantee optimal and reproducible irradiation conditions.

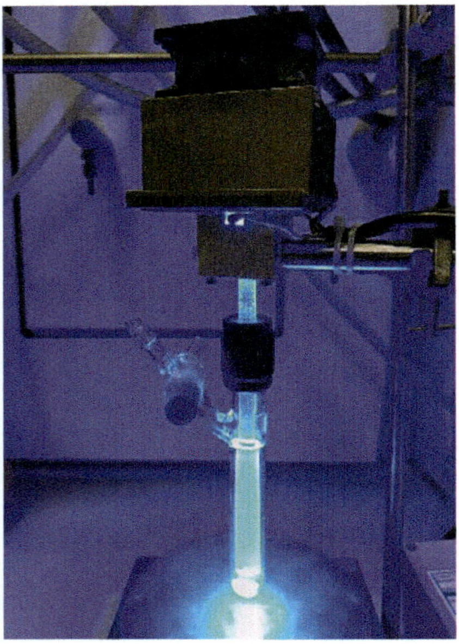

Figure 2.12 Photochemical apparatus employed for the synthesis of **11a-c.** *Source:* Rigotti et al. [68], Reproduced with permission from Royal Society of Chemistry.

2.3.2 Reactions Involving the C=O Double Bond

The Paternò–Büchi (PB) reaction is one of the earliest known photochemical reactions having a synthetic potential and involves the direct excitation of an (aromatic) carbonyl to promote a [2+2] cycloaddition onto an olefin to form an oxetane [12]. Very recently, PB cycloadditions have been applied to dearomatization reactions. The tetrahydrooxeto[2,3-*b*]indole core in compounds **13a-e**, bearing up to three contiguous all-substituted stereocenters, was efficiently built from indoles **12a-e** and aromatic ketones (e.g. benzophenone) upon visible light (405 or 456 nm) photolysis in toluene (Scheme 2.5) or acetone [69]. The procedure was smoothly optimized also under microfluidic flow conditions.

*Typical procedure for the synthesis of **13a-e***: The chosen indole (**12a-e**, 0.1 mmol, 0.1 M) and benzophenone (1.0 equiv.) were dissolved in toluene or acetone (1 mL) and the reaction mixture was bubbled with N_2 for one minute. The vial containing the so-prepared solution was then degassed and irradiated for 16 hours in the photoreactor illustrated in Figure 2.13. The desired product was obtained via purification by chromatography (when needed).

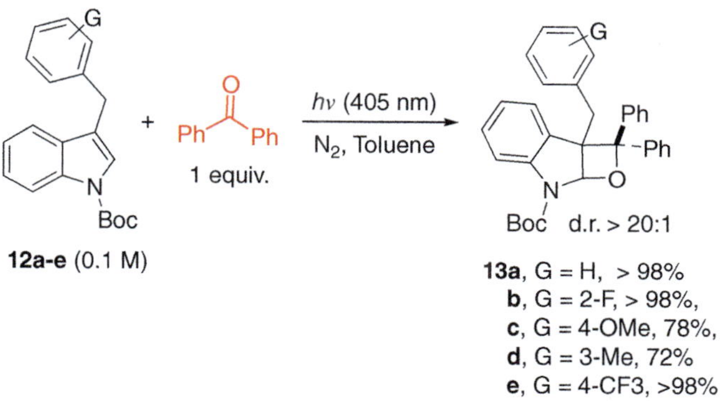

Scheme 2.5 Paternò–Büchi reaction for the construction of the tetrahydrooxeto [2,3-*b*]indole core.

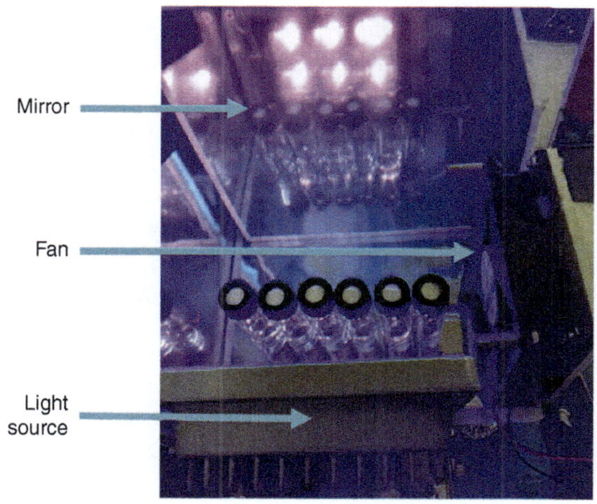

Figure 2.13 Photochemical apparatus employed for the preparation of **13a-e.** *Source:* Rigotti et al. [68], Reproduced with permission from Royal Society of Chemistry (Creative Commons Licence).

Experimental setup: Figure 2.13 shows the general setup of a batch reaction under 405 nm irradiation. The reaction mixture was placed in the middle of the photoreactor (at about 1.5 cm distance from the light source). To maintain a stable reaction temperature, one fan was placed close to the vials (25 ± 2 °C) and the temperature was controlled by a thermometer.

2.3.3 Reactions Involving a Photoinduced Homolysis

N-Chloroamines have been considered as visible light reactive substrates to be used in atom-transfer radical addition (ATRA) reactions [70]. Recently, a continuous flow reactor equipped with a separation membrane was used for the preparation of *N*-chloroamines (e.g. **14**). Subsequently, these species were directly used for the photochemical functionalization of olefins **15a-e** to afford β-chloroamines **16a-e** in up to 91% yield and a throughput of more than 20 g h^{-1} (Scheme 2.6) [71]. All the reactions have been carried out by means of a commercial continuous flow reactor consisting of a compact glass fluidic module (155 × 125 × 8 mm size, 0.4 mm channel size, 2.77 mL internal volume), encased within a high-capacity heat exchange channel (20 mL volume). Interestingly, the light source can be tuned according to the reaction conditions (365 nm for direct irradiation and 405 nm when 9-fluorenone is present as photosensitizer).

*Typical procedure for the flow synthesis of **16a-e**:* All solutions (containing bleach, piperidine, and **15a-e**) were degassed by sparging with an argon balloon. The reactor was turned on (LEDs, pump, and thermostats) and the system was given ~10 minutes to equilibrate. The solution reservoirs (loops of tubings) were charged with the amine and the bleach solutions and injected into the reactor. To account for any dilution effects at the edges of the injected sample, only the central fraction (~1 mL for optimization, 5.54 or 8.31 mL for isolation) was collected.

Experimental setup: Flow reactions were conducted in a commercial continuous flow reactor: Corning Advanced-Flow Lab Photo Reactor.

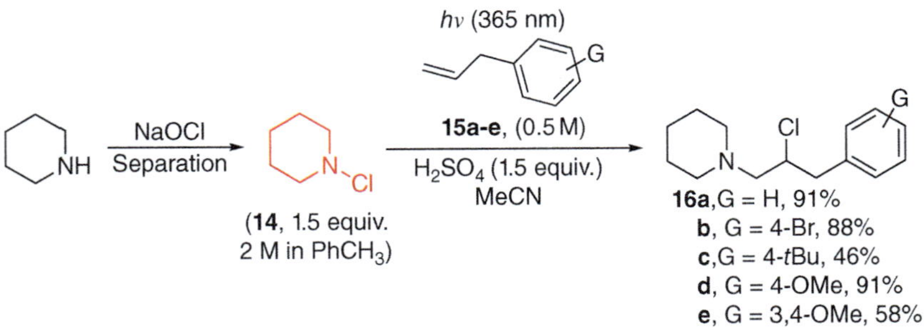

Scheme 2.6 Atom-transfer radical addition (ATRA) reaction between an *N*-chloroamine and olefins.

Arylazo sulfones ($ArN_2SO_2CH_3$, **17a-e**) are compounds bearing a dyedauxiliary group (a substituent responsible for both the color and the photoreactivity of a given starting substrate) and found wide application in the visible light and sunlight-driven formation of both Ar-C and Ar-heteroatom bonds [72, 73]. The arylation of various (hetero)arenes (Figure 2.14, top) was successfully conducted in a lab-made solar microcapillary reactor (the "sunflow" reactor, Figure 2.14, bottom) in three different geographical locations (Germany, Italy, and Brazil) with a short exposition time (one hour) [74].

*Typical procedure for the flow synthesis of **18a-e**:* The chosen arylazo sulfone (**17a-e**, 0.2 mmol, 1.0 equiv., 0.05 M) was flushed with argon in a bottom flask and dissolved in degassed acetonitrile : water (9 : 1, 4.0 mL). The flask was wrapped in aluminum foil and then the chosen arene (furan, thiophene, 10–25 equiv.) added. The stock solution was thus charged in a "sunflow" reactor that was in turn positioned outdoor and exposed to sunlight for one hour. Biaryls **18a-e** were isolated from the raw photolysate by solvent removal followed by column chromatography.

Experimental setup: The "sunflow" reactor (Figure 2.14) was flushed with argon and the stock solution was injected via a Luer–Lock syringe. The reactor was positioned outdoor and exposed to sunlight for one hour.

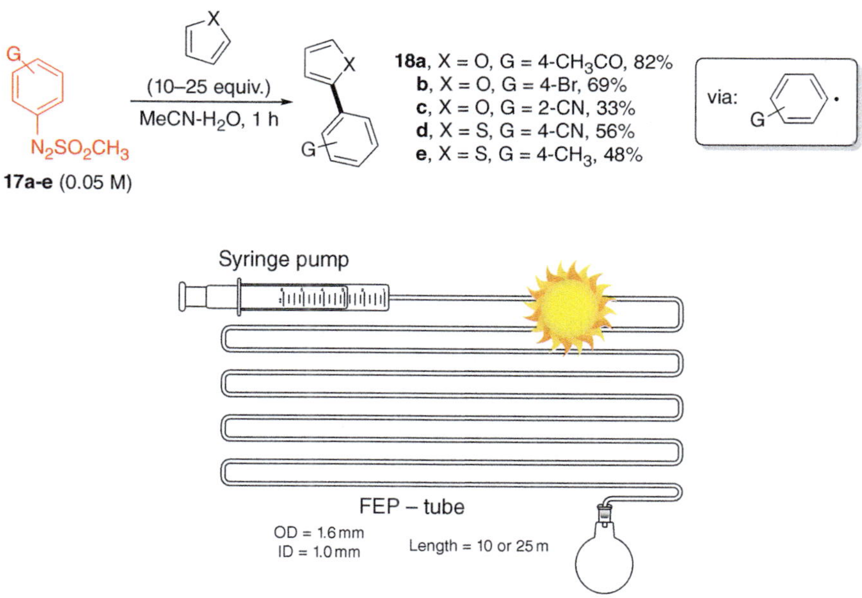

Figure 2.14 Photochemical flow reactor employed for the synthesis of **18a-e**. *Source:* Reproduced with permission from da Silva et al. [74]/John Wiley & Sons.

2.3.4 Reactions Involving Singlet Oxygen

The amount of Artemisinin, which is the most effective antimalarial drug, obtained via extraction from the plant sweet wormwood is obviously not sufficient to supply more than 300 million treatments needed each year, making the cost of such drug unsustainable. However, Artemisinin (**20**) can be prepared from naturally occurring dihydroartemisinic acid (**19**), and the key step in such synthesis is a Schenk-ene reaction with photo-generated singlet oxygen (Scheme 2.7) formed upon photosensitization by using 1,9-dicyanoanthracene (DCA, 0.5 mol%). In a continuous flow protocol, **19** (in turn recovered from the mother liquor accumulated in the commercial artemisinin extraction) is converted into **20** in up to 42% yield with a short residence time (1.5 minutes) and a calculated space-time yield of $3500\,kg\,m^{-3}\,day^{-1}$ [75, 76].

*Typical procedure for the flow synthesis of **20***: The photochemical reactor has been fed with a solution of **19** (0.5 M) and TFA in toluene, in the presence of 9,10-dicyanoanthracene (DCA, 0.5 mol%) as the photosensitizer. An HPLC pump (Knauer, Smartline pump 100) has been used to deliver either substrate solution or pure solvent by means of a two-way switch. In an ETFE T-mixer (IDEX Health and Science), the solution was mixed with oxygen (flow rate, $5\,mL\,min^{-1}$). The oxygen saturated solution was then pumped through the photoreactor that is equipped with a fluorinated ethylene–propylene copolymer (FEP) tubing and the plate cooled down to −20 °C with an immersion cooler. An LED module (420 nm emission wavelength) was employed for the irradiation. The solution was then re-heated to 10 °C and then kept at room temperature. The collected solution was thus washed with aqueous saturated $NaHCO_3$ and the solvent removed under vacuo to afford a yellow solid that was solubilized in acetonitrile and filtered through a PTFE syringe filter with 0.45 μm cutoff. Crystallization of the crude residue gave **20** in 42% yield (based on the starting **19**).

2.3.5 Reactions Involving a Photocatalytic Step

Photocatalysis has recently revolutionized the approach to organic synthesis through radical and radical ion intermediates. These fleeting species can be

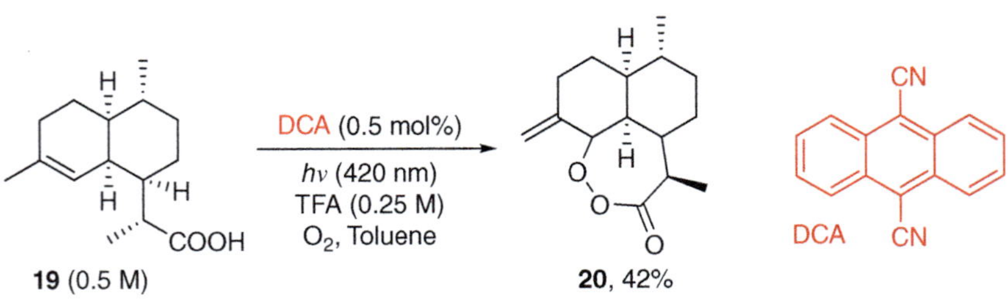

Scheme 2.7 Photosensitized synthesis of Artemisinin **20**.

smoothly generated under extremely mild conditions through two main mechanisms, namely single electron transfer (SET) [23–30] or hydrogen atom transfer (HAT) [31–33].

As for the first mechanism, a seminal and interesting application is represented by the realization of [2+2] cycloadditions occurring via radical ionic intermediates to deliver cyclobutane rings. Indeed, this is a complementary strategy with respect to the above-mentioned photosensitized [2+2] cycloaddition (see Scheme 2.4 above). Intriguingly, the exchange of an electron between the photocatalyst and the substrate, occurring upon irradiation, triggered the reactivity of two double bonds endowed with the same polarity (both nucleophilic or electrophilic). As an example, the crossed intermolecular [2+2] cycloaddition of acyclic enones has been reported to take place in the presence of the well-known $Ru(bpy)_3$ (bpy: 2,2′-bipyridine) photoredox catalyst upon irradiation with visible light (Scheme 2.8a) [77]. A careful choice of the reaction partners was required, with the best performance being obtained by reacting an arylenone (e.g. **21a-d**) with an α,β-unsaturated ketone lacking β-substituents (e.g. the Michael acceptor **22**). In the process, the arylenone was converted by the excited photocatalyst to the

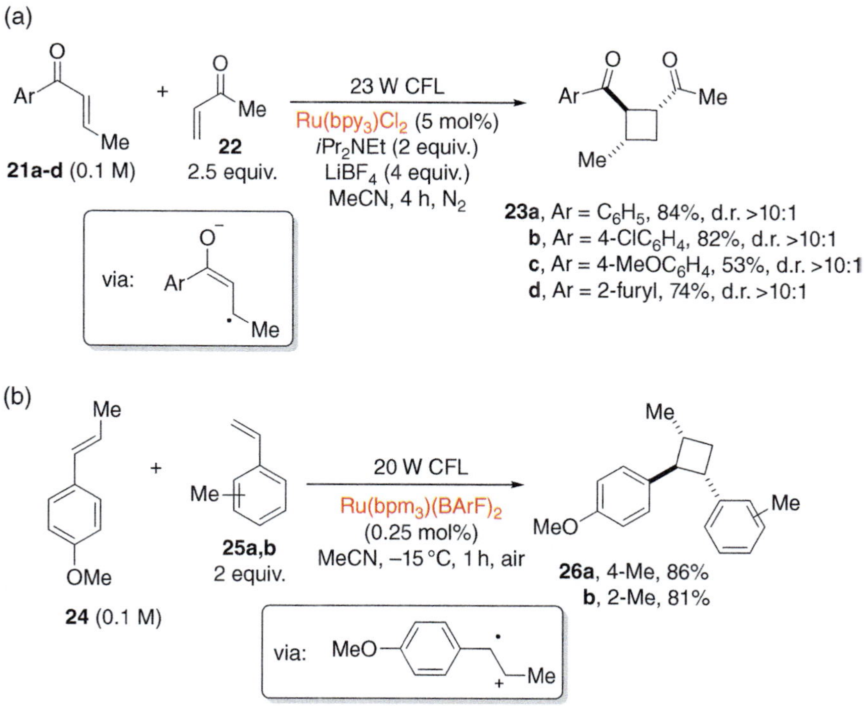

Scheme 2.8 Synthesis of four-membered rings via photoredox catalyzed generation of (a) radical anions, (b) radical cations.

corresponding radical anion that coupled with **22** ultimately delivering cyclobutanes **23a-d**. Interestingly, the reaction could be scaled up to the preparation of 1 g of product **23a** upon irradiation with sunlight (only four hours exposition required, Figure 2.15).

*Typical procedure for the synthesis of **23a-d**:* A dry 25 mL Schlenk tube was charged with the aryl enones **21a-d** (0.1 M), **22** (2.5 equiv.), $Ru(bpy)_3Cl_2{\cdot}6H_2O$ (5 mol%), $LiBF_4$ (4 equiv.), $i Pr_2NEt$ (2 equiv.), and acetonitrile, and degassed by a freeze/pump/thaw cycle (3×) under nitrogen in the dark. The reaction was then allowed to stir and irradiated by a 23 W compact fluorescent lamp at a distance of 30 cm. Upon completion, the solvent was removed in vacuo and the residue purified by column chromatography on silica gel.

Experimental setup: See Figure 2.15.

By reversing the electron flow between the photocatalyst and the substrate, a smooth access to radical cationic intermediates is indeed possible. This strategy has been exploited to realize the crossed intermolecular [2+2] cycloaddition between styrenes (Scheme 2.8b) [78]. Thus, methoxy-substituted styrene **24** and isomeric methylstyrenes **25a,b** delivered cycloadducts **26a,b** in the presence of the $Ru(bpm)_3$ (bpm: 2,2′-bipyrimidine) photocatalyst under visible light irradiation. In the process, the photocatalyst oxidized **24** to the corresponding radical cation, a key intermediate in the process. Furthermore, the choice of the photocatalyst was pivotal to obtain a good reaction yield, since it had to be competent in the mono-electronic oxidation of **24**, but not in that of **26**, possibly causing an undesired cycloreversion event [78]. In a follow-up work, the importance of

Figure 2.15 Photochemical apparatus employed for the synthesis of **23a-d**.
Source: Reprinted with permissions from Du and Yoon [77], Copyright © 2014 American Chemical Society.

radical chain processes in this type of [2+2] photocycloadditions occurring via radical ionic intermediates has been demonstrated thanks to quantum yield measurements [79].

*Typical procedure for the synthesis of **26a,b***: An oven-dried 25 mL borosilicate test tube was charged with **24** (0.1 M), **25** (2 equiv.), and 0.25 mol% $Ru(bpm)_3(BArF)_2$. The test tube was cooled to −15 °C using a controlled-temperature cooling bath and 6.7 mL of anhydrous CH_2Cl_2 was added. The reaction mixture was vigorously stirred under air in front of a 20 W CFL. The solvent was removed by rotary evaporation, and the residue was purified by flash-column chromatography to afford the cycloadducts **26a,b**.

Experimental setup: The reaction vessel is a standard borosilicate test tube immersed in an acetone bath cooled to −15 °C using a Neslab immersion cooler. The light source is a 20 W GE Reveal CFL bulb installed in a clamp light.

Turning to the HAT mechanism, one of the most investigated photocatalysts is undoubtedly the decatungstate anion $[W_{10}O_{32}]^{4-}$, often used as the tetrabutylammonium salt (TBADT). This robust and versatile polyoxometalate derivative has been exploited to cleave the C—H bond in a library of hydrogen donors, including alkanes, ethers, amides and aldehydes, to generate the corresponding C-centered radicals [31–33, 80].

As an example, the nucleophilic acyl radical obtained from heptaldehyde **27** has been conveniently exploited in the hydroalkylation of electron-poor olefins (e.g. esters **28a,b**) via a conjugate radical addition mechanism (Scheme 2.9). Notably, these processes have been carried out in a glass vessel upon solar light irradiation, in an example of the so-called window-ledge chemistry (Figure 2.16) [55]. The desired products 29a,b were indeed obtained on up to *ca.* 10 g scale by merely exposing the reaction mixtures to sunlight for a few days without any need of artificial energy input.

The versatility of decatungstate photocatalysis is further demonstrated by the possibility to develop a flow-version of this chemistry. Thus, as previously

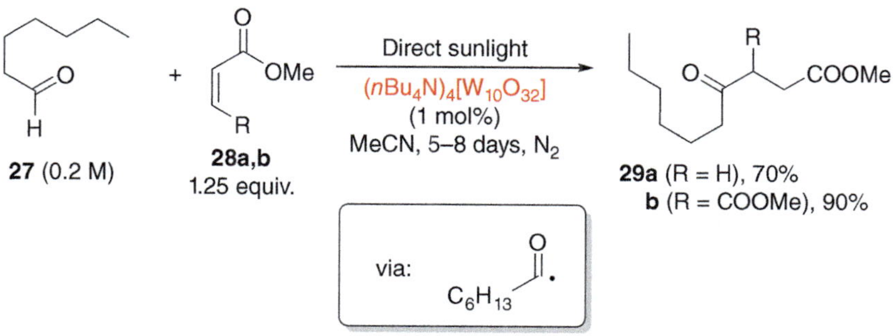

Scheme 2.9 Sunlight-induced formation of unsymmetrical ketones from aldehydes.

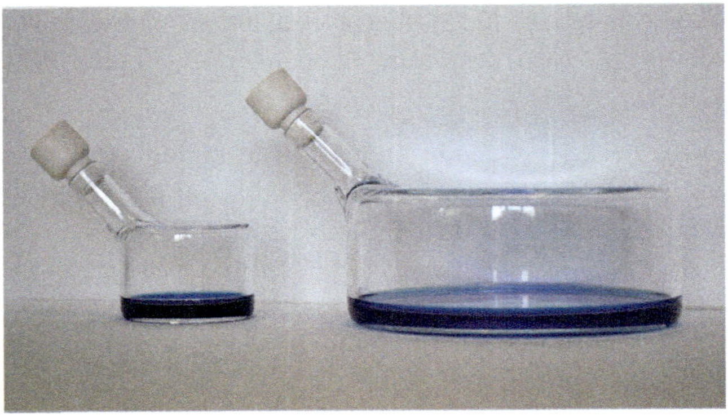

Figure 2.16 Photochemical apparatus employed for the synthesis of **29a-b.**

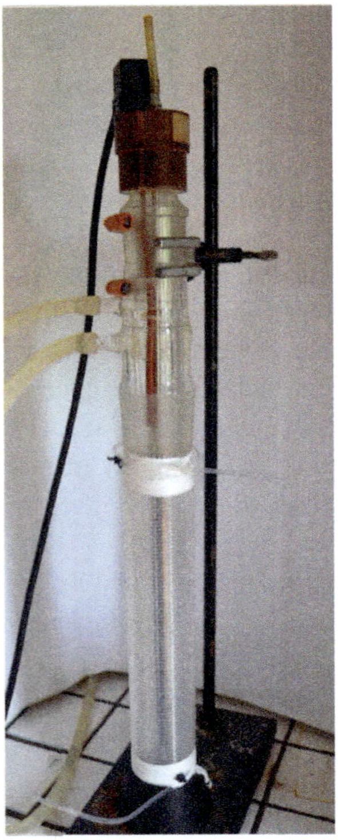

Figure 2.17 Photochemical flow reactor employed for the synthesis of **29a,b.**
Source: Reprinted with permission from Bonassi et al. [81], Copyright © 2016 John Wiley & Sons, Inc.

described [58], a photochemical reactor may be assembled by flowing the reaction mixture in an UV-transparent FEP (fluorinated ethylene propylene) tubing wrapped around a medium pressure Hg-vapors lamp (Figure 2.17).

Thanks to this experimental setup, it was possible to synthesize the same unsymmetrical ketones obtained upon solar light irradiation (e.g. product **29b** was prepared in 79% yield), albeit with a reduced irradiation time, thus increasing the overall productivity [81].

Typical procedure for the synthesis of ***29a,b*** *(batch):* A solution (100 mL) of heptaldehyde **27** (0.2 M), and **28a,b** (1.25 equiv.) in the presence of TBADT (1 mol%) in MeCN was poured into a glass Pyrex vessel and purged for 10 minutes with nitrogen, capped and exposed to sunlight on a window ledge. The solvent was removed in vacuo from the photolyzed solution and **29a,b** isolated by bulb-to-bulb distillation.

Experimental setup: Photocatalyzed reactions were carried out on solutions corresponding to a layer of about 1 cm thick with a surface area of 1 dm^2 in the cylindrical vessel used (Figure 2.16, right side). Vessels of on the left side were used for small-scale experiments for the optimization of the reaction conditions (left side).

*Typical procedure for the synthesis of **29a,b** (flow):* **27** and **28a,b** in the presence of a catalytic amount of TBADT were dissolved in 50 mL of acetonitrile. The solution was charged in a flask and pumped through the flow apparatus (Figure 2.17); **29a,b** were isolated upon column chromatography.

Experimental setup: The photochemical reactor (Figure 2.17) was equipped with a water-cooled 500 W medium-pressure mercury lamp. All the tubings were made of fluorinated ethylene propylene (FEP, outer diameter: 3.18 mm; inner diameter: 2.1 mm; reactor volume = 50 mL).

2.4 Conclusions

Summing up, it is not difficult work to arrange a small photochemical lab to explore the viability of photochemical steps. Lamps operating with an emission centered at 254, 305, and 350 nm in the UV region and LEDs for the visible spectrum are often sufficient for this purpose. As reactor vessels, common glassware (round-bottom flask, vials, etc.) and some quartz tubes are well fitted for this aim. In case of sunlight-promoted reactions, the use of this freely available light source does not require any investment. Exhaustive textbooks and reviews dealing with the basis of photochemistry [82–84], as well as dealing with the synthetic aspects of organic photochemistry [85–87], are available. For references on the use of photochemistry/photocatalysis for undergraduate organic chemistry laboratory experiments, see Refs. [88–90].

Acknowledgment

We thank Miss Alessandra Del Tito for taking some pictures included in this chapter.

References

1 Albini, A. and Fagnoni, M. (2004). Green chemistry and photochemistry were born at the same time. *Green Chemistry* 6: 1–6.

2 Albini, A. and Fagnoni, M. (2008). 1908: Giacomo Ciamician and the concept of green chemistry. *ChemSusChem* 1: 63–66.

3 Hoffmann, N. (2012). Photochemical reactions of aromatic compounds and the concept of the photon as a traceless reagent. *Photochemical & Photobiological Sciences* 11: 1613–1641.

4 Bonfield, H.E., Knauber, T., Lévesque, F. et al. (2020). Photons as a 21st century reagent. *Nature Communications* 11: 804.

5 Protti, S., Ravelli, D., and Fagnoni, M. (2019). Wavelength-dependence and wavelength-selectivity in photochemical reactions. *Photochemical & Photobiological Sciences* 18: 2094–2101.

6 Balzani, V., Ceroni, P., and Juris, A. (2014). *Photochemistry and Photophysics: Concepts, Research, Applications*. John Wiley & Sons, Inc.

7 Noyori, P. (2010). Insight: Green chemistry: the key to our future. *Tetrahedron* 66: 1028.

8 White, J.D. (1998). George Buchi. Obituary. *Organic Syntheses* 77: xxiii–xxvi.

9 Chang, L., An, Q., Duan, L. et al. (2022). Alkoxy radicals see the light: new paradigms of photochemical synthesis. *Chemical Reviews* 122: 2429–2486.

10 Guo, J.-J., Hu, A., and Zuo, Z. (2018). Photocatalytic alkoxy radical-mediated transformations. *Tetrahedron Letters* 59: 2103–2111.

11 Kamijo, S. (2019). Ketones and aldehydes. In: *Photoorganocatalysis in Organic Synthesis* (ed. M. Fagnoni, S. Protti, and D. Ravelli), 1–38. Singapore: World Scientific Publishing Europe Ltd.

12 D'Auria, M. (2019). The Paternò–Büchi reaction – a comprehensive review. *Photochemical & Photobiological Sciences* 18: 2297–2362.

13 Albini, A. and Fagnoni, M. (2013). *Photochemically-Generated Intermediates in Synthesis*. Hoboken: John Wiley & Sons.

14 Crespi, S. and Fagnoni, M. (2020). Generation of alkyl radicals: from the Tyranny of Tin to the photon democracy. *Chemical Reviews* 120: 9790–9833.

15 Ravelli, D., Protti, S., and Fagnoni, M. (2016). Carbon-carbon bond forming reactions via photogenerated intermediates. *Chemical Reviews* 116: 9850–9913.

16 Shields, D.J., Chakraborty, M., Abdelaziz, N. et al. (2020). Review of laser flash photolysis of organic molecules (2015–2018). *Photochemistry* 48: 70–121.

17 Protti, S., Ravelli, D., and Fagnoni, M. (2022). Designing radical chemistry by visible-light promoted homolysis. *Trends in Chemistry* 4: 305–317.

18 Dichiarante, V., Protti, S., and Fagnoni, M. (2017). Phenyl cation: a versatile intermediate. *Journal of Photochemistry and Photobiology A: Chemistry* 339: 103–113.

19 Großkopf, J., Kratz, T., Rigotti, T. et al. (2022). Enantioselective photochemical reactions enabled by triplet energy transfer. *Chemical Reviews* 122: 1626–1653.

20 Pipiri, I., Buscemi, S., Piccionciello, A.P. et al. (2018). Photochemically produced singlet oxygen: applications and perspectives. *ChemPhotoChem* 2: 535–547.

21 You, Y. (2018). Chemical tools for the generation and detection of singlet oxygen. *Organic & Biomolecular Chemistry* 16: 4044–4060.

22 Wau, J.S., Robertson, M.J., and Oelgemöller, M. (2021). Solar photooxygenations for the manufacturing of fine chemicals—technologies and applications. *Molecules* 26: 1685.

23 Narayanam, J.M.R. and Stephenson, C.R.J. (2011). Visible light photoredox catalysis: applications in organic synthesis. *Chemical Society Reviews* 40: 102–113.

24 Xuan, J. and Xiao, W.-J. (2012). Visible-light photoredox catalysis. *Angewandte Chemie, International Edition* 51: 6828–6838.

25 Reckenthäler, M. and Griesbeck, A.G. (2013). Photoredox catalysis for organic syntheses. *Advanced Synthesis & Catalysis* 355: 2727–2744.

26 Xi, Y., Yi, H., and Lei, A. (2013). Synthetic applications of photoredox catalysis with visible light. *Organic & Biomolecular Chemistry* 11: 2387–2403.

27 Stephenson, C.R.J., Yoon, T.P., and MacMillan, D.W.C. (ed.) (2018). *Visible Light Photocatalysis in Organic Chemistry*. Weinheim, Germany: Wiley-VCH.

28 Hossain, A., Bhattacharyya, A., and Reiser, O. (2019). Copper's rapid ascent in visible-light photoredox catalysis. *Science* 364: eaav9713.

29 König, B. (ed.) (2020). *Chemical Photocatalysis*, 2ee. De Gruyter.

30 Pitre, S.P., Overman, L.E. (2022). Strategic use of visible-light photoredox catalysis in natural product synthesis. *Chemical Reviews* 122: 1717–1751.

31 Protti, S., Ravelli, D., and Fagnoni, M. (2015). Photocatalytic C-H activation via hydrogen atom transfer in synthesis. *ChemCatChem* 7: 1516–1523.

32 Cao, H., Tang, X., Tang, H. et al. (2021). Photoinduced intermolecular hydrogen atom transfer reactions in organic synthesis. *Chemical Catalysis* 1: 523–598.

33 Capaldo, L., Ravelli, D., and Fagnoni, M. (2022). Direct photocatalyzed hydrogen atom transfer (HAT) for aliphatic C–H bonds elaboration. *Chemical Reviews* 122: 1875–1924.

34 Fukuzumi, S. and Ohkubo, K. (2014). Organic synthetic transformations using organic dyes as photoredox catalysts. *Organic & Biomolecular Chemistry* 12: 6059–6071.

35 Romero, N.A. and Nicewicz, D.A. (2016). Organic photoredox catalysis. *Chemical Reviews* 116: 10075–10166.

36 Sideri, I.K., Voutyritsa, E., and Kokotos, C.G. (2018). Photoorganocatalysis, small organic molecules and light in the service of organic synthesis: the awakening of a sleeping giant. *Organic & Biomolecular Chemistry* 16: 4596–4614.

37 Fagnoni, M., Protti, S., and Ravelli, D. (ed.) (2019). *Photoorganocatalysis in Organic Synthesis*. Singapore: World Scientific Publishing Europe Ltd.

38 Albini, A. and Fagnoni, M. (ed.) (2010). *Handbook of Synthetic Photochemistry*. Weinheim: Wiley-VCH.

39 Ramamurthy, V. and Venkatesan, K. (1987). Photochemical reactions of organic crystals. *Chemical Reviews* 87: 433–481.

40 Garcia-Garibay, M.A. (2007). Molecular crystals on the move: from single-crystal-to-single-crystal photoreactions to molecular machinery. *Angewandte Chemie, International Edition* 46: 8945–8947.

41 Kuzmanich, G., Natarajan, A., Chin, K.K. et al. (2008). Solid state photodecarbonylation of diphenylcyclopropenone: a quantum chain process made possible by ultrafast energy transfer. *Journal of the American Chemical Society* 130: 1140–1141.

42 Knowles, J.P., Elliott, L.D., and Booker-Milburn, K.I. (2012). Flow photochemistry: old light through new windows. *Beilstein Journal of Organic Chemistry* 8: 2025–2052.

43 Garlets, Z.J., Nguyen, J.D., and Stephenson, C.R.J. (2014). The development of visible-light photoredox catalysis in flow. *Israel Journal of Chemistry* 54: 351–360.

44 Schuster, E.M. and Wipf, P. (2014). Photochemical flow reactions. *Israel Journal of Chemistry* 54: 361–370.

45 Buglioni, L., Raymenants, F., Slattery, A. et al. (2022). Technological innovations in photochemistry for organic synthesis: flow chemistry, high-throughput experimentation, scale-up, and photoelectrochemistry. *Chemical Reviews* 122: 2752–2906.

46 Matsushita, Y., Ichimura, T., Ohba, N. et al. (2007). Recent progress on photoreactions in microreactors. *Pure and Applied Chemistry* 79: 1959–1968.

47 Roberge, D.M., Zimmermann, B., Rainone, F. et al. (2008). Microreactor technology and continuous processes in the fine chemical and pharmaceutical industry: Is the revolution underway? *Organic Process Research & Development* 12: 905–910.

48 Coyle, E.E. and Oelgemöller, M. (2008). Micro-photochemistry: photochemistry in microstructured reactors. The new photochemistry of the future? *Photochemical & Photobiological Sciences* 7: 1313–1322.

49 Basso, A. and Capurro, P. (2021). Recent applications of photochemistry on large-scale synthesis (2015–2019). *Photochemistry* 48: 294–322.

50 Candish, L., Collins, K.D., Cook, G.C. et al. (2022). Photocatalysis in the Life Science Industry. *Chemical Reviews* 122: 2907–2980.

51 As a representative supplier for UV lamps. https://www.heliosquartz.com/ (accessed 11 April 2023).

52 For the applications of Kessil lamps in synthesis. https://kessil.com/products/science_main.php (accessed 11 April 2023).

53 Ravelli, D., Protti, S., and Fagnoni, M. (2016). Application of visible and solar light in organic synthesis. In: *Applied Photochemistry: When Light Meets Molecules* (ed. G. Bergamini and S. Silvi), 281–342. Switzerland: Springer International Publishing.

54 Schultz, D.M. and Yoon, T.P. (2014). Solar synthesis: prospects in visible light photocatalysis. *Science* 343: 1239176.

55 Ravelli, D., Protti, S., and Fagnoni, M. (2016). Decatungstate anion for photocatalyzed "Window Ledge" reactions. *Account of Chemical Research* 49: 2232–2242.

56 For the characteristics of a SolarBox. https://cofomegra.it/en/solarbox-1500-3000/ (accessed 11 April 2023).

57 Protti, S. and Fagnoni, M. (2009). The sunny side of chemistry: green synthesis by solar light. *Photochemical & Photobiological Sciences* 8: 1499–1516.

58 Hook, B.D.A., Dohle, W., Hirst, P.R. et al. (2005). Practical flow reactor for continuous organic photochemistry. *The Journal of Organic Chemistry* 70: 7558–7564.

59 Rehm, T.H. (2020). Reactor technology concepts for flow photochemistry. *ChemPhotoChem* 4: 235–254.

60 For the different types of photochemical flow reactors. https://www.corning.com/worldwide/en/innovation/corning-emerging-innovations/advanced-flow-reactors.html (accessed 11 April 2023)

61 For continuous flow devices. https://www.syrris.com/product/asia-photochemistry-system/ (accessed 11 April 2023)

62 For the design of a falling film reactor. https://photoreactors.com/products/falling-film-photoreactors/ (accessed 11 April 2023).

63 Arias-Rotondo, D.M. and Mc Cusker, J.K. (2016). The photophysics of photoredox catalysis: a roadmap for catalyst design. *Chemical Society Reviews* 45: 5803–5820.

64 Buzzetti, L., Crisenza, G.E.M., and Melchiorre, P. (2019). Mechanistic studies in photocatalysis. *Angewandte Chemie, International Edition* 58: 3730–3747.

65 von Lintig, J., Kiser, P.D., Golczak, M. et al. (2010). The biochemical and structural basis for trans-to-cis isomerization of retinoids in the chemistry of vision. *Trends in Biochemical Sciences* 35: 400–410.

66 Huang, M., Zhang, L., Pan, T. et al. (2022). Deracemization through photochemical *E/Z* isomerization of enamines. *Science* 375: 869–874.

67 Folkmann, M. P. and Hansen, E. T. (2007). Isomerisation of pharmaceutical intermediates. EP 1844010.

68 Rigotti, T., Schwinger, D.P., Graßl, R. et al. (2022). Enantioselective crossed intramolecular [2+2] photocycloaddition reactions mediated by a chiral chelating Lewis acid. *Chemical Science* 13: 2378–2384.

69 Mateos, J., Vega-Peñaloza, A., Franceschi, P. et al. (2020). A visible-light Paternò–Büchi dearomatisation process towards the construction of oxeto-indolinic polycycles. *Chemical Science* 11: 6532–6538.

70 Govaerts, S., Angelini, L., Hampton, C. et al. (2020). Photoinduced olefin diamination with alkylamines. *Angewandte Chemie International Edition* 59: 15021–15028.

71 Steiner, A., de Frutos, O., Rincón, J.A. et al. (2021). N-Chloroamines as substrates for metal-free photochemical atom-transfer radical addition reactions in continuous flow. *Reaction Chemistry & Engineering* 6: 2434–2441.

72 Qiu, D., Lian, C., Mao, J. et al. (2020). Dyedauxiliary groups, an emerging approach in organic chemistry. The case of arylazo sulfones. *The Journal of Organic Chemistry* 85: 12813–12822.

73 Fagnoni, M. (2022). Colored compounds for eco-sustainable visible-light promoted syntheses. In: *Sustainable Organic Synthesis: Tools and Strategies* (ed. S. Protti and A. Palmieri), 150–180. The Royal Society of Chemistry.

74 da Silva Jr, P.E., Amin, H.I.M., Nauth, A.M. et al. (2018). Flow photochemistry of azosulfones: application of "Sunflow" reactors. *ChemPhotoChem* 2: 878–883.

75 Kopetzki, D., Lévesque, F., and Seeberger, P.H. (2013). A continuous-flow process for the synthesis of artemisinin. *Chemistry A European Journal* 19: 5450–5456.

76 Lévesque, F. and Seeberger, P.H. (2012). Continuous-flow synthesis of the anti-malaria drug artemisinin. *Angewandte Chemie, International Edition* 51: 1706–1709.

77 Du, J. and Yoon, T.P. (2009). Crossed intermolecular [2+2] cycloadditions of acyclic enones via visible light photocatalysis. *Journal of the American Chemical Society* 131: 14604–14605.

78 Ischay, M.A., Ament, M.S., and Yoon, T.P. (2012). Crossed intermolecular [2+2] cycloaddition of styrenes by visible light photocatalysis. *Chemical Science* 3: 2807–2811.

79 Cismesia, M.A. and Yoon, T.P. (2015). Characterizing chain processes in visible light photoredox catalysis. *Chemical Science* 6: 5426–5434.

80 Protti, S., Ravelli, D., Fagnoni, M. et al. (2009). Solar light-driven photocatalyzed alkylations. Chemistry on the window ledge. *Chemical Communications* 47: 7351–7353.

81 Bonassi, F., Ravelli, D., Protti, S. et al. (2015). Decatungstate photocatalyzed acylations and alkylations in flow via hydrogen atom transfer. *Advanced Synthesis & Catalysis* 357: 3687–3695.

82 Horspool, W.H. and Armesto, D. (1992). *Organic Photochemistry: A Comprehensive Treatment*. London: Ellis Horwood.

83 Turro, N.J., Ramamurthy, V., and Scaiano, J.C. (2009). *Modern Molecular Photochemistry of Organic Molecules*. Basingstoke: Palgrave Macmillan.

84 Klan, P. and Wirz, J. (2009). *Photochemistry of Organic Compounds*. Wiley-Blackwell.

85 Mattay, J. and Griesbeck, A. (ed.) (2008). *Photochemical Key Steps in Organic Synthesis: An Experimental Course Book*. Wiley Weinheim.

86 Griesbeck, A., Oelgemöller, M., and Ghetti, F. (ed.) (2012). *Handbook of Organic Photochemistry and Photobiology*, 3ee. CRC: Press.

87 Bergamini, G. and Silvi, S. (ed.) (2016). *Applied Photochemistry: When Light Meets Molecules*. Switzerland: Springer International Publishing.

88 Contreras-Cruz, D.A., Cantû-Reyes, M., García-Sánchez, J.M. et al. (2019). Shedding Blue Light on the Undergraduate Laboratory: an easy-to-assemble LED photoreactor for aromatization of a 1,4-dihydropyridine. *Journal of Chemical Education* 96: 2015–2020.

89 Chen, S. (2018). let there be light: hypothesis-driven investigation of ligand effects in photoredox catalysis for the Undergraduate Organic Chemistry Laboratory. *Journal of Chemical Education* 95: 872–875.

90 Cruz, C.L., Holmberg-Douglas, N., Onuska, N.P.R. et al. (2020). Development of a large-enrollment course-based research experience in an Undergraduate Organic Chemistry Laboratory: structure–function relationships in pyrylium photoredox catalysts. *Journal of Chemical Education* 97: 1572–1578.

3

How to Confidently Become an Electrosynthetic Practitioner

Sylvain Charvet[1,*], *Taline Kerackian*[1,*], *Camille Z. Rubel*[1,2,*], *and Julien C. Vantourout*[1]

[1]*Institut de Chimie et Biochimie Moléculaires et Supramoléculaires (ICMBS, UMR 5246 du CNRS), Université Lyon, Villeurbanne, France*
[2]*Department of Chemistry, The Scripps Research Institute, La Jolla, CA, USA*

Glossary

Current Electric current is defined as the flow of electric charge through a conductive material or medium. It is the rate of movement of electric charge in a specific direction, usually measured in amperes (A). The electric charge can be carried by electrons in a metallic conductor, ions in an electrolyte, or any other charged particles in a plasma. The direction of the electric current is defined as the direction of the flow of positive charge, even though in most cases it is the negative charges that move. Electric current plays a fundamental role in a wide range of electrical and electronic devices, including motors, generators, batteries, and power distribution systems.

Current density Current density is defined as the amount of electric current flowing through a unit area perpendicular to the direction of the current flow. It is a vector quantity that describes the magnitude and direction of the flow of electric charge in a material. Mathematically, it can be represented by the symbol J and is defined as $J = I/A$, where J is the current density, I is the current flowing through the material, and A is the surface area of the material through which the current is flowing. The unit of current density is amperes per square meter ($A.m^{-2}$) in the SI system of units. Current density is an important parameter in many areas of electrical engineering, such as the design of electrical conductors, semiconductors, and electronic devices.

* *These authors contributed equally.*

Enabling Tools and Techniques for Organic Synthesis: A Practical Guide to Experimentation, Automation, and Computation, First Edition. Edited by Stephen G. Newman.

It is also used to analyze and optimize the performance of electrical circuits and systems.

Cyclic voltammetry Cyclic voltammetry (CV) is a widely used electrochemical technique. During a typical CV experiment, the voltage is ramped up and down in a repetitive cycle, while the resulting current response is recorded. This technique is commonly used to study the electrochemical behavior of a wide range of chemical and biological systems. Some uses include the determination of the redox potentials of redox-active species, the kinetics of electron transfer reactions, and the characterization of electrode surfaces and electroactive molecules. CV data can also provide a wealth of information about the electrochemical system being studied, including the nature and concentration of the electroactive species present, the number and types of electron transfer reactions occurring, and the rate of these reactions.

Electrochemistry Electrochemistry is the study of electron movement in an oxidation or reduction reaction at a polarized electrode surface. This term can an also refer to the transfer of electrons from one chemical species to another.

Electrolyte An electrolyte is a substance that contains ions and can conduct electricity. When an electrolyte is placed in a solution, the ions in the solution become mobile and can carry an electrical current through the solution. Electrolytes can be either liquids or solids, and they can be either organic or inorganic compounds. In aqueous solutions, common examples of electrolytes include salts such as sodium chloride (NaCl), potassium chloride (KCl), and calcium chloride ($CaCl_2$). In organic solvents, common examples of electrolytes include salts such as tetrabutylammonium tetrafluoroborate ($TBABF_4$), potassium hexafluorophosphate (KPF_6), and lithium bromide (LiBr). Electrolytes are important in electrochemical reactions because they play a crucial role in the transfer of electrons and the conduction of electrical charge. They are often used in batteries, fuel cells, and other electrochemical devices.

Electrosynthesis Electrosynthesis refers to the process of using electrical energy to drive a chemical reaction, typically involving the conversion of a molecule into a different molecule. This process involves the application of an electric potential or current to a reactant solution, causing electrons to be transferred between the electrode and the solution at the electrode surface, which in turn drives the desired chemical reaction. Electrosynthesis has been demonstrated in the production of various substances, such as pharmaceuticals, organic compounds, and electroplating of metals.

Faraday The Faraday is defined as the amount of electrical charge (in coulombs) that is transferred when one mole of electron flows through a conductor during an electrochemical reaction. The term "Faraday" is used to describe the amount of electrical charge transferred during an electrochemical reaction. It is named after the English scientist Michael Faraday who made significant contributions to the field of electromagnetism and electrochemistry in the nineteenth century. It is often used in the form of the Faraday constant, which is approximately equal to 96 485 coulombs per mole of electrons. The Faraday is an important unit of measurement in electrochemistry as it is used to calculate the amount of substance produced or consumed during an electrochemical reaction. It is also used to determine the efficiency of electrochemical processes and to calculate the amount of electric current needed to produce a certain amount of substance in an electrolytic cell.

Metallophotoredox Metallophotoredox is a type of catalytic reaction that involves the use of a photosensitizer and a metal catalyst to facilitate electron transfer reactions. The photosensitizer is typically a molecule that absorbs light and becomes electronically excited, while the metal catalyst serves to mediate electron transfer reactions. Together, they enable a wide range of organic transformations, including cross-coupling reactions, hydrogenation reactions, and oxidation reactions. Metallophotoredox has become an increasingly important area of research in synthetic chemistry and is a powerful tool for the efficient construction of complex molecules with high levels of control and selectivity.

Oxidation and anode An electrochemical oxidation is a chemical reaction that involves the loss of electrons by a substance through the application of an electric current. In this process, the species is subjected to an electric potential, which causes it to undergo an oxidation reaction, resulting in the formation of an oxidized product. During an electrochemical oxidation, the oxidation half-reaction takes place at the anode, which is the electrode connected to the positive terminal of the power source. The anode attracts negatively charged ions (i.e. anions) from the solution, and these anions are oxidized by the loss of electrons. Electrochemical oxidation is used to convert a wide range of organic and inorganic substances into their oxidized forms, which can have a variety of applications in industry, environmental remediation, and energy storage.

Passivation Electrochemical passivation is the process by which a metal surface is made more resistant to corrosion through the formation of a thin, protective oxide layer. This oxide layer is created through an electrochemical reaction between the metal and an electrolyte, such as oxygen or water. The process typically occurs by applying a potential difference between the metal and the electrolyte, which causes the metal to oxidize and form the protective oxide layer. Once the oxide layer has formed, it acts as a barrier to further oxidation and corrosion, effectively "passivating" the metal surface. In the context of electrosynthesis, the process hampers the desired chemical reaction from occurring.

Potential Potential is the potential energy difference between two points in an electrochemical system due to the presence of charged particles or ions, which can affect the direction and rate of chemical reactions. It is expressed in units of volts (V) and is an important concept in fields such as electrochemistry, biology, and physics. Current and potential are linked by

Ohm's law. This law states that the current through a conductor between two points is directly proportional to the voltage across the two points, provided that the temperature and other physical conditions of the conductor remain constant. In other words, as the voltage increases, the current through the circuit will also increase proportionally. Mathematically, Ohm's law can be expressed as: $I = V/R$, where I is the current flowing through the conductor, V is the voltage across the conductor, and R is the resistance of the conductor.

Reduction and cathode An electrochemical reduction is a chemical reaction that involves the gain of electrons by a substance through the application of an electric current. In this process, the species is subjected to an electric potential, which causes it to undergo a reduction reaction, resulting in the formation of a reduced product. During an electrochemical reduction, the reduction half-reaction takes place at the cathode, which is the electrode connected to the negative terminal of the power source. The cathode attracts positively charged ions (i.e. cations) from the solution, and these cations are reduced by the gain of electrons. Electrochemical reduction is commonly used in electroplating, in which a metal is reduced onto a substrate by electrochemical means. It is also used in the production of chemicals and fuels, such as hydrogen gas and ammonia, by electrolysis of water and nitrogen gas, respectively.

Synthon In organic chemistry, a synthon is a hypothetical or real molecular fragment (or intermediate) that can be used as a building block in the synthesis of more complex molecules. An organic synthon is a synthetic equivalent of a functional group or group of atoms, which can be used to prepare a target molecule via a series of chemical reactions. Organic synthons are typically used in retrosynthetic analysis, which is a strategy for planning organic syntheses by breaking down a target molecule into simpler precursors. The use of synthons can simplify retrosynthetic analysis by reducing a complex molecule to a smaller and more easily handled fragment.

Abbreviations

Al	Aluminum
C	Carbon
DMA	*N,N*-Dimethyl acetamide
DMF	*N,N*-Dimethyl formamide
DMSO	Dimethyl sulfoxide
ERCC	Electrophile reductive cross-coupling
Fe	Iron
FTO	Fluorine-doped tin oxide.
GC-MS	Gas chromatography – mass spectrometry
HCl	Hydrochloric acid

LC-MS	Liquid chromatography – mass spectrometry
LiBr	Lithium bromide
Mg	Magnesium
Ni	Nickel
NMR	Nuclear magnetic resonance
Pt	Platinum
Rpm	Rotations per minute
RVC	Reticulated vitreous carbon
THF	Tetrahydrofuran
TLC	Thin layer chromatography
Zn	Zinc

3.1 Introduction

Electrosynthesis is the process of using electricity to induce chemical reactions. The history of organic electrosynthesis started almost 200 years and has led to an extensive breadth of research since 2000 [1]. Electrosynthesis has a wide range of applications in various industries, including chemical synthesis, energy storage, environmental remediation, and pharmaceuticals [2–7]. Electrosynthesis can also enable the production of chemicals and materials that are difficult or impossible to obtain through traditional chemical processes. In the chemical industry, electrosynthesis has the potential to transform the production of specialty chemicals and intermediates such as those in cosmetics, agrochemicals, and pharmaceuticals [2, 3]. Traditional chemical synthesis methods for these compounds are often energy-intensive and result in significant waste. Electrosynthesis can provide a more sustainable alternative, enabling the production of these compounds with high selectivity and purity. In the energy storage sector, electrosynthesis is being explored as a means of converting renewable energy sources, such as solar and wind power, into chemical energy [4, 5]. This can be achieved through the production of fuels, such as hydrogen, from water using electrosynthesis [6, 7]. Electrosynthesis can also be used for environmental remediation, enabling the removal of pollutants from contaminated water and soil [8, 9].

However, electrolysis remains a very under-used procedure for the preparation of organic compounds in both the laboratory and industry. This is because, historically, organic electrosynthesis publications were inaccessible to synthetic organic chemists. Seminal reports were not written with the typical synthetic chemist reader in mind, which led to multiple misunderstandings on why, when, and how one should use electrochemistry in a synthetic laboratory. For example, many electrochemical reactions are reported while performing analytic measurements such as impedance, but these are not necessarily useful at a synthetic scale. Unfortunately, most reports often fail to distinguish if a reported transformation is of practical use or not, leading to a body of literature that is misleading and confusing, especially to non-electrochemists. In addition, too often an attempt to repeat an electrosynthesis in a different laboratory leads to lower selectivity and/or yield due to insufficient detail in the description of the cell (geometry, dimensions, materials of components. . .) and the conditions (solvent, concentration of reagents and electrolyte, pH, cell current, temperature. . .). *Thus, in this introduction, our goal is to clarify why it is important to use electrosynthesis and how to perform electrochemical reactions in the context of organic chemistry.*

3.2 General Definition of Organic Electrosynthesis

As previously mentioned, organic electrosynthesis is a branch of organic chemistry that uses electrochemistry to carry out chemical reactions involving organic compounds. In organic electrosynthesis, an electric current is passed through a solution containing the organic compounds, which causes the transfer of electrons between the reactants and the electrodes. This electron transfer results in the formation of radicals, ions, and other reactive intermediates, which can then undergo further chemical transformations to produce the desired products.

3.3 Why is Organic Electrosynthesis Used?

There are several reasons why organic electrosynthesis is advantageous. Organic electrosynthesis is often a cleaner and more sustainable alternative to traditional chemical synthesis. Indeed, while chemical processes require the use of toxic reagents and generate large amounts of waste, electrochemical reactions are driven only by electrons. In addition, electrochemistry can be highly selective, allowing chemists to target specific chemical bonds in a molecule, which is of particular interest in the context of complex organic molecule synthesis. Moreover, many organic electrosynthesis reactions can be carried out at mild temperatures and pressures, which can reduce the risk of unwanted side reactions and improve overall reaction efficiency. Finally, organic electrosynthesis is a safer alternative to traditional chemical synthesis, as there is often less risk of fire, explosion, or other hazards associated with using highly oxidizing or reducing chemical reagents.

3.4 How is Organic Electrosynthesis Performed?

General steps to perform electrosynthesis are listed below and will be elaborated on in Section 3.6 and 3.7:

1) Determine the reaction you want to carry out and identify the reactants and the product(s).

2) Select the appropriate electrodes for your reaction. Electrodes are typically made of a conductive material such as platinum, graphite, or carbon.
3) Choose an appropriate electrolyte that can conduct electricity. The choice of electrolyte will depend on the nature of the reaction.
4) Setup an electrochemical cell by placing the two electrodes in the electrolyte solution. The two electrodes should be connected to a source of electrical energy, such as a battery or power supply.
5) Apply a voltage or a current to the cell to drive the reaction. The voltage or current required will depend on the nature of the reaction and the electrodes used.
6) Monitor the progress of the reaction by measuring the current passing through the cell, and the change in the concentrations of the reactants and products over time. *All typical analytical techniques like TLC, GC, or HPLC can be used.*
7) Once the reaction is complete, collect the product and purify it if necessary as you would do for an organic reaction.

3.5 Where to Start with Electrosynthesis?

Over the years, many books and reviews have regularly highlighted the diversity of chemistry possible under electrochemical conditions leading to more sustainable and efficient processes. Key general references are listed below, which are excellent starting points for organic chemists to learn about both the practical and fundamental aspects of using electrochemistry for synthesis:

Selected General Reviews

Baran: Synthetic Organic Electrochemical Methods Since 2000: On the Verge of a Renaissance [1]
Nagaki: Modern Strategies in Electroorganic Synthesis [10]
Fronatana-Uriba: Organic electrosynthesis: a promising green methodology in organic chemistry [11]
Little: Redox catalysis in organic electrosynthesis: basic principles and recent developments [12]
Pletcher: Organic Electrosynthesis [13]
Zeng: Use of Electrochemistry in the Synthesis of Heterocyclic Structures [14]
Waldvogel: Electrochemical Screening for Electroorganic Synthesis [15]

Selected General Guides

Hilt: Basic Strategies and Types of Applications in Organic Electrochemistry [16]
Baran: A Survival Guide for the "Electro-curious" [17]
Lam: A practical guide to electrosynthesis [18]

It is important to note that electrosynthesis can be a complex process and may require optimization of various parameters such as electrode materials, electrolyte, voltage, and current/potential. Careful control of these parameters can lead to increased product yield, purity, and selectivity. In addition, being able to confidently reproduce the results is also an important factor as it will favor the uptake of electrochemical reactions. Challenges associated with optimization, parameter control, and reproducibility have been substantially improved by the development of standardized equipment such as the ElectraSyn 2.0 from IKA. Since its commercialization in 2017, the field of electrochemistry has witnessed an increase of interest due to the practicality offered by the instrument.

Given the breadth of recent reviews, we have chosen to make this book chapter a focused and practical "how-to" guide for beginners to the field. The first part of this chapter will thus aim to teach someone who has never conducted an electrochemical experiment, or who has never used an IKA ElectraSyn 2.0, how to do so confidently. It contains pictures and explanations, as well as tips and tricks. It focuses on the laboratory aspect of electrochemical experimentation rather than the conceptual aspect of experiment design, which will be covered in the case study section. The IKA ElectraSyn 2.0 is relatively inexpensive (MSRP US $2,650 for the most basic model at the time of writing) and is an excellent starting point. However, designing custom electrochemical reactors is not particularly difficult, even for beginners. The following discussion on the ElectraSyn 2.0 will still be informative for those designing their own systems thanks to the descriptions of useful parameters including current, potential, electrolytes, and electrodes. Readers are also encouraged to read the reviews by the Hilt [16], Baran [17], and Lam [18] research groups for beginner-friendly discussion on designing custom reactors.

3.6 Electrasyn 2.0

3.6.1 Machine and Consumables

3.6.1.1 Opening the IKA ElectraSyn 2.0 Box

When you buy an IKA ElectraSyn 2.0, you receive a briefcase-like black plastic case with various components (Figure 3.1a). This case contains the ElectraSyn, a 20 mL glass vial and its white cap, a square carrier for connecting vials to the ElectraSyn, the plug-in power and various adapters, and a stir plate. The ElectraSyn can support various attachments including a stir plate, the square carrier used to connect one ElectraSyn vial, or the carousel used to connect six vials at once (Figure 3.1b). One can also attach the Gogo module (Figure 3.1b), which allows you to connect the vial to the ElectraSyn but place it elsewhere for heating, sonication, or cooling. The carousel and the Gogo module are not included in the ElectraSyn 2.0 basic package and are considered accessories.

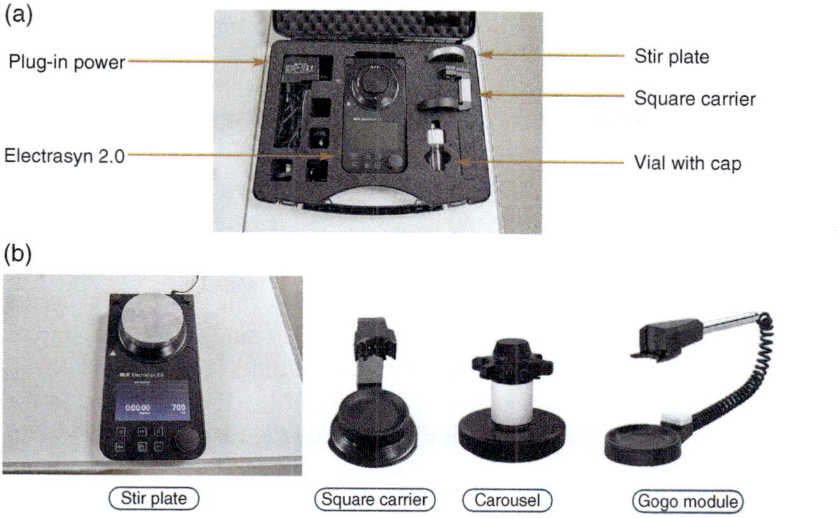

Figure 3.1 Commercially available ElectraSyn 2.0. (a) Commercial box; (b) Equipment.

3.6.1.2 Cell (Vial and Cap)

The reaction cell consists of the ElectraSyn vial and cap.

1) Glassware
   Glassware comes in 1, 2, 5, 10, and 20 mL sizes. These can be purchased on the IKA website or can be made by a glassblower (Figure 3.2).
2) Caps and septum
   You can purchase vials with or without caps. The caps come with a smaller screw-on cap with a septum for introducing reagents into reactions. However, we have found the seal of these to be poor, so if air-sensitivity is an issue, it is recommended to instead cover this opening with a small rubber septum, which can be wrapped in parafilm or tape for even better sealing (Figure 3.2).

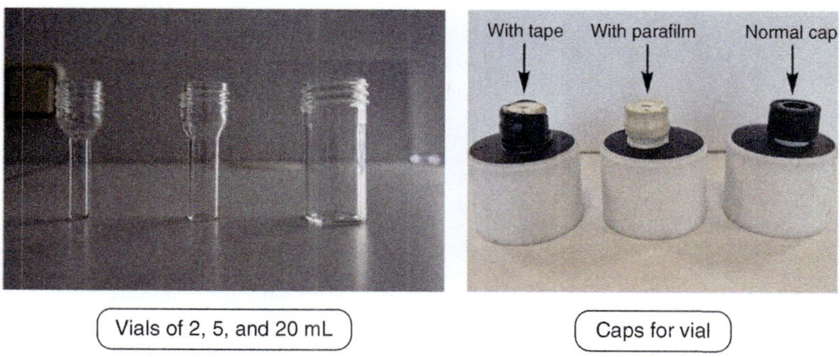

Figure 3.2 Commercially available accessories.

3.6.1.3 Electrodes

Each electrochemical reaction must be run with two electrodes, a cathode and an anode. These come in many varieties. A third electrode can also be used as a reference electrode to properly control the potential but is not mandatory, especially for reactions run under constant current [16, 18].

1) Cathode and anode
   There are various types of electrodes. Any (semi)conductive material can serve as an electrode material, ranging from cheap aluminum foil to expensive platinum plates or boron-doped diamond electrodes. Knowing the reaction occurs at the surface of electrodes, the choice of the material could have a huge impact on the reactivity. The best way to find the right electrodes is typically trial-and-error. To start, carbon-based materials are most of the time used due to the broad variety of surfaces available including graphite, carbon felt, glassy carbon, and reticulated vitreous carbon (RVC) (Figure 3.3). The choice will therefore depend on the type of electrochemical reactions (*for more details on this particular topic the reader should consider reading Refs.* [17, 18]). For example, for a purely reductive process, a sacrificial anode needs to be used (that means the anode will be consumed during the reaction). Once again, a broad range of materials are available including Zn, Mg, Al, and Fe (Figure 3.3). More information can be found in several published reviews or guides.
   In the context of the ElectraSyn 2.0, micro electrodes are used for 1 or 2 mL vials and regular ones are used for other vial sizes.
2) Reference electrode
   When running a reaction with constant voltage, a reference electrode can be used to measure the true potential against a standard reference, which has a stable and well-known potential. Running a constant potential reaction

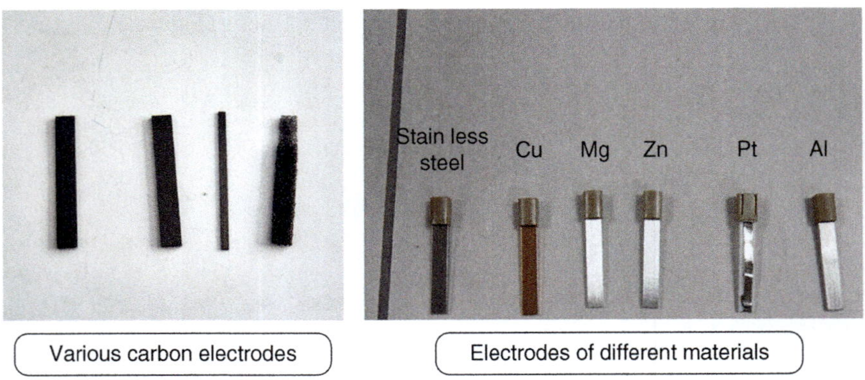

Figure 3.3 Electrodes.

without this reference electrode only measures the non-standardized potential between the cathode and anode [17].

3) Purchasing electrodes
   Electrodes can be purchased from the IKA website. However, an electrode is simply an appropriately sized piece of pure metal that can be connected to an electrical source. Budget-friendly electrodes can be obtained by purchasing metal sheets from Amazon and cutting them to the desired size for electrochemical flow or other applications. We tend to purchase RVC and carbon felt from outside suppliers such as Ultramet.

3.6.2 Interface

3.6.2.1 Hardware

On the ElectraSyn 2.0, there are 6 "touch" buttons and one knob (Figure 3.4). From top left to top right on the first line is the power touch button (1), the cyclic voltammetry button (2), and the home button (4). On the second row, there is the button to lock the screen (3), a button to view a graph of potential over the course of a constant current reaction or current over the course of a constant potential reaction (6), and the return button (5). The knob is used for scrolling within various menus and can be pressed to validate a choice.

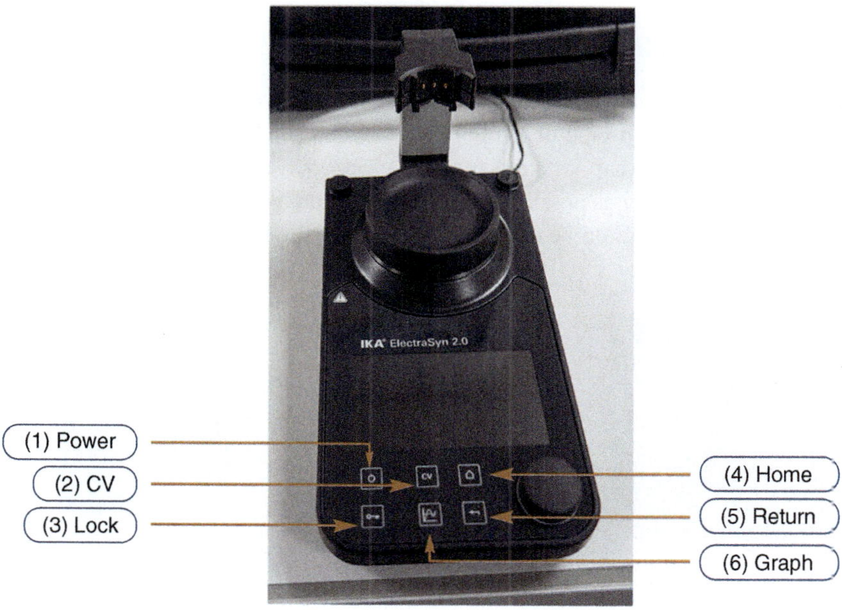

Figure 3.4 ElectraSyn 2.0.

3.6.2.2 Menus

1) Main menu

   Once the ElectraSyn 2.0 has been turned on via the power button, the main menu appears. It consists of three choices:

   - "Experiments" allows the user to choose from a previous experiment that has already been saved to the device by the user.
   - "New Experiments" allows the user to design a new experiment and leads them through the selection of various parameters.
   - "Assist Mode" can be used to help a user determine which current is best for their reaction. The general options "high," "medium," and "low" are listed, prompting the user to answer whether their reaction involves starting materials with a high, medium, or low reduction or oxidation potential.
   - On the top-right corner of the screen is a gear symbol, which is the settings menu.

2) Settings

   Various settings can be modified within this menu. In "Communication," you can change the language of the instrument. Perhaps the most important component of the settings is the voltage limit. Immediately upon receiving a new ElectraSyn, the user should change the voltage limit from 10 V to the maximum 30 V because the user will likely run a reaction (or set of reactions if using the carousel) that requires more than 10 V (Figure 3.5).

3.6.3 How to Set Up the Cell

1) Add a stir bar to the vial. If the reaction is air-sensitive, wrap the threading of the vial with Teflon tape or similar to create a strong seal. Place the desired electrodes into the cap, making sure the electrodes are placed such that they are as close together as possible (Figure 3.6). When looking at the cap from the front, the anode should be on the left and the cathode should be on the right

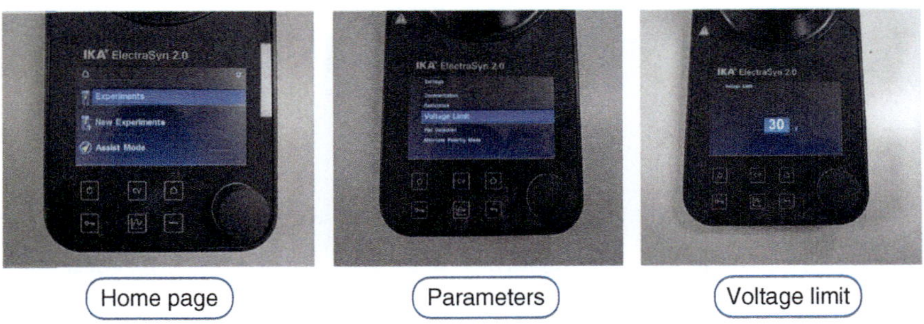

Figure 3.5 Adjusting voltage limit.

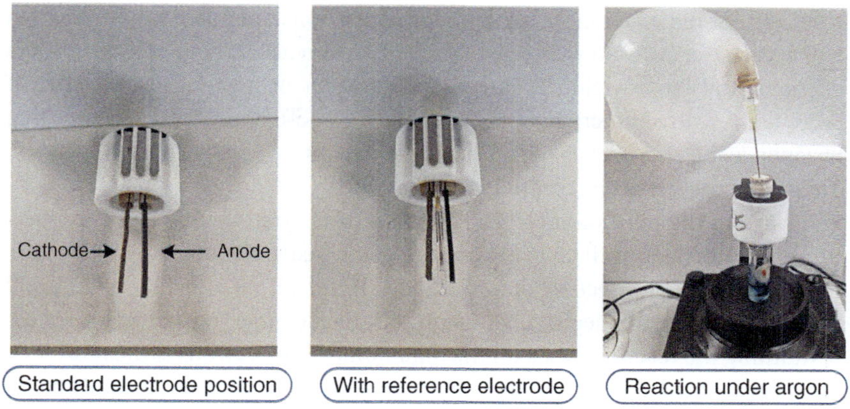

Figure 3.6 Reaction setup.

when using a positive current. If desired, the reference electrode can be screwed into the middle (Figure 3.6).

2) As with any reaction, it is best to add solids first. Then, if working under inert atmosphere, you can place the vial under said atmosphere and add liquids through the septum.
3) Place the vial into the ElectraSyn square carrier or carousel. For various reasons, if the connection is poor, electricity will not flow properly. When placed correctly, the bottom of the vial should touch the black plate and you should not be able to twist the vial left to right, only slide it up and down (Figure 3.6). With older caps, be sure to check that there is not corrosion where the cap meets the ElectraSyn. Lastly, make sure that the square carrier or carousel is rotated all the way to the right. If the equipment is not rotated all the way to the right until it stops, the reaction will not work due to a poor connection (this is probably the culprit if the reaction is started, and the voltage maximizes immediately to 30 V).
4) If the contents are air-sensitive, or a special reaction gas is desired, place a balloon filled with nitrogen, argon, or that gas on the vial (Figure 3.6).

3.6.4 How to Start an Experiment

Select "New Experiment" on the main menu (Figure 3.7). From here, the instrument will prompt you through a few questions.

- *Constant current or constant potential.*
  In general, it is best to choose constant current. This is because as the reaction goes on, redox active species are consumed, resulting in an increase in the

resistance of the reaction. If a reaction is run at constant potential, the current will decrease as resistance increases such that the reaction becomes slower, less efficient, and lower yielding. One would consider using constant potential if there are sensitive functional groups that could be harmed by a reaction that has increasing potential as the resistance increases.

- *Time, total charge, or run continuous.*

  You can run the reaction for a set amount of time, until a set number of electrons are added, or until you as the user tell the instrument to stop. We recommend choosing the second option under almost all circumstances and will continue through this section as though the "total charge" option was selected.
  - Running the reaction based on the total number of electrons is the most logical because the practitioner can determine how much current to apply based on the redox event happening in the vial. They can also decide how much excess to apply in this manner based on reaction efficiency. This parameter can be investigated during the optimization of the reaction.

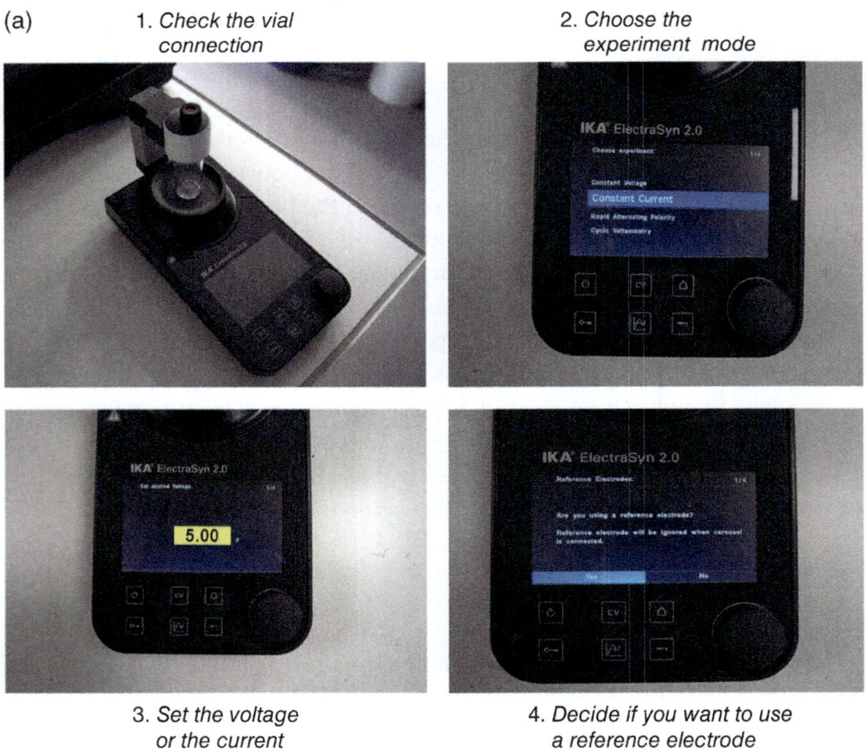

Figure 3.7 (a) Starting an electrochemical experiment. (b) Starting an electrochemical experiment.

(b)

5. *Choose the time of the reaction*

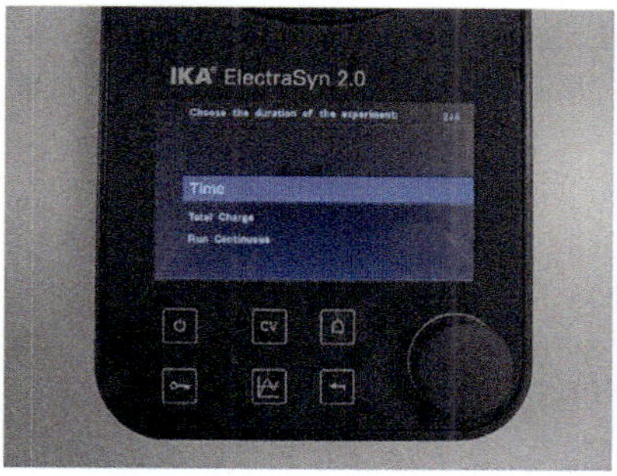

6. *In total charge, put mmol of substrate and the number of faraday*

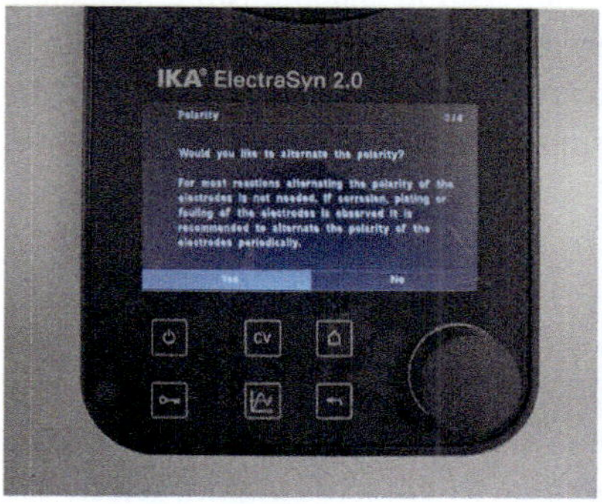

7. *Choose if you want to alternate the polarity*

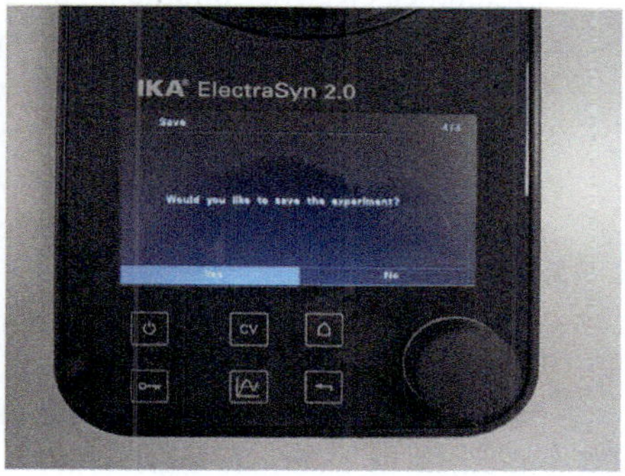

8. *Decide if you want to save your experiment*

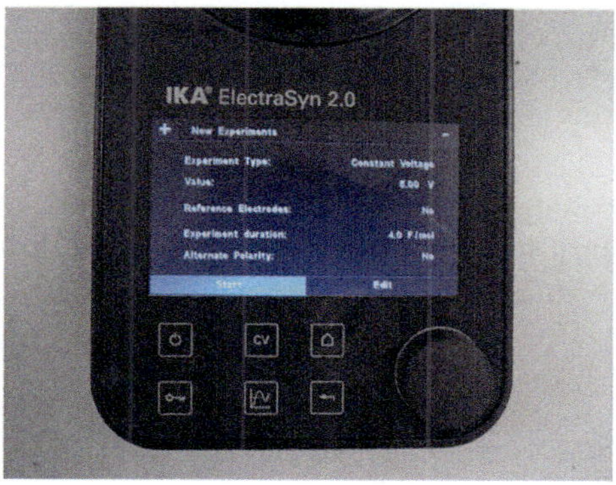

9. *Start the experiment*

Figure 3.7 (Continued)

- *Reaction scale and equivalents of electrons in Faradays per mol (F·mol^{-1}).*
  - Once you enter the reaction scale in mmols, you must select the equivalents of electrons. Thinking about desired changes in oxidation state of the redox-active substances in solution can allow you to easily choose this value. For example, if a single-electron reduction of the substrate should lead to the desired reaction, you can choose 1 F·mol^{-1}. If there might be any inefficiency, you can apply more electrons that are needed, say 1.5 F·mol^{-1}. For reactions in which a catalytic species will be affected by reduction or oxidation, you will want to adjust the F·mol^{-1} according to the change in oxidation state per substrate equivalent.
- *Reference electrode.*
  - A reference electrode should be used if the reaction is being run at constant potential. The reference electrode measures the potential compared to a known standard. The IKA reference electrode is silver and can be used with KCl as a Ag/AgCl redox pair. If a constant potential reaction is run without a reference electrode, the instrument will measure and apply a potential based on the difference between the two electrodes in use.
- *Alternating polarity.*
  - Alternating polarity can only be used when the cathode and anode are made of the same material. When this mode is selected, the direction of the current changes at the frequency indicated by the user. This can be used to prevent buildup at an electrode and has also been used as a tool to enhance selectivity and reactivity.
- *Save experiment.*
- *Summary.*

  If you put something in wrong, you can use the turning nob to adjust or check that parameter at this point.
- *Start.*
  - Stir rate can be adjusted once the reaction is running. The ElectraSyn will automatically run at 400 rpm but increasing the stir rate to at least 1000 rpm is recommended because most electrochemical reactions are mass-transfer limited.

3.6.5 During the Reaction

1) If the connection between the square carrier and ElectraSyn is faulty, the voltage will maximize (the other reason for this could be that the resistance of the solution is too high due to not enough electrolyte or a nonconducting solvent like THF). With the carousel, if the connection to the ElectraSyn is not correct, the electrolysis will not start. If there is a vial connection issue you will observe "–"on the screen at the position of the problematic vial.

Figure 3.8 During the reaction.

2) Check the value of the voltage/current (Figure 3.8). From our experience, for a constant current experiment run at 5–40 mA, the resulting voltage should most likely be lower than 20 V. For a constant voltage experiment run at 2–20 V, the current should be below 40 mA if the reaction is properly set up. When running multiple vials at once on the carousel, the total voltage of the system can reach a maximum of 30 V. This means that to accommodate many vials, the system will automatically lower the applied current of the reaction if the corresponding total voltage of all the vials on the carousel is greater than 30 V. If you are concerned about high resistance in a reaction due to solvents that have a low dielectric constant or low concentration of electrolyte, do not use the carousel or do not fill the entire carousel. You can check for this by removing one of the vials and seeing if the voltage of the other vials increases. It is possible that the potential increases drastically over the course of the reaction. This is likely due to solids depositing on one or both electrodes. You can verify this after the reaction, but it indicates a problem with the reaction conditions such as solubility of products or salts issued from the sacrificial anode, plating-out of metals onto the electrode, or unwanted side reactions.
3) Check electrodes for gas formation and plated metals or accumulation of organic materials.
4) Note changes of color.
5) Note solid precipitation.
6) Take aliquots of the reaction as desired for analysis by NMR, GC-MS, TLC, etc.

3.6.6 After the Reaction

1) Check that the reaction was completed (all $F{\cdot}mol^{-1}$ have passed, or time of reaction is correct).
2) Note electrode state and reaction mixture state.
3) Clean vial like you would with any glassware. Remove the electrodes from the vial cap. Clean the cap with acetone and water if needed. Replace the septum if it is dirty. If connections appear rusted where the cap connects to the

ElectraSyn or where the electrodes connect to the cap, wet a paper towel with 1 M HCl and scrub these areas. Electrodes can be cleaned with water, 1 M HCl, and acetone. Graphite and metal-based electrode can also be cleaned with sandpaper. For more detailed descriptions of cleaning each type of electrode, IKA has a useful guide.[1] In general, if an electrode looks dirty or darkened and cannot be cleaned, you should use a different one. Depending on the reaction being run, electrodes can expire quickly or be reused many times. As a chemist, it is best to use visual inspection to make this determination. If a reaction is suffering from reproducibility issues, it may also be time to use a fresh electrode. Porous materials like graphite and RVC tend to have shorter lifetimes while metallic electrodes tend to have longer ones.

4) Reaction analysis can proceed as desired.

3.7 Case Study

Because electrochemistry requires electrodes and electrolyte as well as specific types of solvents, it can be daunting to translate a traditional chemical reaction into an electrochemical one. This section is meant to explain the logic behind the parameters of an electrochemical experiment, including reaction design and the optimization process. It describes a recent project developed in our group illustrating how a known photochemical reaction was achieved using electrochemistry. Through a combination of chemical intuition, knowledge of relevant precedents, and trial and error, we hope this case study enables the practitioner to confidently learn how to design an electrochemical experiment.

3.7.1 Project Overview

Dialkyl ketones are a motif often found in biologically active molecules (Figure 3.9) [19]. One way to synthesize dialkyl ketones is through reductive cross-coupling (RCC). However, traditional RCC methods require the use of stoichiometric reductants [20–27]. Electrochemical RCC (ERCC) overcomes the use of chemical reductants in the synthesis of dialkyl ketones. Examples of ERCC for the synthesis of dialkyl ketones have been reported but are limited to the formation of benzylic or methyl dialkyl ketones and couple highly activated acylative reagents like acyl chlorides and anhydrides [28–30]. Imides are an alternative to more common reactive acyl precursors, and because of their stability, imides are especially useful as late-stage acyl synthons in multistep syntheses. In the past, our lab has developed the coupling of imides with alkyl bromides using metallophotoredox catalysis to access ketones (Scheme 3.1a) [31], but the need for an

1 https://www.ika.com/ika/pdf/flyer-catalog/202103_Electrasyn%202.0_cleaning%20electrodes_EN.pdf.

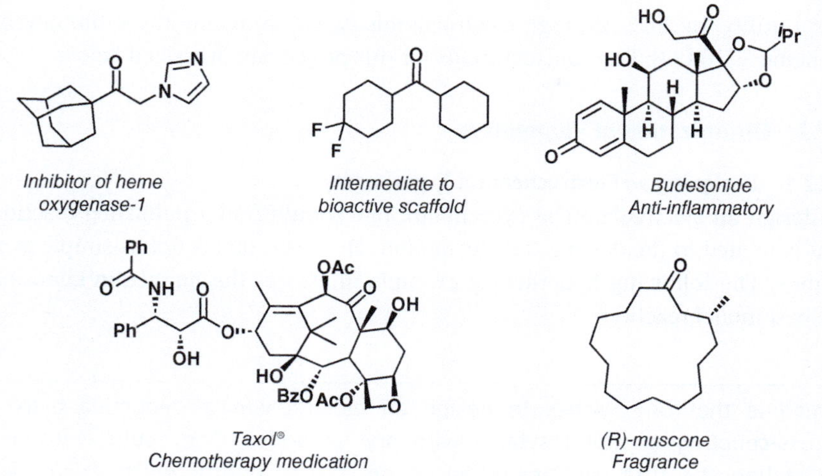

Figure 3.9 Occurrence of dialkyl ketones in bioactive molecules.

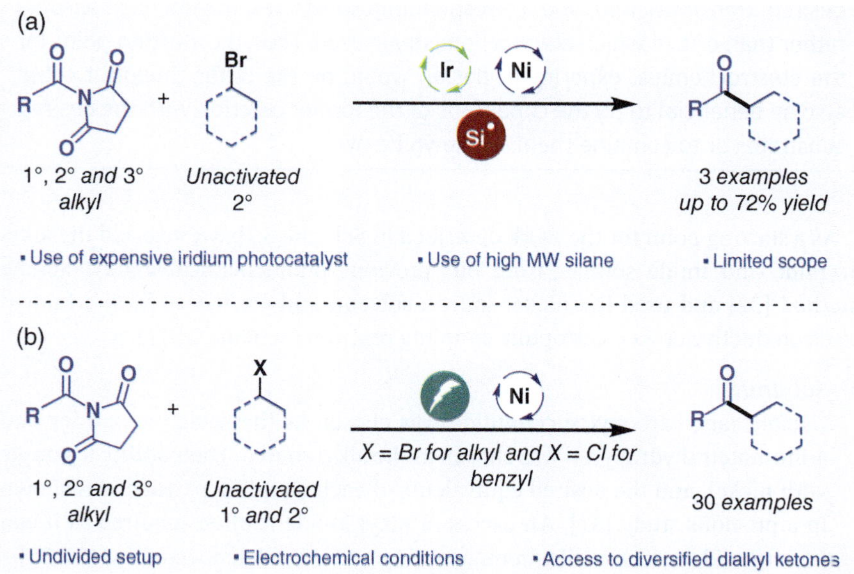

Scheme 3.1 (a) Photochemical Ni-catalyzed formation of dialkyl ketones from imides [31]. (b) Electrochemical nickel reductive cross electrophile coupling of imides with alkyl halides [32].

expensive photocatalyst and stoichiometric silane reductant diminished the reaction's utility. In 2022, through electrochemistry, we overcame these limitations (Scheme 3.1b) [32]. Key considerations for this project are described below.

3.7.2 Optimization of Parameters

3.7.2.1 Designing an Electrochemical Experiment

To design an electrochemical experiment, one should find a published reaction that is related to the desired transformation. However, this is not as simple as it sounds. The following hypothetical example illustrates the nuance in choosing the best model reaction:

> Imagine that one wishes to design an electrochemical reductive biaryl cross-coupling of aryl tosylates with aryl chlorides. One could envision modeling the first experiment based on a hypothetical electrochemical reductive cross coupling of alkyl carboxylic acids and alkenyl bromides. On the other hand, one could model the first experiment based on a hypothetical cross coupling between aryl tosylates and aryl chlorides with zinc as the terminal reductant. Although it might seem easier to transition one electrochemical reaction to another, we recommend using the reaction in which the desired transformation and corresponding substrates match most closely rather than one in which electrochemistry is used. Thus, the starting point for the electrochemical experiment design would be the latter, though it could also be beneficial to try the conditions of the former reaction with the desired substrates or to combine them as shown below.

As a starting point for the work described in Scheme 3.1b, we selected the alkyl bromide and imide sources from our previous photochemical dialkyl ketone method [32] and used the nickel source, solvent, and electrodes from a known electroreductive cross electrophile coupling reaction (Scheme 3.2) [33].

1) *Substrates*
   Cyclohexane carbonyl succinimide was chosen as the acylative partner and 4-bromotetrahydropyran was chosen as the alkyl source. Their ability to couple with nickel, and the desired equivalents of each coupling partner, was shown in a previous study [31]. An excess of alkyl halide is often required as it can undergo side reactions like homocoupling and protodehalogenation [34, 35].
2) *Catalyst*
   A pre-formed nickel(II) bromide 2,2′-bipyridine complex was chosen as the nickel precursor because it is known to avoid plating of nickel at the cathode [36].

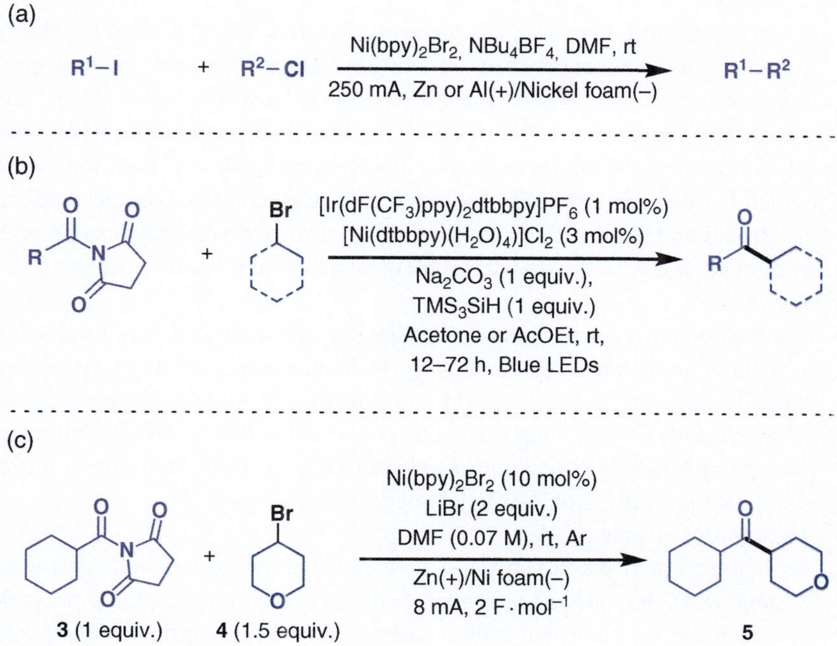

Scheme 3.2 From literature precedents to first electrochemical attempt. (a) Pioneering electrochemical Ni-catalyzed formation of dialkyl ketones from acyl chlorides [33]. (b) Photochemical Ni-catalyzed formation of dialkyl ketones from imides [31]. (c) Electrochemical proof of concept.

3) *Electrolyte*

An electrolyte is a substance that contains ions and can conduct electricity. When an electrolyte is placed in a solution, the ions in the solution become mobile and can carry an electrical current through the solution. In organic solvents, common examples of electrolytes include salts such as tetrabutylammonium tetrafluoroborate ($TBABF_4$), potassium hexafluorophosphate (KPF_6), and lithium bromide (LiBr). Regarding our study, as bromide is already present in the media from alkyl bromide and the nickel source, LiBr was chosen as the electrolyte.

4) *Additives*

No additives were added in the reaction as we could draw a viable catalytic cycle without any.

5) *Solvent*

ERCC often employ polar aprotic solvents like DMF or DMA. Previously reported conditions indicated that DMF could be a suitable solvent for this reaction [33].

6) *Concentration*
   In the 5 mL ElectraSyn vial, a volume of 3–4 mL is recommended to achieve proper surface contact between the solution and the electrodes. We first opted for a concentration of 0.1 M (*which is common in cross coupling reactions*) and a scale of 0.25 mmol of substrates to be engaged in the reaction (*a scale where all analysis can be performed easily including any potential isolation steps*), which corresponds to 2.5 mL of solvent. To have good surface contact between electrodes and the media, an extra milliliter of solvent was added, so the concentration was 0.07 M with 0.25 mmol of substrate and 3.5 mL of solvent.
7) *Electrodes*
   The desired cross coupling is purely reductive, meaning that only a reduction is required to turn over the catalytic cycle. Such reaction needs a setup with a cathode and a sacrificial anode (*for more details regarding the different modes of electrochemical reactions, the reader is directed to Refs.* [16–18]). Following the example of ERCC reactions in electrosynthesis from literature, a nickel foam cathode and a zinc sacrificial anode were chosen [32].
8) *Voltage and current*
   As it is common when starting a new electrochemical reaction (*please see Section 3.6.4*), the constant current mode of electrolysis was adopted. For a scale of 0.25 mmol, we decided to apply a current value of 8 mA as most of the examples we found in the literature were applying currents between 5 and 10 mA.
9) *Number of electrons*
   Theoretically, two electron equivalents should be enough for our reaction to be completed, so 2 $F{\cdot}mol^{-1}$ were applied. *At an early stage, we noticed that the starting material was not totally consumed after applying 2* $F{\cdot}mol^{-1}$, *so we increased the* $F{\cdot}mol^{-1}$ *to 4.*
10) *Temperature*
   Heating and cooling are difficult, though not impossible, to perform using ElectraSyn 2.0 without using the Gogo module accessory. To make screening faster, our reaction was run at room temperature.

3.7.3 Proof of Concept

The final reaction conditions are shown in Scheme 3.3. The desired product was obtained in 36% yield determined by GC-MS with an internal standard, and we proceeded with optimization from this point.

3.7.3.1 Optimization

While reaction parameters can be optimized in any order, depending on the reaction and what is known from literature precedent, there may be a logical order in which to conduct them. For our study, we first optimized the parameters

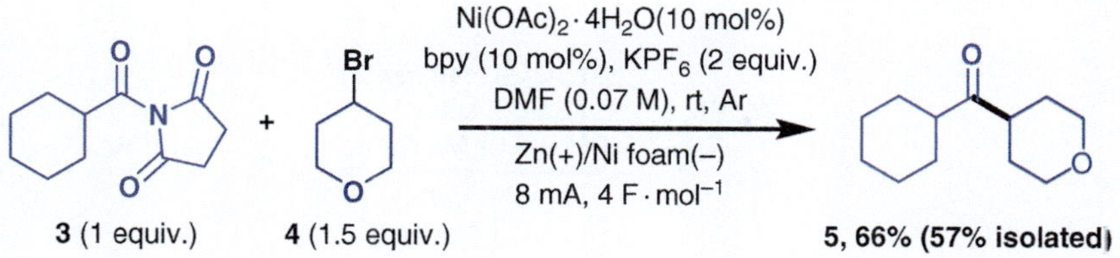

Scheme 3.3 Optimized conditions for the nickel-catalyzed electroreductive cross coupling of imides with alkyl halides.

we expected to have the greatest impact on reaction yield. Thus, we started with electrolyte, followed by nickel catalyst source and ligand, the solvent, additives, equivalents of alkyl bromide, and finally electrochemical parameters including current, number of electrons, and electrodes.

1) Electrolyte
Electrolytes are composed of a cation and an anion that can interfere with the catalyst. Therefore, we decided to first screen electrolytes. After screening a handful of classic salts, several parameters were considered to choose the most fitting one. Lithium perchlorate, tetrabutylammonium tetrafluoroborate, tetrabutylammonium hexafluorophosphate, and potassium hexafluorophosphate were all high yielding, but perchlorates have safety concerns. Tetrabutylammonium salts have a high molecular weight and are thus less atom-economic. In addition, electrolytes such as tetrabutylammonium tetrafluoroborate and tetrabutylammonium hexaflurophosphate are expensive. Therefore, considering safety, molecular weight, and cost, potassium hexafluorophosphate (KPF_6) was used for the rest of the optimization. Noteworthy, we decided to use two equivalents of electrolyte to insure sufficient conductivity that was assessed by looking at the voltage range obtained during the reaction (Table 3.1).
2) Precatalyst and ligand
We screened pre-ligated and non-pre-ligated nickel(II) bromide complexes and found bipyridine to be important (Tables 3.2 and 3.3). Then, other nickel sources were tested with this 2,2′-bipyridine ligand. $NiCl_2$•glyme and nickel(II) acetate tetrahydrate were both high yielding. Nickel(II) acetate tetrahydrate is 1/100th the price of $NiCl_2$·glyme and $NiBr_2$·glyme. Therefore, a pre-ligated complex of nickel(II) acetate 2,2′-bipyridine was attempted but afforded diminished yield. As we were interested in probing the mechanism during the precatalyst screening, bis(1,5-cyclooctadiene)nickel, the pseudo-nickel(I) complex $[(dtbbpy)NiCl]_2$, and a possible reaction intermediate were tested as potential active catalysts (entry 8–10). After this screening, the

Table 3.1 Electrolyte screen.

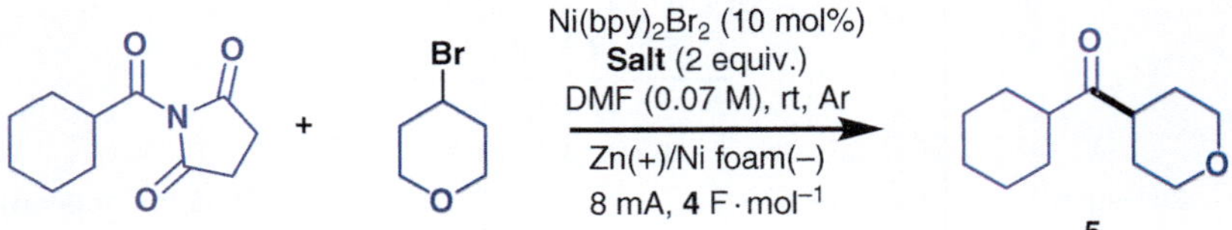

Entry	Salt	Yield of 5[a]
1	$LiClO_4$	60%
2	LiBr	34%
3	$MgBr_2$	17%
4	$TBAPF_6$	65%
5	$TBABF_4$	63%
6	KPF_6	62%

[a] Reactions were performed on a 0.25 mmol scale. Yields determined by GC-MS using benzophenone as an internal standard.

Table 3.2 Pre-catalyst screen.

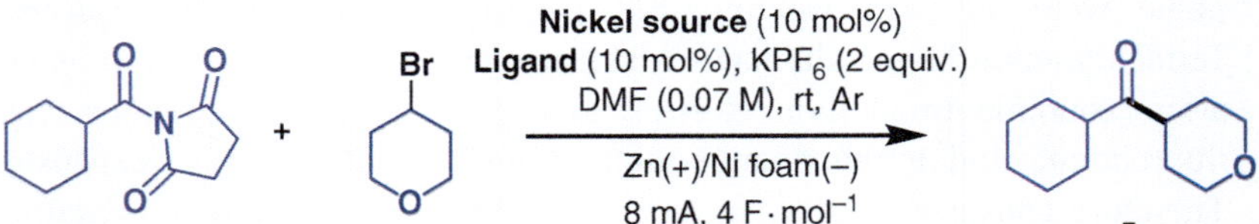

Entry	Nickel source	Ligand	Yield of 5[a]
1	$Ni(bpy)_2Br_2$	—	62%
2	$Ni(bpy)_3Br_2$	—	55%
3	$NiBr_2\cdot$ glyme	bpy	69%
4	$NiBr_2\cdot$ glyme	dtbbpy	36%
5	$Ni(OAc)_2\cdot 4H_2O$	bpy	66%
6	$NiCl_2\cdot$ glyme	bpy	69%
7	$Ni(dtbbpy)(OAc)_2\cdot 2H_2O$	—	50%
8	$[Ni(dtbbpy)Cl]_2$	—	29%
9	$Ni(cod)_2$	bpy	46%
10	Ni(dtbbpy)(COCy)(succ)	—	42%

[a] Reactions were performed on a 0.25 mmol scale. Yields determined by GC-MS using benzophenone as an internal standard.

Table 3.3 Ligand screen.

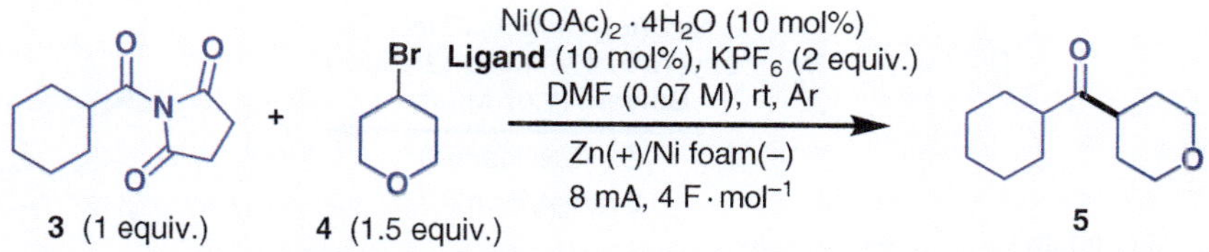

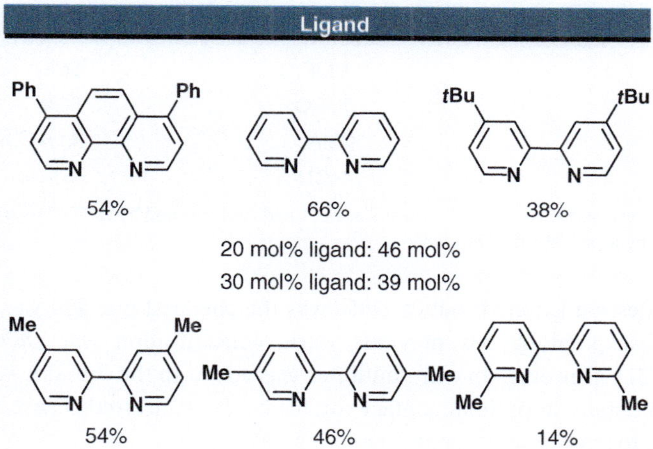

Reactions were performed on a 0.25 mmol scale. Yields determined by GC-MS using benzophenone as an internal standard.

nature of the nickel was found to only have a minor impact on the reaction yield. Thus, nickel(II) acetate tetrahydrate was chosen as a cost-effective nickel source.

On the other hand, the ligand clearly impacted the yield of the reaction. Only nitrogen-based ligands were tested as they are reported to be especially effective for nickel-catalyzed acylative RCC [28–30]. The steric properties of the ligand were significant; 6,6′-dimethyl-2,2′-dipyridyl only gave14% yield. Of the bipyridine ligands tested, 2,2′-bipyridine gave the highest yield and is a cost-effective option. Higher equivalents of ligand compared to nickel afforded diminished yield, so equimolar equivalents of 2,2′-bipyridine and nickel were adopted.

3) Solvent

Polar solvents were explored because resistance in solvents such as THF, hexanes, toluene, and DCM is too high to run the reaction at a reasonable potential due to low solubility of salts in those solvents as well as lower dielectric constant (Table 3.4). Out of all the screened solvents that afforded high yields

Table 3.4 Solvent screen.

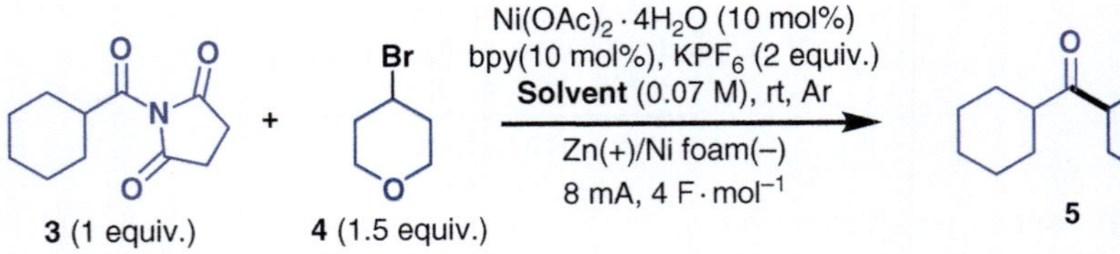

Entry	Solvent	Yield of 5[a]
1	DMF	66%
2	DMSO	60%
3	DMAc	41%
4	AcCN	5%

[a] Reactions were performed on a 0.25 mmol scale. Yields determined by GC-MS using benzophenone as an internal standard.

of the desired ketone product, DMF was the cheapest one and was thus preferred. Inspired by our previous work, concentration was chosen to be 0.1 M [31]. However, an extra milliliter was added to the initial 2.5 mL of solvent to obtain an optimal contact surface between electrodes and the media, leading to a reaction concentration of 0.07 M.

4) Additives

Some previous reports involve additives to enhance the yield, and we tested these additives as well (Table 3.5) [37–40]. First an iodide source was tested to promote halogen exchange with the nickel catalyst, but this diminished the yield. Potassium carbonate, a coordinating base, was added, but it did not significantly impact the yield. Magnesium salt, a coordinating species, and lithium bromide, an additional source of bromide in the media, also afforded diminished yield. Catalytic amount of zinc salts, added at the start of the reaction, were proven to enhance catalyst activity and to increase the yield of RCC reactions [38]. Under our electrochemical reaction conditions, this was not the case. Water can also help to solubilize some ionic species. In our case, catalytic amounts of water were slightly detrimental, showing that the reaction is sensitive to water. Finally, an electron shuttle has been described in ERCC reactions to avoid overreduction of substrates by interacting with the cathode in place of the other reaction components [41]. The shuttle gets reduced at around −1.8 V vs Fc/Fc^+ and can be added to avoid reduction below this potential [41]; 20 mol% of $Ni(MeBPI)_2$ shuttle was added to the reaction. Although it avoided nickel plating at the cathode, showing that it did act as an electron shuttle, no change in yield was observed, which demonstrates that overreduction is not a limiting parameter.

Table 3.5 Additives screen.

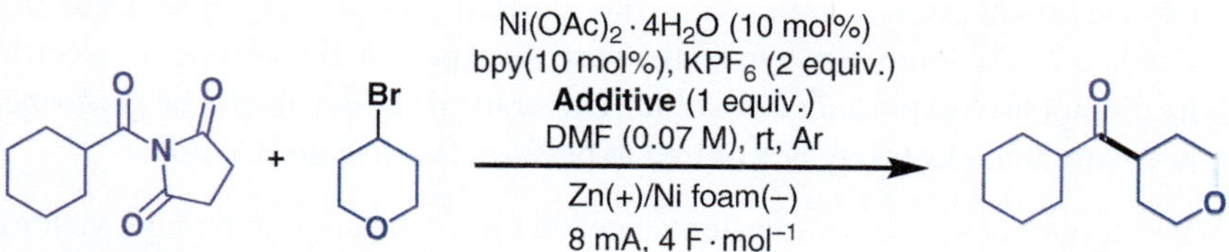

3 (1 equiv.) **4** (1.5 equiv.) **5**

Entry	Addtive	Yield of 5[a]
1	None	66%
2	NaI	28%
3	K_2CO_3	57%
4	$MgCl_2$	30%
5	LiBr	35%
6	$ZnBr_2$ (10 mol%)	53%
7	H_2O (20 mol%)	47%
8	$Ni(MeBPI)_2$ (20 mol%)	61%

[a] Reactions were performed on a 0.25 mmol scale. Yields determined by GC-MS using benzophenone as an internal standard.

5) Substrate equivalents
The importance of the ratio between coupling partners has been highlighted in recent studies (Table 3.6) [42]. The optimal ratio was reached at 1 : 1.5 imide : alkyl bromide.

Table 3.6 Equivalents screen.

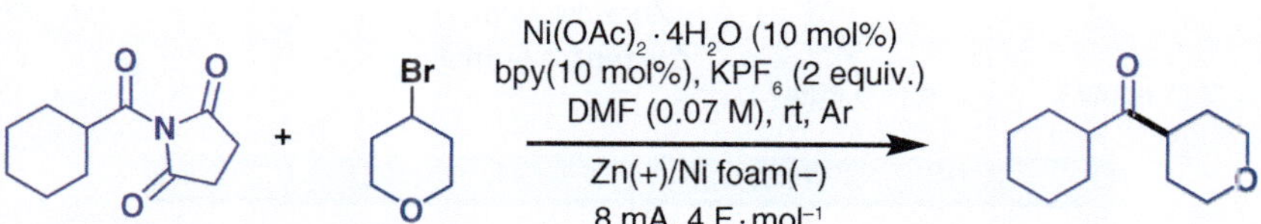

3 (1 equiv.) **4** (**X** equiv.) **5**

Entry	X equiv. of 4	Yield of 5[a]
1	1	43%
2	1.5	66%
3	2	48%

[a] Reactions were performed on a 0.25 mmol scale. Yields determined by GC-MS using benzophenone as an internal standard.

6) Electrochemical parameters

   Of electrochemical parameters, the current should be optimized first (Table 3.7). We found 8 mA to be the best. As expected, the absence of electricity did not lead to product formation. This control should always be performed to ensure that electricity is required to perform the desired reaction.

The number of electron equivalents needed for the reaction to be high yielding was 4 $F \cdot mol^{-1}$ (Table 3.8). Fewer equivalents give a lower yield and greater equivalents did not improve the yield.

Finally, different electrode pairs were tested (Table 3.9). First the zinc sacrificial anode was replaced by magnesium as such electrode was successfully used in the context of reductive cross coupling reactions [33]. Unfortunately, this sacrificial anode only provided trace amount of desired product. This is certainly due to the formation of undesired by-products issued from the reaction between Mg cations and DMF as demonstrated by the group of Périchon in one of their electrochemical studies [43]. An attempt to replace the sacrificial anode by a combination of amine (*added as a sacrificial reductant, see Ref.* [44] *for more details*) with a non-sacrificial anode of graphite, only gave traces of product. Then, different cathodes were explored. Graphite gave very low yield, probably due to its low surface area. An RVC cathode gave good yields, but nickel foam was better. As such, nickel foam cathode and zinc anode provided the best results.

Table 3.7 Current variations.

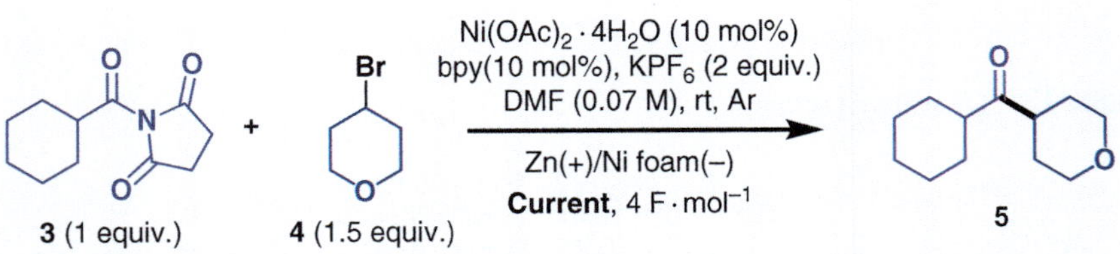

Entry	Current	Yield of 5[a]
1	4 mA	54%
2	6 mA	53%
3	8 mA	66%
4	10 mA	64%
5	12 mA	60%
6	No electricity	0%

[a] Reactions were performed on a 0.25 mmol scale. Yields determined by GC-MS using benzophenone as an internal standard.

Table 3.8 *F·mol^{-1}screen.*

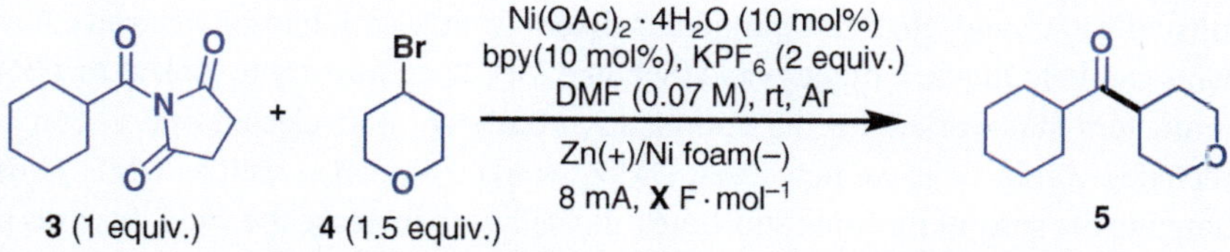

Entry	X F·mol^{-1}	Yield of 5[a]
1	3	42%
2	4	66%
3	5	60%

[a] Reactions were performed on a 0.25 mmol scale. Yields determined by GC-MS using benzophenone as an internal standard.

Table 3.9 Electrode screen.

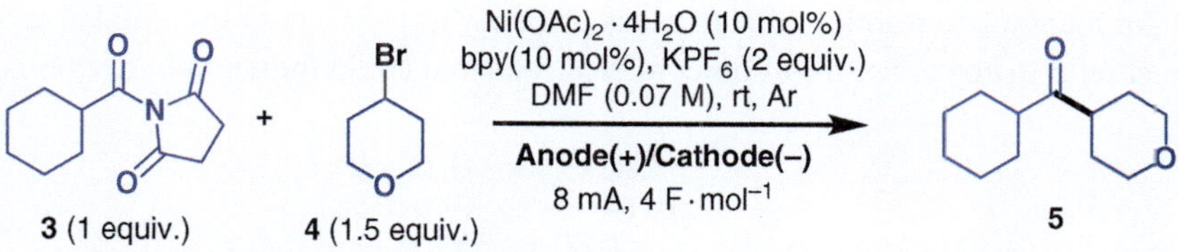

Entry	Anode(+)/Cathode(–)	Yield of 5[a]
1	Zn(+)/Ni foam(–)	66%
2	Mg(+)/Ni foam(–)	<5%
3	Zn(+)/RVC(–)	48%
4	Zn(+)/Graphite(–)	14%
5	Graphite(+)/Ni foam(–), NEt_3 (2 equiv.)	<5%

[a] Reactions were performed on a 0.25 mmol scale. Yields determined by GC-MS using benzophenone as an internal standard.

Overall, the optimization process led us to a high yielding and cost-effective electrochemical transformation (Scheme 3.3).

During the optimization, it is important to pay attention to possible decomposition of substrates, precipitate formation, or gas release. In our reaction, a change of color from colorless to dark orange seemed to indicate a successful reaction. We believe the color originated from the formation of (bpy)Ni(COCy)(succ) complex. While a change of color does not guarantee a good outcome, no color change meant it would fail.

3.7.3.2 Substrate Scope

During the substrates scope investigation, if some desired products are obtained in unusually low yield, changing parameters like solvent, electrolyte, or electron equivalents can help to reach higher yield (Scheme 3.4). Therefore, try to look at reaction parameters that worked during optimization but were not selected due to cost or efficiency. In our case, we noted that DMSO worked almost as well as DMF. Thus, changing the solvent for some substrates allowed us to increase the yield. In cases in which starting material was still present, greater electron equivalents were passed in order to favor the conversion by increasing catalyst turnover (i.e. cathodic reduction of the nickel catalyst). Finally, if the limiting reagent was still observed at the end of the reaction, more equivalents of alkyl bromide were added.

This case study should give the practitioner a good start on how to optimize an electrochemical reaction. Synthetic parameters optimization is difficult to generalize as it depends on the transformation design. Electrochemical parameters however can be generalized as the following:

1) Current optimization can lead to an optimum value.
2) Electron equivalents can be important, as an underestimated value will lead to an incomplete reaction (due to a lack of electrons to react). One should also be careful to any excessive number of electrons that could induce side reactions.

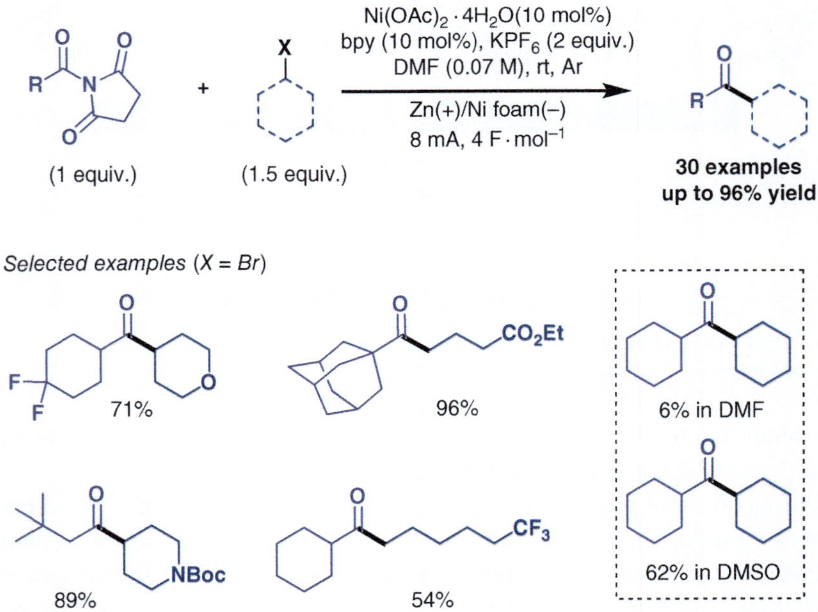

Scheme 3.4 Selected scope.

3) Electrodes should first be chosen depending on the reaction mode (paired electrolysis, only reduction, or only oxidation) [16–18]. Then the porosity of the electrode material might affect the reaction significantly. Ultimately, cost of electrodes should be considered, as some materials can be expensive.

Naturally, the order of synthetic or electrochemical parameters variations can be interchanged depending on the reaction being studied.

3.8 Conclusion

As demonstrated throughout this book chapter, electrosynthesis is a field that can mistakenly be perceived as complex and challenging to outsiders. However, the development of standardized equipment started changing this perception as exemplified by the increase in the number of electrochemistry papers published since 2017. Thus, we really hope that this practical step-by-step guide on how to use ElectraSyn2.0 and to perform an optimization campaign will motivate organic chemists to confidently become electrosynthetic practitioners. If the reader wishes to learn more about the basic aspects of electrochemistry, we suggest carefully reading Ref. [17] from the Baran research group. If the reader wishes to learn more about practical aspects, we recommend reading Refs. [16, 18] from the Hilt and Lam laboratories, respectively.

References

1 Yan, M., Kawamata, Y., and Baran, P.S. (2017). Synthetic organic electrochemical methods since 2000: on the verge of a renaissance. *Chemical Reviews* 21 (117): 13230–13319.

2 Stork, W. (2004). Specialty chemicals. *Chemical & Engineering News supplement* 82: 35–39.

3 Bahuguna, A. and Jandhyala, H. (2022). Specialty Chemicals Market Size, Share & Trends Analysis Report, 2030.

4 Badwal, S.P., Giddey, S.S., Munnings, C. et al. (2014). Emerging electrochemical energy conversion and storage technologies. *Frontiers in Chemistry* 2 (79): 1–28.

5 Yan, Z., Hitt, J.L., Turner, J.A., and Mallouk, T.M. (2019). Renewable electricity storage using electrolysis. *Proceedings of the National Academy of Sciences* 23 (117): 12558–12563.

6 Terlouw, T., Bauer, C., McKenna, R., and Mazzoti, M. (2022). Large-scale hydrogen production via water electrolysis: a techno-economic and environmental assessment. *Energy & Environmental Science* 15: 3583–3602.

7 Shiva Kumar, S. and Himabindu, V. (2019). Hydrogen production by PEM water electrolysis – a review. *Materials Science for Energy Technologies* 2 (3): 442–454.
8 Xia, R., Overa, S., and Jiao, F. (2022). Emerging electrochemical processes to decarbonize the chemical industry. *Journal of the American Chemical Society Au* 2 (5): 1054–1070.
9 Wen, D., Fu, R., and Li, Q. (2021). Removal of inorganic contaminants in soil by electrokinetic remediation technologies: a review. *Journal of Hazard Mater* 401 (5): 123345–123364.
10 Yoshida, J., Kataoka, K., Horcajada, R., and Nagaki, A. (2008). Modern strategies in electroorganic synthesis. *Chemical Reviews* 108 (7): 2265–2299.
11 Frontana-Uribe, B., Little, R.D., Ibanez, J.G. et al. (2010). Organic electrosynthesis: a promising green methodology in organic chemistry. *Green Chemistry* 12: 2099–2119.
12 Francke, R. and Little, R.D. (2014). Redox catalysis in organic electrosynthesis: basic principles and recent developments. *Chemical Society Review* 43: 2492–2521.
13 Pletcher, D. (2014). Organic electrosynthesis. In: *Developments in Electrochemical Science Inspired by Martin Fleischmann* (ed. D. Pletcher, Z.-Q. Tian, and D.E. Williams), 77–94. Chichester: Wiley.
14 Jiang, Y., Xu, K., and Zeng, C. (2018). Use of electrochemistry in the synthesis of heterocyclic structures. *Chemical Reviews* 118 (9): 4485–4540.
15 Gütz, C., Klöckner, B., and Waldvogel, S.R. (2016). Electrochemical screening for electroorganic synthesis. *Organic Process Research and Development.* 20 (1): 26–32.
16 Hilt, G. (2020). Basic strategies and types of applications in organic electrochemistry. *ChemElectroChem* 7 (2): 395–405.
17 Kingston, C., Palkowitz, M.D., Yusuke, T. et al. (2020). A survival guide for the "electro-curious". *Accounts of Chemicals Research* 53 (1): 72–83.
18 Leech, M.C. and Lam, K. (2022). A practical guide to electrosynthesis. *Nature Reviews Chemistry* 6: 275–286.
19 Foley, D.J. and Waldmann, H. (2022). Ketones as strategic building blocks for the synthesis of natural product-inspired compounds. *Chemical Society Review* 51: 4094–4120.
20 Wotal, A.C. and Weix, D.J. (2012). Synthesis of functionalized dialkyl ketones from carboxylic acid derivatives and alkyl halides. *Organic Letters* 14 (6): 1476–1479.
21 Cherney, A.H., Kadunce, N.T., and Reisman, S.E. (2013). Catalytic asymmetric reductive acyl cross-coupling: Synthesis of enantioenriched acyclic α,α-disubstituted ketones. *Journal of the American Chemical Society* 135 (20): 7442–7445.
22 Zhao, C., Jia, X., Wang, X., and Gong, H. (2014). Ni-catalyzed reductive coupling of alkyl acids with unactivated tertiary alkyl and glycosyl halides. *Journal of the American Chemical Society* 136 (50): 17645–17651.

23 Shi, R. and Hu, X. (2019). From alkyl halides to ketones: nickel-catalyzed reductive carbonylation utilizing ethyl chloroformate as the carbonyl source. *Angewandte Chemie International Edition* 58 (22): 7454–7458.

24 Ni, S., Padial, N.M., Kingston, C. et al. (2019). A radical approach to anionic chemistry: synthesis of ketones, alcohols, and amines. *Journal of the American Chemical Society* 141 (16): 6726–6739.

25 He, J., Song, P., Xu, X. et al. (2019). Migratory reductive acylation between alkyl halides or alkenes and alkyl carboxylic acids by nickel catalysis. *ACS Catalysis* 9 (4): 3253–3259.

26 Wang, J., Cary, B.P., Beyer, P.D. et al. (2019). Ketones from nickel-catalyzed decarboxylative, non-symmetric cross-electrophile coupling of carboxylic acid esters. *Angewandte Chemie International Edition* 58 (35): 12081–12085.

27 Wang, J., Hoerrner, M.E., Watson, M.P., and Weix, D.J. (2020). Nickel catalyzed synthesis of dialkyl ketones from the coupling of N-alkyl pyridinium salts with activated carboxylic acids. *Angewandte Chemie International Edition* 59 (32): 13484–13489.

28 Jiao, K.-J., Ma, C., Liu, D. et al. (2021). Nickel-catalyzed electrochemical reductive relay cross-coupling of alkyl halides with alkyl carboxylic acids. *Organic Chemistry Frontiers* 8: 6603–6608.

29 Troupel, M., Ocafrain, M., Dolhem, E. et al. (1998). Electrosynthèses de composés carbonylés catalysées par des complexes du nickel: Activation du monoxyde de carbone ou de composés acylés. *The Canadian Journal of Chemical Engineering* 76 (6): 1013–1019.

30 Marzouk, H., Rollin, Y., Folest, J.C. et al. (1989). Electrochemical synthesis of ketones from acid chlorides and alkyl and aryl halides catalysed by nickel complexes. *Journal of Organometallic Chemistry* 369 (3): C47–C50.

31 Kerackian, T., Reina, A., Bouyssi, D. et al. (2020). Silyl radical mediated cross-electrophile coupling of N-acyl-imides with alkyl bromides under photoredox/nickel dual catalysis. *Organic Letters* 22 (6): 2240–2245.

32 Kerackian, T., Bouyssi, D., Pillet, G. et al. (2020). Nickel-catalyzed electro-reductive cross-coupling of aliphatic N-acyl imides with alkyl halides as a strategy for dialkyl ketone synthesis: scope and mechanistic investigations. *ACS Catalysis* 12 (19): 12315–12325.

33 Durandetti, M., Nédélec, J.-Y., and Périchon, J. (1996). Nickel-catalyzed direct electrochemical cross-coupling between aryl halides and activated alkyl halides. *Journal of Organic Chemistry* 61 (5): 1748–1755.

34 Zhang, W., Lu, L., Zhang, W. et al. (2022). Electrochemically driven cross-electrophile coupling of alkyl halides. *Nature* 604: 292–297.

35 Xue, W., Jia, X., Wang, X. et al. (2021). Nickel-catalyzed formation of quaternary carbon centers using tertiary alkyl electrophiles. *Chemical Society Review* 50: 4162–4184.

36 Kawamata, Y., Vantourout, J.C., Hickey, D.P. et al. (2019). Electrochemically driven, Ni-catalyzed aryl amination: scope, mechanism, and applications. *Journal of the American Chemical Society* 141 (15): 6392–6402.

37 Moragas, T., Correa, A., and Martin, R. (2014). Metal-catalyzed reductive coupling reactions of organic halides with carbonyl-type compounds. *Chemistry A European Journal* 20 (27): 8242–8258.

38 Weix, D.J. (2015). Methods and mechanisms for cross-electrophile coupling of Csp^2 halides with alkyl electrophiles. *Accounts of Chemical Research* 48 (6): 1767–1775.

39 Gu, J., Wang, X., Xue, W., and Gong, H. (2015). Nickel-catalyzed reductive coupling of alkyl halides with other electrophiles: concept and mechanistic considerations. *Organic Chemistry Frontiers* 2: 1411–1421.

40 Richmond, E. and Moran, J. (2018). Recent advances in nickel catalysis enabled by stoichiometric metallic reducing agents. *Synthesis* 50 (03): 499–513.

41 Truesdell, B.L., Hamby, T.B., and Sevov, C.S. (2020). General C(sp2)–C(sp3) cross-electrophile coupling reactions enabled by overcharge protection of homogeneous electrocatalysts. *Journal of the American Chemical Society* 142 (12): 5884–5893.

42 Everson, D.A. and Weix, D.J. (2014). Cross-electrophile coupling: principles of reactivity and selectivity. *The Journal of Organic Chemistry* 79 (11): 4793–4798.

43 Troupel, M., Nedelec, J.Y., and Perichon, J. (1991 and 2023). Electrochemical carboxylation of organic halides in the presence of consumable magnesium anode and modified cathode. In: *Electroorganic Synthesis*, 355–360. Routledge.

44 Perkins, R.J., Hughes, A.J., Weix, D.J., and Hansen, E.C. (2019). Metal-reductant-free electrochemical nickel-catalyzed couplings of aryl and alkyl bromides in acetonitrile. *Organic Process Research & Development* 23 (8): 1746–1751.

4

Flow Chemistry

Yosuke Ashikari and Aiichiro Nagaki

Department of Chemistry, Faculty of Science, Hokkaido University, Sapporo, Japan

Glossary

Batch chemistry Organic chemistry carried out in conventional reaction vessels such as flasks and beakers.

Collection time The time over which a reaction solution is collected from the output of a flow reaction. This is not related to the terms "reaction time" or "residence time," which is used to describe the time during which reacting chemicals are exposed to each other (see Section 4.1.1.1). In flow reactions, the reaction mixture is continuously delivered. Collection time in flow is comparable to reaction volume in batch – it is a measure of the quantity of material obtained.

Competitive sequential reaction a chemical reaction in which multiple reactions occur sequentially. Reactions where a primary product is generated from a starting material; however, they may react again with each other, which is termed the overreaction. Typical examples include the reduction of esters by metal hydrides and reactions of Grignard reagents with esters (ketones are primarily generated and then converted to alcohols). Further details are provided in Section 4.1.2.1.

Disguised chemical selectivity A principle for low reaction selectivity occurring with fast reactions propounded by Paul Rys in 1976 [1], which claims that it is difficult to control the selectivity of fast competitive sequential reactions, even if the first reaction is sufficiently faster than the latter. Section 4.1.2.1 describes the details.

Flow rate The velocity of a moving fluid or a volume of a fluid sent from pumps expressed per unit time, in $mL\,min^{-1}$.

Enabling Tools and Techniques for Organic Synthesis: A Practical Guide to Experimentation, Automation, and Computation, First Edition. Edited by Stephen G. Newman.

Homogeneous In a solution, all the dissolved materials or compounds are uniformly distributed at all points.

Laminar flow The condition of flow where fluids move parallel to the wall without vortexes. Antonym: turbulent flow.

Micromixer A device with a micrometer-sized inner diameter to mix two or more fluids.

Pressure drop The energy required to push fluid inside the tube. Generally, tubes with a longer length and thinner inner diameter or more viscous fluids tend to exhibit a higher pressure drop. To pass fluid in a high-pressure drop condition, suitable pumps that can steadily send fluids under higher pressures are required.

Reaction integration When several reactions are sequentially carried out without isolating the intermediates. The absence of the isolation step can reduce chemical waste and operation time, although the reaction conditions tend to be dirty, making the transformation difficult.

Residence time The length of time that a fluid exists in a reactor or the length of the fluid that passes from the inlet to the outlet. Residence time is calculated as follows:

$$t_{\mathrm{R1}} = \frac{V_{\mathrm{R1}}}{F}$$

where t_{R1} is the residence time in tube reactor R1, V_{R1} is the volume of tube reactor R1, and F is the flow rate of the fluid introduced into tube reactor R1 (see Figure 4.1). The linear velocity is distributed within the tube reactor, and the inside flow speed is faster than that near the outer wall. However, because the difference between these speeds is negligible, for ease of calculation, the average time is taken as the residence time.

Theoretical yield The amount of product generated by an ideal reaction. In batch theoretical yields, it can be calculated from the amount of added reagent. The product yield is then calculated from the amount of the obtained product divided by the theoretical yield. In the flow reactions, the theoretical yield is calculated using the flow rate and concentration of the reagent. The general equation is as follows:

$$\text{mole of theoretical yield} = C_{\text{reactant}} \times F_{\text{reactant}} \times t_{\text{collect}}$$

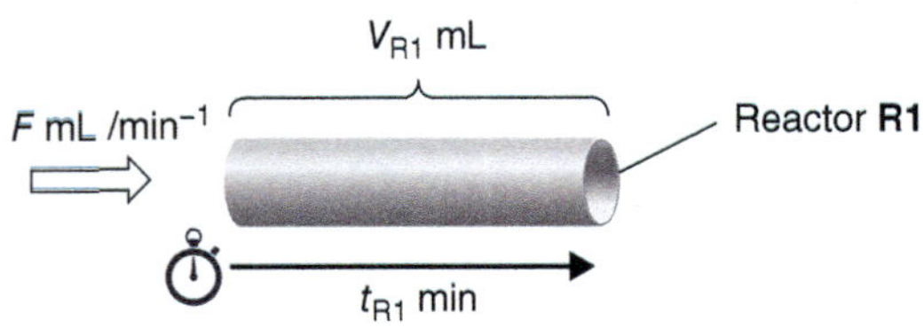

Figure 4.1 Schematic for simple tube reactor R1 and its variable. Units are representative.

where C_{reactant} is the concentration of the reactant for the standard, F_{reactant} is the flow rate of the reactant for the standard, and t_{collect} is the collection time. This equation indicates that the concentration, flow rate, and collection time must be recorded accurately to precisely calculate the product yield. Section 4.3 provides several examples.

Turbulent flow The condition of flow where fluids move irregularly. Many sizes of unsteady vortices are generated leading to accelerate the mixing of fluids. Antonym: laminar flow.

4.1 Introduction

Organic chemistry has primarily been developed by "batch chemistry." Historically, batch chemistry has provided a wide range of chemical reactions, and chemists have established numerous theories to predict the results of these reactions. For example, chemists can predict a specific product, even if the reaction can produce several possible species. However, high selectivity, including stereoselectivity, regioselectivity, and chemoselectivity, is not always accomplished. In particular, if the reactions are extremely fast, as is common when highly reactive reagents are used, controlling the selectivity becomes more difficult. In another challenging case, species with extremely short lifetimes are difficult to use in conventional reactions. These reactive intermediates can quickly decompose after they are generated. As a third challenge, when chemists need to increase the quantity of product produced, they cannot always adopt identical reaction conditions as the lab-scale reaction in manufacturing. Large-scale synthesis can lead to undesirable results such as low yield and low selectivity when macro-batch reactors are used.

In this chapter, we introduce an alternative class of reaction vessels called flow microreactors, and we provide general information about the chemistry carried out in flow to perform reactions that are difficult or impossible in batch chemistry. In particular, we focus on our group's expertise carrying out fast, exothermic chemistry such as organometallic reactions in a controlled manner. This chapter aims to provide readers with sufficient comfort and understanding of these technique to enable them to perform simple experiments, reproduce literature results, and consult more specialized literature. Section 4.1 describes general advantages of using flow microreactors. Section 4.2 provides practical information about performing experiments such as choosing equipment and designing experiments. Section 4.3 describes a detailed case study.

4.1.1 What is Flow Microchemistry

Flow chemistry is a type of synthetic chemistry that focuses on the use of flow reactors, especially flow microreactors. Flow microreactors have a flowing reaction mixture within a microscale reaction volume. This contrasts conventional batch reactors, which accumulate all reaction components at the same time and have a macro-scale (larger than a centimeter) reaction volume. Most of the advantages of flow microreactors arise from the two aforementioned characteristics. In this section, these key advantages are described, with a particular focus on organic synthesis.

4.1.1.1 Reaction Time Controllability

Most published chemical reactions involves stirring a reaction mixture some time, and then adding another reagent (or a quenching agent). Chemists can operate this procedure manually because such a time scale is manageable. However, what if they need to add a reagent just after one second of the addition of the primary two reagents? In addition, how can they control the exact 1.0 second, neither 0.9 nor 1.1? How can they repeat identical operation with no difference of the timing?

When using a flow microreactor, the reaction time is equivalent to the residence time, which can be controlled by the volume of the reaction space and the flow rate. Figure 4.2 shows a simple schematic of the precise control of reaction time in the flow microreactor. In a batch reactor, reactant **R1** is added to a flask containing a solution of substrate **S**, and after reaction time *t*, reagent **R2** is added. In a flow reactor, a solution of **S**, **R1**, and **R2** are introduced from inlet-1, inlet-2, and inlet-3, respectively. The respective flow rates of these solutions re F_S, F_{R1}, and F_{R2}, respectively, and the volume between inlet-2 and inlet-3 is V. When applying the

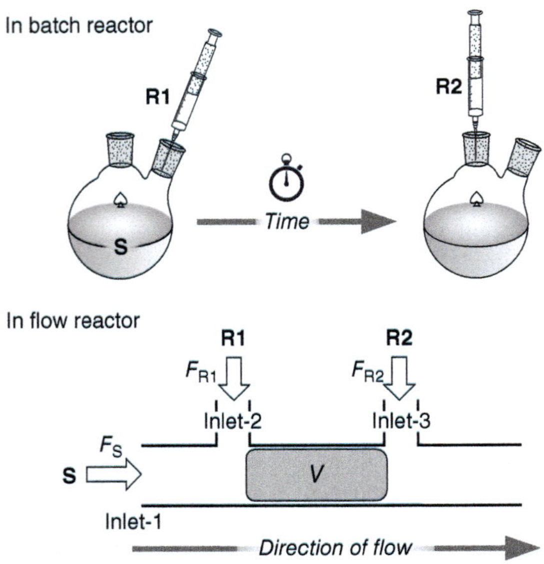

Figure 4.2 Schematic for residence time control.

flow system, the interval time t between the addition of **R1** and **R2** is equal to the residence time between inlet-2 and inlet-3, which can be calculated as follows:

$$t = \frac{V}{F_S + F_{R1}}$$

In batch, **R2** is added after t, following the addition of **R1**, to **S**. However, in flow **R2** is added when the mixture of **S** and **R1** passes through inlet-3. The interval between the additions is equivalent to the residence time in the flow microreactor. Because F_1 and F_2 can be controlled mechanically and V can be accurately determined by the length and cross-sectional area of the tube, the reaction time can be precisely controlled. Theoretically, with a higher flow rate and a smaller tube volume, the residence time can be shortened to milliseconds. For example, we demonstrated a thin and short-tube reactor with a residence time of 13 ms [2].

4.1.1.2 Fast Mixing

The time for diffusion can be written as:

$$t \sim \frac{d^2}{D}$$

where t is the diffusion time, d is the diffusion length, and D is the diffusion coefficient. As reaction scale is increased, so is the time required for the mixture to become homogeneous if relying only on diffusion. To overcome this, chemists stir the solution, although achieving homogeneity still relies on molecular diffusion in small segments. For example, common centimeter-sized magnetic bars create a large vortex that will not perfectly mix a solution. The time scale in which the mixtures become homogeneous can be sufficiently long that issues with reaction selectivity can arise.

In flow microreactors, solutions can be divided into small segments, the biggest being micrometer-sized. The size of the segments depends on the micromixer and the segment's inner structure. In one approach, the solvent flow is separated and then integrated to reduce the segment scale (Figure 4.3). When the flow is laminar, the width of the laminar flow is the length of the diffusion path. This reduces the stream thickness, resulting in a shorter mixing time. When the flow is turbulent, the segment scale can be further reduced, contributing to significantly fast mixing. A representative of such multi-lamination-type micromixers is manufactured by the Institute of Microtechnique Mainz (IMM), Germany. Another type of micromixer, the collision-type T-shaped micromixer, is described in Section 4.2.1.

If the mixing is slow, it is illogical to set a shorter residence time. For example, if it takes several seconds to homogenize the mixed solution, controlling the residence time within a millisecond scale will have a negligible effect. The addition of

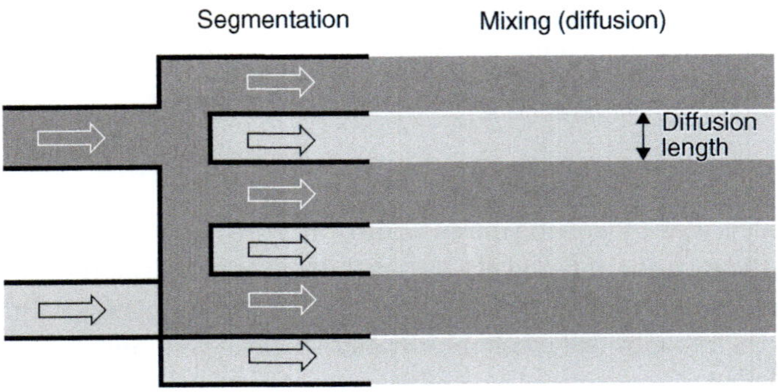

Figure 4.3 Schematic for mixing by multi-lamination-type micromixer.

a second reagent, such as **R2** in Figure 4.2, must occur after the solution becomes homogeneous.

4.1.1.3 Temperature Controllability

Reaction temperatures are primarily controlled by heat transfer from an outer source, such as an ice bath for cooling or an oil bath for heating. Because the heat transfer occurs over the surface of the reactors, a higher specific surface area directly affects the efficiency of heat transfer. The volume of an object is proportional to the length cubed, whereas the surface area is proportional to the length squared, which is known as the square-cube law. According to this mathematical principle, when the length of an edge of an object increases 10-fold, its surface area and volume increase by 100- and 1000-fold, respectively. This results in 1/10 of the specific surface area (Figure 4.4). Thus, the efficiency of the heat transfer differs according to the reactor scale. This is one key reason why a large-scale reaction, using identical conditions to the smaller-scale reaction, can result in failure. Therefore, a flow microreactor, which has a significantly high specific surface area, is suitable for controlling the reaction temperature. Several temperature-controlled exothermic reactions, including anionic polymerization, have been demonstrated in flow microreactors [3].

4.1.2 Reactions Enabled by Flow Microreactors

In Section 4.1.1, key characteristics and principles are summarized. In this section, we focus on certain types of reactions enabled by the use of flow microreactors.

4.1.2.1 Competitive Sequential Reactions

The most important benefit of fast mixing is overcoming disguised chemical selectivity (see Glossary). The following example shows a typical sequential competitive reaction with disguised chemical selectivity: Assume that the reaction of

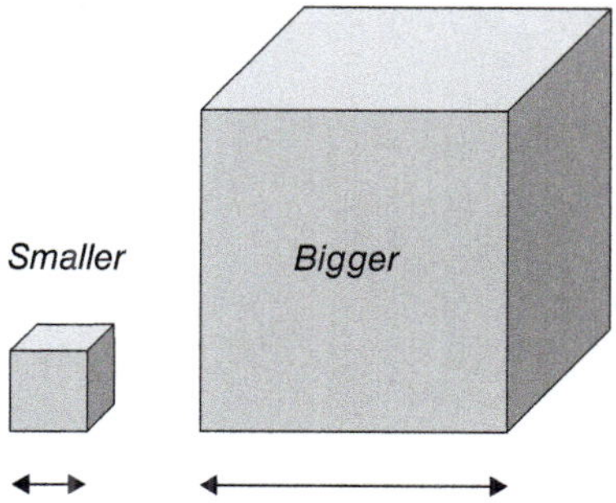

	L	$10 \times L$	Multiplication
Length	L	$10 \times L$	10
Surface area	$6 \times L^2$	$6 \times (10 \times L)^2$	100
Volume	L^3	$(10 \times L)^3$	1000
Specific surface area	$\frac{6 \times L^2}{L^3}$	$\frac{6 \times (10 \times L)^2}{(10 \times L)^3}$	0.1

Figure 4.4 Scaling merit for specific surface area.

starting material **S** with reagent **R** generates the primary product $\mathbf{P_1}$ (Figure 4.5a). $\mathbf{P_1}$ is a reactive molecule that can react with **R** to be converted into the by-product $\mathbf{P_2}$. This type of reaction is known as a competitive sequential reaction. If the kinetic constant of the primary reaction ($k1$) is sufficiently higher than that of the latter reaction ($k2$), the reaction using a one-to-one ratio of **S** to **R** is assumed to provide $\mathbf{P_1}$ selectively. However, in practice, as is often the case with fast reactions, the predicted selectivity does not always appear. This is because a rapid reaction may occur before the reactants become homogeneous.

A brief selectivity schematic is shown in Figure 4.5b. In an ideal situation, the reaction begins after the reaction components are homogeneous. However, if the reaction is faster than homogenization, the reaction starts at the interface between **S** and **R** (Figure 4.5b-(1)). At this moment, the generated $\mathbf{P_1}$ separates **S** from **R**, thereby preventing the reaction (Figure 4.5b-(2)). However, because $\mathbf{P_1}$ can contact **R**, the side reaction can generate by-product $\mathbf{P_2}$ (Figure 4.5b-(3)). Thus, although **S** is more reactive than $\mathbf{P_1}$, $\mathbf{P_1}$ can react with **R**, lowering selectivity. This is because the mixing is slower than the reaction, and the selectivity depends not on the kinetics but on the mixing efficiency, which is called the disguised chemical selectivity.

A typical example is the reaction of alkyl lithium with acyl chlorides [4]. Most textbooks state that acyl chlorides have a higher reactivity than the corresponding

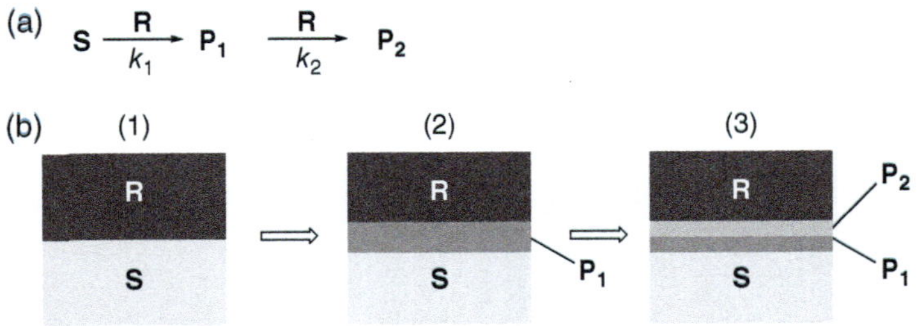

Figure 4.5 (a) Typical example of a competitive sequential reaction. (b) Schematic for disguised chemical selectivity.

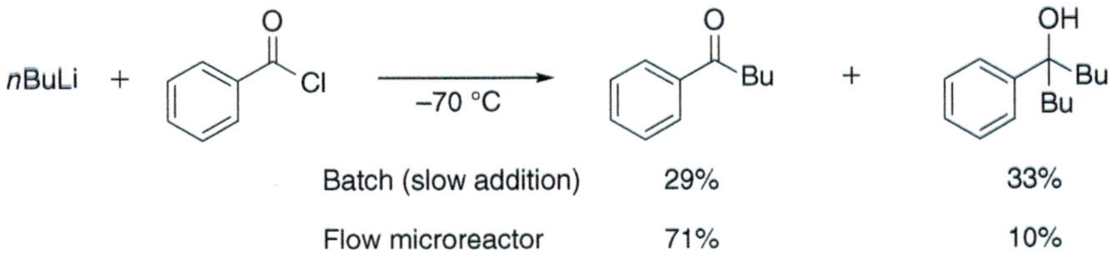

Scheme 4.1 Reaction of *n*-butyllithium with benzoyl chloride.

ketones. This indicates that the reaction of nucleophiles with acyl chlorides is faster than that with ketones. Thus, many would assume that the selective generation of ketones from nucleophiles with acyl chlorides can be achieved under optimized conditions. This claim is not always correct, especially for nucleophiles with high reactivity, which may cause poor selectivity. The reaction shown in Scheme 4.1 is a good example, where the reaction of *n*-butyllithium (*n*BuLi) with benzoyl chloride results in a mixture, with the alcohol by-product being derived from the over-reaction of *n*BuLi with the ketone product. This low selectivity seems to be due to disguised selectivity.

Conversely, the fast mixing of the flow microreactor has the potential to enhance the reaction selectivity, thereby potentially achieving ideality. An improvement in the selectivity of the aforementioned reaction of *n*BuLi with benzoyl chloride was demonstrated using flow microreactors. Several fast competitive-sequential reactions have been controlled by flow microchemistry (see Section 4.3.1 case study) [5].

4.1.2.2 Reactions Mediated by Unstable Intermediates

Flow microreactors can sequentially add reagents so that multiple reactions can be continuously carried out. Adding reagents with an adequate interval is called "reaction integration." In particular, they apply to reactions mediated by short-lived active species because of the precise controllability of the reaction time to as low as a millisecond scale. Scheme 4.2 shows an example of this. In this reaction, the starting material **S** is mixed with reactant **R** to be converted to reactive intermediate **I**. The quenching reagent **Q** is then added to convert **I** to the desired product **P**. Because of the high reactivity, **I** can be decomposed to form by-product **B**.

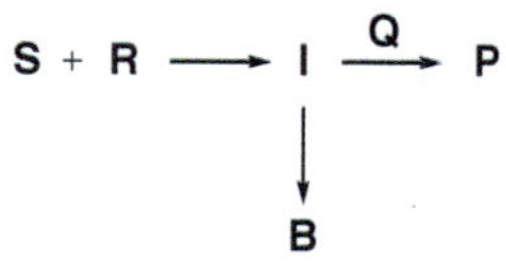

Scheme 4.2 Schematic for reaction mediated by unstable intermediate.

In a batch reactor, if intermediate **I** has a sufficiently long lifetime, **Q** can be added before **I** decomposes to **B**. However, if the lifetime of **I** is short and within seconds or milliseconds range, the manual operation is not applicable for the high-speed addition of **Q**. In a typical batch reactor, the diffusion time could be longer than the lifetime of **I**, which makes it difficult to trap **I** using **Q** in a step-by-step reaction.

In contrast, a flow microreactor can help control such difficult reactions. Because of the precise control of the residence time, quencher **Q** can be added before the decomposition of **I**. With fast mixing, **Q** can then be quickly distributed throughout the reaction solution to capture **I** before it decomposes.

In such experiments, it is important to consider both the yield of products and the conversion of starting materials. Figure 4.6 shows a theoretical trend of the yield of **P** and the recovery of **S** versus the time interval for the addition of **Q**. As the interval time increases, the conversion and yield first increase and then decreases. The decrease in yield indicates an increase in the decomposition of **I** to **B**. The time when the yield starts to decrease is approximately the lifetime of intermediate **I**. For synthetic utility, we can choose the residence time before the lifetime of **P** to maximize the amount of obtained **P**.

Based on this idea, reactions mediated by a range of highly reactive intermediates can be achieved using flow microreactors. A typical experiment of a reaction mediated by an unstable intermediate is described in Section 4.3.2.

The precise control of the residence time enables the substitution of reactive intermediates. For example, we reported the generation of (*E*)-1,2-dichlorovinyl

Figure 4.6 Theoretical trend of yield and recovery against reaction time in a reaction mediated by an unstable intermediate.

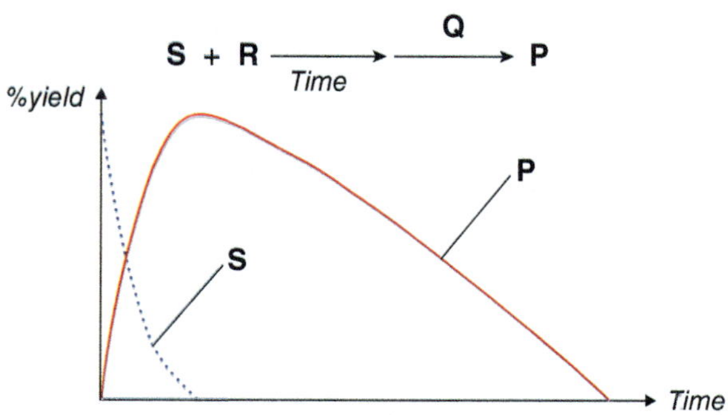

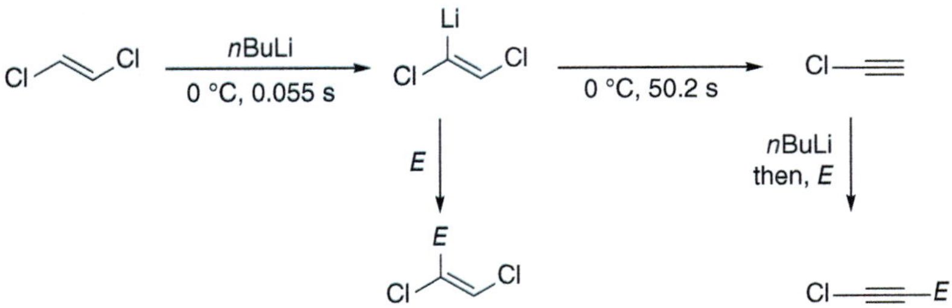

Scheme 4.3 Residence time controlling for the structure of highly reactive intermediate by flow microreactor. "*E*" indicates the electrophiles.

lithium and its reaction with electrophiles [6]. This reactive intermediate could be converted to chloroacetylene with the liberation of lithium chloride within 50 seconds (Scheme 4.3). We performed the flow reaction at different residence times to demonstrate that (*E*)-1,2-dichlorovinyl lithium can be used as a reactive intermediate with a shorter residence time (55 ms), whereas chloroacetylene, whose terminal proton can be deprotonated by additional n-butyllithium, can be used with a longer residence time (50.2 seconds). Taking advantage of the flow microreactor, millisecond-to-second-scale treatment of the intermediate can be achieved. For further examples, see Ref. [7].

The word "unstable" does not solely indicate the tendency for decomposition. Several reactive intermediates quickly lose their stereochemistry, diminishing the reaction's diastereoselectivity or enantioselectivity. For such stereochemically meaningful intermediates, the flow microreactor provides a solution; for example, trapping intermediates before their racemization [8].

Assembling multiple tubes and mixers allows several reactions to be performed sequentially. Scheme 4.4 provides a typical example of subsequent reactions mediated by multiple intermediates, beginning when the starting material **S** reacts with reagent $\mathbf{R_1}$ (reaction-1) to form intermediate $\mathbf{I_1}$. After reaction-1 completes, $\mathbf{I_1}$ can react with reagent $\mathbf{R_2}$ (reaction-2) without the isolation of $\mathbf{I_1}$. The intermediate $\mathbf{I_2}$ obtained was then mixed with reagent $\mathbf{R_3}$ to convert to product **P** via reaction-3. This type of reaction system, in which multiple reactions occur sequentially, can also be called telescoping or reaction integration [9].

In batch chemistry, these reactions would be carried out by adding the reagents after a period elapsed (Figure 4.7a). Thus, these reactions can be referred to as being separated or integrated by time. Alternatively, the sequential reactions can be achieved in flow reactors composed of multiple mixers (Figure 4.7b). In this

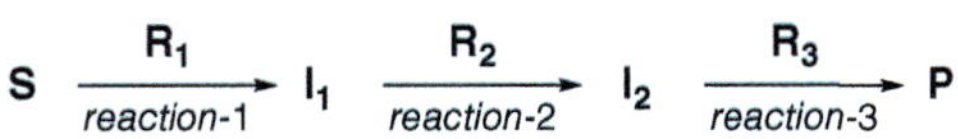

Scheme 4.4 Schematic for reaction integration.

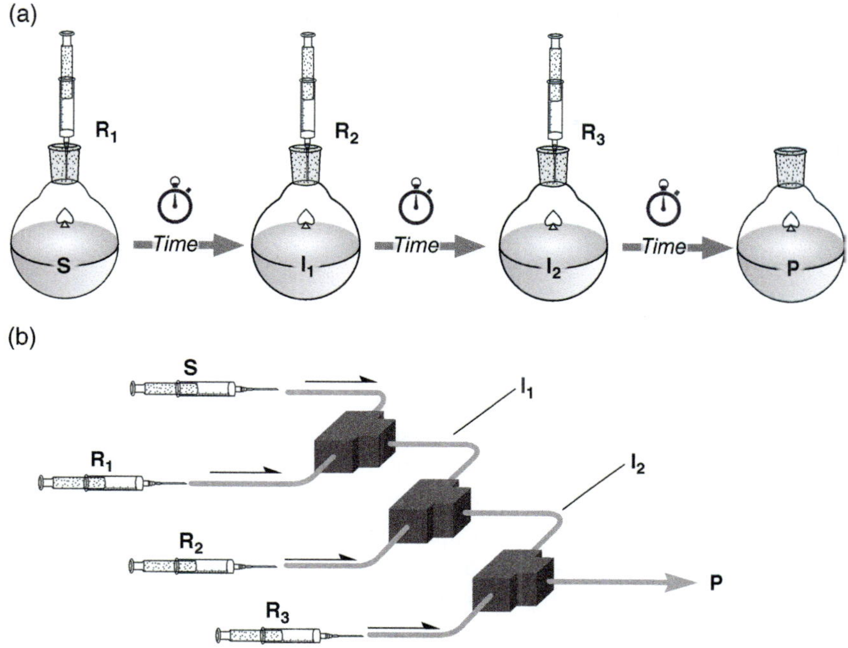

Figure 4.7 Reaction integration in (a) batch reactor and (b) flow reactor.

case, because reagents are added to different mixers, the reactions would occur in different reactors. Thus, these reactions can be regarded as separated by space. Because reagents are continuously introduced into the flow reactor, all reactions can occur at an identical time across the flow system. This contrasts with the reaction integration in batch, where reactions occur in an identical space but at different times. Section 4.3.3 shows a typical example.

4.1.2.3 Reactions Occurring at the Surface: Two-Phase Reactions, Electrochemical Reactions, and Photoreactions

Similar to the principle of temperature controllability, flow microreactors are also suitable for two-phase reactions, such as gas–liquid and solid–liquid reactions. Because two-phase reactions occur at the interfacial surface, micrometer-sized reactors are beneficial due to their higher specific surface area. One of the most important types of multi-phase reactions are electrochemical reactions, in which an electron moves between the reagents and the electrode. Electron transfer occurs on the electric double layer at a site close to the surface of the electrode.

Similarly, photoreactions can also be enhanced by flow microreactors. According to the Lambert–Beer law, the energy of light exponentially attenuates as it passes the medium, such that the photoreactions primarily occur close to the light source or the surface of the reactor. Thus, the flow microreactor can play a crucial role in these photoreactions by taking advantage of its high surface area. Many two-phase reactions [10], electrochemical reactions [11], and photoreactions [12] have been demonstrated using flow microreactors.

4.1.3 Further Applicability of Flow Microsynthesis

4.1.3.1 Scalability

In flow microreactors, although reactions occur at the micrometer scale, the reaction solution can continuously flow into the reaction space, increasing the total reaction scale. In our laboratory, it is common to use flow rates of approximately 10 mL min^{-1}, which allows 100 mL scale synthesis by a 10-minute operation, and 1 L scale synthesis by a 100-minute continuous operation. Because the reagents are steadily introduced into and removed from the reactor, the product quality is consistent over long-term operation. This means reoptimization of reaction conditions for scale-up can be more easily avoided in flow compared to batch synthesis. There are many examples of long-time operation of a flow reactor providing a larger amount of the product without loss of yield, selectivity, and purity [13].

4.1.3.2 Safety Operation

An important characteristic of the flow microreactor is the safety of reactions that can generate dangerous intermediates. Several reactions are mediated by toxic, explosive, or corrosive intermediates, and the risk of their exposure increases as the reaction scale increases. However, because intermediates only exist in the restricted space in the flow microreactor, which is a small subset of the total throughput, the risk can be reduced. Reports have demonstrated that this safe method applies to reactions that generate toxic gases as intermediates [14]. Another aspect of the safety of flow synthesis is the ease in scaling up, as mentioned above. As there is no requirement for a large vessel for large-scale synthesis, the risk of uncontrolled reactions that could lead to hazardous accidents can be minimized [15].

4.2 General Information for Flow Microreactors

In this section, the information for those who want to start flow experiments is described. There are several examples of flow reactions carried out in channel-type reactors, which are plate-shaped reactors with partial grooves on the surface.

However, our lab usually utilizes tube-type reactors due to their ease of performing reactions, especially those with higher flow rates. In this section, we introduce what is used in our laboratory for carrying out fast reactions where careful control over temperature and time is crucial.

4.2.1 Tools and Equipment for Flow Chemistry

4.2.1.1 Micromixer

Many companies, such as Sanko Seiki (Japan), IMM (Germany), YMC (Japan), and Mikroglas (Germany), sell micromixers. We utilize confliction-type T-shaped micromixers manufactured by Sanko Seiki, as depicted in Figure 4.8. As indicated by the gray arrows in the schematic shown in Figure 4.8b, the solutions are introduced from the side inlets and collide with each other in the micro space. Then they scatter into tiny droplets and mix together. Smaller droplets shorten the diffusion length, resulting in a shorter homogenization time of the two solutions, i.e., a shorter mixing time. Thus, increasing the collision energy can aid faster homogenization. A high flow rate and a short inner diameter are therefore important variables in faster mixing.

Micromixers vary depending on the shape (the angle of the introduced solutions) and the inside diameter. The mixing efficiencies vary between Y-, T-, and V-shaped micromixers, which are named for ease of descriptive purposes. The T-shape has a square angle between the inlet and outlet, the Y-shape has an obtuse angle and the V-shape has a sharp angle [16]. Additionally, there is a wide choice of materials for micromixers such as resins, plastics, and metals. Although they each have advantages, stainless steel is preferred because of its high chemical resistance and thermal conductivity.

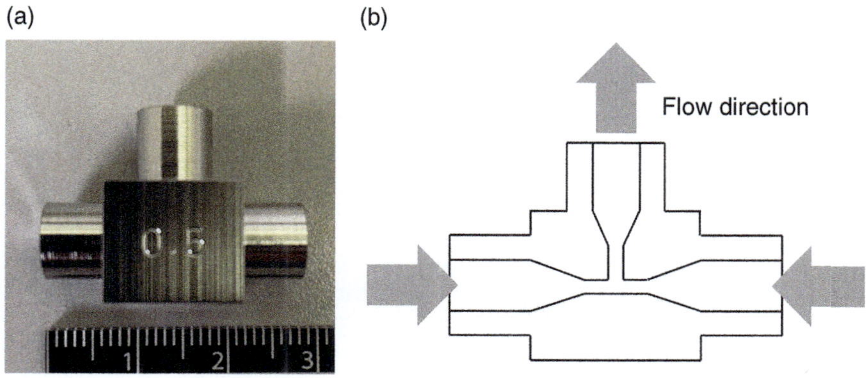

Figure 4.8 Photography of T-shaped micromixer (Sanko Seiki) (a) and its schematic (b).

4.2.1.2 Tube Reactor

To control the reaction temperature, tubes made of stainless are a favoured choice in our lab, similar to commercial HPLC tubes. An inner diameter of 1000 μm is a good choice, as smaller diameters are easily blocked. Accordingly, when setting up our lab, we purchased 10 m stainless tubes with outer diameters of 1/16 inches and inner diameters of 1000 μm. After smoothing the edges of the cut tube by tool for finish after cuts (GL Science, catalog number: 3001-31704), it was wound in a spiral configuration to aid the process of dipping the flow reactor in a bath (Figure 4.9). We then connected the fittings at either end of the tube. There are several tube makers (GL science, Swagelok, SMC, etc.) and their fitting parts cannot be an alternative for each other, so specific fittings must be chosen to connect tubes to a micromixer. In our laboratory, the one-ring ferrule (catalog number: 6010-41200) and stainless steel joint (catalog number: 6010-43010) manufactured by GL Science were used as they can be connected to the mixers manufactured by Sanko Seiki.

4.2.1.3 Pump

Pumps are classified as nonpositive and positive displacement types, and the latter is used in flow reactions because they can flow solutions constantly at a fixed speed. The most common are syringe pumps, in which the pusher block pushes the plungers of the set syringes (Figure 4.10). The use of a gastight syringe

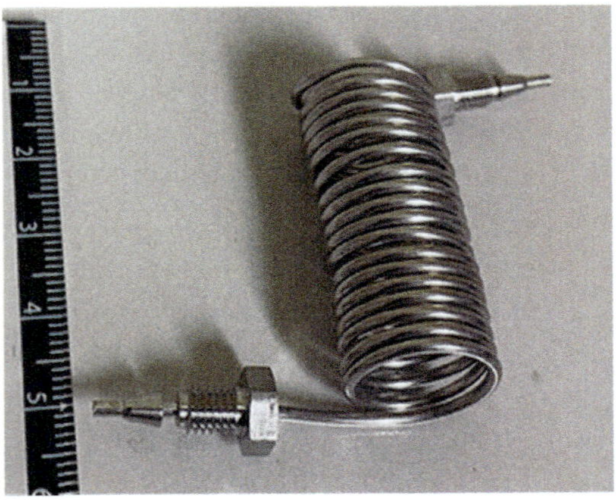

Figure 4.9 A tube reactor made of stainless steel.

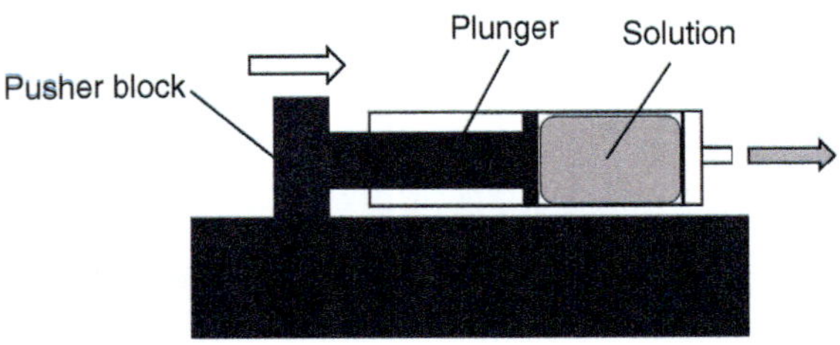

Figure 4.10 Schematic for a syringe pump.

is recommended to prevent solution leakage from the gap between the plunger and outer syringe. The pushing part of the syringe pumps is usually moved by a screw so that the pulsation of the flow was completely diminished. Several syringe pump makers are available, such as Harvard Apparatus, KD Scientific, and YMA. We use them manufactured by Harvard Apparatus (Harvard PHD ULTRA) because of their stability in pushing solutions with a higher back pressure.

4.2.1.4 Pre-Cooling Tubes

Pre-cooling tubes are recommended to control the temperatures of the introduced solutions by dipping them in a cooling or heating bath. Because the solutions are stored at room temperature in gastight syringes, they must be cooled or heated before they are mixed in the micromixer; otherwise, the temperature of the reaction, which must be initiated in the micromixer, is not accurate. Stainless tubes are suitable as pre-cooling tubes due to their good heat-transfer ability. Thus, we use stainless tubes with an inner diameter of 1000 μm with GL Science fittings, which are almost identical to the tube reactors described in Section 4.2.1.2. Because the tubes need to maintain the temperature of the solutions inside a specific temperature (the bath temperature), they should be long enough to complete the heat transfer. A length of 50 cm is chosen for volumetric flow rates lower than $4\,\mathrm{mL\,min^{-1}}$, whereas a length of 100 cm is usually chosen for volumetric flow rates higher than $4\,\mathrm{mL\,min^{-1}}$. When applying much higher flow rates, such as $20\,\mathrm{mL\,min^{-1}}$, longer pre-cooling tubes are recommended.

4.2.1.5 PTFE Tubes

Long (50 or 100 cm) PTFE (or Teflon) tubes are recommended to connect the gastight syringes with the pre-cooling tubes (Figure 4.11) because the stainless pre-cooling tubes are less flexible. Because they are used for transporting the solutions from the syringe to the stainless tube, any length, inner diameter, and fitting are applicable. In our experiments, we purchased PTFE tubes from GL Science (inner diameter: 1.00 mm, outer diameter: 1/16 in.) because of their diameter tolerances.

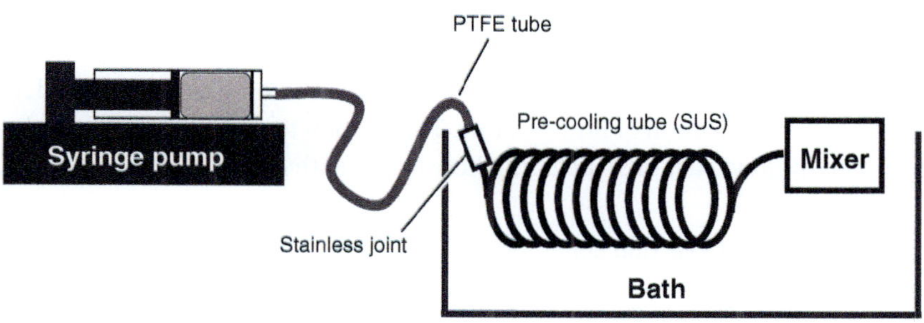

Figure 4.11 Schematic for the use of PTFE tube and pre-cooling tube.

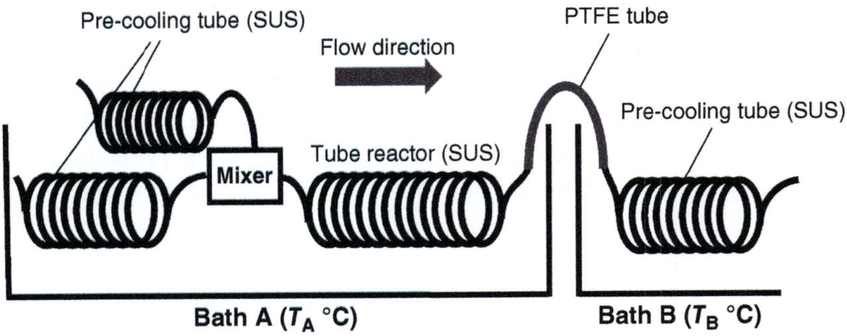

Figure 4.12 Schematic for the use of PTFE tube as a connection for two baths.

After cutting the PTFE tube to a certain length, both edges of the tube were connected with PEEK fittings purchased from GL Science (PEEK fingertight fitting, catalog number: 6010-48600). Stainless joint purchased from GL Science (catalog number: 6010-43120) was used to connect the PTFE tube to the pre-cooling tube.

Because PTFE tubes are structurally flexible, they are also used to connect two reactors in different baths, for example, to carry out sequential reactions at different temperatures. In such cases, we use a short (10–12.5 cm) PTFE tube to connect the reactors. Figure 4.12 shows a schematic of typical equipment used to change the reaction temperature. After the reaction takes place in a tube reactor dipped in bath A at T_A °C, the solution passes through the PTFE tube and the pre-cooling tube dipped in bath B at a temperature of T_B °C. Because the temperatures T_A and T_B are different, the solution must pass through the pre-cooling tube to change the temperature from T_A to T_B. Although the residence time in the PTFE tube is short, the temperature of the solution cannot be easily controlled while passing through the tube. Thus, the reaction, especially those including unstable intermediates, should be terminated before the solution reaches the PTFE tube; otherwise, the residual highly reactive species could cause uncontrolled side reactions.

4.2.2 How to Perform Experiments

A guide for operating these reactions is necessary in aiding the reader's understanding of the application of flow reactors. This section focuses on recipes for the preparation of flow reactions, including how to set up the flow reactors.

4.2.2.1 Selection of Reaction Conditions

Flow rate. In contrast to conventional batch reactions, flow reactions have a unique reaction parameter: flow rate (see Glossary). When using syringe pumps, the syringe diameter must be known. The diameter provided by the vendors can be used or manually measured using calipers. In practice, to maintain a steady flow, we recommend the flow rate should range from 0.5 to 15 mL min^{-1} if plausible. It is noteworthy that when mixing two solutions with the aforementioned confliction-type mixers, a flow rate with a certain difference is recommended because, experimentally, it leads to better mixing than mixing with solutions with identical flow rates. We usually choose a 1 : 4 ratio for the flow rates of two mixing solutions.

Equivalent amount. The idea of using equivalent amounts of reagents is somewhat different from that in batch chemistry. In the batch reactor, the equivalent amount of reagents was considered as the molar ratio of the added reagents. Because the reagents are introduced into the reactor continuously in a flow system, the equivalent ratio of reagent **A** to **B** is calculated as follows:

$$(\text{equivalent ratio of } \mathbf{A} \text{ to } \mathbf{B}) = C_A \times F_A : C_B \times F_B$$

where C_A and C_B are the concentrations of **A** and **B** and F_A and F_B are the initial flow rates of **A** and **B**, respectively. When considering the molar ratio of multiple reagents, one can directly compare the number of moles of reagents introduced into the flow reactor per unit time (molar flow rate). Figure 4.13 shows a case study in which a solution of starting material **S** (0.10 M) was mixed with a solution of reagent **R** (0.42 M) in M1. The resulting solution was mixed with a solution of quenching agent **Q** (0.60 M) to produce the desired product **P**. The molar flow rates of the chemicals were 0.60 mmol min^{-1} (**S**), 0.63 mmol min^{-1} (**R**), and 1.8 mmol min^{-1} (**Q**). According to these numbers, it appeared that **R** was added in a 1.05 equivalent of **S** and that the **Q** added was equal to 3 equivalents of **S**. These equivalent ratios depend only on the initial concentrations and flow rates of the reagent solutions, and not on any other parameters such as the total flow rate, reaction temperature, residence time, and collection time.

Residence time. Once the flow rate is selected, the residence time (see Glossary) can be chosen, which is comparable to the reaction time in batch reactions. We use stainless tubes with a 1.0 mm inner diameter as a tube reactor and

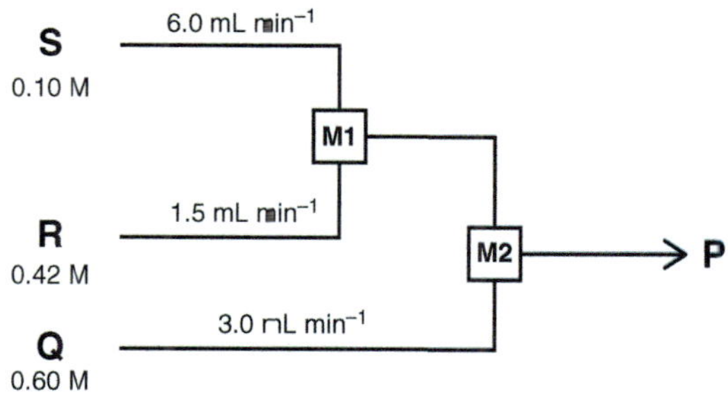

Figure 4.13 Case study for calculating equivalent amounts.

cut them into arbitrary whole numbers such as 10, 50, and 100 cm. The volume of the 100 cm reactor 0.785 mL. A flow rate of 5.0 mL min^{-1} would result in a residence time of 9.4 seconds. The difference in the length of the tubes must be distinct and/or labelled; otherwise, the exact volume of the tube or the residence time cannot be ensured. Varying the residence time by increasing or decreasing the flow rate is possible, but is not always suitable because the flow rate affects not only the residence time but also the mixing efficiency. The influence of mixing efficiency on the reaction, particularly the selectivity, cannot be ignored in the reaction results. Thus, we changed the tubes to change the residence time (Section 4.3.2 displays a practical example).

Stabilizing time. The reaction must be evaluated after reaching a steady state. The situation immediately after the pump started could be different from that the steady-state condition after the solution has been pumping for some time. If the reaction conditions, such as the temperature of the bath, are changed, the residual solution inside the tube reactor must be removed; otherwise, it affects the reaction or contaminates the passing solution. We assume that twice the residence time of the reactor is sufficient for the removal of the residuals. We refer to this interval time, i.e. for the time the flow takes to reach a steady state, as the stabilizing time. If we carried out a polymerization reaction in microreactors, we set a longer stabilizing time, for example three times that of the residence time or possibly even larger. This is presumably because the chain-like shape of the polymers aids their retention in the tubes. A longer residence time helps decrease the risks of undesired errors; however, excessive loss of substrates results in a decrease in the number of available samples in an experiment.

Collecting time. In the experiment, following the stabilizing time point, you can subsequently start collecting the solution flowing from the outlet of the reactor. The collection time is used to calculate the reaction yield; thus, accurate timing is important. As long as you can ensure the exact time, any length of the collection time is applicable. To exclude experimental error, we set it as 30 seconds or longer, or as 10 seconds to conserve the reagent solution. It must be noted that a shorter collection time may lead to some experimental errors. For example, if the collection time is 30 seconds, the addition of 0.1 second due to handling is negligible (less than 0.5%), while it can be significant for a short collection time.

Number of samples. Once solutions of reagents are prepared, experiments can be continued under different reaction conditions as long as there is sufficient solution. Because the type of solvent and concentration of reagents cannot be changed after preparation, we primarily change the residence time and the reaction temperature during one experiment. The most time-consuming step is the preparation of solutions, so we tend to perform multiple flow reactions under varying conditions. For example, the protocol described in Section 4.3.2 aims to carry out eight reactions simultaneously using identical solutions. This depends on the importance of the reagents, such as the difficulty of their synthesis.

Volume of the prepared solutions. Once the reaction conditions (the flow rate, collecting time, stabilizing time, and a number of samples) are known, the amount of reagent solution needed to complete the series of experiments can be determined. The simplest way is to multiply the flow rate by the sum of the stabilizing time and collecting time. However, events such as blocked tubes may occur during the experiments, which wastes reagent solutions. Thus, it is recommended, especially for beginners, to prepare double the volume than is needed.

4.2.2.2 Preparation of Reagent Solution

Because the reactions are performed in flow, all reagents must be fully dissolved in solvents; otherwise, the flow reactors will become blocked. Moreover, because the reactions described in this chapter utilize highly reactive reagents such as organolithiums, the reagent solution must be completely dried. This section describes the preparation of these reagent solutions.

General. Reaction solutions should be prepared in flame-dried flasks, and the internal atmosphere must be substituted with dried inert gases (nitrogen or argon). After cooling the flask to room temperature, the reagent and solvent can be added.

Reagents. If the reagent is oil, it can be added via a syringe, whereas if the reagent is solid, the flask should be opened and the solid added, followed by capping the flask with the septum as soon as possible. After the addition of a solid reagent, inert gas can substitute the inside atmosphere.

Solvents. Dried solvent can be purchased from vendors and used without further drying. In addition, homemade dried solvents can be used.

Alkyl lithiums. If using any alkyl lithium reagents, they should be diluted with *n*-hexane for ease of reaction. Higher concentrations of lithium reagents tend to produce insoluble lithium salts, which result in a reliable blocked flow reactor. In most cases, the concentration of the lithium reagent is approximately 0.4 M. Because of its high reactivity, the solution of *sec*-butyllithium (*s*BuLi) should be diluted just before starting the flow reaction. Following a period, the solution may became clouded, suggesting that the *s*BuLi is partially inactive; however, in most cases, we find the loss of *s*BuLi is relatively small and negligible.

Concentration. The reagent concentration must be precisely known in order to accurately determine the product yield. Solutions can be made by recording the weight of reagent and adding an appropriate volume of the solvent. If the reagent is a liquid, the volume of the reagent should not be ignored, especially if high concentrations are used. The sum of the volume of the solvent and reagent is regarded as the total volume of the solution, although there must be some difference (Section 4.3.1 as a case study). If the reagent is solid, its volume can usually be ignored during the preparation of the solution as long as the concentration is low (less than ~0.50 mol L^{-1}). If the reagent is a solution (e.g. *n*-butyllithium), it can be diluted with dry solvents.

4.2.2.3 Preparation for Reactions

The setup for flow reactions varies depending on the reaction. In Section 4.3, several types of reactions are described with schematics of the settings. Section 4.3.1 shows the detailed manual for operating the flow reactions. Before these practical examples are described, specific techniques should be noted.

Setup for the syringe pumps. When using syringe pumps, there are two parameters to be set. One is to determine the size/inner diameter of the syringe. Some syringe pumps, such as the HARVARD PHD ULTRA, have preset parameters with manufactured gastight and glass syringes. Thus, you can select the preset parameter, or the inner diameter of the syringe can be measured using a slide gauge and then input. Another parameter to be set is the flow rate. Occasionally, as maintenance, we determine the exact flow rate of the pumps by collecting the ejected solutions for one minute and confirming whether the collected volume is identical to the set flow rate.

Before turning on the pumps. One feature of flow reactions is that syringes are directly used for the reaction, whereas in batch reactions, they are used only for delivering solvents and reagents to flasks. This is important, especially if the reaction must be operated without water. First, when drawing the solutions into the syringe, the air in the needle must be substituted with a dry and inert gas, such as nitrogen or argon gas. Typically, we draw an amount of such inert gas in a flame-dried flask using a syringe equipped with a needle and eject it to the outside of the flask, which is repeated more than three times. Moreover, when using gastight syringes, for example, by introducing solvents to the flasks for the preparation of the reaction solution or drawing the prepared solution to start the flow reaction, the syringes should be prewashed with a small amount of these solutions. This is because gastight syringes cannot be flame-dried, and the residual moisture inside the syringe cannot be fully removed.

After drawing the reagent solution into the gastight syringe, the plunger must be drawn to inject inert gas into the syringe. This should help separate the solution inside the syringe from the outside atmosphere when changing the needle to the tube reactor (Figure 4.14a). After connecting the tubes with the syringe, push the plunger to exclude the internal atmosphere until drops appear from the outlet (Figure 4.14b). This aids the cleaning of the inside of the tubes, which confirms that the tubes are not blocked.

After filling the solutions inside the tubes, the gastight syringes are allowed to be set on the syringe pumps. The syringes are placed horizontally on the pumps and fixed using clips and/or screws (Figure 4.15). After placing the syringes, we

(a) (b)

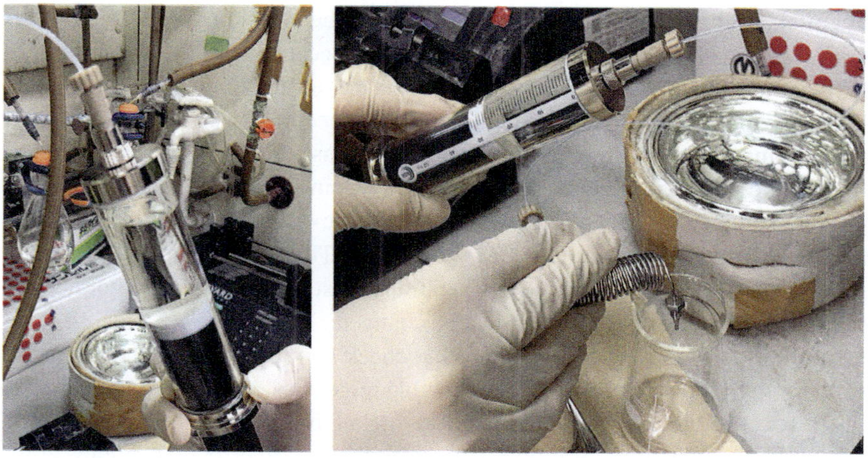

Figure 4.14 Tips for syringe techniques. (a) Filling the syringe with an inert gas. (b) Filling the tubes with the solution.

recommend to start pumping to confirm that the droplets are ejected from the tube (similar to the previous technique for filling the solutions inside the tubes). The pumps do not always tightly contact to the plunger of the syringe, or sometimes contact with the pushing block is forgotten, which results in an inefficient pump that does not start pumping and therefore backflow of solutions from other pumps. Therefore, before connecting the tubes to the mixers, we confirm that the pumps were ready to push the solutions.

Figure 4.15 Gastight syringe set on syringe pump horizontally.

During operation of the pumps. One of the most important aspects to pay attention to during the pumping is to confirm the solution flow correctly; for example, checking the solutions are not leaking. Because most flow reactions are carried out in heating or cooling baths, if leakage occurs, the media in the baths become cloudy. Sometimes, leaking from the connecting parts is difficult to find; thus, seeing the momentum of the ejecting solutions would be beneficial. In addition to the connecting parts, there is another point where leakage occurs: the gastight syringe. Because of the high back pressure or the blocked tube, the solutions leak between the plunger and the syringe. Thus, the pumps are carefully monitored. If there is a leak, all syringe pumps must be turned off, the connectors tightened or the gastight syringe must be changed.

Another point to consider is the need to check and resolve blockages. Because of the generation of undesired and insoluble molecules, tubes and mixers sometimes become blocked. In some cases, a higher back pressure caused by the increased viscosity of the solutions can result in a pump error. The pumps automatically stop when the system is blocked or the back pressure becomes high, and in that case, other pumps must be turned off; otherwise, the solutions might backflow to ruin the reagents in the syringe. After you stop the pumps, you can wash the inside of the tubes and mixers to remove any stock. When organolithium reagents are used, the stock solution should be a lithium salt. The inside is washed with methanol or acetone and pushed by syringes. After the blockage is removed, the flow system can be assembled again and the experiment can be restarted. Once the flow reactor clogs, an extra amount of reactant solution is required to continue the experiment; thus, we recommend that beginners prepare more reagent solution than needed. Because the volumes needed for a single operation can be easily calculated, we continue the reaction with blocked conditions or perform further reactions with another condition as long as the solutions of the reagents remain.

It is important to note that all the stainless tubes were fully dipped in the bath. Although the temperature of the media (mainly water or acetone) is controlled, any equipment outside of the media has varying temperatures. Thus, to maintain the reaction temperature, all stainless tubes, including the pre-cooling tubes and tube reactors, must be placed under the surface of the medium.

4.2.2.4 Preparation for Reaction Evaluation

In flow chemistry, after preparation of the reagents and device settings, each experiment was completed within 1 minute (30 seconds stabilizing and 30 seconds collection). Thus, during a single experiment, multiple samples can be obtained. To reduce the evaluation time, we recommend using gas chromatography (GC) or liquid chromatography (LC) with an internal standard. Before starting the experiment, the desired product (and starting materials) is analyzed by GC or LC.

Additionally, after confirmation, it is recommended to prepare a calibration line for these compounds using an appropriate internal standard.

4.2.2.5 Cleaning Up

In contrast to batch reactors, because of the micrometer-sized structure, we cannot brush the inside of a flow microreactor. Thus, as soon as the experiments are finished, you should start washing them. Washing with acetone is sufficient for most organic reactions. If the reaction possibly generates inorganic salts, such as reactions with organolithium reagents that might generate some lithium salts, we can use pure water for washing inside the reactors and mixers. When washing, it is recommended to detach all tubes and mixers for the thorough removal of residues. After flowing sufficient acetone and water, the detached tubes and mixers are dipped in a 50% acetone–water bath, which is then placed in an ultrasound bath to remove any residue. After ultrasound washing for 30 minutes, the tubes and reactors are briefly washed with acetone again to substitute the residual water with acetone, and then dried using a heating drier. Finally, the samples were placed in a heated oven at 50 °C for further drying. Because of the small and elaborate inside structure of these reactors and mixers, excess heating should be avoided, although heating higher than 100 °C is essential for drying the glassware. Since the stabilizing time in the next experiment aids the removal of residual water, it is not necessary to consider residual water in the tubes and micromixers. The gastight syringes were dried in an identical oven after the removal of residual organic and inorganic compounds.

4.3 Case Studies

In this section, some simple flow microreactions that cannot be easily performed in batch systems are described.

4.3.1 Competitive Sequential Reaction (General Procedure)

As described in Section 4.1.2.1, one of the most important characteristics of flow microsynthesis is the control of fast competitive-sequential reactions. A typical and important example is the reaction of organolithium reagents with trialkyl borates to produce boronic esters [17]. Boronic esters are ubiquitous in organic synthesis and are especially used as reagents for Suzuki–Miyaura cross coupling. In most cases, the corresponding organometallic reagents, such as organolithium and organomagnesium, react with trialkyl borates to generate boronic esters (Scheme 4.5). This reaction is carried out at a cryogenic temperature because multiple additions of the organometal to the trialkyl borate can occur, forming undesired by-products such as borinic esters (di-adduct in Scheme 4.5).

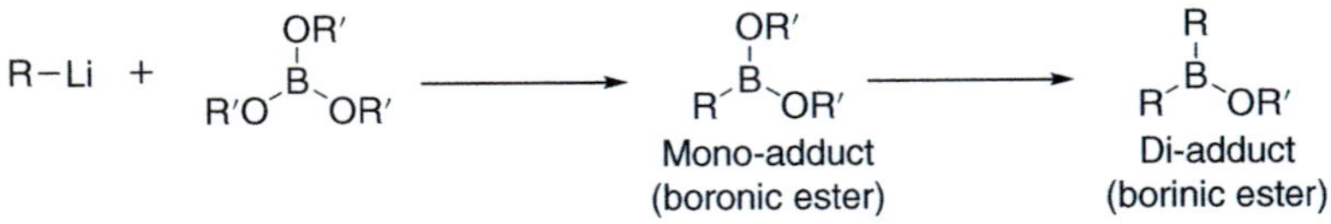

Scheme 4.5 Schematic for addition of organolithium to trialkyl borate. R and R′ are alkyl and aryl group.

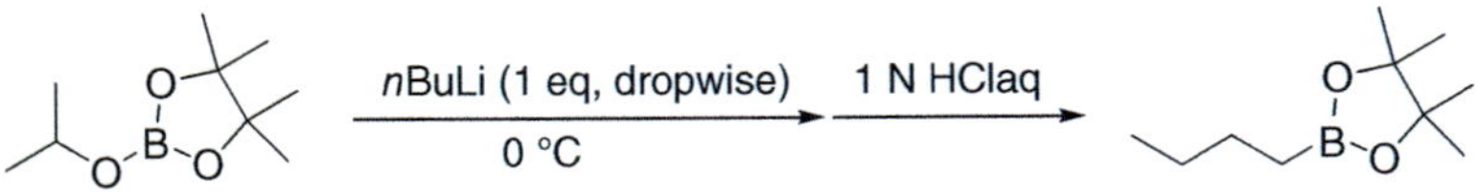

1 **2**, 63%

Scheme 4.6 Borylation of lithium reagent (model reaction).

To demonstrate the selective control of such competitive sequential reactions, a simplified reaction using *n*-butyl lithium with isopropoxy pinacolborane (**1**) is described. When these chemicals were mixed 1 : 1 ratio at 0 °C in batch, the yield of the desired product (*n*-butylborane pinacol ester, **2**) was 63% (Scheme 4.6), with the remainder of the mass balance proposed to be from multiple additions of *n*-butyllithium [18]. This relatively poor selectivity can be evaluated and improved using flow microreactors according to the following procedure.

4.3.1.1 Preparation

Table 4.1 lists the typical equipment used in this experiment and Figure 4.16 shows a photograph of equipment (a) and its assembly (b). Two gastight syringes (purchased from SGE) were used to introduce the reagents: one was used for the solution of isopropoxy pinacolborane (**1**) and the other for the solution of *n*-butyllithium. As described, we used PTFE tubes with a length of 50 cm, inner diameter of

Table 4.1 List for equipment for an initial trial.

Equipment	Size (volume, length, inner diameter, etc.)[a]	Quantity
Gastight syringe (SGE)	Volume: 50 mL	2
Pre-cooling tube	Length: 100 cm, ID: 1000 μm, OD: 1/16 in.	2
PTFE tube	Length: 50 cm, ID: 1000 μm, OD: 1/16 in.	2
Connector (GL Science)	OD: 1/16 in.	2
Micromixer (Sanko Seiki)	T-shape, ID: 500 μm	1
Tube reactor	Length: 100 cm, ID: 1000 μm, OD 1/16 in.	1

[a] ID = inner diameter, OD = outer diameter.

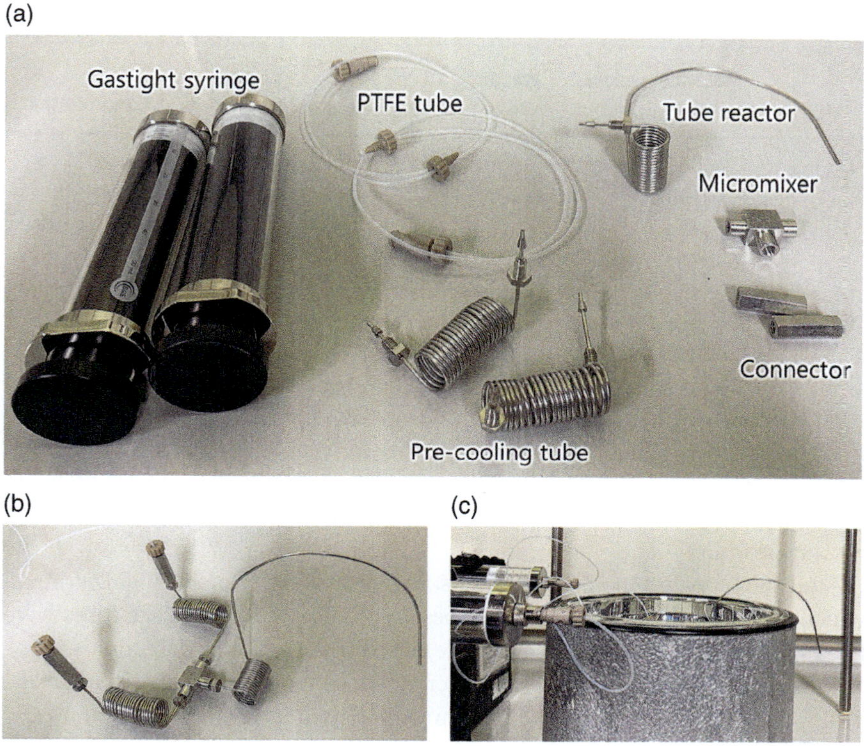

Figure 4.16 (a) Equipment used. (b) Reactor assembly. (c) Reactor in cooling bath.

1000 μm, and outer diameter of 1/16 in. The length of the pre-cooling tubes was chosen to be 100 cm because the residence time within this tube, typically longer than a few seconds, was sufficient to make the solution temperature equal to the preset temperature. To connect the external screw of the PTFE tubes with that of the pre-cooling tubes, a connector (stainless joint purchased from GL Science), which has two internal screws at both ends with a negligible inner volume was used. After connecting each piece of equipment, as shown in Figure 4.16b, they were tightened using a wrench to prevent the flowing solution from leaking. The assembled flow system, connected to the gastight syringes, was dipped into a bath during the experiment, as shown in Figure 4.16c. The stainless tubes were kept in a cooling or heating medium, with only the outlet of the flow reactor emerging.

The following table (Table 4.2) presents a flowchart of this flow experiment. As described in Section 4.2.2.2, a flame-dried round-bottomed flask filled with inert gas was used to prepare the solutions. Because **1** is a liquid reagent, its volume was recorded and the total volume was counted. The total volume of the solution was 50.0 mL so that the concentration was 0.080 M. The weight of **1** was checked using

Table 4.2 Recipe for the reaction of **1** with *n*-butyl lithium.

Reagent	Molecular weight density	Weight volume	Amount	Solvent volume	Concentration
1	186.06 $0.92\,g\,mL^{-1}$	744 mg 0.8 mL	4.00 mmol	THF 49.2 mL	0.0800 M
*n*BuLi	1.60 M in *n*-hexane	4.00 mL	6.40 mmol	*n*-hexane 12.0 mL	0.400 M

a precision balance, and the volume of the added solvent was recalculated to accurately obtain a concentration of 0.0800 M. Each concentration of the commercial *n*BuLi solution varied; thus, the total volume was changed. Although not shown in Table 4.2, quenching agents should also be prepared. In this experiment, the reaction was quenched with 1 N HCl aq, and the aqueous solution was prepared prior to the experiment and added to a sample tube for collection of the eluted reaction solution.

After the preparation of the solutions, they were drawn into gastight syringes, which were connected to the PTFE tubes and pre-cooling tubes. The tubes were filled with the solutions, and syringes were set on the syringe pumps (Section 4.2.2.3). Before connecting the tubes to the mixer, it is recommended to confirm that the solution starts passing from the outlet immediately after turning on the pump and that the droplets are continuously emitted. This ensure the pumps are not loose, which could lead to the backflow of one solution to another. After confirmation of flow, the tubes were connected to the mixers and tightened using wrenches. A large bath was used to set the reaction temperature (Figure 4.16c). For the reaction at 0 °C, iced water was used, whereas at cryogenic temperatures, acetone with dry-ice or other cooling media was used.

4.3.1.2 Experiment

After preparing the flow pumps and reactors, the experiments were started. The experimental conditions are presented in Table 4.3. Additionally, a schematic is given in Figure 4.17. Because the concentration of *n*BuLi was fivefold higher than that of **1**, the flow rate of **1** was set fivefold faster than that of *n*BuLi, setting the molar ratio of these two reactants as 1 : 1 (Section 4.2.2.1 contains a brief explanation). After confirming that the settings were all correct, especially the settings of the pumps (the inner diameter and flow rate), the flow reactor was dipped in the ice-water bath, as shown in Figure 4.16c. The pumps were then started, and after a while, the eluted droplets could be found. The droplets then steadily flowed, and the timer was started to count the stabilizing and collection times. After the stabilizing time passed, the outlet of the reactor was subsequently moved to a sample tube that included 1 mL of

Table 4.3 Reaction conditions.

Condition	Value
Flow rate of **1** (F_1)	12.5 mL min^{-1}
Flow rate of *n*BuLi (F_{BuLi})	2.5 mL min^{-1}
Bath temperature	0 °C
Quencher	1 N HClaq (1 mL)
Residence time in the reactor	3.1 s
Stabilizing time	30 s
Collection time	30 s

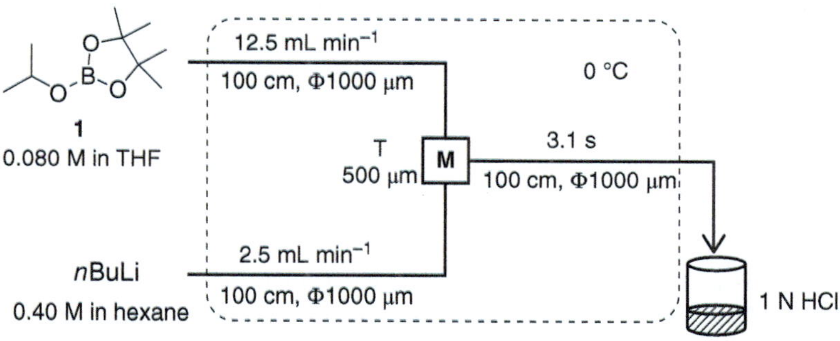

Figure 4.17 Schematic for the flow reaction. The lengths of the PTFE tube are not recorded because they do not affect the reaction.

1 N HCl aq as a quencher, and the solution was collected during the collection time. After the collection time passed, the outlet of the reactor was quickly moved from the sample tube to the wasting bottle, and the pumps were turned off.

4.3.1.3 Screening of Reaction Conditions

Provided the amount of the reagent solution is sufficient for further experiments, several experiments can be run. For example, one can envision installing a new reactor with longer length and adjust the flow rates on the pumps to keep the same residence time. This allows study of the impact of flow rate on the reaction outcome, which is related to mixing efficiency. Table 4.4 shows one option for how flow rates could be screened.

It is recommended to plan a series of reaction screening experiments at one time and immediately perform them. The experiments described in Table 4.4 can be organized in one's lab book as described in Table 4.5, filling in the specific collection time (approximately 30 seconds). The columns of the "internal standard (mg)" and "yield (%)" can be filled after the experiments are analyzed.

Table 4.4 Typical conditions for changing the flow rates.

F_1 (mL min^{-1})	F_{BuLi} (mL min^{-1})	Length of reactor (cm)	Residence time (s)
3.125	0.625	25	3.1
6.25	1.25	50	3.1
12.5	2.5	100	3.1
25	5	200	3.1

Table 4.5 Prepared table in experiment notebook.

Entry	F_1 (mL min^{-1})	F_{BuLi} (mL min^{-1})	Length of reactor (cm)	Collecting time (s)	Internal standard (mg)	Yield (%)
1	3.125	0.625	25			
2	6.25	1.25	50			
3	12.5	2.5	100			
4	25	5	200			

In this experiment, the tube was changed after each entry. After the pumps were turned off, the flow reactor was removed from the ice–water bath, and the tube reactor was detached from the micromixer. Another tube of a different length was then attached to the mixer, and the newly assembled flow reactor was dipped into the bath. After adjusting the set flow rates of the pumps, they were turned on to start the next experiment. Before starting a new reaction, it is recommended to confirm the temperature of the bath, although in this case, the temperature of the bath can be easily confirmed as long as ice is present in the bath. To test the effect of the mixing efficiency, changing the micromixer, for example the shape (T or V) and the inner diameter, is also a good choice.

Another option for further experiments is to change the temperature. Changing the ice–water bath to other baths, such as those filled with acetone–dry ice medium, is sufficient for this entry. As there is no need to detach and connect the tube reactors, this operation is relatively easy.

4.3.1.4 Analysis

As described in Section 4.2.2.4, the reactions can be efficiently analyzed using GC or LC. For this case study, and in most cases, we prefer using GC. After collecting all the samples listed in Table 4.5 and consuming the prepared reagent solutions, the experiment was completed. Then, an internal standard (mainly *n*-tetradecane) was added, and its weight was recorded in the table of the notebook (see Table 4.6).

Table 4.6 Example of the table in experiment notebook after adding internal standard.

Entry	F_1 (mLmin^{-1})	F_{BuLi} (mLmin^{-1})	Length of reactor (cm)	Collecting time (s)	Internal standard (mg)	Yield (%)
1	3.125	0.625	25	30	15.1	
2	6.25	1.25	50	30	14.9	
3	12.5	2.5	100	30	15.2	
4	25	5	200	30	15.0	

To avoid adding any salts into the GC, ethyl acetate (4 mL) and water (2 mL) are added to each of the samples, which is then briefly shaken. Then, an aliquot of the upper organic phase is collected and added it to a GC sample tube.

The theoretical yield of this reaction must be calculated in moles. For example, in entry 3 of Table 4.6, the theoretical yield of **2** can be calculated as follows:

$$\text{theoretical yield of } \mathbf{2}(\text{mol}) = 0.080(\text{mol L}^{-1}) \times 1.25(\text{mL min}^{-1}) \times 0.50(\text{min}) \\ = 0.50(\text{mmol})$$

For this calculation, the concentration and flow rate of **1** was used. The same theoretical yield can be calculated using *n*BuLi because a 1 : 1 ratio was used, and thus both components are limiting reagents. According to the similar calculation, the theoretical yields of **2** in each entry of Table 4.6 are 0.125 mmol for entry 1, 0.25 mmol for entry 2, and 1.0 mmol for entry 4.

With a known theoretical yield and known quantity of internal standard, analyses and calculations can be carried out to determine the yields (Table 4.7) [18]. All yields obtained were higher in this flow experiment compared to batch, highlighting the value of flow microreactors in controlling competitive sequential reactions. Additionally, the yield increased with an increasing total flow rate, highlighting the effect of mixing efficiency on such competitive reactions.

Table 4.7 Completed table for the flow reaction.

Entry	F_1 (mLmin^{-1})	F_{BuLi} (mLmin^{-1})	Length of reactor (cm)	Collecting time (s)	Internal standard (mg)	Yield (%)
1	3.125	0.625	25	30	15.1	68
2	6.25	1.25	50	30	14.9	79
3	12.5	2.5	100	30	15.2	83
4	25	5	200	30	15.0	82

Source: Adapted from Ashikari et al. [18].

4.3.1.5 Clean Up

The general method for cleaning the microreactors is described in Section 4.2.2.5. After detaching the tubes from the mixer, acetone was introduced into the tubes using a small syringe. The samples were then washed with water and acetone. The tubes and mixers were then dipped in a 50% acetone–water bath and washed in an ultrasound bath for more than 30 minutes. After ultrasound washing, the tubes and mixer were briefly washed with acetone again to substitute the residual water with acetone and then dried using a blow drier. Finally, the samples were placed in an oven at 50 °C for further drying. The gastight syringes were first washed with acetone, and water was used to remove salts. After washing with acetone again, the gastight syringes were dried using a heating drier and heated in an oven at 50 °C. The glassware, including the glass syringe and round-bottomed flasks, was washed as per normal.

4.3.2 Reactions Mediated by Short-Lived Intermediates

In Section 4.3.1, fundamental techniques for carrying out simple, single-step flow reactions with short residence times were discussed. In this section, a guide for a more complicated flow system is provided, specifically, where the reaction proceeds by a short-lived intermediate (see Section 4.1.2.2. for the principle). In this example, transiently stable cyano-substituted aryl lithiums are harnessed [19]. Due to their electrophilicity, cyano groups can easily react with aryl lithium reagents. Thus, reactions mediated by aryl lithiums bearing cyano groups must be carried out at cryogenic temperatures, else the intermediates will decompose before they can react with electrophiles. In a batch reactor, the reaction of 4-bromobenzonitrile (**3**) with *n*-butyl lithium, followed by the addition of methanol, resulted in a low yield of the desired product (benzonitrile, **4**) at 0 °C, although an identical reaction carried out at −78 °C gave **4** in a good yield (Scheme 4.7). The high conversion (higher than 90% in both cases) indicates that the generated intermediate (4-cyano phenyl lithium) is destroyed before the addition and reaction with methanol as an electrophile.

In this case study, the improvement of this reaction by screening residence time and reaction temperature was the primary focus. Thus, the flow reactor depicted

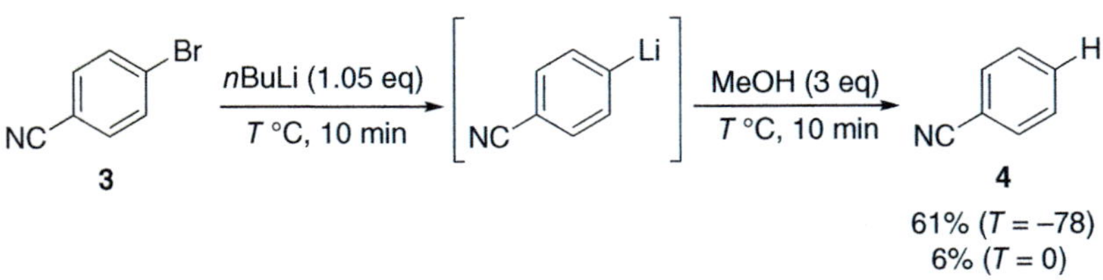

Scheme 4.7 Generation and reaction of cyano-substituted aryl lithium in a batch reactor.

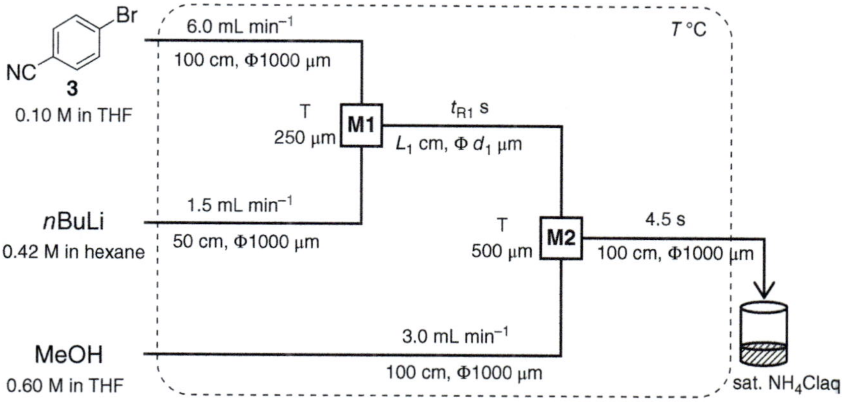

Figure 4.18 Schematic for the flow reaction of 4-bromobenzonitrile with *n*-butyl lithium.

in Figure 4.18 was used. To change the residence time, four types of stainless-tube reactors (their length and inner diameter are described below) were used, whereas for the screening of temperature, baths of 0 and −48 °C were used. The quantities of reagents (Table 4.8) were selected to be sufficient to enable running all eight possible experiments (Table 4.9) run in sequence.

The preparation and operation of the experiment were similar to that described in Section 4.3.1. The volumes of the prepared solutions were sufficient for all eight experiments. It is easier to change the reaction temperature than changing the residence time, as the latter requires detaching the tubes from the mixer, whereas the former simply requires changing the bath. Thus we collect the sample of entry 5 after collecting that of entry 1 because they have identical reaction conditions, except for the temperature. After, entry 2 and 6 are collected, and so on.

If the experiment did not proceed ideally, for example if the system becomes blocked, the reactant solution could be unexpectedly consumed. In such a case,

Table 4.8 Quantities for the reaction of **3**, *n*-butyl lithium, and methanol.

Reagent	Molecular weight density	Weight volume	Amount	Solvent volume	Concentration
3	182.02	910 mg	5.00 mmol	THF 50.0 mL	0.100 M
*n*BuLi	1.60 M in *n*-hexane	4.00 mL	6.40 mmol	*n*-hexane 12.0 mL	0.400 M
MeOH	32.04 0.792 g mL^{-1}	634 mg 0.80 mL	19.8 mmol	THF 32.2 mL	0.600 M

Table 4.9 Prepared table to summarize the reaction[a].

Entry	T (°C)	d_1 (μm)	L_1 (cm)	t_{R1} (s)	Collecting time (s)	IS (mg)	Recovery of 3 (%)	Yield of 4 (%)
1	0	250	3.5	0.014				
2	0	1000	3.5	0.22				
3	0	1000	12.5	0.76				
4	0	1000	100	6.3				
5	−48	250	3.5	0.014				
6	−48	1000	3.5	0.22				
7	−48	1000	12.5	0.76				
8	−48	1000	100	6.3				

[a] d_1 is the inner diameter of the first tube reactor and L_1 is its length. Figure 4.18 shows the determination of d_1 and L_1. The column "IS" is for the weight of the added internal standard.

the quantity of remaining reagents should be considered to determine whether it is sufficient for collecting all planned samples. If the remaining samples are not enough, the collection time could be reduced, such as by 20 seconds, to save solutions and to achieve the eight experiments with a comfortable margin.

Table 4.10 shows our reported results with varying residence times and temperature [19], showing that the product yields in the flow reactor were higher than those in batch at 0 °C (entries 1–8). At this temperature, with a longer residence time t_{R1}, the yield decreased, indicating the gradual decomposition of the

Table 4.10 Summary of results for the flow reaction under the conditions shown in Table 4.9.

Entry	Residence time t_{R1} (s)	Temperature T (°C)	Conversion of 3 (%)	Yield of 4 (%)
1	0.014	0	100	88
2	0.22	0	100	85
3	0.76	0	100	84
4	6.3	0	100	48
5	0.014	−48	50	33
6	0.22	−48	100	85
7	0.76	−48	100	89
8	6.3	−48	100	86

Source: Adapted from Nagaki et al. [19].

intermediate. The results at −48 °C show no decomposition of the intermediate, indicating this low reaction temperature helped prolonging the lifetime of the intermediate. At this temperature, an increase in the conversion was observed. Continuing such investigations, the stability of the intermediate can be evaluated by drawing a contour map. Thus, these series of experiments can afford both the practical optimized condition as well as fundamental insight into the rate of formation and decomposition of the intermediate.

Once the conditions for the generation and reaction of the reactive species were optimized, the scope of the electrophiles was then tested. Alkyl halides, aldehydes, trialkylsilyl chlorides, trialkylstannyl chlorides, and boronic esters were used as general electrophiles. It is worth mentioning that electrophiles that are less reactive than the sensitive functional group on the aryllithium (in this case – a nitrile) will not be in scope.

Following a similar process, a range of short-lived aryllithiums, such as those bearing alkoxy carbonyl [20], nitro [21], formyl [22], and ketone [23] groups, as well as alkyl lithiums [24], can be generated and utilized for reactions with diverse electrophiles [25]. These reactive species were investigated by screening the reaction temperature and residence time to determine their stability and optimized conditions.

4.3.3 Reaction Integration

In both Sections 4.3.1 and 4.3.2, reactions using two or three reagents were demonstrated. In this section, a reaction involving five reagents, as shown in Scheme 4.8, is cited as an example of more complex multistep synthesis. The difunctionalization of 1,2-dibromobenzene (**5**) is an example of reaction integration, where four reactions occur sequentially: (i) a halogen–lithium exchange reaction, (ii) a reaction with an electrophile, (iii) a second halogen–lithium exchange reaction, and (iv) a reaction with another electrophile [26]. Notably, 2-bromophenyl lithium (**6**), the first intermediate, can quickly decompose to form

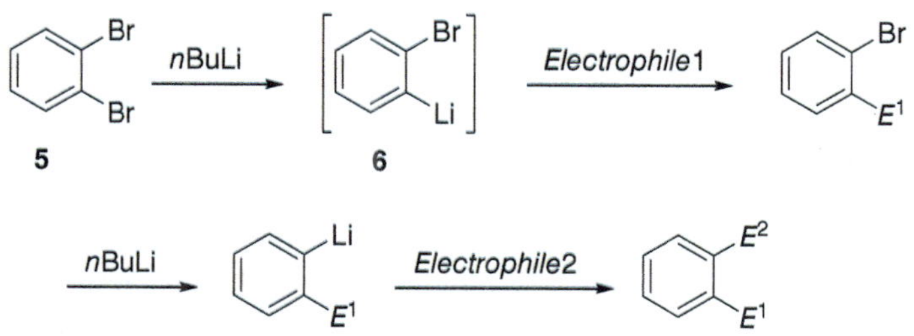

Scheme 4.8 Introduction of two electrophiles to 1,2-dibromobenzene. E^1 and E^2 are functional groups.

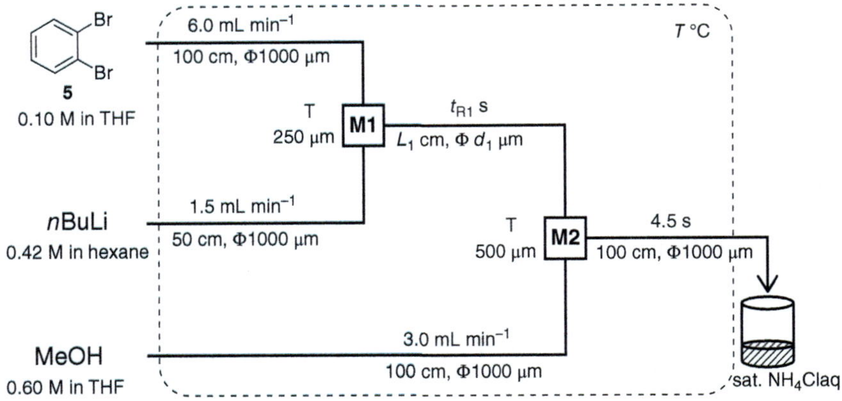

Figure 4.19 Optimization of two aforementioned reactions of the integrated reaction.

a benzyne, making this another example of a reaction mediated by unstable intermediates.

Because there are four reactions in this transformation, it is difficult to simultaneously optimize the reaction conditions. Thus, we separate the reactions to individually optimize each step. Figure 4.19 shows a typical example of the optimization of the generation and reaction of **6**. This experiment is almost identical to that described in Section 4.3.2. The temperature and residence time can be varied to determine the best conditions to obtain the highest yield of the desired product (bromobenzene). The temperature (T) range studied was 0 to −78 °C, and the residence time (t) studied were identical to those in Table 4.9.

Once these conditions were been optimized, the scope of electrophiles was then investigated. Based on our previous results [25], methyl triflate, trimethylsilyl triflate, tri-n-butylstannyl chloride, and benzaldehyde are good electrophiles in this transformation. Using one of the aforementioned electrophiles, the integration of this lithiation-substitution reaction with additional reactions was investigated. Because the best conditions for these reactions are not always identical to those of the original reaction, it was necessary to optimize the conditions of the latter reaction, especially its reaction temperature and residence time. Two baths were used to change the temperature to the optimized conditions. Section 4.2.1.5 shows the reactor's connections to different baths.

A schematic of the flow reactor is shown in Figure 4.20. In this case, we adopted the condition of −78 °C temperature T and 100 cm L_1 (residence time t_{R1}: 6.3 seconds) for the first step. With the condition of the former reaction fixed, the parameters being optimized in the latter reaction were the residence time and temperature. A second bath was necessary to study temperature in the second step. The temperature (T_2) was varied from 0 to −78 °C, and the length of the tube reactor was

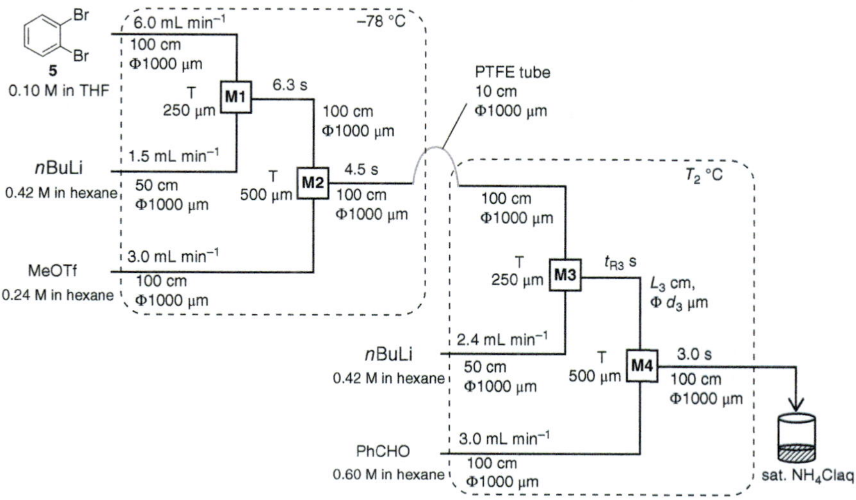

Figure 4.20 Schematic for flow reactor system for the sequential reaction of **5**.

varied from 5 to 200 cm. Other parameters, such as the flow rate of the reagent solutions and their concentrations, were found to have less of an effect on the reaction. Thus, we set the flow rate of the second *n*BuLi entering micromixer M3 as 2.4 mL min^{-1} whose concentration is the same as the first *n*BuLi solution (0.42 M). The ratio of flow rates of solutions entering micromixer M3 is 1 : 4.4. Similarly, we chose 3.0 mL min^{-1} as the flow rate of the second electrophile (benzaldehyde in Figure 4.20) to maintain the ratio of flow rates entering the same micromixer around one to four or five (1 : 4.3 in M4 in Figure 4.20).

The solution entering the latter mixer will include undesired compounds, such as residual reagents and byproducts from the first step, which must be considered. For example, residual methyl triflate, the electrophile shown in Figure 4.20, may react with *n*-butyllithium introduced in M3. Thus, the amount of *n*-butyllithium added to M3 must be sufficient to tolerate its loss based on the side reaction with methyl triflate. To account for this, the amount of methyl triflate added in M2 was lowered relative to the amount of methanol that was used in Figure 4.19. It is necessary to confirm that this decrease does not affect the reaction yield. The equivalent of the reagents is higher than that added to the former mixers (MeOTf: 1.2 eq, *n*BuLi: 1.68 eq, and PhCHO: 3.0 eq. Section 4.2.2.1 shows the calculation for the molar equivalent in flow reactions) to capture the unreacted reagents. We sometimes doubted whether the residence time of the final reaction tube is sufficient, and thus investigated the length of the tubes. If the final reaction (the reaction with benzaldehyde) appeared too slow to be completed in the tube reactor, we additionally stirred the collected reaction solution in a flame-dried flask.

4.4 Further Expertise

During the case studies, fundamental techniques for carrying out the flow microreactions were described to familiarize the readers with this difficult but powerful synthetic methodology. In this section, we more briefly describe other examples of how flow reactions can be used for efficient organic synthesis.

4.4.1 Reaction Integration

Figure 4.21 shows two types of reaction integration: linear integration and convergent integration [27]. The characteristics of the former involve multiple reactions in a single-flow process, as shown in Section 4.3.3. In convergent-type integration, reactions are allowed to occur in parallel in a different reaction space, and the generated intermediates can be mixed once they have been formed. Because the residence times can be controlled, multiple unstable reactions can occur in this type of integration.

Examples of convergent integration

- A. Nagaki, D. Ichinari, J. Yoshida, Three-component coupling based on flash chemistry. Carbolization of benzynes with functionalized aryllithium followed by reactions with electrophiles. *J. Am. Chem. Soc.* 2014, **136**, 12245–12248.
- A. Nagaki, Y. Takahashi, J. Yoshida, Generation and reaction of carbamoyl anions in flow: Applications in the three-component synthesis of functionalized α-ketoamides. *Angew. Chem., Int. Ed.* 2016, **55**, 5327–5331.
- Y. Takahashi, Y. Ashikari, M. Takumi, Y. Shimizu, Y. Jiang, R. Higuma, S. Ishikawa, H. Sakaue, I. Shite, K. Maekawa, Y. Aizawa, H. Yamashita, Y. Yonekura, M. Colella, R. Luisi, T. Takegawa, C. Fujita, and A. Nagaki, Synthesis of biaryls with a piperidylmethyl group based on space integration of lithiation, borylation, and Suzuki-Miyaura coupling. *Eur. J. Org. Chem.* 2020, 618–622.

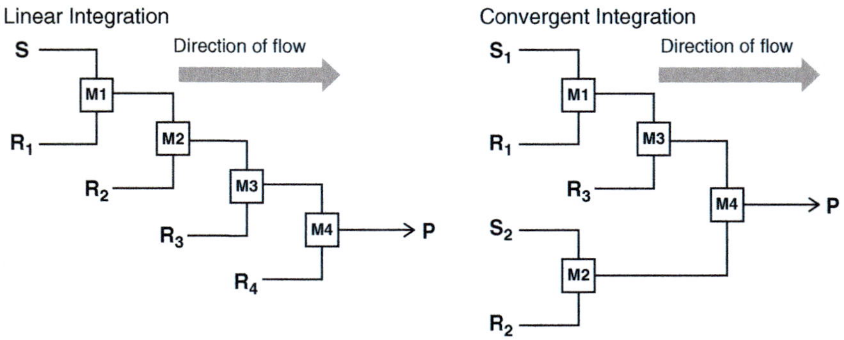

Figure 4.21 Schematic of flow reactors for linear and convergent reaction integration.

4.4.2 Chemoselective Reactions

As described in Sections 4.1.2.1 and 4.3.1, an important feature of flow microreactors is the provision of highly selective reactions. The reaction in Section 4.3.1 demonstrates selectivity by preventing undesired over-reactions. Another type of selectivity to be controlled is chemoselectivity, which is the selectivity of reactions toward multiple functional groups.

Based on an idea similar to disguised selectivity, it becomes difficult to control the chemoselectivity of the reaction using highly reactive reagents, although there may be a sufficient difference in the reaction kinetics. This is because the reaction kinetics are only the dominant factor when homogeneous, and extremely fast reactions can be initiated before the mixed reagents become homogeneous. Thus flow microreactors enabling fast mixing can lead to highly chemoselective reactions. The following references are examples of reactions using organolithiums, which are highly reactive reagents, with reagents bearing multiple functional groups:

- A. Nagaki, K. Imai, S. Ishiuchi, and J. Yoshida, Reactions of difunctional electrophiles with functionalized aryllithium compounds: Remarkable chemoselectivity by flash chemistry. *Angew. Chem. Int. Ed.* 2015, **54**, 1914–1918.
- A. Nagaki, S. Ishiuchi, K. Imai, K. Sasatsuki, Y. Nakahara, J. Yoshida, Micromixing enables chemoselective reactions of difunctional electrophiles with functional aryllithiums. *React. Chem. Eng.* 2017, **2**, 862–870.
- M. Colella, A. Tota, Y. Takahashi, R. Higuma, S. Ishikawa, L. Degennaro, R. Luisi, A. Nagaki, Fluoro-substituted methyllithium chemistry: external quenching method using flow microreactors. *Angew. Chem. Int. Ed.* 2020, **59**, 10924–10928.

4.4.3 Heterogeneous Catalytic Reactions

An additional benefit of flow reactions is that reactants are passed through the specific reaction space, which has good applicability with heterogeneous catalysis. Because heterogeneous catalysts are recyclable and easy-to-remove, the use of a cartridge filled with insoluble catalysts, called a "column reactor," is a powerful application for flow chemistry. Several synthetic reports take advantage of column reactors [28], and the followings are simple examples of the heterogeneous reactions in flow:

- A. Nagaki, K. Hirose, Y. Moriwaki, K. Mitamura, K. Matsukawa, N. Ishizuka, and J. Yoshida, Integration of borylation of aryllithiums and Suzuki–Miyaura coupling using monolithic Pd catalyst. *Catal. Sci. Technol.* 2016, **6**, 4690–4694.
- Y. Ashikari, K. Maekawa, M. Ishibashi, C. Fujita, K. Shiosaki, H. Bai, K. Matsuyama, and A. Nagaki, Stille, Heck, and Sonogashira coupling and hydrogenation catalyzed by porous-silica-gel supported palladium in batch and flow. *Green Process. Synth.* 2021, **10**, 722–728.

- Y. Ashikari, K. Maekawa, M. Takumi, N. Tomiyasu, C. Fujita, K. Matsuyama, R. Miyamoto, H. Bai, A. Nagaki, Flow grams-per-hour production enabled by hierarchical bimodal porous silica gel supported palladium column reactor having low pressure drop. *Catal. Today* 2022, **388–389**, 231–236.

4.5 Summary and Outlook

In this chapter, characteristics of flow chemistry were described. Based on the key advantages of flow microreactors, including reaction time controllability, fast mixing, and temperature controllability, several kinds of that are not applicable in the conventional batch reactors were achieved. In particular, very fast reactions, e.g. those done in milliseconds, can be controlled, which allows high selectivities. In Section 4.3, some typical experimental techniques are described to operate those flow reactions.

Since the reactants in microreactors must be in flow, reactions with precipitate are not easily achieved. Reactions performed below the freezing point of the solvent are also not suitable.

Once these reactions, which cannot be easily done in batch, are achieved by using flow chemistry, a major advantage of the flow system is that scaling up is more readily achieved. However, continued development and understanding of flow chemistry is necessary to accelerate reaction development and our society to quickly access new materials and molecules.

References

1 Rys, P. (1976). Disguised chemical selectivities. *Acc. Chem. Res.* 9: 345–351.
2 Colella, M., Tota, A., Takahashi, Y. et al. (2020). Fluoro-substituted methyllithium chemistry: external quenching method using flow microreactors. *Angew. Chem. Int. Ed.* 59: 10924–10928.
3 (a) Nagaki, A., Tomida, Y., Miyazaki, A., and Yoshida, J. (2009). Microflow system controlled anionic polymerization of alkyl methacrylates. *Macromolecules* 42: 4384–4387. (b) Nagaki, A., Miyazaki, A., Tomida, Y., and Yoshida, J. (2011). Anionic polymerization of alkyl methacrylates using flow microreactor systems. *Chem. Eng. J.* 167: 548–555. (c) Petersen, T.P., Becker, M.R., and Knochel, P. (2014). Continuous flow magnesiation of functionalized heterocycles and acrylates with TMPMgCl·LiCl. *Angew. Chem. Int. Ed.* 53: 7933–7937.
4 Nagaki, A., Sasatsuki, K., Ishiuchi, S. et al. (2019). Synthesis of functionalized ketones from acid chlorides and organolithiums by extremely fast micromixing. *Chem. Eur. J.* 25: 4946–4950.
5 (a) Suga, S., Nagaki, A., and Yoshida, J. (2003). Highly selective Friedel–Crafts monoalkylation using micromixing. *Chem. Commun.* 354–355. (b) Nagaki, A.,

Togai, M., Suga, S. et al. (2005). Control of extremely fast competitive consecutive reactions using micromixing. Selective Friedel–Crafts aminoalkylation. *J. Am. Chem. Soc.* 127: 11666–11675. (c) Webb, D. and Jamison, T.F. (2012). Diisobutylaluminum hydride reductions revitalized: a fast, robust, and selective continuous flow system for aldehyde synthesis. *Org. Lett.* 14: 568–571. (d) Nagaki, A., Yamashita, H., Takahashi, Y. et al. (2018). Selective mono addition of aryllithiums to dialdehydes by micromixing. *Chem. Lett.* 47: 71–73.

6 Nagaki, A., Matsuo, C., Kim, S. et al. (2012). Lithiation of 1,2-dichloroethene in flow microreactors: versatile synthesis of alkenes and alkynes by precise residence-time control. *Angew. Chem., Int. Ed.* 51: 3245–3248.

7 (a) Asai, T., Takata, A., Ushiogi, Y. et al. (2011). Switching reaction pathways of benzo[*b*]thiophen-3-yllithium and benzo[*b*]furan-3-yllithium based on high-resolution residence-time and temperature control in a flow microreactor. *Chem. Lett.* 40: 393–395. (b) Nagaki, A., Takahashi, Y., Henseler, A. et al. (2015). Flash chemistry using trichlorovinyllithium: switching the reaction pathways by high-resolution reaction time control. *Chem. Lett.* 44: 214–216. (c) Nagaki, A., Kim, S., Miuchi, N. et al. (2016). Switching between intermolecular and intramolecular reactions using flow microreactors: lithiation of 2-bromo-2'-silylbiphenyls. *Org. Chem. Front.* 3: 1250–1253.

8 (a) Nagaki, A., Takizawa, E., and Yoshida, J. (2009). Oxiranyl anion methodology using microflow systems. *J. Am. Chem. Soc.* 131: 1654–1655. (b) Tomida, Y., Nagaki, A., and Yoshida, J. (2011). Asymmetric carbolithiation of conjugated enynes: a flow microreactor enables the use of configurationally unstable intermediates before they epimerize. *J. Am. Chem. Soc.* 133: 3744–3747.

9 Some examples of reaction integration in flow: (a) Nagaki, A., Imai, K., Kim, H., and Yoshida, J. (2011). Flash synthesis of TAC-101 and its analogues from 1,3,5-tribromobenzene using integrated flow microreactor systems. *RSC Adv.* 1: 758–760. (b) Asai, T., Takata, A., Nagaki, A., and Yoshida, J. (2012). Practical synthesis of photochromic diarylethenes in integrated flow microreactor systems. *ChemSusChem* 5: 339–350. (c) Nagaki, A., Yamada, D., Yamada, S. et al. (2013). Generation and reactions of pyridyllithiums via Br/Li exchange reactions using continuous flow microreactor systems. *Aust. J. Chem.* 66: 199–207. (d) Nagaki, A., Tokuoka, S., and Yoshida, J. (2014). Flash generation of α-(trifluoromethyl) vinyllithium and application to continuous flow three-component synthesis of α-trifluoromethylamides. *Chem. Commun.* 50: 15079–15081. (e) Ichinari, D., Ashikari, Y., Mandai, K. et al. (2020). A novel approach to functionalization of aryl azides through the generation and reaction of organolithium species bearing masked azides in flow microreactors. *Angew. Chem. Int. Ed.* 59: 1567–1571. (f) Ashikari, Y., Kawaguchi, T., Mandai, K. et al. (2020). A synthetic approach to dimetalated arenes using flow microreactors and the switchable application to chemoselective cross-coupling reactions. *J. Am. Chem. Soc.* 142: 17039–17047.

10 (a) Nagaki, A., Takahashi, Y., and Yoshida, J. (2014). Extremely fast gas/liquid reactions in flow microreactors: carboxylation of short-lived organolithiums. *Chem. Eur. J.* 20: 7931–7934. (b) Mandai, K., Yamamoto, T., Mandai, H., and Nagaki, A. (2022). Rapid gas–liquid reaction in flow. Continuous synthesis and production of cyclohexene oxide. *Beilstein J. Org. Chem.* 18: 660–668.

11 (a) Horii, D., Fuchigami, T., and Atobe, M. (2007). A new approach to anodic substitution reaction using parallel laminar flow in a micro-flow reactor. *J. Am. Chem. Soc.* 129: 11692–11693. (b) Elsherbini, M. and Wirth, T. (2019). Electroorganic synthesis under flow conditions. *Acc. Chem. Res.* 52: 3287–3296. (c) Noël, T., Cao, Y., and Laudadio, G. (2019). The fundamentals behind the use of flow reactors in electrochemistry. *Acc. Chem. Res.* 52: 2858–2869. (d) Hu, P., Peters, B.K., Malapit, C.A. et al. (2020). Electroreductive Olefin–Ketone coupling. *J. Am. Chem. Soc.* 142: 20979–20986. (e) Takumi, M., Sakaue, H., and Nagaki, A. (2022). Flash electrochemical approach to carbocations. *Angew. Chem. Int. Ed.* 61: e202116177.

12 (a) Cambié, D., Bottecchia, C., Straathof, N.J.W. et al. (2016). Applications of continuous-flow photochemistry in organic synthesis, material science, and water treatment. *Chem. Rev.* 116: 10276–10341. (b) Sambiagio, C. and Noël, T. (2020). Flow photochemistry: shine some light on those tubes! *Trends Chem.* 2: 92–106. (c) T. H. Rehm (2020). Flow photochemistry as a tool in organic synthesis. *Chem. Eur. J.* 26: 16952–16974.

13 Nakahara, Y., Furusawa, M., Endo, Y. et al. (2019). Practical continuous-flow controlled/living anionic polymerization. *Chem. Eng. Technol.* 42: 2154–2163.

14 (a) Mastronardi, F., Gutmann, B., and Kappe, C.O. (2016). Continuous flow generation and reactions of anhydrous diazomethane using a Teflon AF-2400 tube-in-tube reactor. *Org. Lett.* 15: 5590–5593. (b) Pieber, B. and Kappe, C.O. (2016). Generation and synthetic application of trifluoromethyl diazomethane utilizing continuous flow technologies. *Org. Lett.* 18: 1076–1079. (c) Strauss, F.J., Cantillo, D., Guerra, J., and Kappe, C.O. (2016). A laboratory-scale continuous flow chlorine generator for organic synthesis. *React. Chem. Eng.* 1: 472–476.

15 Skrotzki, E.A., Vandavasi, J.K., and Newman, S.G. (2021). Ozone-mediated amine oxidation and beyond: a solvent-free, flow-chem. *Flow-Chemistry Approach. J. Org. Chem.* 86: 14169–14176.

16 Perez, K., Picard, B., Vuluga, D. et al. (2020). Bromine–lithium exchange on a gem-dibromoalkene, part 2: comparative performance of flow micromixers. *J. Org. Process Res. Dev.* 24: 787–791.

17 Nagaki, A., Moriwaki, Y., and Yoshida, J. (2012). Flow synthesis of arylboronic esters bearing electrophilic functional groups and space integration with Suzuki–Miyaura coupling without intentionally added base. *Chem. Commun.* 48: 11211–11213.

18 Ashikari, Y., Maekawa, K., and Nagaki, A. Unpublished data.

19 Nagaki, A., Kim, H., Usutani, H. et al. (2009). Generation and reaction of cyano-substituted aryllithium compounds using microreactors. *Org. Biomol. Chem.* 8: 1212–1217.

20 (a) Nagaki, A., Kim, H., and Yoshida, J. (2008). Aryllithium compounds bearing alkoxycarbonyl groups: generation and reactions using a microflow system. *Angew. Chem. Int. Ed.* 47: 7833–7836. (b) Nagaki, A., Kim, H., Moriwaki, Y. et al. (2010). A flow microreactor system enables organolithium reactions without protecting alkoxycarbonyl groups. *Chem. Eur. J.* 16: 11167–11177.

21 Nagaki, A., Kim, H., and Yoshida, J. (2009). Nitro-substituted aryl lithium compounds in microreactor synthesis: switch between kinetic and thermodynamic control. *Angew. Chem. Int. Ed.* 48: 8063–8065.

22 Nagaki, A., Tsuchihashi, Y., Haraki, S., and Yoshida, J. (2015). Benzyllithiums bearing aldehyde carbonyl groups. A flash chemistry approach. *Org. Biomol. Chem.* 13: 7140–7145.

23 Kim, H., Nagaki, A., and Yoshida, J. (2011). A flow-microreactor approach to protecting-group-free synthesis using organolithium compounds. *Nat. Commun.* 2: 264.

24 (a) Nagaki, A., Yamashita, H., Hirose, K. et al. (2019). Alkyllithium compounds bearing electrophilic functional groups: a flash chemistry approach. *Angew. Chem. Int. Ed.* 58: 4027–4030. (b) Nagaki, A., Yamashita, H., Hirose, K. et al. (2019). Generation and reaction of functional alkyllithiums by using microreactors and their application to heterotelechelic polymer synthesis. *Chem. Eur. J.* 25: 13719–13727.

25 Nagaki, A. (2019). Recent topics of functionalized organolithiums using flow microreactor chemistry. *Tetrahedron Lett.* 60: 150923.

26 (a) Usutani, H., Tomida, Y., Nagaki, A. et al. (2007). Generation and reactions of *o*-bromophenyllithium without benzyne formation using a microreactor. *J. Am. Chem. Soc.* 129: 3046–3047. (b) Nagaki, A., Tomida, Y., Usutani, H. et al. (2007). Integrated micro flow synthesis based on sequential Br–Li exchange reactions of *p*-, *m*-, and *o*-dibromobenzenes. *Chem. Asian J.* 2: 1513–1523.

27 Nagaki, A., Ashikari, Y., Takumi, M., and Tamaki, T. (2021). Flash chemistry makes impossible organolithium chemistry possible. *Chem. Lett.* 50: 485–492.

28 Kobayashi, S. (2016). Flow "fine" synthesis: high yielding and selective organic synthesis by flow methods. *Chem. Asian J.* 11: 425–436.

5

Reaction Optimization Using Design of Experiments

Laura Forfar and Paul Murray

Paul Murray Catalysis Consulting Ltd, Yate, Bristol, UK

Glossary

- A **factor** (also called a variable, parameter, input) is an experimental setting, which can be controlled and whose levels are believed to affect the outcome of the experiment, for example temperature, concentration, reagent stoichiometry.....
  - Factors can be *discrete* (categorical, qualitative, e.g. solvent, equipment, reagent), *continuous* (numeric, quantitative, e.g. temperature, rate of addition), or *mixture* (a component of a blend, e.g. binder, lubricant, surfactant)
- A **response** is a measured outcome of the experiment, for example conversion, purity, or selectivity.
- A **level** is the exact value of the factor used, e.g. 50 °C is a level of temperature, a continuous factor, and toluene is a level of a solvent, a discrete factor.
- **OFAT/OVAT** is one variable/factor at a time, where one factor is changed while the remaining factors of the reactions are held constant.
- A **design** is a precisely structured set of experiments that are analyzed together to assess the effect of changing the factors on the response.
- A **model** is an equation that approximately relates the factors to the responses.
- A **factorial design** is one which consists of two or more factors investigated at different levels, where every combination of the levels across all factors is interrogated.
- A **fractional factorial design** is based on a full factorial design but only a fraction of the experimental runs is included.

Enabling Tools and Techniques for Organic Synthesis: A Practical Guide to Experimentation, Automation, and Computation, First Edition. Edited by Stephen G. Newman.

- A **confounding** variable is a factor whose presence affects the factor(s) being studied so that the results you get do not reflect the actual relationship between the factor(s) being studied and the response. It arises when not enough experiments have been carried out to determine whether a change in the response is due to one factor or another.
- **Correlation** between two variables is when an observed change in the level of one variable is accompanied by a change in the level of another variable.
- The **resolution** of a design describes the amount of confounding/aliasing in a design.
- An **interaction** in a process or reaction is where the combined change in 2 factors produces a greater or lesser effect than the sum of the effects expected from either factor alone, i.e. 2 + 2 does not equal 4.
- **Curvature** describes when the response is not linear i.e. you see a curved line rather than a straight line. A quadratic term is required to explain the curvature.
- **Response surface designs** assess factors at more than two levels and are used to explain curvature
- A **centre point** is an experiment with all factors set at the mid-point between the low and high level. This is generally repeated (at least three times) to get a measure of experimental error for the process.
- **Main effect**: a measure of how changes in the level of a factor affect the response.
- **Statistical significance**: a model is statistically significant if the probability (p) of obtaining the model by chance is less than the significance level (α, commonly 0.05, 5%), $p \leq \alpha$.
- **R^2**: percent of the variation of the response explained by the model, it shows the model fit. A value close to 1 (100%) indicates that the model fits the data very closely.
- **Q^2**: Percent of the variation of the response predicted by the model according to cross validation and expressed in the same units as R^2. Q^2 estimates the predictive ability of the model. Q^2 should be greater than 0.1 for a significant model and greater than 0.5 for a good model. The difference between R^2 and Q^2 should also be smaller than 0.4 for a good model.
- **Replication**: the variation of the response under the same conditions (pure error), often at center points.
- **Reproducibility**: the variation of the response under the same conditions (pure error), compared to the total variation of the response.
- To **Complement** the design is to add new experiments to resolve interactions or nonlinearities in the current design.

5.1 Introduction

Have you ever had a result you have been unable to repeat? Early in one's education and career, it is disappointing and frustrating when you follow a literature prep to the letter and do not achieve the stated outcome; however, if this happens later in your career, for example during the large-scale manufacture of a product, this can be catastrophic in terms of safety, productivity, and financial losses. This is what can happen when a process is not robust – the results are not reproducible and the process is not as well understood as the scientist originally thought.

Design of experiments (DoE) is a technique that can be used to gain more understanding of your chemical reaction and eventually deliver an efficient, well-understood, and robust process [1]. DoE has several names: statistical experimental design, factorial experimental design, or simply experimental design. The topic has specific terminology and consists of many acronyms. It is a statistical approach to experimentation, which investigates several variables simultaneously in a set of experiments, referred to as a "design," to assess their effect on the experimental outcome. Experimental design is commonly used in industry but is rarely mentioned in chemistry education, possibly as the theory is extensive and many think of it as an optimization technique rather than a useful tool for novel research.

Our aim for this chapter is to explain the background of DoE with enough math as necessary to explain how the process works, but not so much as to alienate the readership. We hope it will give the reader enough information and tips to add the technique in your chemistry toolkit, making your experimentation more ordered and efficient, delivering valuable information in as few experiments as possible. The chapter will work through an example that has been adapted from a real industrial project to give a view of how DoE can be used. Despite the efforts of the authors, it is unlikely that you will fully understand all of the contents of this chapter from reading along. The concept will likely fall into place once you're practically using design of experiments. This will require specific software (see Section 5.7), trials of which are available for personal use.

5.1.1 How Do We Experiment and DoE Terminology

As a synthetic chemist you may describe the way you currently conduct research as designing experiments to test your theories. Traditionally in chemistry, experimentation is taught to focus on changing one variable at a time (OVAT, or one factor at a time, OFAT) to assess the effect of one specific variable on the experimental outcome. This seems logical: how would you be able to see the effect of temperature on a reaction if both temperature and concentration are changed in a set of experiments? But DoE does just that. It is a structured approach that considers several factors together and varies them simultaneously in a particular set of experiments. These sets of experiments are referred to as "designs" and when this word is used throughout the chapter, it is specifically referring to a group of experiments that are a package of work and the data from which will be analyzed together to deliver increased information about a process. DoE is commonly performed with the assistance of specialist software that can aid in selecting which experiments to perform and the interpretation of results (see Section 5.7).

When you perform an experiment, you are aiming for a target, whether that is the highest purity, the maximum yield, the cheapest process or something fit for your current purpose, and to this end you change input settings (factors or variables) to see what effect their change has on the outcome (response). If you do not get a satisfactory result with an experiment, you would generally change the level of the input factors and see if that change makes a difference to the response.

- A **factor** (or variable, parameter, input) is an experimental setting, which can be controlled and whose levels are believed to affect the outcome of the experiment, for example, temperature, concentration, reagent stoichiometry.....
  - Factors can be *discrete* (categorical, qualitative, e.g. solvent, equipment, reagent), *continuous* (numeric, quantitative, e.g. temperature, rate of addition), or *mixture* (a component of a blend, e.g. binder, lubricant, surfactant)
- A **response** is a measured outcome of the experiment, for example, conversion, purity, or selectivity.
- A **level** is the exact value of the factor used, e.g. 50 °C is a level of temperature, a continuous factor, and toluene is a level of a solvent, a discrete factor.

We can introduce the concept of factors and responses by discussing an everyday "reaction." The readership will have an extremely varied chemical knowledge, but the vast majority will have made a hot drink and many students require caffeine to power them through the PhD. Making a coffee sounds like a simple task but when broken down into steps, there are actually many things to consider. Will we use instant or ground coffee? Which brand? Is it single origin or a blend, caffeinated or decaffeinated? How much coffee to add, how much water to add, the brewing time, brewing temperature, roasting temperature, agitation. Are we adding milk, cream, sweetener (artificial or sugar) and what is the order of addition? Thinking back further, what was the roasting temperature and time of the beans (or in fact any other "pre-treatment" of the coffee itself). I'm sure if you go and make a coffee now you can think of even more factors, but it is clear that there are many factors to consider in a seemingly simple process.

Having thought about all the factors that may influence the coffee making process, we can now think about how we will assess the resulting coffee. We cannot simply say whether it was good or bad as this is subjective and not measurable; I like my coffee with milk and no sugar, while you may prefer a sweetened black coffee. Your experiments will only give you useful and accurate information about your process if you have a quantifiable response and an accurate way of measuring this. To evaluate the coffee, we could measure the caffeine content, calculate the cost of the coffee, and even assess the color if we used a proper color chart. There *are* ways to evaluate more subjective responses such as taste, which would involve panels of professional tasters but objective responses can offer more detailed process understanding [2].

5.1.2 OVAT vs. DoE

With so many variables, it can seem overwhelming to decide on where to start with an approach for optimizing a procedure, especially when the OVAT method is adopted. You would have to start with the factor you thought was most important and vary that first while holding everything else constant. Then when the best result from varying the first factor is found, hold at this setting and vary the second factor, and so on (see below).

5.1.2.1 A Simple Chemical Example

We can take a simple substitution reaction as an example (Scheme 5.1). Adding methanol to 1-chloro-1-phenylethane will result in the formation of the ether product, but how much of the alcohol is necessary? Is an excess required to deliver the best possible yield? What temperature is required? We want to prioritize formation of the substitution product and minimize the elimination product. We can start by looking at 1 equivalent of alcohol (SM2) and varying the temperature, finding that an increase in temperature results in an increase in product but then this drops off above a certain threshold. If the best yield is observed at 80 °C, the next step is to hold the temperature at 80 °C and vary the amount of SM2 to observe the effect of changing the second factor.

There are several flaws with this method and the main one is that many factors are interdependent and there is no way of finding out reliably how the differing levels of the factors combine to change the response.

- An **interaction** in a process or reaction is where the combined change in 2 factors produced a greater or lesser effect than the sum of the effects expected from either factor alone, i.e. 2 + 2 does not equal 4.

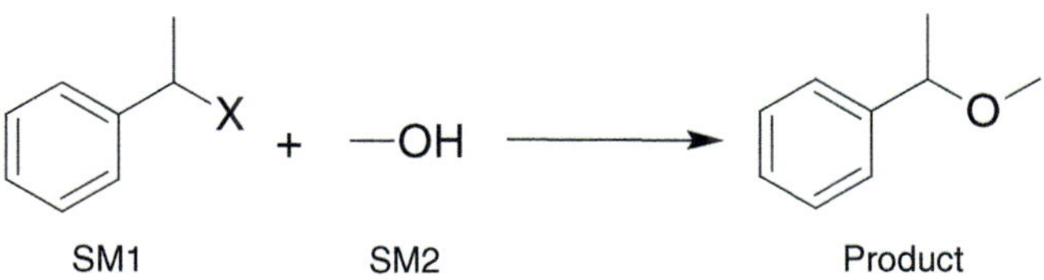

SM1 SM2 Product

Scheme 5.1 Substitution reaction.

In a set of experiments investigating the effect of temperature and catalyst (the factors) on conversion (the response), it was observed that catalyst A gave the best conversion at 50 °C, whereas catalyst B gave the best conversion at 30 °C. These results show there is an interaction between catalyst and temperature.

The OVAT approach is imperfect because it only looks at cross-sections of chemical space (as illustrated in Figure 5.1 when testing two factors). Furthermore, it could be impossible to uncover the best possible conditions for the process as

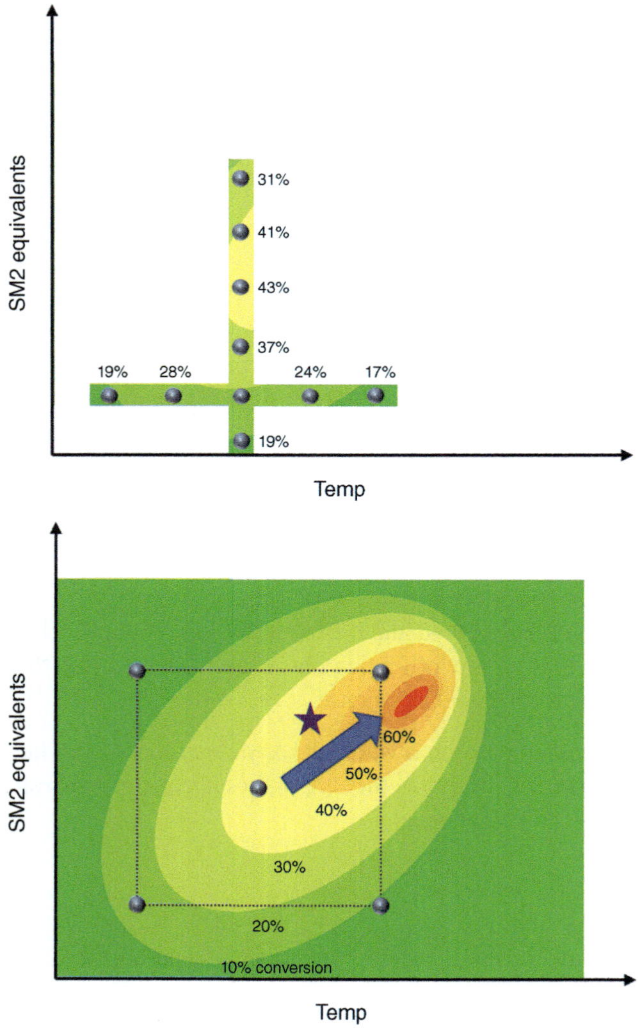

Figure 5.1 OVAT screening (top) vs. DoE screening (bottom).

there is no way to know if improvements are possible (all you have are the individual results) and limited information about which direction to take to find improved results. If faced with this set of results in the lab, you may take the current best setting and try to improve by varying other factors such as pressure or solvent, but it is unlikely you would try further combinations of temperature and SM2. OVAT is also inefficient; in Figure 5.1 (top), 10 experiments have been carried out using the OVAT approach. The results are plotted against the temperature and amount of SM2 used, with the results listed next to them. The results range from 17 to 43% but from this dataset it is unclear if there is a different combination of SM2 equivalents and temperature, which could deliver a higher yield.

Finally, OVAT does not have a measure of inherent variability. Each experiment is generally carried out once, the results are interpreted, and the experimenter moves on. If the result was inaccurate (for example if SM2 was incorrectly dosed due to an issue with the balance, or there was an analytical error which was not detected), it could lead the scientist to draw incorrect conclusions about the reaction. There will always be experimental error which can be calculated by repeating experiments and comparing the results, however, it is unlikely that the scientist will carry out their experiments in duplicate, or even better triplicate, as this would be too costly in terms of time and resource.

The alternative approach to OVAT experimentation is DoE. This approach investigates a reaction or process by considering several factors simultaneously in a rationally chosen set of experiments (a "design"). The chosen experiments are structured in a particular way to enable the use of statistics to create a mathematical model, which can relate the factors to the response and see how the response changes as a function of changing the factors. A well-performed design of experiments can tell you which factors are key to your process (i.e. what effect changing the factors has on the response), if there are any interdependent factors (interactions), the reaction conditions that will bring about acceptable performance, and if the process is robust. Well executed DoEs will allow an estimation of error, enabled by repeated experiments (commonly a "center point" experiment carried out at least three times to give an idea of replication). Having a value for experimental error will provide a confidence interval for the model and predictions, and will give significance to the results, i.e. tell you whether an effect is real and above noise or not.

In the simplest design, each factor is examined at a high and low level. Several experiments are carried out in one package of work where the order is structured in a particular way to ensure each factor is investigated equally. If the substitution reaction was investigated using DoE, you would set a low and high level for equivalents of SM2 and a low and high temperature. Four experiments are then carried out: low SM2 and low temperature, low SM2 and high temperature, high SM2 and low temperature, and high SM2 and high temperature, forming a square of chemical space (Figure 5.1, bottom). Ideally, a center-point experiment is also run with SM2 and temperature investigated halfway between the high and low point. The results of these experiments can be used to create mathematical models that relate the response to the changing factors and can explain how changing the factors changes the response. For simple designs, this can be done with simple maths by hand, but generally DoE software is used. From this design investigating two factors you can find out how the factors impact the response, if there is an interaction between the factors, and how to change your factor settings to improve the result or reach your target. If you do not hit your target response level in the first set of experiments, the model can be used to predict different factor settings to reach the desired target. Figure 5.1 (bottom) shows a contour plot that has been generated by DoE software. In this illustrative example, five experiments have been carried out with varying temperatures and equivalents of SM2 and the best result from these five gives a conversion of 55%. Using the model generated by the software, we are able to predict the estimated conversion from any combination of temperature and SM2 equivalents, even if that experiment has not been physically conducted. For an experiment carried out under the conditions at the purple star (a mid-high temperature and a mid-high SM2), the model predicts a conversion of approximately 51%. The model will ultimately guide the chemist toward the optimal 80% yield, following the direction of travel (the blue arrow).

5.1.3 A Note on Error, Accuracy, and Precision

If you carry out an experiment once and get a measurement of 55%, how can you know that this value is accurate and a true representation of the reaction? You are assuming that the reaction and the analysis have been carried out properly and accurately. But it is easy to make a mistake in a research lab and when a reaction or experiment is only tested once there is no way to account for mistakes. If you perform the reaction twice, you will end up with an average of the two results. This is better than just the one result; however, there is a chance that a mistake occurred with one of the reactions, measurements, sample collections, or interpretation of the results and you will not know which result is less accurate. With three repeats, you can begin to assess whether you have a true reflection of the response – the three results could all be very similar or you could have two results that are almost identical and one which is quite different. If this happened, you could look into the different result and either try to rationalize the difference or determine whether something went wrong.

Even if the experimenter carries out the steps as precisely and accurately as possible, there are many points in which error can be introduced into a design. In order to reduce this error, all apparatus and equipment should be calibrated, environmental conditions should be as controlled as possible, and samples and measurements should only be taken when the system has reached equilibrium.

It is important to consider how accurately factor settings can be achieved and if the difference between the low, mid, and high factor settings is large enough to see a true change. If the low temperature setting is 5 °C and the high setting is 10 °C, this leaves a mid-point of 7.5 °C. The measurement error on a thermometer could be ±1 °C; therefore, the difference in the low and mid-point could potentially only be 0.5 °C (6–6.5 °C). Will you see an effect of temperature when it is only changing by half a degree? See more about this in Section 5.5.2.

Nevertheless, the most important sources of experimental error are the unknown factors that are not controlled in the experiment. Thorough factor selection, a clear and detailed reaction protocol, and careful experimentation should minimize this error. Before carrying out a design, it may be worth repeating an experiment a few times to check the reproducibility of the system. If your error is small, you have control. If the experiment is not reproducible, then you are not considering an important factor and should return to factor selection. You will not get the desired information from DoE if you do not have good reproducibility.

5.2 When and How Can DoE Be Used?

There are many different ways in which experimental design can be used – the simplest way and therefore the most common is final optimization of a process. This is commonly done in process development chemistry where a team of scientists have a target molecule, have developed the desired route, and fixed their starting materials. They can then run a small DoE to find out the best levels of the starting materials along with a couple of conditions such as temperature and concentration. A DoE at this stage in development will give valuable information in confirming the best operating conditions for a reaction. However, if DoE is used earlier in the development process, it is possible to uncover exciting possibilities by looking at a larger number of factors. Early designs can provide answers to questions such as "what is the effect of a different solvent on the yield?" You could potentially achieve dramatically different yields or generate fewer impurities by changing the solvent, but this should be investigated early in the development process.

The area where DoE could perhaps make the biggest difference to chemists, research scientists, and developers is in route scouting and reaction development. The coffee example showed us that there can be large numbers of factors that can affect the response and the only way to determine what that effect is, is to experiment. There will be some factors that have a large effect (i.e. changing that factor has a significant impact on the response) and some which have a minimal or no effect. There may be some interactions or on rare occasions, none. Using experimental design in the early stages of development is called screening, and designs used at this point are screening designs. The purpose of a screening design is to determine which factors are important (i.e. which factors have a significant effect on the response), in what way are they important (i.e. does an increase in the factor increase or decrease the response), and whether there are interactions between the factors. In a screening design, not all the investigated factors will impact the response and they are referred to as insignificant or unimportant factors. An initial screening design near the beginning of a project can help to build process understanding while trying to minimize the cost and time associated with gathering this information by reducing experiment numbers. We are not suggesting that the first experiments run in a project are part of a screening design. Indeed, some preliminary reactions will be required to check specifics of a reaction, e.g. solubilities and the selection of appropriate factor ranges. However, our experience of working with industrial clients has shown that many leave DoE until quite late in their process, after hundreds of experiments have already been carried out. Once the initial sighting experiments reach numbers above 10–15, you should consider reviewing the information you have to date and start planning an experimental design.

It requires confidence to perform a design at the start of a project before you fully understand what is going on, but this is the point of the screening design - to estimate what is and is not important and point you in the right direction. We must note that another problem with a large number of sighting experiments is the experimenter may be led to false conclusions about the significance of a factor as these experiments will have been performed OVAT and therefore you cannot tell if the factor is important in an interaction.

5.3 What Information Can I Get from a DoE and How Is It Obtained?

When it comes to discussing the maths, which is the basis of experimental design, some synthetic chemists can lose interest or be overwhelmed by the thought of statistics and complicated equations. The approach of this chapter is DoE for chemists, written by chemists, so we will give you enough information so that you understand how the technique works and what the numbers in the statistical software means, but we will not go further than simple addition, subtraction, division, and multiplication. For those who are interested in the maths, chapter 3 of *Design and Optimization in Organic Synthesis* by Carlson [3] is a good place to start, and *Statistics for Experimenters* by Box and Hunter [4] is very thorough. We will discuss the maths in the context of a simple two-factor design.

The nucleophilic substitution reaction discussed in Section 5.1.2 has many potentially significant factors (solvent, concentration, temperature, stoichiometry of the alcohol, halide...), but to keep it simple we will start with varying the temperature and the equivalents of SM2. A comprehensive way to investigate the effect of both factors on the conversion to product (the response) is to run four experiments at combinations of high- and low-factor settings. This is called a factorial design. A factorial design at two levels (i.e. high and low temperature) requires 2^n runs for "n" factors. Ideally, in addition to the 2^n experiments, a center point experiment would also be run in triplicate to assess reproducibility.

For this example, temperature is varied between 50 and 80 °C, while SM2 equivalents is varied between 1.0 and 2.0 equivalents.

After a chemist runs these experiments and measures the response (i.e. % conversion, Table 5.1), we are able to determine which factors have an effect on conversion, if the relationship between temperature and SM2 equivalents is important (i.e. if there is an interaction), if the factor settings used deliver an acceptable conversion, or if different settings could deliver improvements, for example if a temperature >80 °C would result in higher conversions. Using simple arithmetic, it is possible to create a mathematical model showing how the response changes as a function of the changing factors. See Section 5.3.2.

5.3.1 Which Factors Are Important?

With the data collected from a DoE, a lot of information can be extracted. Generally, this will be done using computer software, but some of it can be done by hand using simple maths. A simple but critical metric is the **main effect**. This is a measure of how changes in the level of a factor affects the response, i.e. how much the response value will change when moving from a low setting of a factor to the high setting. It is calculated by taking the difference between the average of the response values at the high setting and the average of the response values at the low setting.

From Table 5.1, the main effect of temperature $= [(93+80)/2-(73+60)/2]$ $= 86.5-66.5 = 20$.

This means that, on average, changing the temperature from 50 to 80 °C increased the conversion by 20%.

Table 5.1 Two-factor design table.

Experiment number	Temperature (°C)	SM2 (equivalents)	Conversion (%)
1	50	1	73
2	50	2	60
3	80	1	93
4	80	2	80
5	65	1.5	81

From Table 5.1, the main effect of SM2 equivalents $= [(80+60)/2-(93+73)/2] = 70-83 = -13$

This means that, on average, changing the SM2 equivalents from 1 to 2 decreases the conversion by 13%, i.e. the lower level of 1 equivalent is better for conversion. If we compare the two factors, the effect of temperature is greater than the effect of SM2 equivalents over the ranges investigated.

In designs of many factors, several of these will have no effect or a very minimal effect on the response. However, a factor cannot be discounted as insignificant unless it is also not present in an interaction.

As described in Section 5.1.2.1, an **interaction** is where the combined change in two factors produces a lesser or greater effect than the sum of effects expected from either factor alone. An interaction exists when differences in the response from one factor depend on the level of another factor (Figure 5.2). In an interaction plot, one factor is varied along the *x*-axis and the other factor is presented as two separate lines, one for the high level and one for the low level. To construct an interaction plot for a two factor, two level design, choose one factor (F1) to be the *x*-axis and have the response as the *y*-axis. Plot the results for the low and high level of

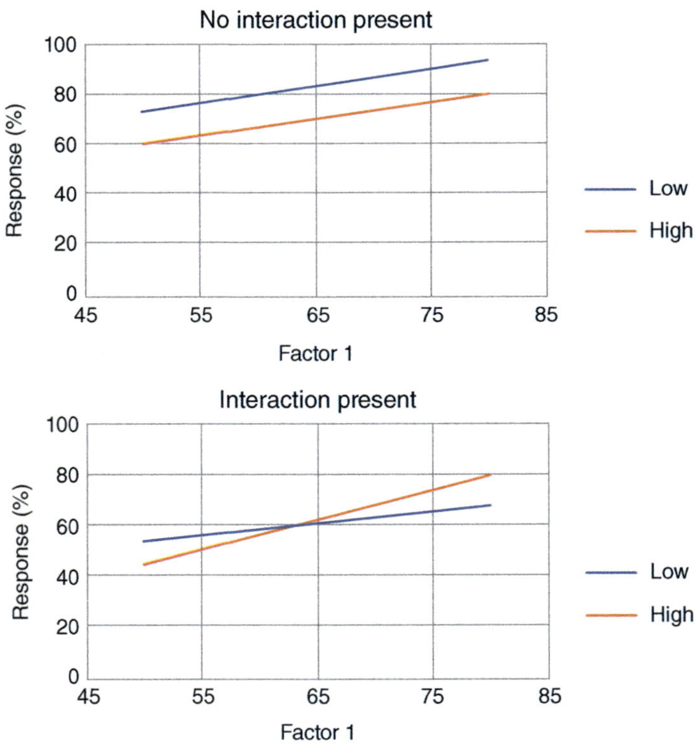

Figure 5.2 Interaction plots.

F1 when the other factor is at the same setting (e.g. factor 2 is low) and draw a line between them. Next, plot the results for the low and high level of F1 when factor 2 is at the opposite setting (e.g. factor 2 is high) and draw a line between them.

If the lines are parallel, there is no interaction but if the lines deviate from parallel, there is an interaction. You will only be able to find out about interactions between factors by using DoE, OVAT experiments will not provide this information.

5.3.2 How Are the Models Generated?

A model is an equation that approximately relates the factors to the response. Specific computer software is available to aid in the generation of models and their interpretation and you will not have to calculate these by hand; however, we would like you to understand how they are generated.

To experiment is to perform a series of tasks or to follow a recipe, then look at a response (y). When developing a process, you change the input factors (x^i) in the hope that they will change the response. Therefore, we can say that the response is a function of the input factors and it is fair to assume there is a functional relationship between the experimental inputs and the measured response. Experimental design uses modeling to try and determine this functional relationship.

$Y = f(x....)$ where x is the experimental variable and y is the response.

With a simple experiment investigating two factors, x^1 and x^2, we presume that both factors will have an effect on the response ($y = f(x^1, x^2) + e$), where (e) is experimental error. But there could also be an effect from an interaction between the two factors ($y = f(x^1, x^2, x^1x^2) + e$).

Each factor will affect the response in different ways, either causing a small or large change in the reaction or be part of different interactions. We can expand the equation to quantify these effects:

$$Y = b^0 + b^1x^1 + b^2x^2 + b^{12}x^1x^2 + b^{11}\left(x^1\right)^2 + e$$

- b^0 is a value calculated from the center point experiment as an estimate of the response with all factors set to their midpoint (0)
- b^1 is a direct measure of the linear dependence of the factor x^1 on the response, i.e. how much changing factor x^1 directly changes y
- b^{12} is a measure of the interaction between factors x^1 and x^2
- $b^{11}(x^1)^2$ is a quadratic term to describe nonlinearity (also referred to as curvature) in the response. This can occur when changing the factor from a low level to a high level initially changes the response in one direction, then as the factor level increases further the response changes back in the original direction, for example increasing the temperature may initially increase conversion of

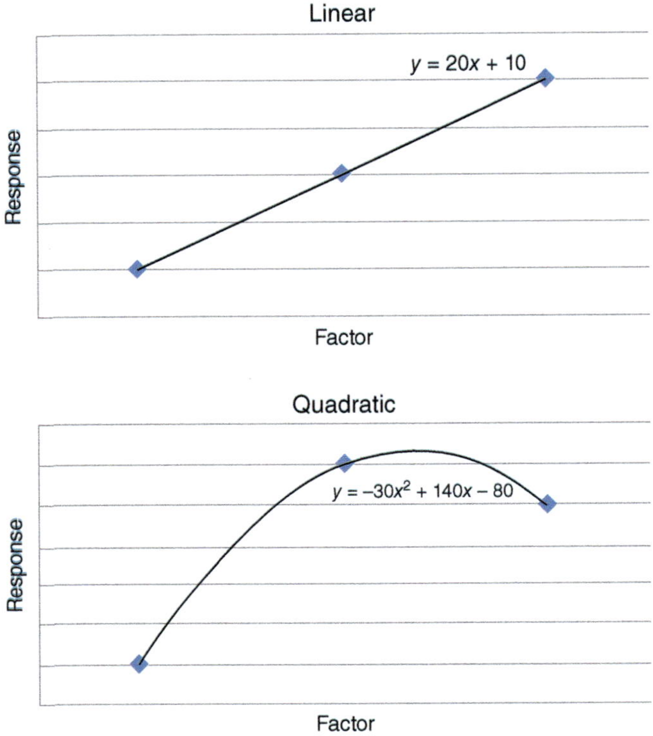

Figure 5.3 Linear line equation and quadratic line equation.

a reaction; however, past a certain temperature the response will begin to decrease. For example, above the best temperature setting the starting material begins to decompose or a new reaction pathway opens up, consuming the starting material to another product or further consuming the product. Figure 5.3 shows a linear increase and a nonlinear change and shows why a quadratic term is required to fit a model to such a response.

Table 5.2 is adapted from the 2-factor design originally shown in Table 5.1. The temperature is denoted factor 1, and the equivalents of SM2 are denoted factor 2. Mathematically, the levels for factors are normalized to −1 for the low level and +1 for the high level, with the center point set to 0. For the interaction between factors 1 and 2, use the product of the factor setting. In experiment 1 when the low value for temperature (50 °C) and equivalents of SM2 (1 equivalent) are selected and normalized, the interaction between them is $x^1 * x^2 = -1 * -1 = +1$. Three experiments [5–7] have been added at the center point value. The inclusion of these experiments and their measured responses will enable the calculation of b^0 and the experimental error and can help to solve the equations below.

Table 5.2 Results table with factor and interaction codings.

Experiment	Factor 1: Temperature	Factor 2: SM2 equivalents	Interaction (1*2)	Response
	x^1	x^2	$x^1 x^2$	y
1	−1	−1	+1	73
2	−1	+1	−1	60
3	+1	−1	−1	93
4	+1	+1	+1	80
5	0	0	0	81
6	0	0	0	82
7	0	0	0	81

Considering the data in Table 5.2, we can write an equation for each of the experiments, then use different statistical techniques to solve the equations that will give values for the main effects and interaction effects.

$$73 = b^0 - b^1 - b^2 + b^{12} \quad (5.1)$$

$$60 = b^0 - b^1 + b^2 - b^{12} \quad (5.2)$$

$$93 = b^0 + b^1 - b^2 - b^{12} \quad (5.3)$$

$$80 = b^0 + b^1 + b^2 + b^{12} \quad (5.4)$$

$$81 = b^0 \quad (5.5)$$

$$82 = b^0 \quad (5.6)$$

$$81 = b^0 \quad (5.7)$$

Equations (5.1)–(5.4) above represent the experiments 1–4. Equations (5.5)–(5.7) use the center point experiments. We could also take the average of these and include an error (Eq. 5.8). This equation can be rearranged (Eq. 5.9) to get a value for b^0, showing that if we have an experimental error (which is calculated from the repeated experiments), we can work out b^0.

$$(81 + 82 + 81)/3 = b^0 + e \quad (5.8)$$

$$b^0 = 81.33 - e \quad (5.9)$$

The remaining unknowns are b^1, b^2, and b^{12} and these can be revealed by using matrix algebra. Multiple linear regression (MLR) is a statistical approach to modeling the relationship between a response (y) and two or more factors (x^n). The data is modeled using linear functions and unknown model parameters are estimated from the data. While it is possible to work this out by hand, statistical software does this easily, even for designs of 8+ factors. We saw in Section 5.3.1 how the main effect could be calculated. The main effect is the difference between the high and low setting. The b^1 and b^2 values are from the mid-point to the high or low and therefore $b^i = (\text{main effect}^i/2)$.

The take home message from this section is that from a factorial design, you can find out how the investigated factors affect the response and how the two-way interactions between all combinations of factors affect the response. These metrics can easily be calculated by hand for two-factor designs but for higher factor numbers, statistical software can be used.

The previous section focused on a design of two factors but we know that chemistry is complicated and it is likely that you will want to investigate many more factors. With a two-factor design, it is easy to get information about the way the total system works (main effects and interactions) in relatively few experiments. However, with factorial designs the number of experiments required increases exponentially with the number of factors you want to investigate, i.e. 5 factors is $2^5 = 32$ experiments, 8 factors = 256 experiments. There comes a point when you will need to balance the information gained with the number of experiments you are prepared to, or able to carry out. Limits on how many experiments you are able to perform arise from expensive materials, time pressures, and many other reasons, meaning that although you have 8 potentially significant factors that you wish to investigate, it is not possible to run 256 experiments. Indeed, it is possible to investigate 8 factors in fewer than 256 experiments but in reducing the number of experiments you will no longer be able to understand everything about the process.

5.4 What Types of Design Are Available?

There are different design options depending on what you want to know about your process. Obviously, the more information the better, but when investigating several factors simultaneously, there will be a limit on how many experiments you can carry out. If you are at the start of development, it could be more beneficial to you to find out what factors are significant in a reaction, without knowing all the details of if and how they are interacting. If you already know that only three factors are important and you want to understand everything about these factors, you may choose a design which gives more information but requires higher experiment numbers. Sections 5.4.1, 5.4.2, and 5.4.3 below discuss a few of the design types that are commonly used by synthetic chemists.

5.4.1 Screening Designs

Screening designs are used to quickly "screen" the factors to find out which ones are having a significant influence on the response and an idea of in what way. This may help you to understand the mechanism and gain new information on your process while looking at lots of factors. Factorial designs (see Section 5.3), where each factor is examined at a low and high level, are thorough and information rich but the number of required experiments quickly escalates as the number of factors increases.

5.4.1.1 Fractional Factorial Designs

For screening chemical reactions, fractional factorial designs can be used to gain balance between the number of experiments with the amount of information obtained. These designs run a carefully selected subset of the experiments in a full factorial design - i.e. a fraction of the experiments, as the name implies. In these designs, every factor is still investigated at a high level and a low level, but unlike in a full factorial design you do not run every single combination.

However, in reducing the number of experimental runs, you reduce the power of the design and the result of this is the inability to detect all main effects and interactions which may be affecting the response. There are different resolutions of fractional factorial designs and they can be grouped into (i) interaction designs, those which inform you of two-factor interactions, and (ii) linear designs, those which inform you of the main effects only. If you are planning a screening design investigating 6 factors, you will have 3 main options to consider: a resolution III design that requires 8 runs, a resolution IV design that requires 16 runs, or a resolution V design that requires 32 runs.

Resolution III designs are linear designs. They only provide information about main effects. You have not performed enough experiments to assess the two-factor interactions.

Resolution V designs are interaction designs. They provide information about which main effects are important as well as all combinations of two-factor interactions.

Resolutions IV designs are the middle ground. They will tell you if there are two-factor interactions, but will not give the detail of the exact interactions; the two-way interactions are said to be confounded (see below). If interactions are present, further experiments will be required to confirm what those interactions are. These designs are useful when you have several factors (6+) and think there is the possibility of interactions but are not able to run the resolution V or full factorial design due to the higher number of experiments required.

A **confounded** term is one which cannot be estimated independently of other terms. Confounding arises when not enough experiments have been conducted to resolve one term from another, i.e. you cannot confidently assign the effect on the response to one term over another. Table 5.3 shows the confounded terms from resolution III, IV, and V designs investigating 3 or 4 factors. For the resolution III design, the main effects are confounded with the two-factor interactions. For the resolution IV design, the main effects are free from confounding but the two-factor interactions are confounded with each other. In a resolution V design, there is no confounding in the main effects or two-factor interactions.

Table 5.3 Resolution of fraction factorial designs and confounding terms.

Resolution III		Resolution IV		Resolution V	
Term	**Confounded with**	**Term**	**Confounded with**	**Term**	**Confounded with**
A	B*C	A		A	
B	A*C	B		B	
C	A*B	C		C	
A*B	C	D		D	
A*C	B	A*B	C*D	A*B	
B*C	A	A*C	B*D	A*C	
		A*D	B*C	A*D	
				B*C	
				B*D	
				C*D	

We expect to see significant interactions in the majority of designs. Resolution III designs should only be used in certain circumstances: to screen for important factors when there are lots of factors or if the experiments are expensive in terms of cost, time, or resources, as they require fewer experiments. A follow up design is recommended and you may still miss potential interactions.

We recommend fractional factorial designs are used for building-up process understanding at the beginning of a new project, where approximations are adequate rather than having to know the precise conditions for the perfect reaction.

5.4.1.2 Definitive Screening Designs

In Section 5.3.2, we introduced squared or quadratic terms that are required to explain a non-linear response (curvature). These terms cannot generally be estimated from factorial or fractional factorial designs as three levels (e.g. low, medium, and high) are required to assess curvature. When a fractional factorial design has center points, the presence of curvature can be detected, but it is not possible to say which factor or factors cause this and further experimentation (to a response surface design, see Section 5.4.3) is required to fully resolve these terms. Definitive screening designs (DSDs) are three-level designs that were created to estimate quadratic terms in similar experiment numbers to fractional factorial designs [5]. The number of runs required is approximate to $2n + 1$ for n factors.

This design type can be used to identify which factors are important and an estimation of the quadratic effect of each factor. Two-factor interactions are correlated with other two-factor interactions and the quadratic terms. From a definitive screening design of at least six factors, it is possible to estimate all possible full quadratic models involving three or fewer factors. In our experience, two-way interactions are more common than square terms and in doing this design you risk losing information about these.

These designs are useful for screening, when you are confident there will be quadratic behavior but still have 6+ factors to screen. On the other hand, they can make analysis more difficult, it is easier to assign significance to some terms while missing some terms due to correlation.

Correlation means it is not possible to completely separate the variables from each other. Variables are said to be correlated if an observed change in the level of one variable is accompanied by a change in the level of another variable.

Initially, these designs could only be used with continuous factors, but they have now been developed to include two-level discrete factors [6].

5.4.2 Designs for Optimizing Reactions

Optimization is the art of finding the optimal settings for significant factors taking into account potentially conflicting demands of different responses. Following an initial screening design that has revealed the important factors and whether interactions are present, further designs can be used to improve the process.

A follow-on screening design is an additional design (generally factorial) created with the best result from the previous design either as the center or at the corner of the design space, which excludes the insignificant factors from the first design. This design is intended to identify if the maximum can be achieved within the design space or if there is further room for improvement and therefore highlight a new direction of travel. The process is iterative until a maximum is obtained, but this should happen within less than three designs in our experience, provided the ranges or levels investigated are sensible.

If the initial screening design is very close to the target, or the interpretation of the initial design was simple (i.e. without curvature and interactions), the method of steepest ascent can be used to follow the direction of travel indicated in the first design (blue arrow). This involves performing single experiments along the line of increase, until a maximum is reached. Once the new high point is reached, a further smaller design can be carried out to confirm process understanding (Figure 5.4).

5.4.3 Response Surface Designs

The designs introduced so far have focused on looking at main effects and two factor interactions with the definitive screening design exploring quadratic terms. Response surface designs are necessary to explore the more complex relationships

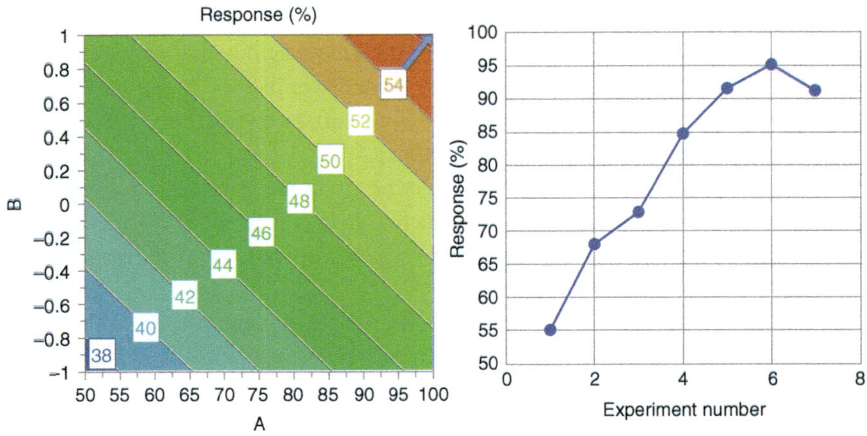

Figure 5.4 Left, results from initial design. Right, sequential experiments to reach optimum.

between several factors and one or more responses for more detailed process understanding. These designs study the factors at more than two levels, enabling the model to include quadratic terms. There are several design options to choose from when looking to understand curvature in a response; however, before you embark on one of these designs please be sure you need to run one. Since three levels are investigated for each setting, the number of experiments required is much larger and therefore they will take longer to run and analyze. We commonly see scientists run response surface designs "to be thorough" and upon analysis of the results and fitting of the model realize they do not have a quadratic term and in some instances do not even have interactions. Our preferred way of working is to run a (resolution IV) factorial design, then if the model for the response includes a quadratic term, additional experiments are carried out to estimate the quadratic terms.

A central composite design is constructed from a resolution V fractional factorial design with each factor at two levels (the orange dots in Figure 5.5), a set of center points and a further set of axial points (also known as star points, shown in red in Figure 5.5), which have all factors set at their center point (the third level) except one factor that is then at a fourth (low) and fifth (high) level. The axial points can be outside of the factorial range making a circumscribed central composite design (CCC). These designs provide high-quality predictions over the entire design space.

A face-centered composite design (CCF) has the axial points on the faces of the cube (for a three-factor design). These designs can give much more information than a factorial design due to the addition of the third level but give poorer precision on the pure quadratic coefficients when compared with a CCC. However, they are a good option if you have limits on the design space.

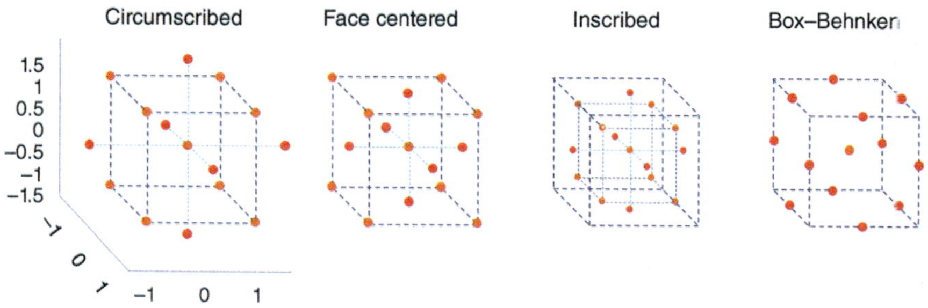

Figure 5.5 Different response surface designs.

A central composite inscribed (CCI) design also uses points within the factor ranges originally specified but does not provide the same high-quality predictions over the entire space as all the experiment runs, including the axial points, fit within the cube of the pre-stated factor levels. Like the CCC, this design has five levels however the explored design space is smaller, imagine a sphere fitting inside a cube, and therefore the accuracy of the quadratic term predictions is over a smaller range.

Finally, a Box–Behnken design can be used when there are certain combinations of factor settings that should be avoided. The "corners" of the design, i.e. the factorial points, are not run. This design required three levels and as such they contain regions of poorer prediction quality. They do, however, require fewer treatment combinations than a central composite design in cases involving 3 or 4 factors.

Choosing the correct design is always a balance between the amount of information you need and the number of experiments you can afford. Response surface designs require higher experiment numbers, and therefore, if you do not know if this design is required, we suggest you start with a fractional factorial design and if, upon analysis, it becomes clear that a quadratic term is necessary for the model, either the additional experiments are carried out, or a new response surface design looking at the significant factors in more refined ranges is carried out.

Note: factorial and fractional factorial designs should look at factors at only two levels. If you have a discrete factor such as solvent or acid that you wish to investigate at more than two levels (i.e. you want to run a design exploring three solvents), you will either need to perform a different design type (i.e. a d-optimal design, see Ref. [3], p. 183 for details), or use principal component analysis (PCA) to map and compare your reagents [7, 8].

5.5 The DoE Process

Although we have mentioned at length the disadvantages of experimenting by changing one variable at a time, one practical benefit of this method is you do not waste a great deal of time if an experiment is carried out incorrectly, or you have not thought of something. When conducting a design, many experiments (normally somewhere between 11 and 35) will be carried out consecutively in a short space of time before the results are collated and statistically analyzed. This means that significant effort is put into experimentation without knowing if the experiments are going to give you good results that can answer your questions. The experimental design process requires the scientist to think of everything that needs to be investigated before performing the design; therefore, it is suggested that you follow an eight-step protocol to guide you through the process. In our experience, a properly planned design is more likely to meet your expectations and deliver useful information. The validity of a design is directly affected by its construction and execution, so attention to detail is very important.

To see how DoE works, we are going to use a Wittig reaction (Scheme 5.2) as a case study and we will follow this through the whole DoE process. We will use whatever knowledge we have about the specific reaction to look at each step of the DoE process. This example is modified from a real industrial project, therefore some changes have been made but the general process remains the same.

5.5.1 Aim and Objective

Define the problem. You will not get an answer to your question if the question is not properly defined. Different people have different ideas about what the problem is, so it is a good idea to bring together a multidisciplinary team to try and set the initial aim. Think about whether you fully understand the problem. What do you want to achieve?

Determine the objective. If your aim is to make a target molecule there could be different ways of breaking it down into components with different disconnections. One objective could follow one route, whereas a different pathway would be a separate objective; each would require a different design. When deciding your objective, you should consider any limitations that you could face regarding equipment, time, and money. For example, while a high temperature may be required for the reaction, your upper limit of temperature will be restricted by the

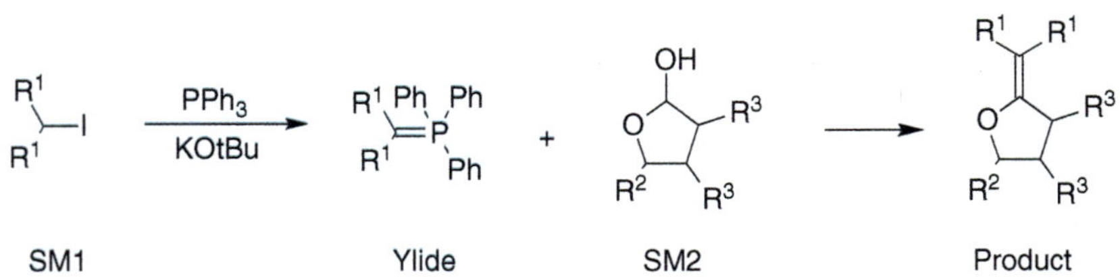

Scheme 5.2 A Wittig reaction for optimization using the DoE process.

equipment you have available. Think about where the gaps are in your knowledge - is there something you do not yet understand? This is your opportunity to tailor the design to deliver that new understanding.

The aim and objective for this Wittig reaction are the same: to maximize the yield of product while minimizing impurities, and ensure the reaction is robust for large scale manufacture.

5.5.2 Selecting Factors and Ranges

5.5.2.1 Factors

When deciding which factors to investigate in a design, it is good practice to list all the potentially significant factors (PSFs) that may affect the outcome, then choose which to control and which to vary. If all PSFs are captured, they can be referred to later on if a problem arises. Again, factor selection is best done as a team and it is likely that there will be different opinions on what factors are the most important. Even if a factor does not initially seem important, it should be noted down as that may become a consideration later. It is good practice to explain or justify why you think something will not be important. This usually helps to clarify your thinking about what you really know. Previous observations for the exact chemistry are most valuable. Who has carried out the reaction before? What does past data show? If the chemistry is new to you, exhaust the literature to understand the factors (and their ranges) that might be important. If it is novel research, as much relevant research on related chemistry should be gained to ensure you consider all the relevant factors. Consider the mechanism of the reaction and all the steps in the process. If you already have a batch process sheet (a detailed experimental procedure), it is useful to work through this step by step and think of all factors relating to each procedure. Things like the order of addition can have a big impact on the mechanism, as well as the equipment, reagent, additive, and even the person carrying out the reaction. It is more difficult to control scale-dependent factors on a small or lab scale when carrying out large numbers of experiments, and so practically it is easier to start by investigating the "fundamental" scale independent factors, making sure to minimize variation from the scale-dependent factors by having good mixing and rapid heat transfer. The scale-dependent factors such as heat transfer, mixing, and mass transfer can be investigated on a larger scale after the important factors are understood.

During factor selection, consider all steps in the process but keep the reaction and work-up separate. Once you have optimized the reaction, then you can focus on the work-up, and hopefully be working up a cleaner process with fewer impurities to remove as the reaction was optimized. The reaction and work-up need to be brought together at the end, but considering them together may not be ideal, especially when the factor levels end up at very different settings from where they started. This is also important when you are investigating different reagents or solvents because these may require different work-up techniques.

In general, an understanding of the reaction mechanism, as well as the cause and fate of impurities, can help to choose the appropriate order of addition. For the Wittig reaction, 20 potential factors were identified, which are shown in Scheme 5.3.

Once all the PSFs are identified, you must opt to set each factor at a fixed level or investigate the factor by selecting a range of levels. Be careful of factors that are not independent. Some will be connected to others and you will be unable to set both at different levels simultaneously. For example, if a process is exothermic as a reagent is added, you can either control the addition rate or the temperature, but depending on the extent of the exotherm and the equipment used, it might not be possible to set both. A key deliverable of experimental design is a measure of reproducibility and experimental error, which gives confidence in and significance to the results. Controlling the factors that are not being investigated should minimize unexpected variation. The experimental procedure should be explicitly written and closely followed by whoever is conducting the design to minimize any uncontrolled variation and therefore error.

Experimental factors can be continuous, as is the case for temperature, stoichiometry, rate of addition, and for these, you can choose any level to be investigated. Alternatively, there may be discrete factors whose selection is constrained by

Scheme 5.3 Potentially significant factors considered for the Wittig reaction.

availability or existence, such as reagent, solvent, equipment. Discrete factors are often more problematic. A chemical may not exist at the point in between the chemicals you have chosen so how do you know the center point?

In our experience as consultants in experimental design, we commonly see scientists doing designs which look at few factors (between 2 and 4) and when we get involved, we typically identify 15 PSFs for the process. That is not to say we want all 15 to be investigated, but by choosing not to investigate a factor, you are choosing to lose the information on the effect, and potentially restrict the scope that changing this factor will have on the response. Generally, we believe that more factors give more opportunities to understand and improve your reaction.

We recommend an exercise: think about a process you are working on and try to list all the factors that may affect it. You may find it useful to use a fishbone diagram to list all the PSFs.

More detail of the factors considered for the Wittig reaction is given below:

- Phosphine: Triphenylphosphine has successfully been used in the process. Where triphenylphosphine oxide can be difficult to remove the process has been developed to separate the by-product during the reaction work-up. This factor was fixed.
- Phosphine, base, and iodide equivalents could affect the formation of ylide and product; these factors should be considered.
- Temperature: the ylide is light and heat sensitive so the reaction should be carried out at low temperature. At −25 °C, ylide formation occurs at a slow rate with lower assay yield in subsequent steps. It is postulated that the formation of the ylide can be conducted at a warmer temperature to drive ylide formation, then for subsequent steps a lower temperature can be used. Temperature will be investigated.
- Volume of solvent: the solubility of triphenylphosphine, base, and SM1 should be considered as this will support the reduction of solvent volumes.
- Addition rates could affect the formation of ylide and product, addition rates will be investigated.

5.5.2.2 Ranges

Selecting the levels for the factors is as much an art as a science and can take a few attempts to get right. The range should be as wide as is practical so a good amount of chemical space is covered and maximum information can be obtained, but not so wide that it would change the response by more than two orders of magnitude. It is ineffective to consider ranges that are impractical in relation to the experimental objective (e.g. using a reaction temperature that is below the activation temperature, or investigating a temperature that is higher than the boiling point of the solvent) although it is acceptable to go further than the preferred reaction domain to understand an effect.

The chosen ranges need to be meaningful to the process, so a low setting of 0.9 equivalents of a reagent that you need 1 equivalent to deliver 100% conversion will result in several experiments, which cannot yield complete conversion. Additionally, choosing a low setting of 0 effectively turns the factor into a discrete one, either it is on or off/absent or present. Instead, try to look at a very low level such as 0.01 equivalents.

The whole range is also important to balance in terms of how much information can be gained from the design. Varying quantities from 1.02 to 1.2 equivalents is a very narrow range for an initial design, and there may be little change in the response. Conversely, 1.02 to 10 equivalents is a wide range, which can result in little difference between the mid and high settings. If just over 1 equivalent is the optimum setting, between 2 and 10 equivalents you are likely to see little effect from that factor as it is always present in an excess. It is important to remember that DoE explains the effect of changing a response from the low to the high setting, and if there is no change, the factor is deemed to be unimportant in the range investigated.

To ensure the ranges set for all the factors are suitable, we recommended to initially run three test reactions with all factors set (i) at their low setting, (ii) at their high setting, and (iii) the center point. If no variation is observed between high and low, then the levels are not set far enough apart. If the variation is too large, then the ranges may be too wide. This allows you to check that the factor ranges are suitable before you carry out the rest of the design. If these ranges are deemed appropriate, these three data points may be used as part of the design.

Do not pick ranges where a change of mechanism may be brought on, e.g. by running out of reagent.

Do not include high temperature or pressure if you cannot achieve it with the equipment available to you. It may give you some information, but do not base the whole design around it if it will be difficult to implement.

For the Wittig reaction, knowledge of the general reaction mechanism as well as prior experiments were used to set the ranges for the factors as per Table 5.4.

Table 5.4 Factors and settings for the Wittig reaction.

Factor	Low setting	High setting	Units
Ylide equivalents	2.2	3	Mol equivalent
Base equivalents relative to ylide	0.9	1.1	Mol equivalent
Temperatures	0	20	°C
Rates of addition	1	3	Hrs

5.5.3 Selecting Responses

A well-defined and well-measured response will help in delivering data, which can then be modeled to give the anticipated process understanding. Responses do not add to the number of experiments, so apart from any additional time to gather the analytical data, they are free information.

While it may be that you want a more sustainable reaction or generally a better reaction output, this must be defined more specifically to enable you to accurately measure the impact of changing the input variables. Specific and measurable responses can help you define in which way your reaction is better and whether your aims and objectives are being met. The measured response should give accurate and consistent results, closely replicate the actual experimental outcome and be as close to the change in event as possible.

Accurate response measurement is needed so the variance seen in the response is due to the changing factors rather than error. To understand how factors affect individual impurities, it is important to look at reaction data before work-up so that a true measure of impurity levels is obtained as well which factors and ranges are best able to control the impurities. Excessive measurement error gives an unreliable response.

Experimental error can come from all steps in the process. This includes variability in starting materials, in measurement, in sampling, and in setting the experimental factors. The error that comes from the response should be minimized so that the overall error is predominantly experimental. The response should vary more than the "noise" of the measurement area as a result of the changes.

There are some types of responses that should be avoided and these include those that are subjective, e.g. taste or color (unless using a standardized chart) and on/off responses. Using the coffee example from Section 5.1.1, we could potentially say that the response is either white coffee or black coffee, but if we did this all the information of the in-between would be lost. To use a light switch to further illustrate the point, the switch is set to on and the light comes on or the switch is set to off and the light turns off. However, the switch could be set to on but there is no light. With an on/off response, there is no information about why this may be the case. There are several reasons why the light may be off; the bulb has blown, the switch is faulty, the fitting is faulty, there is a loose wire, but we cannot find out the cause if we simply look at light on or light off. Additionally, it is difficult to get a strong model with an on/off response and any experiment that delivers an unexpected response or an incorrect response could result in no model or extreme outliers (experiments which the model is unable to explain).

For the Wittig reaction, the responses were solution yield (conversion), impurity, and isolated yield (Table 5.5). Solution yield and the impurity were measured from samples taken immediately from the reaction mixture, whereas the isolated yield was measured after an additional reaction step and work-up.

Table 5.5 Responses, abbreviations, and units for the Wittig design.

Response	Abbreviation	Units
Solution yield (conversion)	Conv	%
Isolated yield	Yield	%
Impurity	Imp	%

5.5.4 Select a Design to Answer the Objective

There are many decisions that can influence the design you carry out. It is dependent on your aim and what you want to achieve, the point you are at in your project, the number of factors you wish to investigate, the type of factors, and the number of experiments you are able to perform.

Whenever possible you should use factorial, fractional factorial, or other "classical" designs. These are the most straightforward designs to construct and analyze. However, there are times when these designs are not suitable, in which case so-called "optimal designs" can be used. These are a class of experimental design that are optimal with respect to a specific criterion. When classical designs are impossible to apply, d-optimal designs are the preferred choice.

Occasions where classical designs are not possible include: investigating discrete factors at more than two levels, investigating mixture factors, having a constrained design space (i.e. certain combinations of factor settings are not possible).

Table 5.6 details the numbers of experiments required for specific design categories between 4 and 16 factors. The number of design runs is given first with the number of minimum center points after (+3). In some circumstances, the number

Table 5.6 How experiment numbers change with factors and design types.

Design	4 factors	6 factors	10 factors	12 factors	16 factors
Fractional factorial resolution III	–	8 + 3	16 + 3	16 + 3	–
Fractional factorial resolution IV	8 + 3	16 + 3	32 + 3	32 + 3	32 + 3
Fractional factorial resolution V	16 + 3	32 + 3	128 + 3	256 + 3	256 + 3
Full factorial ($2n$)	16 + 3	64 + 3	1 024 + 3	4 096 + 3	65 536 + 3
Full factorial ($3n$)	81 + 3	729 + 3	59 049 + 3	531 441 + 3	43 mil
Response surface design	24 + 3	44 + 3	148 + 3	280 + 3	288 + 3

of experiments will not increase even when adding additional factors. For example, 10 factors can be investigated in 35 experiments with a resolution IV fractional factorial design, but you can investigate 16 factors in the same number of experiments. It can be more economical to examine a few extra factors in a screening design than to carry out initial screening, then decide later than another factor could be investigated.

If in any doubt, start with a resolution IV fractional factorial design with as many factors as you expect to be important as well as any factors you are unsure of their effect for more "bang for your buck."

For the Wittig reaction, initially 6 factors were highlighted by the development team as being interesting to investigate but due to time constraints, 19 experiments was considered too large a number for the initial screen. In this case, the number of factors was reduced to four by investigating the equivalents of ylide instead of the separate equivalents of iodide and phosphine. Additionally, the charging temperature was investigated at the same level for both steps and the rate of addition was investigated by using the same level for both the base addition and the SM2 addition. This resulted in a 4-factor design of 11 experiments (8 runs and 3 repeated center points). This is a resolution IV design, so there will be confounding in the design between the two-way interactions, see confounding interactions in Table 5.7 (i.e. from this design, you can tell which main effects have an important impact on the response and if there are significant interactions, but not the exact nature of the interaction).

5.5.5 Carry Out Design and Analyze Samples

In experimental design all your results are used multiple times to build models to represent what is happening in your process. Therefore, it is important to carry out the experiments as carefully as possible. When running the experiments, take note of any observations that may be useful during analysis and try to be precise, but if the factor setting is hard to set precisely, one should record the actual level used. For example, the high level of temperature should be 75 °C, but actually the probe reads 71 °C, record this number, and use it in the DoE software during analysis.

Table 5.7 Confounded interactions in Wittig design.

Interaction	Confounded with
Ylide*base	Temp*rate
Ylide*temp	Base*rate
Ylide*rate	Base*temp

Some points to consider to minimize error:

- Have a clear protocol to follow closely noting any observations and deviations.
- Use a single batch of materials throughout.
- Use identical equipment (as much as possible) to minimize unexpected variation.
- Experiments should be carried out on the same scale.
- Efficient mixing can be important. Higher mixing speeds are generally recommended.
- Apart from the three experiments to check the ranges are correct (see Section 5.5.2.2), the remaining experiments should be run as close to the randomized run order as possible. Ideally, center point experiments are done at the beginning, middle, and end. Do not group experiments for one temperature or one time for easier operation.

Other reaction parameters should be retained as close to the same conditions as possible to minimize experimental error.

The Wittig design experiments were carried out and samples were analyzed by HPLC. The reaction mixture was homogeneous and replication for sampling was acceptable, giving analytical errors of <5% for solution yield, impurity, and yield.

5.5.6 Check Results

To quote Carlson, "The results of a well-planned experiment are often evident from simple graphical analyses, but the world's best statistical analysis cannot rescue a poorly planned experimental program" [3], and Box and Hunter "It is not unusual for a well-designed experiment to analyse itself" [4]. Once you have carried out the experiments and analyzed the samples it is a good idea to give them a visual check in excel as a replicate plot (Figure 5.6). Here you will be able to spot potentially anomalous results. All results from a properly executed and measured experimental design must be included in the dataset even if they appear to be outliers on the residuals plot. The typical reason one of the results is a significant outlier is that either one factor of the experiment was not carried out at the right setting or a significant factor has not been considered. Including all results in the analysis will allow identification of such issues and allow for correction and a more robust design. Only experiments that have a justified reason should be excluded from any dataset. Discounting genuine results risks generating a model that gives a false sense of security in accurate predictions.

Note: how to create a replicate plot in excel.

For each response, add the data and corresponding experiment number into excel. Separate the experiments into the design runs and the repeats. Change the experiment numbers for the repeats to make them all the same number. Insert a scatter

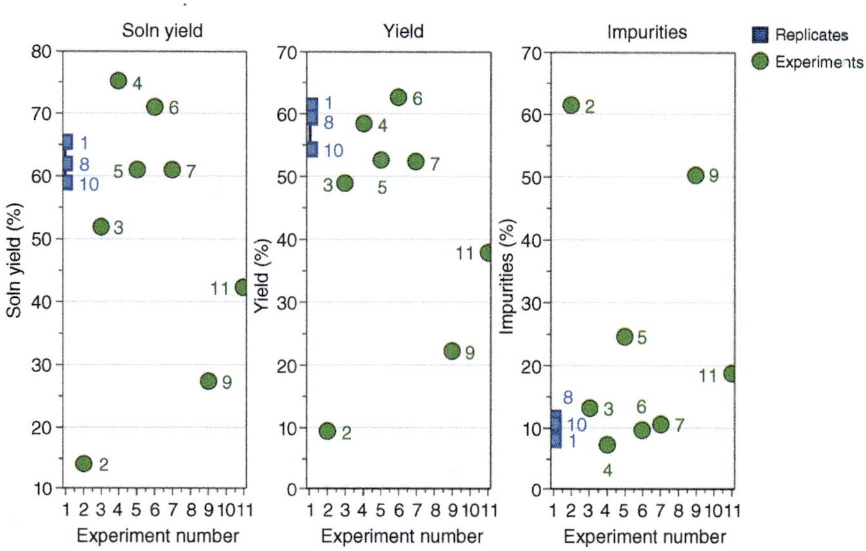

Figure 5.6 Replicate plots.

plot. In "select data," add a series for the design runs with the experiment number on the x-axis and the response results on the y-axis. Add a new series for the repeat experiments and add the data with the same experiment number on the x-axis and the response results on the y-axis. To add a line between the repeat data points, select a data point and select to change series chart type, change to "scatter with straight lines and markers." You can now see the spread between the repeat experiments and how they compare to the rest of the data.

For the Wittig design, the results were checked visually and by looking at replicate plots in Figure 5.6. These show the variation in the results for all experiments for a quick raw data inspection. Repeated experiments appear on the same stick (in blue). The ideal outcome is that the variability of repeated experiments is much less than the overall variability. The replicates for impurities is good with a small range in results. For solution yield, the spread in data is 60%, and the spread in the repeats is 6.7%. The spread in the results for isolated yield is 53%, and the spread in the repeats is around 7%.

5.5.7 Model Data

Statistical analysis of the data can be aided by specific software. The following case study used Modde software to analyze the results of the design and present the models (other available software are described in Section 5.7). Three responses were modeled (i.e. a model equation relating the response to the factors was generated for three individual responses). The analysis and models are presented fully, with the associated metrics and plots for isolated yield, and a shorter summary is given for solution yield and impurities.

How you model the data will depend on the statistical software you use, so it is difficult to provide a complete guide here on how to generate the model equation. The software you use will have a user guide that can be followed for assistance. This section presents the statistical details to check you have a good model whose predictions and information you can be confident in. For a guide and tips on how to get a model look at the user guide of the software you have.

5.5.7.1 General Steps for Developing a Model

1) Check the replicates plot for any obvious problems or conduct a cursory visual check of data for problem experiments, outliers, etc. Check the reproducibility of the center points. What is the experimental error? Order the results from low to high for each response in turn, this can highlight patterns of factor settings that are grouped together and can help you see if one result is vastly different.
2) Look at the confounding variables (if resolution IV or III design) and consider the terms that should/could be in the model.
3) Assess the model and refine by removal of insignificant interactions and quadratic terms first, then by removing insignificant factors.
4) We recommend analyzing the model two or three times (in different ways) to make sure you get to the same model. With confoundings, if there is a confounded interaction made up of two insignificant factors, then this is quite unlikely, it is more likely to be one of the other confounded interactions.
5) Be careful of making assumptions (e.g. with adding squared terms, confoundings)
6) Do you have a good model? Check the fit statistics (definitions below)

Below are some definitions of the statistic checks you can use to assess the model.

ANOVA: Analysis of variance (ANOVA) partitions the total variation of the response (from the sum of squares) into a component due to the regression model and a component due to the residuals (error or unexplained variation). When there are replicated observations, the residual sum of squares is further partitioned into pure error (from the repeats) and lack of fit (model error). It assumes there is random sampling, independent measurements, normal distribution, and equal variance.

ANOVA definitions (see Statistics for Experimenters Ref. [4] for more detail on this).

Sum of squares: the sum of the squared deviations from the mean due to the effect of this term.

Degrees of freedom: the numerator degrees of freedom used to compute the *F*-statistic for the terms and subsets of the model.

Mean squares: the variance associated with the term (sum of squares/degrees of freedom).

F-value: the ratio between explained variation and unexplained variation, the higher value the better.

***p*-Value**: the probability of obtaining the model by chance, or the probability that there is or is not a relationship between the factors and the response based on the null hypothesis.

Null hypothesis: an assumption there is no relationship between the factors and the response.

In the ANOVA table the regression should have a p-value of <0.05, as should the significant coefficients (factors and interactions). Any term with a p-value of >0.05 is not significant, based on an alpha value of 5%. The lack-of-fit should have a p-value >0.05, i.e. the lack-of-fit is not significant.

Observed vs. predicted: this plot should show points on or close to the line. Any outliers are not being predicted well by the model.

Fit statistics: R^2 is a measure of how well the model explains the data. Greater than 75% is expected for a good model. Q^2 (or predicted R^2) explains how well the model predicts the data. Greater than 50% is expected for a predictive model. Adjusted R^2 is a measure of the amount of variation about the mean explained by the model. It's the R^2 adjusted for the number of parameters in the model to make sure the model isn't overfitted.

Residuals normal probability plot: this indicates whether the residuals, anything which is not explained by the model, follow a normal distribution and for a good model should follow a straight line with some scatter. A definite pattern in this plot, such as an s-shaped curve, indicates a transformation (see Ref. [4], chapter 8) could help with data analysis. Be aware of outliers beyond the control limits of four standard deviations – these should be investigated. Outlying experiments must not be excluded without good justification as they are often an indication of something not under control or something not well explained. Outlying experiments may come within the limits when the model is improved.

5.5.7.2 Wittig Reaction

Following factor selection, creating the design, assigning the responses, running the experiments and collecting the data, the results were logged in Modde, a design of experiments software. Modde was used for statistical analysis to develop models to relate the investigated factors to the responses.

The model for isolated yield (Figure 5.7) was developed by starting with a saturated model then removing insignificant interactions and factors, as well as checking the need for a quadratic term to explain a nonlinear response to changes in the factor setting. Upon analysis of the model, the terms which influence the isolated yield are ylide amount, temperature, and a squared term (as seen in the coefficients plot, Figure 5.7c). This software used for this case study (Modde) enables you to put one factor in as a squared term to say whether curvature is significant or not, it does not mean that the input factor is causing the non-linear response, additional experiments are required to determine this. The summary of fit plot (Figure 5.7b) suggests a good model with an R^2 (green bar) of 86% and a Q^2 (dark blue bar) of 64%. The residuals normal probability plot (Figure 5.7d) does not show any outliers and the point follow a straight line with some scatter.

The ANOVA table (Table 5.8) can also be viewed to check the model. It has a regression p-value of 0.002, which means it is statistically significant and a not significant lack-of-fit (0.152).

Two contour plots were generated (Figure 5.8) showing the predictions from the model which includes temperature along the x-axis, ylide equivalents on the y-axis, and the response as a color gradient. The plot on the left contains ylide as the quadratic term with the plot on the right including temperature as the quadratic term. At this point, we do not know which of the factors is causing the curvature; however, it is most likely to be ylide equivalents or temperature as they are significant as main effects. The general trend is similar: high temperature and high ylide equivalents lead to lower isolated yield; however, further experiments are required to refine the model and determine which conditions are optimum for increasing isolated yield.

The observed vs. predicted plot (Figure 5.9) shows some predictions that are quite accurate and some of which are further from the observed values.

The models for solution yield and impurities were developed in the same way as isolated yield, by starting with a saturated model then removing insignificant interactions and factors, as well as checking the need for a quadratic term to explain a nonlinear response to changes in the factor setting. Table 5.9 contains the significant coefficients (i.e. which factors and potential interactions are in the model equation) as well as the important metrics for these responses.

The contour plot for solution yield (Figure 5.10) shows the predictions from the model. It is apparent that to increase solution yield a lower temperature and lower equivalents of ylide are required.

The model for impurities contains an unconfirmed quadratic term and a confounded two-factor interaction. The interaction could either be between ylide*temperature or base*rate. Considering ylide and temperature are present in the model as main effects, whereas neither base or rate are significant, it is more likely that ylide*temp is the important interaction. The software used for this case

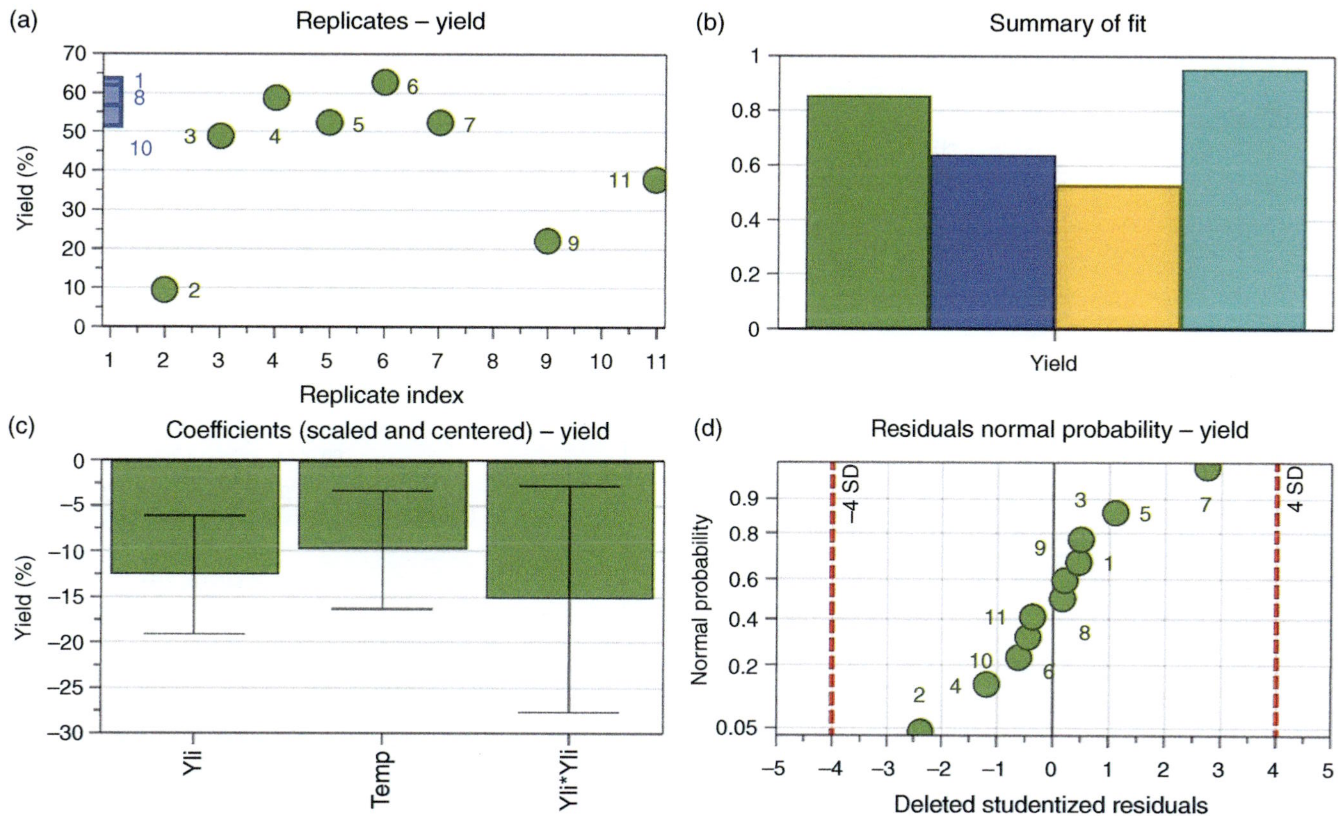

Figure 5.7 Overview plot for the model for isolated yield as generated by Modde software. This includes a replicates plot (a: top left), summary of fit plot (b: top right), coefficients plot (c: bottom left), and residuals normal probability plot (d: bottom right).

Table 5.8 ANOVA table, isolated yield.

Yield	DF	SS	MS (variance)	F	p	SD
Total	11	27 526.4	2502.4			
Constant	1	24 562.9	24 562.9			
Total corrected	10	2963.53	296.353			17.2149
Regression	3	2541.1	847.034	14.0361	**0.002**	29.1038
Residual	7	422.426	60.3466			7.76831
Lack of Fit (Model error)	5	395.406	79.0812	5.85352	**0.152**	8.89276
Pure error (Replicate error)	2	27.0201	13.51			3.6756
	$N = 11$	$Q2 = 0.638$		Cond. no. = 3.601		
	DF = 7	$R2 = 0.857$		RSD = 7.768		
		$R2$ adj. = 0.796				

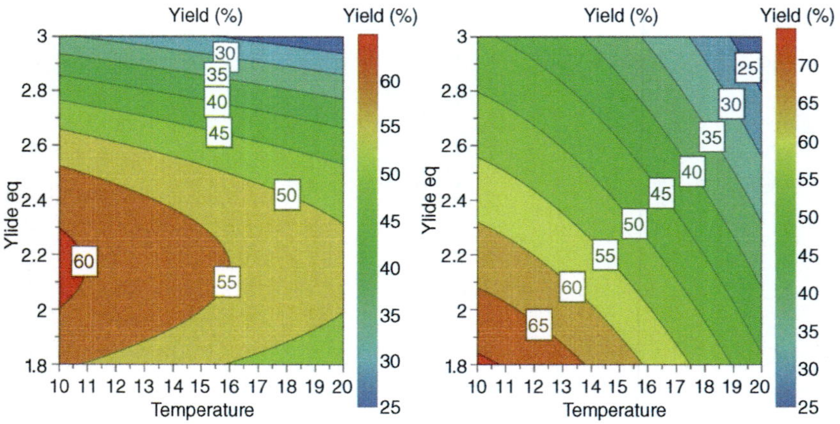

Figure 5.8 Contour plot for the model for yield including ylide^2 (left) or temp^2 (right).

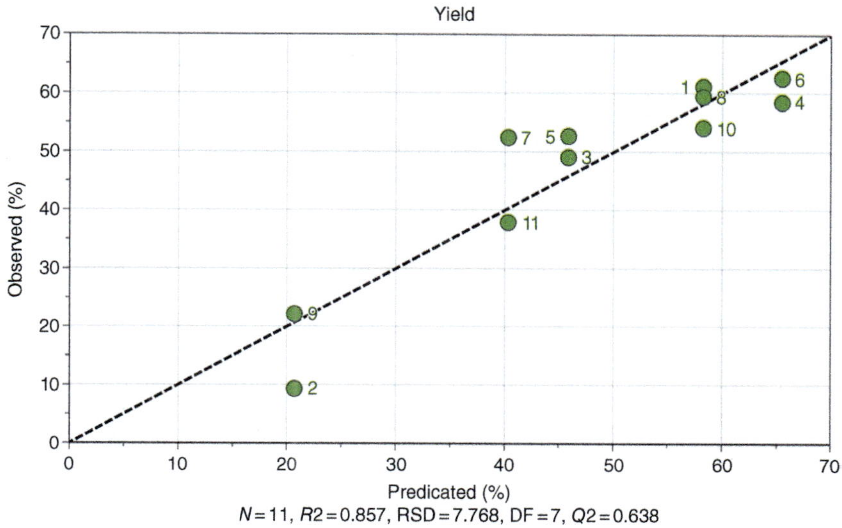

Figure 5.9 Observed vs. predicted plot for the model for isolated yield.

Table 5.9 Model statistics summary for solution yield and impurities.

Response	Coefficients	R^2	Q^2	Regression/ model *p*-value	Lack-of-fit *p*-value
Solution yield	Ylide, temperature	79%	60%	0.002	0.09
Impurities	Ylide, temperature, quadratic term	95%	80%	0.001	0.079
	Ylide*temperature or base*rate interaction				

study (Modde) enables you to put one factor in as a squared term to say whether curvature is significant or not, it does not mean that the input factor is causing the nonlinear response, additional experiments are required to determine this.

The interaction plot for the suspected interaction between ylide equivalents and temperature (Figure 5.11) shows that at low temperature, there is little difference in the level of impurity between low equivalents of ylide and high. However, at high temperature, high equivalents of ylide generate a much higher impurity level.

Two contour plots were generated using the model for impurity (Figure 5.12): one with the model containing ylide as the squared term and one with

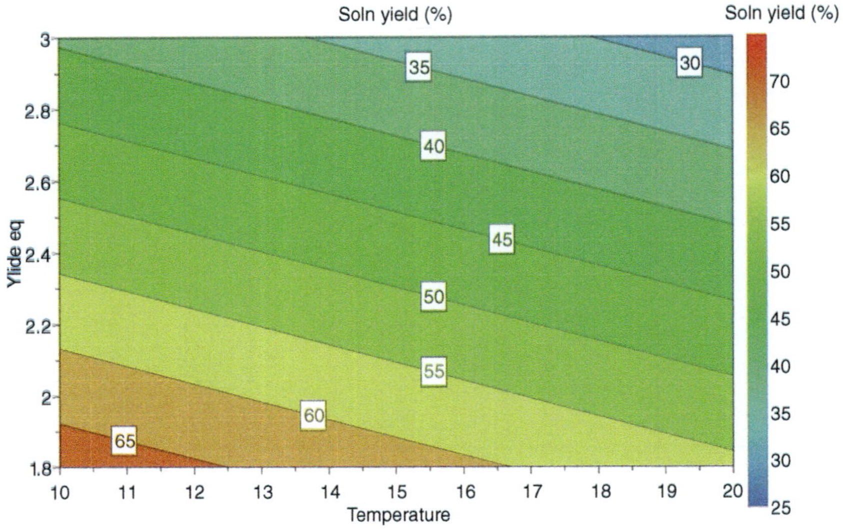

Figure 5.10 Contour plot for the model for solution yield.

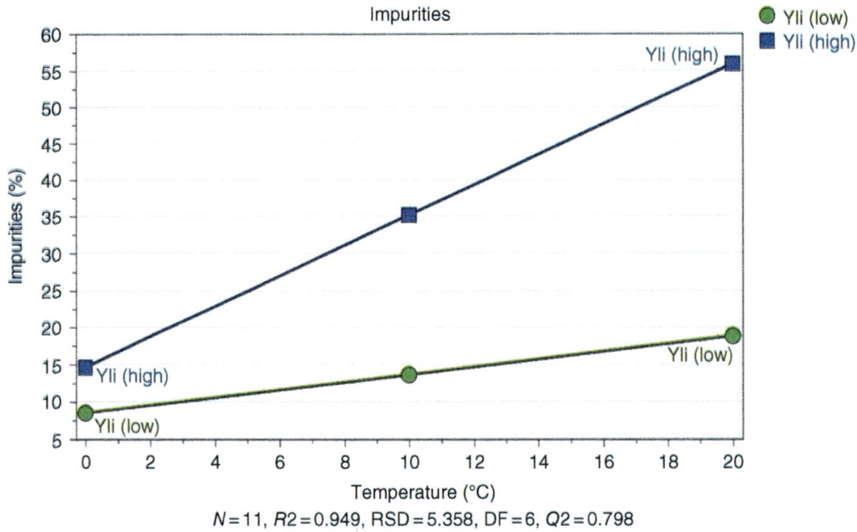

Figure 5.11 Interaction plot for the ylide*temperature interaction in the model for impurities.

temperature as the squared term. The general trend is similar: high temperature and high ylide equivalents leads to higher impurity levels, however further experiments are required to refine the model and determine which conditions are optimum for reducing the impurity.

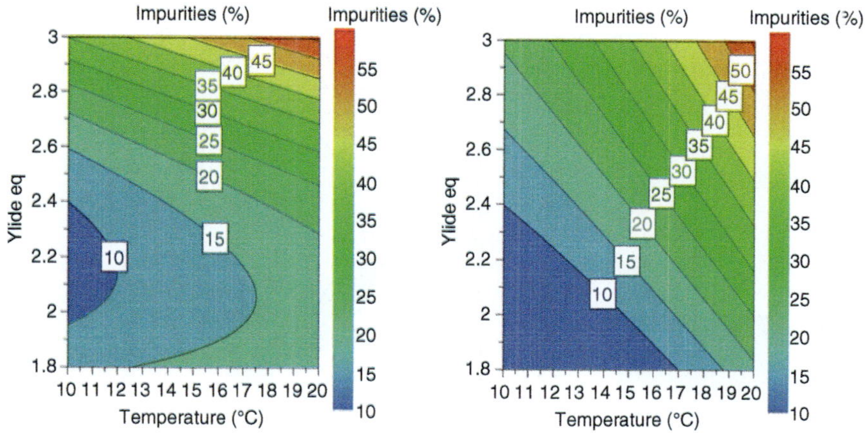

Figure 5.12 Contour plots for the model for impurities including ylide^2 (left) or temp^2 (right).

5.5.7.3 Complementing the Design

The next step is to understand what is causing the curvature in the response. To do this, extra experiments are needed. Because there are only two significant terms in the models for all three responses: ylide and temperature; the additional star point experiments (Section 5.4.3) should be carried out to get more data about the effect of ylide and temperature (Table 5.10). An additional center point experiment is also run to give a point of comparison to the first set of experiments.

The results from the additional five experiments were added to the initial results, and the models were generated again following the procedures mentioned above.

The data for isolated yield is expected to be the most reliable for measuring the output of this reaction as the analytical method is the most robust. For this reason, the model for isolated yield was examined further. The additional runs confirmed that temperature is the quadratic term (Figure 5.13). The model is good, with an

Table 5.10 Additional experiments to complement the design.

Exp no.	Ylide eq	Ratio of base to ylide	Temperature	Rate of addition
12	2.4	1	10	2
13	1.5	1	10	2
14	3.3	1	10	2
15	2.4	1	−5	2
16	2.4	1	25	2

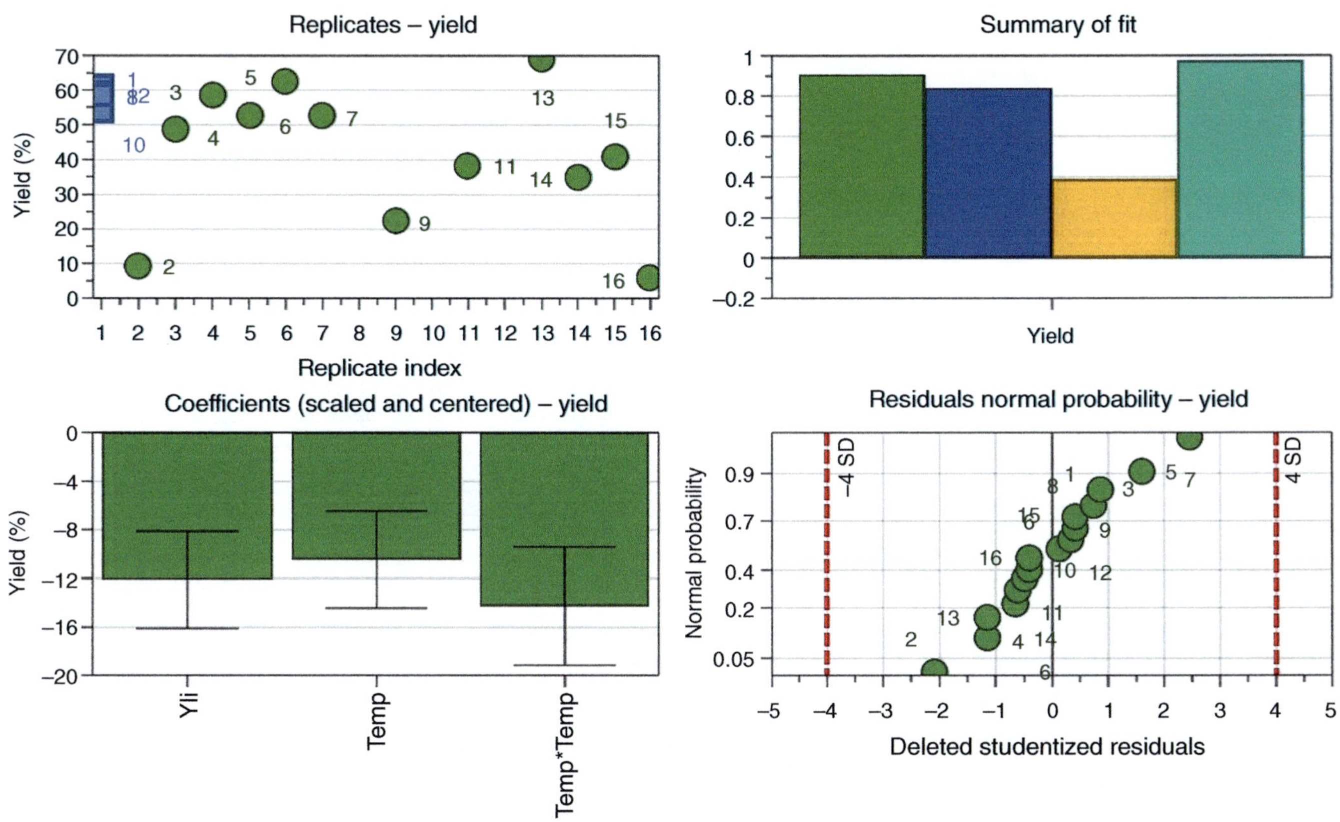

Figure 5.13 Overview plot with all data for the model for isolated yield.

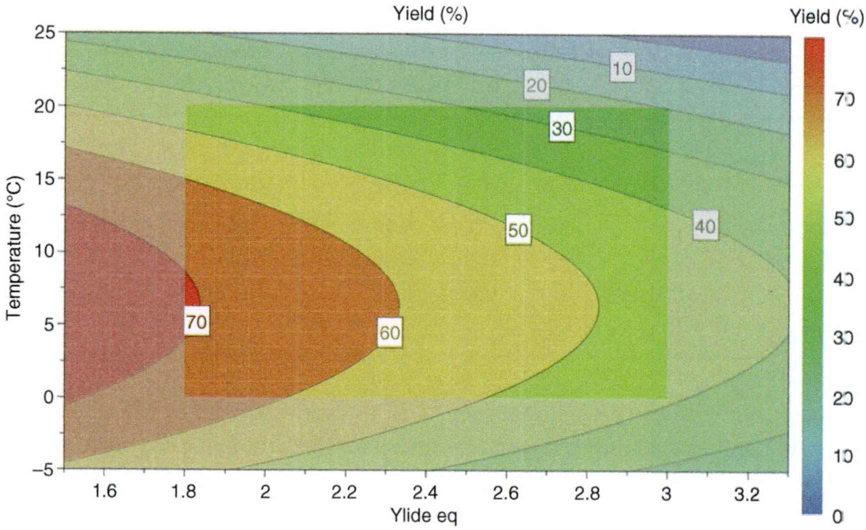

Figure 5.14 Extended contour plot for the model for isolated yield.

R^2 of 91% and a Q^2 of 84%. The model has a regression p-value of 0.000, which means it is statistically significant, and a not significant lack-of-fit (0.087).

The contour plot (Figure 5.14) shows that the highest yield can be achieved at around 7 °C with 1.5 equivalents of ylide. The grayed areas of the contour plot are those that are outside of the original design space. A more accurate prediction was obtained using the "optimiser" feature of the Modde software. A yield of 77% was predicted using 1.5 equivalents of ylide, 0.96 equivalents of base with respect to ylide, at 6 °C.

5.5.8 Validate Predictions

We all want a robust process that generates 100% product with no waste that costs pennies, ideally using sustainable and readily available starting materials, but that is not realistic. A more reasonable aim will be to maximize product and minimize impurity and waste, but as you will see in the Wittig example, the models to achieve these aims can be conflicting and you may have to make compromises with competing demands. The generation of a summary table (Table 5.11) will help to view the required conditions for each response and the software can be used to find optimized conditions. To produce this table, use the models calculated for each response. For each factor (A–H) input the setting required to get the best response (i.e. if % yield is a response and you wish to maximize the yield, input the factor settings predicted to give the highest yield), then populate the far

Table 5.11 Example response summary table.

Response	A	B	C	D	E	F	G	H	Prediction
1	Mid	—	—	—	High	—	—	—	−4 (0)
2	Low	High	Low	Any	Low	X	Low	Low	8
3	—	—	—	High	High	Y	—	—	6.5
4	—	—	—	High	Low	X	High	—	5.2
5	—	High	—	Low	Low	—	—	—	85%
Summary	?	High	Low	?	?	?	?	Low	

right column with the predicted response. In the summary, there will be factors where the required setting is clear as each response has the same setting (e.g. factor B always requires a high setting), but other factor settings will be dependent on which response is the most important as one response requires the factor to be low and another requires a high setting. These are represented with a question mark in the table.

Not all responses will be equally important; therefore, more weight should be given to the more important responses when finding the operating conditions. Most DoE software (Section 5.7) have tools for optimizing responses which assist with this. A validation experiment under the conditions found to meet the targets should be carried out and alongside this is it good practice to run a repeat of the center point. This will give you a point of comparison between the two sets of experiments.

The summary table for the Wittig design (Table 5.12) is presented using information from the refined models, including the additional experiments. To maximize isolated yield requires 1.5 equivalents of ylide, 0.96 equivalents of base, a temperature of 6 °C and as rate was not a significant factor in the model, any rate can be used. The model predicts under these conditions an isolated yield of 77% is achievable. For this process, the most important response is to maximize the yield, however, the secondary goal is to keep the impurity below 10%. The optimizer tool in Modde has been used to target an isolated yield of 77% and a impurity level

Table 5.12 Response summary table for Wittig design.

Response	Ylide (eq)	Base (eq)	Temp (°C)	Stir rate	Prediction
Isolated yield	1.5	0.96	6	Any	77%
Impurity	2	1	−5	Any	6.3%
Summary	1.5	1	6	Any	Yield 77% Imps 10.5%

<10%. The predicted conditions to meet these criteria are 1.5 equivalents of ylide, 1 equivalent of base (with respect to ylide) at 6 °C.

Validation experiments were carried out to check the predictions of the model on a 1 g scale and a 50 g scale. The predictions were slightly higher than the confirmation runs, especially on the larger scale; however, the design uncovered the best conditions to maximize the yield and enabled a reduction in reagents and processing time.

5.6 Combining DoE with Other Screening and Optimization Techniques

There are many things to consider when thinking about using DoE. How complicated is your problem? How much do you already know? What are you hoping to achieve? Sometimes design is not the answer, it may not give you more information than you already have and could cost more in terms of time and experimentation. Sometimes a simple optimization will suffice and sometimes the problem is so complex that a more advanced approach is required.

DoE can be even more powerful in combination with complementary techniques such as Principal Component Analysis (PCA) [7, 8]. Several of the factors that can affect a reaction are discrete; solvent, reagents, catalysts. But discrete factors are harder to screen than continuous, and to assess more than two in a design requires higher experiment numbers. PCA relies on a database of chemicals and their properties to create "principal properties" as way of finding similarities and differences in the data. Patterns can be identified in multivariate datasets and "maps" of chemical space are generated. This technique has been used for screening solvents [9] and ligands [10–12] in combination with experimental design [13]. The principal properties of the chemical are treated as a factor to be investigated with a low and high level. With one principal component (PC), three solvents can be screened (−1, 0 and +1), with two PCs, five solvents can be screened (the four corners and center of a square) and with three PCs, nine solvents can be screened (the eight corners and center of a cube). There are examples of using PCA with DoE to screen solvents, ligands, selecting columns for HPLC [14], selecting fillers for pharmaceutical formulations [15] and selecting surfactants [16].

The 2016 OBC paper by Sheppard and Murray provides a list of 136 solvents and their Principal Components in the SI [9]. Additionally, a table is included in the main paper which makes solvent selection for screening with DoE available to the synthetic chemist.

More recently, enhanced automation in the chemical industry has enabled the coupling of DoE with digital technologies to develop "self-optimizing reactions." There are several publications detailing the specifics, but the general process uses

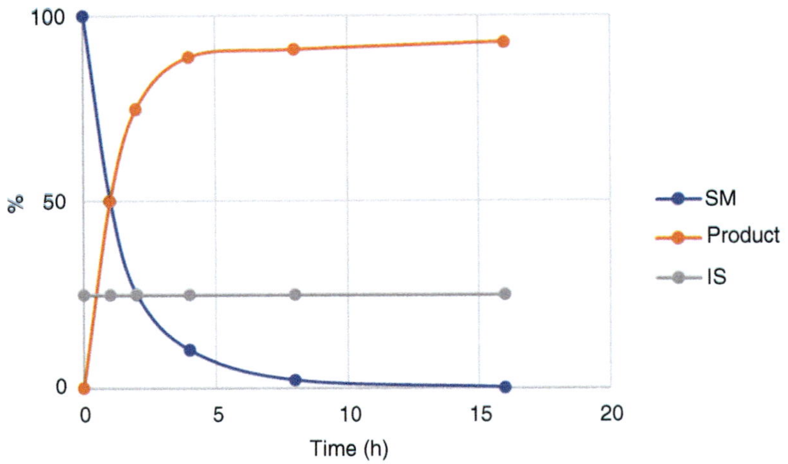

Figure 5.15 An example reaction profile showing consumption of starting material (sm), generation of product and the presence of an internal standard.

a reactor with an automated process control and online process analytical technology which is linked to a feedback algorithm. The system controls the input of reactants and process conditions, analyses a response, uses a simplex-based algorithm to refine the settings used and optimizes automatically [17, 18].

An additional simple technique which can deliver increased process understanding is the use of reaction profiling. Knowing when and how impurities are formed is important to help understand the mechanism of the reaction and combining this knowledge with DoE analysis can lead to better processes. Combining the data for starting material, product, impurities and intermediate in a plot vs. time (the reaction profile, Figure 5.15) can be helpful to visualize the reaction data. From a reaction profile you can see (i) if the product is stable, (ii) when impurities are generated and if they are related to starting materials, reaction initiation or product, and (iii) if there is missing information such as the presence of intermediates. Regular sampling and analysis of the reaction mixture will enable you to generate these profiles.

5.7 Software

As quoted in the last section "the results of a well-planned experiment are often evident from simple graphical analyses, but the world's best statistical analysis cannot rescue a poorly planned experimental program." There are several software options for carrying out experiment design and in our opinion, some are more user friendly and intuitive than others.

It is worth noting that some come with wizards, both for the set up and analysis of designs. While these appear useful, it can often lead into running the wrong designs, or allowing someone to carry out a design without the required forethought. The more carefully a design is planned and carried out, in theory, the more straightforward the analysis will be. Below is a summary of several available software. We will not go into further detail as most providers run software training or have user guides online.

R is a free software for statistical computing and graphics. It requires computer language (S) and advanced users can write C code within R.

Design Expert is DoE software which enables you to create designs for screening, optimization, response surface and formulation. It has wizards for set up and automated model-reduction tools as well as diagnostic and prediction graphs.

Modde is DoE software which enables you to create designs for screening, optimization, response surface, and formulation (depending on the level, there are two options: Go and Pro). It has wizards for set up and automated model-reduction tools as well as diagnostic and prediction graphs. Modde's interface presents the models in a very visual way.

Minitab is DoE software which enables you to create designs for screening, optimization, response surface, and formulation. It also has many additional features including additional multivariate data analysis tools that are not available in every software package.

JMP is a suite of computer programs for statistical analysis including DoE, which enables you to create designs for screening, optimization, response surface and formulation. JMP has many additional features that are used by statisticians.

Most of the software provide a free trial and we suggest trialing these to see which one has the features you are more likely to use and the best interface to work with.

5.8 "I Tried Experimental Design But It Did Not Work"

We often hear from people who are reluctant to use design because they have tried it before and it did not work for them. The probable cause for this is they did not follow the DoE process rigorously and therefore did not get a useful model. There are many reasons why a design might fail, the common ones are listed below.

- *Poor measurement of the response*. This can end in modelling problems. The variance that arises from measurement or sampling error should be much smaller than the variance from changing the settings in the design. Excessive measurement error shows an unreliable response and you can end up with incorrect conclusions or no model.

- *No center points/repeats*. Running a design without center points will mean you cannot check for curvature in your response. You will still be able to get a model but you will not know if you are missing process understanding as you have not tested a third level. However, running a design without repeats means you will not have an estimation of pure error and therefore you cannot say with confidence how different your results are and if your model is significant. When a center point is included, this is the setting that is commonly repeated.
- *No understanding of the scientific or engineering principles in the system*. The scientist still needs to understand the chemistry to choose the correct factors to investigate and how to interpret the models to make chemical sense. If a design is carried out properly (i.e., all un-investigated factors are held constant), a missed factor is just a missed opportunity, but if the factor is missed because it was not thought of, this could be uncontrolled and cause variation in the response.
- If you missed an important factor when using OVAT, you would not know.

There are other mistakes that chemists can make with DoE which will not necessarily mean they will fail to get a model, but the information gained might be minimal or inaccurate. Reasons for this include:

- *Not investigating a significant factor in the required region*. This comes back to selecting the right range to investigate. If a factor is insignificant to the model, it is not causing a change in response under the chosen settings. It is possible that the factor is critical for the process (and you would see this if you took it away), but the range may have been too narrow to see an effect, or too high or low to see the effect.
- *Using the design at the wrong time*. Designs are often used late on in the development of a process, when routes are reasonably fixed which does not allow the flexibility or opportunity to change factors such as starting material or solvent. This means you can use DoE to find the optimum conditions from a potentially imperfect route, but you will not see other opportunities. "The benefits of applying a designed approach diminish as time goes on."
- *Using the wrong design*. Oftentimes a higher power design is used than is needed. For example, response surface designs are run as standard even without quadratic terms. This will not necessarily give you the wrong information but it will result in much higher experiment numbers than necessary and defeats the purpose of using design for economy of experimentation.

5.9 Conclusion

The aim of this chapter was to introduce the topic of design of experiments in a practical sense, focusing on how to implement the technique while describing enough of the maths to ensure DoE and the software is not a black box. It can be an incredibly powerful tool and is being applied in many different industries. It can also be combined with different technologies, e.g. flow chemistry, automation, and PCA for further optimization, where it is a powerful tool for delivering enhanced process understanding to enable informed decision making.

However, DoE is not a magic bullet. There are occasions when DoE is not the right approach and a simpler optimization strategy may work. If you do run a design, ensure you have thought about the process, your aims, and factors carefully before commencing. Senior chemists who have not been in the lab for a while may not have experimented in this way before and may be skeptical about its use. Lack of support can be tricky to deal with and you may have to prove the use of the technique with a smaller, simple design where you can achieve process understanding but not have a complicated model with many model terms. If the DoE protocol is followed, you should have no problems in developing a model, which will aid in convincing any skeptics it is a useful tool to use.

As a final key points: it is critical to justify your factor selection. You do not have to look at everything, but you must understand why you choose to look at something, and/or why you choose to not look at something. Do not limit your selection to literature precedent; consider solvents, reagents, and substrates that may work.

DoE should enable you to work smarter not harder, gaining more information in the same or less time. The benefits include faster time to market (or publication), lower development costs, lower operating costs, lower cost of poor quality, and is applicable in any situation where you need to manipulate several independent variables to optimize one or more dependent variables.

References

1 Fisher, R.A. (1935). *The Design of Experiments [Internet]*. New York: Hafner.

2 Everitt, M.A. (2010). 1 – Designing a sensory quality control program. In: *Woodhead Publishing Series in Food Science, Technology and Nutrition [Internet]* (ed. D. Kilcast), 3–16. Woodhead Publishing https://www.sciencedirect.com/science/article/pii/B9781845694760500010.

3 Carlson, R. and Carlson, J.E. (2005). *Design and Optimization in Organic Synthesis*, 2ee, 1–574. Elsevier.

4 Box, G.E.P., Hunter, J.S., and Hunter, W.G. (2005). *Statistics for Experimenters*, 2ee. John Wiley & Sons, Ltd.

5 Jones, B. and Nachtsheim, C.J. (2011). A class of three-level designs for definitive screening in the presence of second-order effects. *J. Qual. Technol.* 43 (1): 1–15. https://doi.org/10.1080/00224065.2011.11917841.

6 Jones, B. and Nachtsheim, C.J. (2013). Definitive screening designs with added two-level categorical factors. *J. Qual. Technol.* 45 (2): 121–129. https://doi.org/10.1080/00224065.2013.11917921.

7 Carlson, R., Carlson, J., and Grennberg, A. (2001). A novel approach for screening discrete variations in organic synthesis. *J. Chemometrics* April 2000: 455–474.

8 Carlson, R. and Carlson, J.E. (2005). Principal properties and designs for discrete variations. *Org. Process Res. Dev.* 9 (5): 680–689.

9 Murray, P.M., Bellany, F., Benhamou, L. et al. (2016). The application of design of experiments (DoE) reaction optimisation and solvent selection in the development of new synthetic chemistry. *Org. Biomol. Chem.* 14 (8): 2373–2384. https://doi.org/10.1039/C5OB01892G.

10 Fey, N., Harvey, J.N., Lloyd-Jones, G.C. et al. (2008). Computational descriptors for chelating P,P- and P,N-donor ligands1. *Organometallics* 27 (7): 1372–1383. https://doi.org/10.1021/om700840h.

11 Jover, J., Fey, N., Harvey, J.N. et al. (2010). Expansion of the Ligand Knowledge Base for Monodentate P-Donor Ligands (LKB-P). *Organometallics* 29 (23): 6245–6258. https://doi.org/10.1021/om100648v.

12 Jover, J., Fey, N., Harvey, J.N. et al. (2012). Expansion of the ligand knowledge base for chelating P,P-donor ligands (LKB-PP). *Organometallics* 31 (15): 5302–5306. https://doi.org/10.1021/om300312t.

13 Murray, P.M. and Forfar, L.C. (2017). The application of advanced design of experiments for the efficient development of chemical processes. *Chem. Informatics.* 3 (2): 1–15.

14 Forlay-Frick, P., Fekete, J., and Héberger, K. (2005). Classification and replacement test of HPLC systems using principal component analysis. *Anal. Chim. Acta* 536 (1): 71–81. https://www.sciencedirect.com/science/article/pii/S0003267004016666.

15 Gabrielsson, J., Lindberg, N.-O., Pålsson, M. et al. (2003). Multivariate methods in the development of a new tablet formulation. *Drug Dev. Ind. Pharm.* 29 (10): 1053–1075. https://doi.org/10.1081/DDC-120025864.

16 Yao, H.-L., Shi, Y.-C., Yuan, S.-L., and Li, G.-Z. (2009). Quantitative structure-property relationship on prediction of cloud point of surfactants. *J. Dispers Sci. Technol.* 30 (8): 1223–1230. https://doi.org/10.1080/01932690802598531.

17 Clayton, A.D., Manson, J.A., Taylor, C.J. et al. (2019). Algorithms for the self-optimisation of chemical reactions. *React. Chem. Eng.* 4 (9): 1545–1554. https://doi.org/10.1039/C9RE00209J.

18 Taylor, C.J., Baker, A., Chapman, M.R. et al. (2021). Flow chemistry for process optimisation using design of experiments. *J. Flow Chem.* 11 (1): 75–86. https://doi.org/10.1007/s41981-020-00135-0.

6

Introduction to High-Throughput Experimentation (HTE) for the Synthetic Chemist

Stephanie Felten, Michael Shevlin, and Marion H. Emmert

Process Research & Development, Merck & Co., Inc., Rahway, NJ, USA

Glossary

Array The two-dimensional layout of an HTE experiment. HTE experiments are typically performed in plates with an array of 24 (4x6), 48 (6x8), 96 (8x12), or 384 (16x24) "wells" or positions.

Chemical reaction space With chemistry being the science of transforming one compound to another, the chemical reaction space contains all possible chemical reactions [1]. As it pertains to HTE, the relevant space of chemical reactivity is the space that can be explored by the combinations of reagents, solvents, and other reaction partners, with particular focus on a transformation of interest. To map the chemical space for a specific reaction, design of experiments (DoEs) [2] approaches can be used in concert with HTE. DoE employs statistical methods and multifactor optimization to allow a broader survey of the reaction space, limiting the number of experiments that have to be performed to a manageable number.

Custom HTE experiments In contrast to kits (see below "Predispensed reagents, HTE platforms, and kits"), custom HTE experiments use a *de novo* design approach for all or most wells in a plate. Custom HTE experiments require very little or no automation, as all features of the design (reagents, solvents, quantities, reagent ratios, dispensing sequence, reaction scale) are defined *ad hoc*. Therefore, custom HTE experiments are the ideal experiments for HTE novices to deepen their knowledge of HTE design. In advanced HTE laboratories, custom HTE experiments are most common for reaction screening to establish novel reactivity, i.e. in cases where no suitable preplated

Enabling Tools and Techniques for Organic Synthesis: A Practical Guide to Experimentation, Automation, and Computation, First Edition. Edited by Stephen G. Newman.

(see below) HTE platforms are available. Furthermore, custom HTE experiments are common as follow-up experiments for reaction optimization of initial reactivity hits.

Data analysis In the context of HTE, data analysis includes the quantification of analytes (substrates, products, intermediates) and the comparison of these numbers across the array of an HTE experiment. This process often also includes visualizing data, which in turn allows the visual detection of reactivity trends in a reaction array.

Enabling techniques Laboratory techniques that allow synthetic chemists to become more efficient and achieve a synthetic goal faster than with traditional workflows. Examples are HTE, flow chemistry, automation, and biocatalysis, among others.

HTE (or HTS) Abbreviation for high-throughput experimentation (or high-throughput screen); the execution of many experiments in a parallel format, using multi-well plates.

Miniaturization The design and execution of chemical experiments on smaller scale (volume, substrate loading). The typical scale of HTE experiments (50–100 μL solvent, approximately 1 μmol of substrate) is often referred to as "micro-scale" in the HTE community. Miniaturization is a key feature of HTE experiments and allows acquiring more data points per mol of reactant consumed than in traditional experimental chemistry setups on larger scales.

Predispensed (preplated) reagents, HTE platforms, and kits In HTE laboratories, it is common to predispense important components such as bases, ligands, or catalysts into a multi-well plate that can later be used by adding reagents and solvents. The process of predispensing is also known as "preplating." In some cases, all reagents for a certain class of reaction can be predispensed; in those cases, the resulting plates are often referred to as a reaction platform or kit. Examples for kits that are widely available are the KitAlysis® High-Throughput reaction screening kits distributed by Sigma Aldrich.

Stock solution Solution of a reagent or reactant prepared for dispensing into an HTE array. In a typical HTE experiment, many such solutions are prepared and then mixed at different ratios to prepare the reaction solutions in the wells of the HTE plate. Each well typically represents a unique combination of stock solutions and/or preplated reagents/reactants, thus allowing the evaluation of many reaction conditions or reactants in parallel.

Stock slurry and slurry dispensing Some reagents (e.g. inorganic bases) have limited solubility in organic solvents. Such solids can be dispensed into HTE vials using automated solid-dispensing equipment. Alternatively, small quantities can be dispensed into HTE vials as suspensions (commonly also called slurries in the HTE community), followed by evaporating the organic solvent. For further info, see *vide infra*.

Workflow A term describing the typical process of an experiment from start to finish. Within the realm of HTE, this includes experiment design, translation to an HTE array, determining quantities of stock solutions/slurries required, plating all reagents/reactants, heating/cooling/stirring, chromatographic analysis, data analysis, and data visualization. All these topics will be covered in more detail *vide infra*.

6.1 What Is HTE?

High-throughput experimentation (commonly abbreviated as HTE) is generally understood to be the execution of many experiments in a parallel format using multi-well plates. Initially pioneered in biological research, HTE has slowly been adopted into synthetic chemistry over the last decades. Types of HTE experiments include *parallel synthesis* (also called library synthesis), in which a large number of compounds are synthesized using one reaction condition; *HTE for reaction discovery*, which surveys a diverse reaction space to identify novel chemical reactivity; and *HTE for reaction optimization*, which transforms an initial reaction hit into a high-yielding, robust reaction protocol.

Even though the goals of the different types of HTE experiments are distinctly different, all of them are typically executed in an array format (e.g. 24 wells in a 4x6 array or 96 wells in a 8x12 array). The most common way to analyze large numbers of reactions from HTE experiments are automated chromatographic methods. As described in detail further below, HTE allows the rapid surveying of reaction space (reaction discovery or optimization) or chemical space (library synthesis). HTE thus provides large datasets, leading to faster reaction optimization in synthetic chemistry or property optimization for molecules with specific desired functions (e.g. for pharmaceutical or agrochemical discovery chemistry). Since parallel synthesis efforts are widely known and have been reviewed elsewhere [3], this chapter focuses on HTE for reaction discovery and optimization.

6.2 Why HTE and What Can It Achieve?

Unlike other innovative techniques with their origins in academia, HTE and its use in organic chemistry has its roots in the desire to use the time of a highly trained industrial workforce in a more efficient manner. Thus, it is not surprising that the uptake of HTE techniques has been more rapid in industrial than in academic settings [4].

A key reason for the limited uptake of HTE in synthetic academic groups may be the engineering challenges associated with adapting a workflow initially conceived in biology-focused laboratories to the realities of synthetic chemistry: volatile organic solvents, heterogeneous reaction mixtures (i.e. the need for agitation), and varying reaction temperatures, which are all key to the development of chemical reactions. However, many of these reaction conditions common to organic chemistry research are incompatible with HTE equipment used in biological HTE experiments made from polymers (e.g. polystyrene).

This somewhat unique development story for HTE in synthetic chemistry is likely one of the reasons why many misconceptions about this technique have proliferated widely. This chapter aspires to address some of the commonly perceived barriers and provides practical tools for chemists interested in HTE to increase their synthetic efficiency [5].

In the following paragraphs, we will first address common misconceptions and key advantages of HTE as a technique in synthetic chemistry. In the second part of this chapter, we will discuss key considerations and practical resources for typical HTE workflows, which are geared toward synthetic chemists with little or no experience with this technique. To illustrate how HTE workflows can be generalized and used for new discoveries, we will discuss two case studies in detail: The first study describes the development of a screening platform for Ni-catalyzed Suzuki–Miyaura couplings; the second study is a story of reaction discovery and optimization based on the authors' own experience. Finally, perspectives for the future of HTE will be outlined.

6.2.1 Commonly Perceived Barriers to Employing HTE in Synthetic Chemistry

6.2.1.1 Cost

Many synthetic chemists perceive HTE techniques to be out of their reach, particularly if their laboratory is located at a smaller institution. This impression is likely created by high-profile reports on remarkable workflows that heavily utilize automation, programming, and robotics [6–10]. As a result, HTE workflows are perceived to require those types of highly specialized techniques and thus outside of the "comfort zone" of classically trained synthetic chemists.

6.2.1.2 Availability of Dedicated HTE Facilities

Dedicated HTE facilities have been established at many universities and research centers across the globe and many allow external users to access their equipment, using a fee-based model. However, this model requires significant investments in advance to align experimental design, travel time, reagent orders, and the availability of trained staff scientists. These barriers often make it unreasonable and/or untimely to use such dedicated facilities, especially when located physically remote from academic core laboratories. The goal of this chapter is to lay out an alternative approach: The use of HTE in typical synthetic chemistry laboratories, using analytical and other equipment that is already present. Such an approach can, in our opinion, complement collaborations with dedicated facilities.

6.2.1.3 Access to Knowledge and Training

Enabling techniques in chemical laboratories are seldom intuitive and have to be learned. This occurs most often during laboratory courses and research training at universities. Typically, a combination of initial training and continued hands-on use of a technique renders scientists experts in the laboratory techniques of their chosen discipline. In the case of HTE, access to HTE training during the trainee phase of a chemist's career is relatively rare. Currently, the main pathways of acquiring HTE knowledge for synthetic chemists are collaborations with an end-user facility or an industrial laboratory. This chapter aims to broadly share HTE knowledge with the synthetic community, thus to contributing to lowering the barrier to access HTE techniques for interested synthetic chemistry laboratories.

6.2.1.4 Perception of HTE as Antithesis of Hypothesis-driven Research

A more philosophical barrier that remains prevalent among chemists unfamiliar with HTE is the perception that the use of HTE precludes hypothesis-driven research and that its use is an exercise in luck: As long as you can run enough reactions, something interesting will be found in any case. This is a common misconception, likely caused by the frequent reference to serendipitous discoveries made in the HTE space.

All these issues can be perceived as critical hurdles that limit the usefulness of HTE techniques to most synthetic chemists. However, we and others [11, 12] argue that *advanced robotics or dedicated facilities are not needed for the successful use of HTE techniques*, particularly not for the discovery of novel chemical reactivities. This book chapter highlights how HTE workflows can be implemented into a typical synthetic chemistry laboratory and provides practical tips and tricks to achieve this goal.

We further postulate that access to HTE facilities is typically most impactful when performing routine screens in a repetitive fashion. Such scenarios rarely arise in academic laboratories, but are more frequent in industrial settings. For example, for standardized solvent/base screens (e.g. 12 bases and 8 solvents on a 96-well pate), access to a collection of bases already plated out into suitable reaction vials and to a standing collection of solvents is exceedingly helpful. Maintaining such collections is associated with costs to build up stock and secure long-term storage.

Screens that are designed on demand without access to preplated reagents are overall less expensive, as only the needed reagents are added to the reaction array and storage costs are irrelevant. However, such HTE experiments designed "on demand" require more time to execute, as all reagents have to be plated out. This often comes with delays in collecting all chemicals required, but saves costs on reagents. This type of screening can be designed and executed easily in the absence of a dedicated facility. Designing new types of HTE experiments without the use of preplated reagents is often the fastest pathway to success for innovative chemistry developments. Novel reactivity frequently requires novel approaches to uncover it; as such, access to preplated reagent kits may or may not be useful in reaction discovery applications. For this reason, custom or "on-demand" HTE screens are a suitable tool for academic laboratories interested in discovering novel reactivity; those types of screens do not require dedicated facilities.

As we describe in more detail *vide infra*, custom HTE experiments only require access to a glovebox or inertion box, nitrogen, a stirrer/heater (if desired), manual pipettes, and reagents to be tested. Such equipment is widely available in most synthetic chemistry laboratories. To state it clearly: *While robots, automation, or dedicated HTE facilities can be extremely useful, access to such tools is not a requirement for taking advantage of HTE workflows and principles that accelerate synthetic discovery and optimization.*

With access to HTE techniques (as described in further detail below), considering the question if it is "good and hypothesis driven science" to implement such techniques in a synthetic laboratory becomes important. The impact of HTE techniques on the speed of reaction discovery and optimization is widely documented, especially in publications by industrial authors. The more difficult question to answer is if hypothesis-driven research and HTE are incompatible in the first place.

As the readers will see in the case studies that are part of this chapter, HTE design typically begins with a proposed catalytic cycle or a similar hypothesis, which defines the initial conditions to be evaluated. Then, experiment design diverges from more classical approaches in that it aims to cover a range of different reactants (e.g. solvents, ligands, or bases); as such, a good understanding of the physical organic attributes of reactants (e.g. pK_a values, bite angles of bidentate ligands, dielectric constants of solvents) is often key to success. In the data analysis stage of the HTE workflow, discovering patterns in the data, relating these patterns back to the reaction conditions tested, and designing logical follow-up experiments that further elucidate initial hits require significant analytical skill. Both design and data analysis thus benefit from a sound knowledge of the literature, a testable working hypothesis at the beginning of the project, and a deep understanding of how reaction parameters affect transformations. To provide more details and examples for these parts of the workflow, the authors would like to refer readers to the subchapters on HTE design and analysis in the next section and to the case studies in the latter part of this chapter. Those sections provide more support for the above stated fact that designing HTE plates most often starts from hypotheses, especially in cases where the target is to discover new reactivity. *Therefore, we would like to emphasize that HTE can be a simple screening exercise; however, in the right hands, it becomes a powerful tool for testing hypotheses and thus perform hypothesis-driven research in an accelerated fashion.*

6.2.2 Advantages of HTE Workflows vs. Traditional Reaction Setup

In contrast to the perceptions described above, HTE (in the authors' opinion) distinguishes itself from other types of more classical screening approaches by a fundamentally different approach to experimental design. Several key principles are unique to designing and setting up HTE experiments and are responsible for its success across both industrial and academic laboratories. These differentiating factors will be described in detail in the following paragraphs.

6.2.2.1 Setup Time per Reaction

In the hands of experienced synthetic chemists, designing experiments (especially complicated experiments) often consumes more time than the actual experimental setup. When considering this, setting up only one experiment at a time may be a missed opportunity: Weighing out one reagent for an experiment requires the same amount of time as weighing out the same reagent for a stock solution, which could supply 2, 24, 48, 96, or more reactions. Therefore, HTE provides the opportunity to decrease the preparation time per reaction, resulting in more efficient use of the experimentalists' time.

6.2.2.2 Miniaturization and Efficient Reagent Use

In the spirit of green chemistry, miniaturizing reactions from a typical volume of 100 mL to a 100 μL volume (a common reaction volume used in a 1 mL HTE vial; see Figure 6.1) leads to more efficient use of starting materials and reagents. This is particularly impactful if reagents or starting materials are expensive or sparse. Thus, HTE allows for the exploration of a larger reaction space while using less material. This can have a significant impact on success in the early stages of drug discovery and development, as well as in complex molecule synthesis.

6.2.2.3 Multivariable vs. Sequential Optimization

The traditional way of describing reaction optimizations in a synthetic manuscript provides a table of optimizations (Figure 6.2a), which elucidates the effect of changing one variable (e.g. solvent choice, concentration, catalyst precursor) at a time as

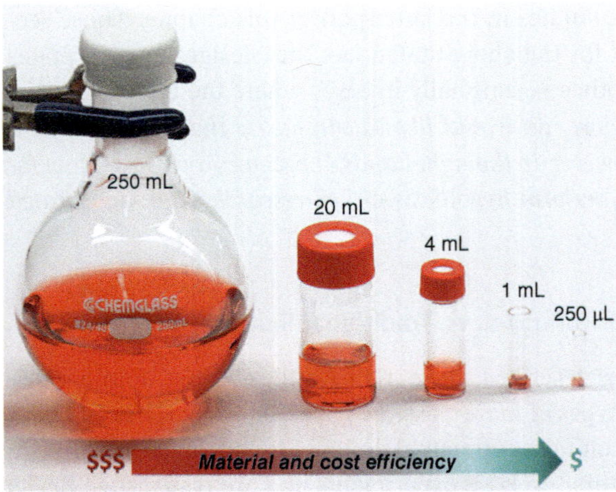

Figure 6.1 **Visualization of reaction miniaturization.** Sizes of reaction vessels with typical fill levels are visualized.

compared to a set of "standard" conditions. In some cases, authors also include information on how the change of two variables impacts reactivity, but that is not always the case. Therefore, such tables are most often limited regarding the type of information provided and do not present a wholistic view of the reactivity space: For example, what happens if two variables are changed in various combinations (e.g. equivalents of reagent and solvent choice)? HTE workflows (Figure 6.2b) address this limitation by designing reaction arrays that elucidate *in the same experiment* how changing two, three, or more variables affect reaction outcomes.

6.2.2.4 Visualizing Reactivity Patterns

With data generated in an array format, a 2D visualization of reaction results can be easily prepared, using selectivity, conversion, and yield data, for example. In the resulting graphics, reactivity patterns are easily spotted. Typically, such visual representations clearly suggest the next set of experiments to be performed, leading to a well-defined workflow, in which one experiment informs the next in a hypothesis-driven fashion.

6.2.2.5 Serendipity in Reaction Discovery

Success in accessing desired reactivity is frequently accelerated by serendipitous discoveries. This idea does not minimize the impact of hypothesis-driven reaction discovery; instead, it reflects the fact that even after more than 100 years of chemical research, even well-known reactions still harbor secrets and can be improved [13]. HTE is particularly prone to accessing serendipitous discoveries,

(a) Example of a traditional sequential optimization.

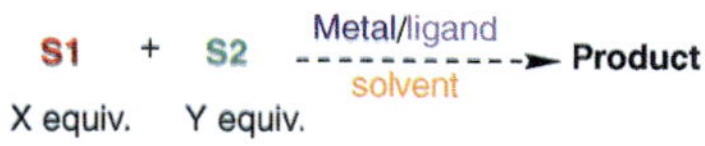

	Entry	Substrate 1	Substrate 2	Solvent	Catalyst/ ligand	Yield (%)
S1 : S2 *Stoichiometry*	1	**1.0 equiv**	**1.0 equiv.**	THF	M_1/ L_1	55
	2	**3.0 equiv.**	**1.0 equiv**	THF	M_1/ L_1	35
	3	**1.0 equiv.**	**3.0 equiv.**	THF	M_1/ L_1	40
Solvent *screen*	4	1.0 equiv.	1.0 equiv.	**DMF**	M_1/ L_1	45
	5	1.0 equiv.	1.0 equiv.	**Toluene**	**M_1/ L_1**	50
M : L *Identity*	6	1.0 equiv.	1.0 equiv.	THF	**M_2/ L_1**	65
	⋮	⋮	⋮	⋮	⋮	⋮

(b) Examples of a multivariable optimization via HTE.

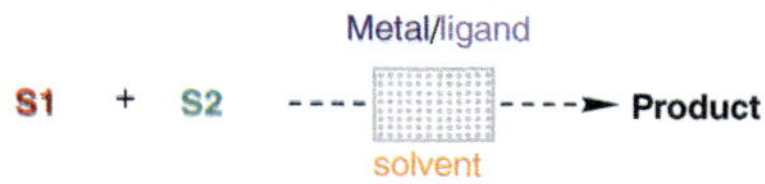

HTE option 1: 12 solvents × 8 substrate ratios

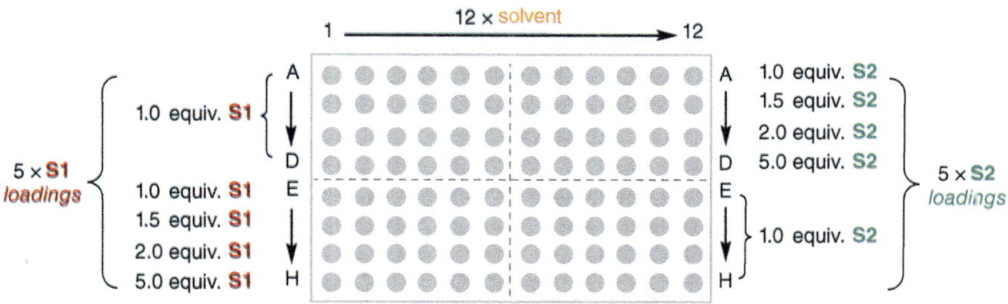

HTE option 2: 12 solvents × 4 substrate ratios × 2 ligands

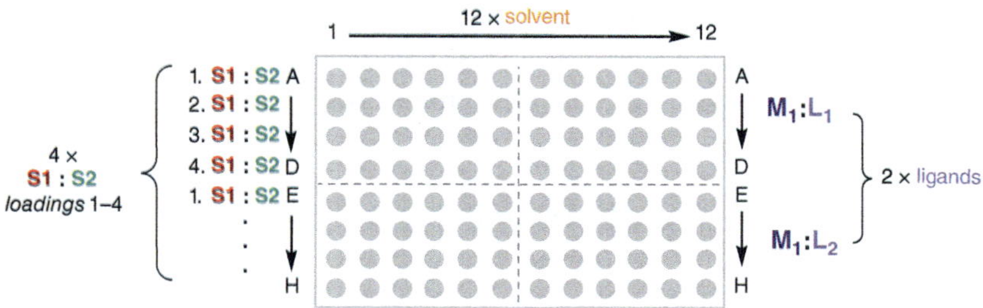

Figure 6.2 Visualization of sequential reaction optimization (a) vs. multivariable reaction optimization via HTE (b).

as practitioners find themselves often in the situation of "having to fill out another column or row" on a design. This can lead to testing conditions that are very different than those originally designed or precedented in the literature because the barrier to accessing novel reaction space is low. Thus, HTE is a valuable tool for scientific creativity, simply by expanding the number of reaction conditions that can be tested in parallel.

6.2.2.6 Avoiding Cross-contamination

Particularly in catalytic reactions, avoiding contamination from unknown sources such as used equipment can be the key to reproducible reactivity. HTE consumables available today are relatively inexpensive (including stir bars), so that one-time use of a reaction vial is common practice. Furthermore, solid reagents are typically dispensed into HTE vials as slurries or solutions, further limiting cross-contamination. Thus, reaction "hits" catalyzed by trace bases, acids, or metals are less likely to occur than in a situation where reaction vessels and stir bars are reused. This in turn limits the time that is needed to follow up on unexplainable results and allows chemists to focus on "real" results instead. As we will describe in the next section, reaction vials can in some cases also directly be used as analytical vials, thus limiting the cost associated with replacing reaction vials continuously and removing one more source of potential contamination.

Overall, we believe that the aforementioned advantages provide a strong case to consider implementing HTE into reaction discovery and optimization workflows. We would like to re-emphasize our earlier statement: HTE is more akin to a mindset that allows efficient experimental design than a technologically enabled, hypothesis-averse screening exercise. If it becomes easy to set up 24, 48, or 96 reactions instead of one without a counter-productive strain on resources, HTE frees scientific creativity to focus on the important problems instead of spending hours weighing out reagents.

This leads us to two key questions: What equipment is needed to set up HTE plates and analyze them? What does a typical workflow look like? These questions form the backbone of this chapter, together with a critical description of two case studies that highlight applications of HTE in the development of new reactivity and screening platforms. We hope that the following practical examples will illustrate the more philosophical advantages of HTE described above and inspire the reader to join the ever-growing field of HTE enthusiasts.

6.3 Practical Considerations and Tools for HTE

Investment in HTE platforms requires a general understanding of what tools are available and which of those tools are applicable for a given environment. For example, an industrial drug design laboratory will have different needs than an academic laboratory starting on the journey of enabling HTE workflows for the discovery and optimization of novel reactivities. The following subchapter provides an overview of available tools and resources. This summary is designed to address crucial questions laboratories face when starting to invest in HTE equipment.

6.3.1 Outline of a Typical HTE Workflow

From beginning to end, a typical HTE workflow consists of several steps shown in Figure 6.3. Each step of the workflow will be discussed in detail below.

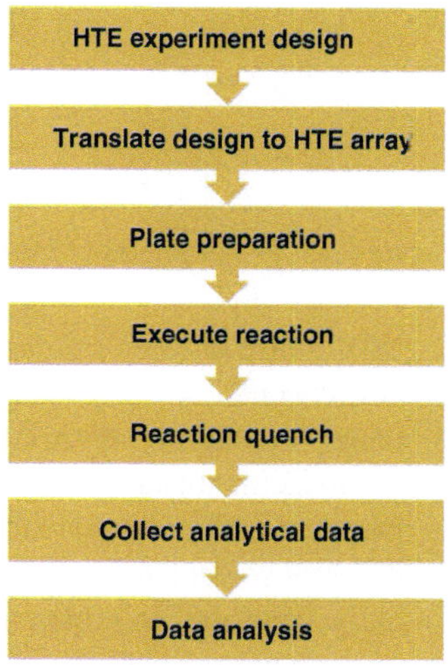

Figure 6.3 Visualization of a typical HTE workflow.

1) HTE experiment design (Figure 6.4): What are the reaction variables to be evaluated? Are there literature conditions for similar reactivity that should be included? Are preplated reagents available that may influence the design?
2) Translate design to HTE array (Figure 6.5): Can the variables of interest be selected such that an appropriate array can be designed? Which other conditions may be of interest (e.g. design for diverse solvents, additives, ligands)? Do the variables chosen fill up all positions available on the array (e.g. on a 24-, 48-, 96-, or 384-well plate)? If not, can additional valuable experiments be envisioned? Does it make sense to run the same plate at two different temperatures? If too many experiments are envisioned, can some experiments be cut, or should an extra plate be designed? Does it

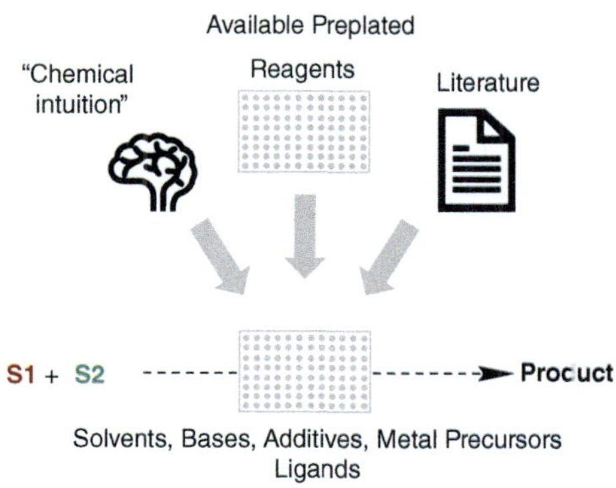

Figure 6.4 Visualization of HTE experiment design process, incorporating information from the literature, available HTE platforms, and chemical intuition.

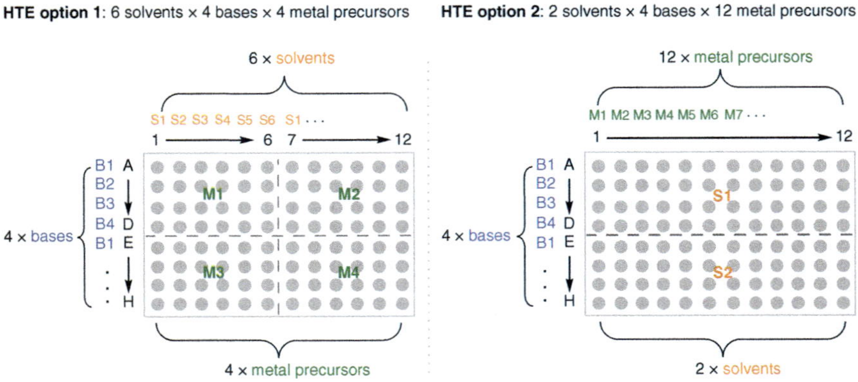

Figure 6.5 Two examples of how to translate a design to an HTE array.

make sense to run a similar plate side by side, e.g. with a different substrate or key reagent?

3) Plate preparation (Figure 6.6): What is the desired order of addition? Are any of the reagents unstable and thus need to be added last? Can stock solutions be prepared in the desired reaction solvents or is it more beneficial to preplate in a different solvent, followed by a solvent switch? What is the desired atmosphere of the plate and is equipment to set up plates under inert atmosphere (glovebox, inertion box) available?

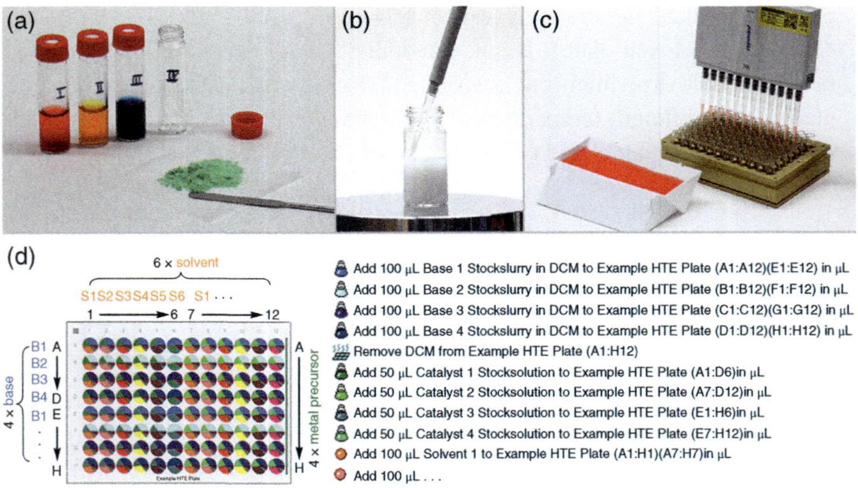

Figure 6.6 **Example of how to prepare an HTE plate.** (a) Preparation of stock solutions, (b) Preparation of stock slurry, (c) Addition of stock solution in reservoir via multi-channel pipette, (d) Outline of addition order, using library studio software.

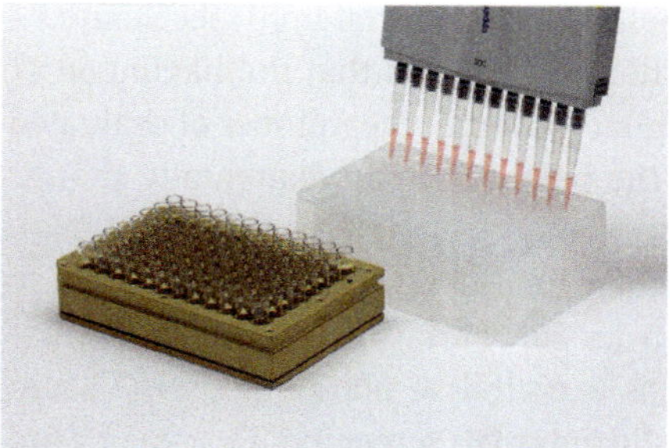

1. Unscrew HTE plate
   -Remove lid and PFA film
2. Dilute reactions
   -Add 500 μL MeCN: DMSO (1:1)
   -Add 50 μL MeOH (quench)
3. Re-seal plate and mix
4. Sample & dilute
   (e.g. 20 μL diluted mixture in 700 μL MeCN)
5. Seal dilution plate
6. Collect analytical
   (e.g. UPLC, UPLC/MS, GC, GC/MS etc.)

Figure 6.7 Example of how to quench and prepare HTE reactions for analysis.

4) Execute the reaction: Is a temperature profile needed? Is cooling/heating/stirring available? Is manual/automated reaction profiling desirable?
5) Reaction quench and analysis (Figure 6.7): What is the desired target concentration for analysis? Are high-throughput analytical methods and instruments available? Which internal standards should be used?
6) Data analysis and visualization: Which software solutions are available? How can the data be exported into a format that allows visualization and informs next steps? How are these data documented and retained?

None of these steps are fundamentally different to running single experiments, but if they were treated with a traditional organic chemistry workflow (e.g. setting up one reaction at a time and NMR analysis after workup and isolation), several days would be required to complete even a 24-reaction screen. Therefore, the HTE field has developed custom solutions for each step of the HTE workflow, which will be detailed in the next section.

6.3.2 Types of HTE Designs

As mentioned in the Introduction, high-throughput reactions can be performed with different goals in mind: Reaction discovery, reaction optimization, and parallel synthesis. The most well-known among these is parallel synthesis (also called library synthesis), as this type of workflow has been applied broadly for medicinal chemistry purposes. Significant progress has been made in this field in the last decades from its infancy in combinatorial chemistry [14] and has been reviewed elsewhere at length [3]. Therefore, the next several paragraphs will focus on reaction discovery and optimization, areas that have been less widely discussed in the community.

6.3.2.1 HTE for Reaction Discovery

What is the best way to discover novel chemical reactivity? The answer to this question is likely as diverse as chemists themselves. Approaches toward discovering new chemical reactions span from hypothesis-driven to serendipity-accelerated – most likely lie somewhere in between. The good news is that the likelihood of discovering new and useful reactivity is higher the larger the number of evaluated reaction conditions. Therefore, HTE has the potential to fundamentally change the pace of reaction discovery.

Designing HTE plates to test *diverse* reaction conditions in parallel is the key to success in reaction discovery. This may mean designing arrays that include (i) diverse sets of ligands for metal catalyzed reactions (e.g. diverse phosphines; N/O-based ligands + phosphines; mono/bi/tridentate ligands); (ii) screening the same ligand plate with different metal precursors; (iii) evaluating basic, acidic, and neutral reaction conditions side by side; or (iv) diversifying the tested solvents. One key design principle is to always include a "background" reaction: As synthetic chemists of all training backgrounds know, reactions in the absence of intended reaction promoters (e.g. catalysts or additives) often provide key insights. Finally, the choice of suitable reactants is important for two reasons: (i) Availability of sufficient amounts of starting materials and desired product can both enable faster follow-up when desired reactivity is identified; and (ii) designing starting materials and products to enable rapid analysis can favor high reaction throughput and thus overall success (e.g. chromatographic analysis is faster with UV-active reagents and products).

6.3.2.2 HTE for Reaction Optimization

Once desired reactivity has been identified, reaction optimization via HTE can significantly accelerate the development of initial reactivity toward a useful methodology by evaluating reaction variables in parallel.

For example, if an initial reactivity hit is promoted by a base in MeOH as solvent, it is prudent to design a series of follow-up screens for optimization (Figure 6.8): (i) A screen that tests a small number of alcoholic solvents (e.g. trifluoroethanol, MeOH, EtOH, *tert*-amyl alcohol) that span a range of different acidities and nucleophilicities together with a large array of bases (inorganic and organic bases with a range of pK_a values and nucleophilicities); (ii) a screen that combines a small number of bases at different equivalents with a larger number of solvents, including at least one aprotic polar and one nonpolar solvent. Patterns emerging from these two HTE experiments will narrow parameters to be screened next.

For reaction optimization, it is important to consider two types of variables: Discrete variables (e.g. identity of base or solvent) versus continuous variables (e.g. temperature, equivalents of reagents added). It is often prudent to first focus on the discrete variables to identify suitable reagent combinations affording moderate yields, with a subsequent focus on optimizing continuous variables.

Using physical organic techniques, it is in principle possible to convert discrete variables into continuous variables. For example, solvents can be classified by their dielectric constants (computed dielectric constants are available for solvent

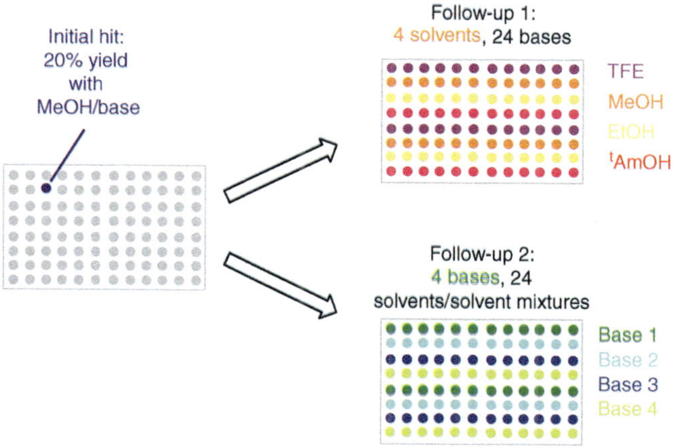

Figure 6.8 Example of initial hit and follow-up screens.

mixtures) [15]; X-type ligands [16] or bases (X^-) can be classified by the pK_a values of their corresponding acids H–X; and Hammett parameters [17] are available for substituted aromatic reactants. Choosing reactants and reagents across a range of these parameters increases the chances of a successful reaction optimization. Furthermore, analyzing the obtained reactivities across a spectrum of physical organic parameters provides mechanistic insights that can aid in designing the next rounds of optimization.

6.3.3 HTE Design Software: Tools for Building Arrays

The next step in an HTE workflow consists of designing the screening array (i.e. the layout). In general, an HTE design incorporates the identity and quantities of all chemicals used plus the reaction procedure, which in this case consists of the sequence of reagent addition and the mode of reagent addition (as a solid, slurry, solution, or neat). A simple and cost-efficient way of outlining high-throughput screens is the utilization of the Microsoft Excel software (see Figure 6.9 for an example). Excel and similar programs are widely available and commonly employed by scientists, which correlates typically with a high comfort level and low barriers to access. Indeed, for academic trainees, designing addition sequences using Excel may provide the opportunity to learn an additional skill.

A more cost intensive, but also highly enabling alternative is to invest in specialized HTE planning software such as LEA Library Studio from Unchained Labs[1] or Katalyst from ACD Labs.[2] These programs are specifically designed to aid

1 https://www.unchainedlabs.com/lea/.
2 https://www.acdlabs.com/products/spectrus/katalyst/index.php.

	96 well plate												
	3 bases (50 µmol/well), 1 row blank												
	solvent: THF as dispense solvent												
Ligand	L1	L2	L3	L4	L5	L6	L7	L8	L9	L10	L11	none	
	1	2	3	4	5	6	7	8	9	10	11	12	Bases
A													NaOAc
B													Cs2CO3
C													KOtBu
D													none
E													NaOAc
F													Cs2CO3
G													KOtBu
H													none

Dispensing quantities						
50 µL/well of base slurry in THF						
10 µL/well of stock solution of all other reagents						
Stocksolution/slurry calculation NaOAc						
Total needed (2 plates)	2400	µL		Total µmol needed	3196.8	µmol
To be made:	4000	µL		Total µmol needed for 4 mL	5328	µmol
				Total needed fo 40 mL	**437.1**	**mg**
				Mw of NaOAc	82.03	g mol^{-1}
Stock solution/slurry calculation CsCO3						
Total needed (2 plates)	2400	µL		Total µmol needed	3196.8	µmol
To be made:	4000	µL		Total µmol needed for 4 mL	5328	µmol
				Total needed fo 4 mL	**1735.9**	**mg**
				Mw of NaOAc	325.8	g mol^{-1}
Stock solution/slurry calculation KOtBu						
Total needed (2 plates)	2400	µL		Total µmol needed	3196.8	µmol
To be made:	4000	µL		Total µmol needed for 4 mL	5328	µmol
				Total needed fo 4 mL	**597.8**	**mg**
				Mw of KOtBu	112.2	g mol^{-1}

Figure 6.9 Example of Excel sheet used for creating an HTE layout (top) and calculating needed quantities of stock solutions/slurries (bottom). The HTE layout summarizes the overall experimental design, such as where certain ligands or bases will be located on the plate, the size of the plate, and the quantities of reagents for each well. The calculation part depicts the logic of preparing stock solutions or stock slurries (for further information on slurry dispensing, see *vide infra*). It shows the overall quantities that need to be prepared (2400 µL = 2 plates x 24 wells per plate for each solution x 50 µL of stock solution dispensed). At a concentration of 1.33 M for each base, 3196.8 µmol of base would be needed for 2400 µL stock solution. However, more stock solution than needed (4000 µL) is prepared to prevent issues stemming from solvent evaporation (especially important for volatile solvents such as CH_2Cl_2 or THF), dripping, or solution sticking to the outside of pipette tips or vial walls. Therefore, the required µmol are adjusted for a factor of 1.67 to prepare 4000 µL of stock solution. Using the molecular weight of each base, the actual amount to be weighed out (in mg) is then calculated.

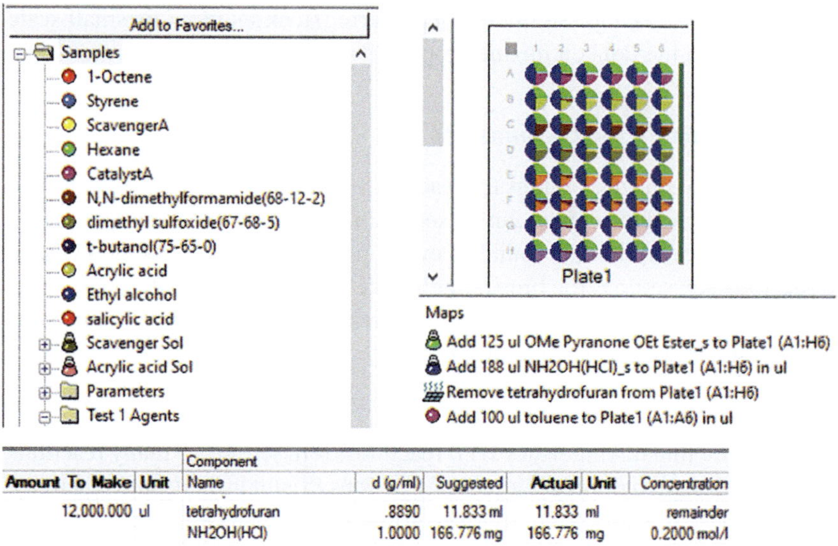

Figure 6.10 Examples of reagent library (top left), plate design and recipe (top right), and stock solution calculations (bottom) in Library Studio®. The reagent library contains all reagents and solutions to be dispensed. The recipe contains the sequence of stock solutions and slurries to be dispensed, resulting in a two-dimensional map of the HTE plate, in which each color represents the volume of one solution added. The image of each well can be selected, resulting in a detailed description of the content. The stock solutions spreadsheet allows selection of the amount of stock solutions to be prepared (Amount to Make) together with a summary of reagent quantities in suitable units (Actual) that need to be mixed to create each stock solution with the desired concentrations.

in creating reaction arrays. They can be used to create reagent libraries, simplify screen design by providing a "drop-and-drag" interface, automate calculations for preparing stock solutions, and provide a visual interface that allows the user to control the composition of the reagent mixtures in each well. Examples are shown in Figure 6.10.

These programs clearly have benefits when utilized regularly and efficiently, particularly when part of an "end-to-end" workflow that includes automated laboratory equipment, as the "recipe" obtained can be directly used to program robotic tools. However, in more resource-constrained environments, a widely available software such as Excel is sufficient.

Regardless of the software used, densities, solubilities, and molecular weights of all reagents, as well as the type of mixing need to be considered to ensure accurate calculations. If densities of organic solids are unknown, they can often be approximated as ~1 g mL^{-1}; for inorganic compounds including salts, an approximation of ~3 g mL^{-1} can be generally used. In addition, ideal mixing [18] is typically a good working assumption for preparing stock solutions for small-scale screening, if no other data on mixing properties are available.

6.3.4 HTE Reactors and Consumables

Next, we want to consider options for reaction vessels in which HTE screens can be performed. Reaction screens can be set up in disposable vials (equipped with suitably sealing caps) that are found in every modern laboratory. The only issue with this low-tech setup is the time it might take to screw all vials open and close. If that is not a concern – voilà, your HTE reactor is already present in your laboratory, and you didn't even know about it!

6.3.4.1 Reaction Blocks

One key feature that distinguishes HTE reactions from setting up many reactions in vials is the design that utilizes arrays of reactions to gain insights into reactivity. For that purpose, it is beneficial to utilize array-type (4x6, 6x8, or 8x12) heating plates (with or without sealing top plates; see below for options), which are readily available from many vendors, such as Chemglass[3] or Analytical Sales & Services[4] (Figure 6.11).

6.3.4.2 HTE Vials

Finding a suitable reaction vial to fit into the chosen HTE reaction blocks is essential; 2 mL (HPLC vials), 4 mL (1 dram vials), and 8 mL (2 dram vials) glass vials are available from a wide range of vendors, including ThermoFisher, VWR,

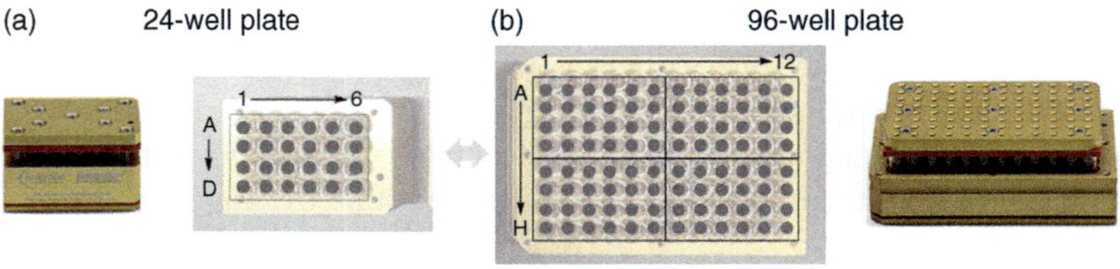

Figure 6.11 **Examples of commonly utilized high-throughput plates.** (a) 24-well plate for 1 mL vials (~50–150 μL reaction volume; left: side view with top sealing plate; right: top view without sealing plate). (b) 96-well plate for 1 mL vials (left: top view without sealing plate; right: side view with top sealing plate).

---

3 OptiBlock Parallel Synthesis Reaction Block Start Up Kit, OP-5000-K1. https://chemglass.com/optiblock-parallel-synthesis-reaction-blocks.

4 Analytical Sales and Services: **24-well plate** – (i) Volume: 250 μL, SKU: 24250; (ii) Volume: 1 mL, SKU: 24249; (iii) Volume: 4 mL, SKU: 24015; (iv) Volume: 8 mL, SKU: 24017. **48-well plate** – Volume: 2 mL, SKU: 48012. **96-well plate** – (i) Volume: 250 μL, SKU: 96970; (ii) Volume: 1 mL, SKU: 96960. https://www.analytical-sales.com/.

Chemglass, Analytical Sales & Services, and many others. Most of those screw cap vials can be used in combination with the delivered vial caps, as long as those are stable to the solvents and reagents to be tested. Alternatively, Analytical Sales & Services manufactures vial blocks for each of those size vials, which can be sealed with a top plate[5]. Vials for 24- or 96-well plate reactions with vial volumes of 250 μL and 1 mL can be purchased from Analytical Sales & Services[6] and other vendors.

6.3.4.3 Reaction Blocks with Sealing Top Plate

The use of a sealing top plate for all vials in a reaction block is a key feature of HTE reaction setup, as this results in significant time savings as compared to sealing every reaction mixture separately. Such high-throughput plates are commercially available as 24-, 48-, and 96-well plates of varying volumes (250 μL, 1 mL, 2 mL, 4 mL, and 8 mL) from different vendors (Unchained Labs, Chemglass [see footnote 3], or Analytical Sales & Services [see footnote 4]; Figure 6.11 shows plates for 1 mL vials; Figure 6.13a shows a plate for 250 μL vials). In addition to reaction plates designed for HTE applications, some vendors also offer "HTE starter kits,"[7] which include other tools and some consumables (e.g. screws, sealing mats, and vials).

6.3.4.4 Special Reactors for Photochemistry, Electrochemistry, and High-Pressure Reactions

Besides HTE reactors designed for parallel synthesis under thermal conditions, other types of HTE blocks have been designed to enable reactions with special engineering requirements such as photoredox catalysis [19], electrochemistry [20], or reactions requiring pressurized gases [21]. Photoredox HTE blocks and accessories (e.g. LEDs with different wavelengths, LED control panels) are commercially available for 24- or 96-well reaction blocks.[8,9] Specialized

---

5 SKU 48012 (2 mL vials/HPLC vials); SKU 24015 (1 dram/4 mL vials); SKU 24017 (2 dram/8 mL vials); all at https://www.analytical-sales.com/.

6 HTE vials. (i) Volume: 1 mL, Dimension: 8 x 30, SKU: 9984001; (ii) Volume: 250 μL, Dimension: 84 x 21, SKU: 96342. https://www.analytical-sales.com/.

7 **HTE Starter Kits**. *Sigma Aldrich*: KitAlysis™ 24-well Reaction Block and Screwdriver Set – SKU: Z742107-1KT. https://www.sigmaaldrich.com/US/en. *Analytical Sales and Services*: (i) 24-well screening – SKU: HTEKit1; (ii) 24- and 96-well screening – SKU: HTEKit2; (iii) 24-, 48- and 96-well screening – SKU: HTEKit3. https://www.analytical-sales.com/.

8 Photoredox HTE plate sources: *Sigma Aldrich*: Photo KitAlysis™ Starter Kit — SKU: Z742612-1KT. https://www.sigmaaldrich.com/US/en. *Analytical Sales and Services*: **24-well plate** – (i) Volume: 250 μL, SKU: 24250; (ii) Volume: 1 mL, SKU: 24253; **96-well plate** – (i) Volume: 50 μL, SKU: 96970; (ii) Volume: 1 mL, SKU: 96973. LED boards with different wavelengths are also commercially available. https://www.analytical-sales.com/.

9 Commercialization of the HTe-Chem reactor described in reference 31 is underway by Analytical Sales & Services.

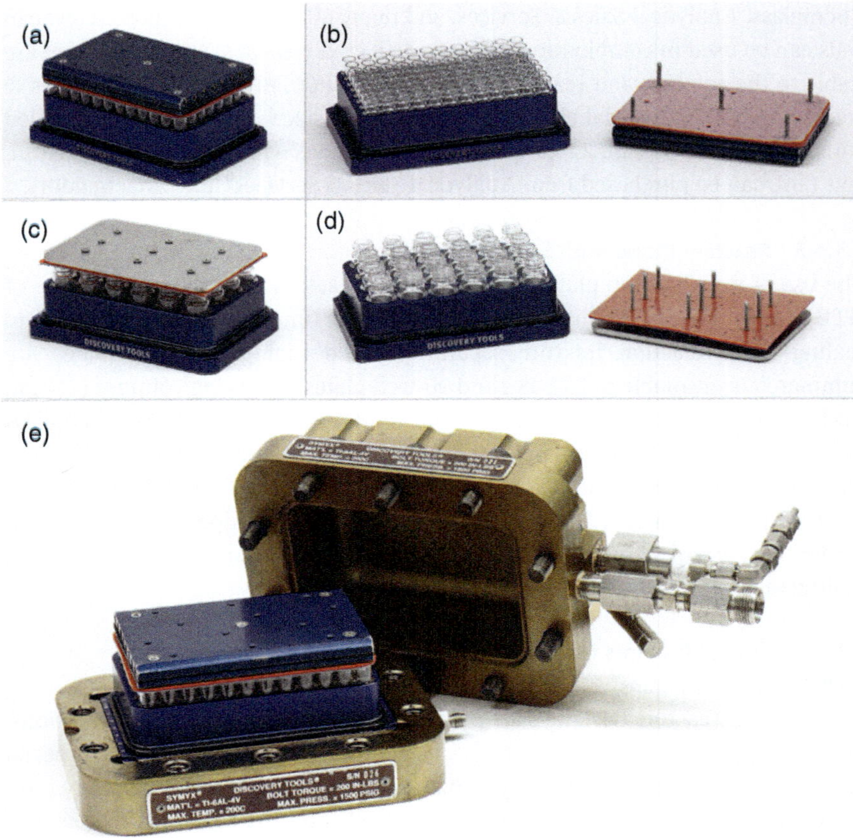

Figure 6.12 **Examples of high pressure HTE blocks from unchained labs.** (a) Sealed 96-well plate for 1 mL vials. (b) Open 96-well plate with lid for 1 mL vials. (c) Sealed 24-well plate for 4 mL vials. (d) Open 24-well plate with lid for 4 mL vials. (e) High-pressure HTE block in gas manifold reactor.

high-throughput reaction blocks for gas reactions (e.g. hydrogenation, carbonylation, carboxylation; see Figure 6.12) are available from Unchained Labs[10] or Swagelok. These reaction blocks are equipped with sealing plates that allow external pressurized gases to be connected (Figure 6.12a, c); contact between the gases and the external gas supply is made via holes in the rubber mats (see Figure 6.12b, d).

10 Gas HTE Block. *Unchained Labs*: (i) **96-well plate** – Volume: 1 mL, Serial No: S14085 Rev A; (ii) **24-well plate** – Volume: 4 mL, Serial No.: S113415 Rev B. For more information quote contact support service on https://www.unchainedlabs.com/.

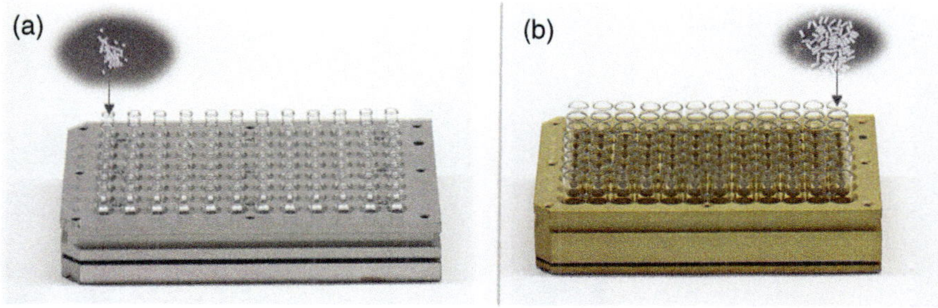

Figure 6.13 Small stir bars for 250 μL vials (a) and 1 mL vials (b).

6.3.4.5 Reaction Stirring and Temperature Control

Stir bars for 4 mL and larger vials are widely available from all common manufacturers. VP Scientific 5 mm x 2 mm stir bars can be utilized for 1 and 2 mL reaction vials.[11] Stir bars for smaller scale high-throughput screens are available from Analytical Sales & Services (Figure 6.13).[12] For simple transfer of small stir bars into a whole plate, a drop dispenser[13] is recommended. Alternatively, the utilization of a straw-like, plastic spatula can also be helpful to place small stir bars into the desired reaction vials.

Related to the discussion around suitable stir bars is the need to identify stirring solutions that can reliably stir at any scale. For 24- and 48-well plates as well as relatively large reaction vials (larger than 2 mL), the use of a traditional heater/stirrer found in most chemistry laboratories is a budget-friendly solution to stirring HTE experiments. However, once 96-well reaction blocks and vials with 250 μL or 1 mL volume are involved, stirring becomes inconsistent on traditional stirrer/heaters and has been shown to influence reaction outcomes, especially in heterogeneous reaction mixtures. As such, using a tumble stirring unit (commercialized by Unchained Labs or V&P Scientific)[14] is key to the most reproducible HTE data and reaction outcomes (Figure 6.14).

A more accessible alternative to tumble stirrers can be the use of shakers and/or incubators, which are broadly used in biological laboratories.[15] However, the

11 SKU: VP 774-5 http://vp-sci.com/.

12 Stir bars. (i) Stir Bars for Photoredox & Parallel Synthesis (gold) Block Systems, Dimension: 1.98 mm diameter x 4.80 mm length, SKU: 13258; (ii) Stir Bars for Photoredox Micro (silver) Block Systems, Dimension: 1.32 mm diameter x 1.57 mm length, SKU: 13257. https://www.analytical-sales.com/.

13 http://vp-sci.com/chemistry/reaction-vials-stir-bars.html.

14 Tumble stirrers or shakers: *V&P Scientific* or *Unchained/Freeslate Core Module*, cost will depend on vendor and will require a quote.

15 Temperature controlled shakers: *Eppendorf* — Eppendorf ThermoMixer® C, Catalog No. 5382000023. https://www.eppendorf.com. (ii) Incubator: EchoTherm™, cost will depend on vendor and model.

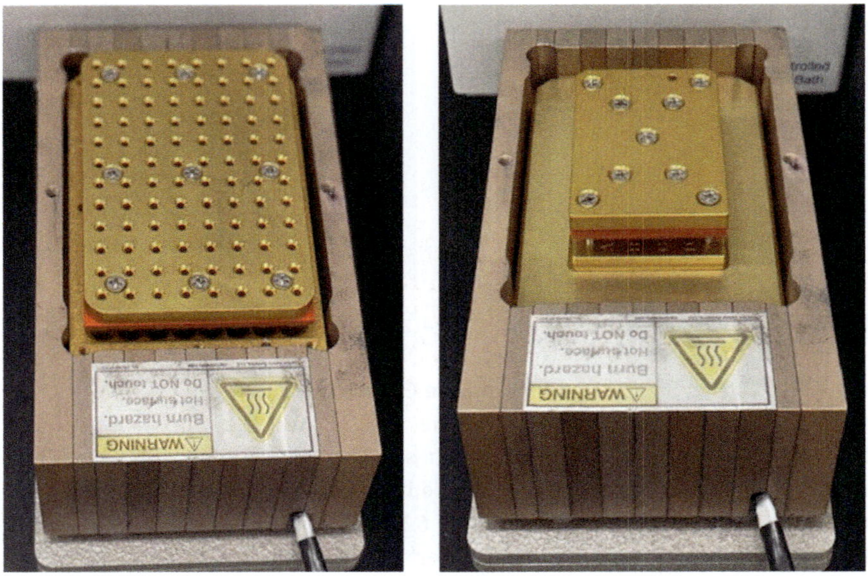

Figure 6.14 Temperature regulated tumble stirrer with a 96-well plate (left) and a 24-well plate (right).

reproducibility of reactions with solid reagents often depends on the ability of the chosen mode of agitation to break up those solids, which suggests that the homogeneity or heterogeneity of the reaction mixture needs to be considered when choosing the mode of agitation. Glass beads as alternative to stir bars are also available for use in a shake reactor[16]; in practice, those beads are most often used for reactions with gases (e.g. hydrogenations) to improve gas transfer into solution.

Closely associated with the mode of stirring is the mode of temperature control. Tumble stirrers can typically be connected to circulating temperature control solutions that allow cooling or heating in the same reaction setup. Such temperature control solutions are available from many manufacturers (e.g. Huber[17]). For "traditional" stir plates, a heating element is common, but cooling is not often part of the solution. For this purpose, Cambridge Reactor Design has provided a simple solution: The polar bear, a chilled stir plate that can be equipped with different reactor blocks for parallel synthesis and HTE.[18]

---

16 Glass Beads. *Fisher Scientific*: Solid Borosilicate Glass Beads – Size: mm, Catalog No.: 10-310-1. https://www.fishersci.com/us/en/home.html.
17 For example, the Huber Unistat 705 series: https://www.huber-online.com/us/product_datasheet.aspx?no=1068.0001.01.
18 Polar Bear Product Family: https://www.cambridgereactordesign.com/polarbearplus/index.html

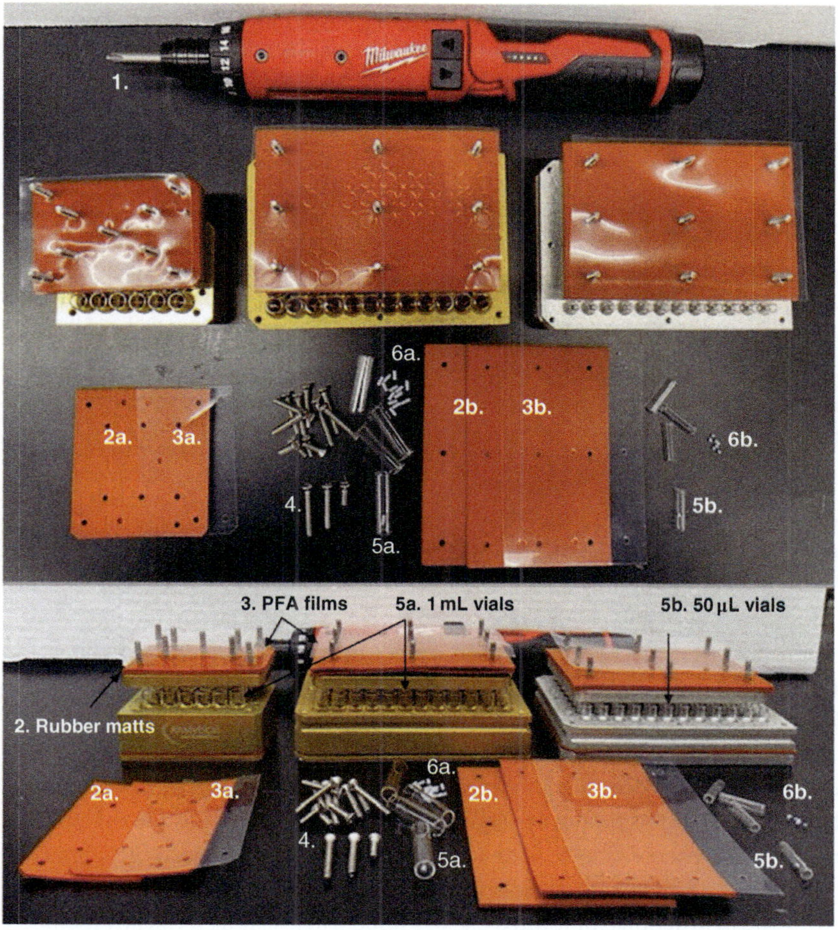

Figure 6.15 **Example of consumable items for HTE.** 1. Electric screwdriver, 2. rubber mats of different sizes, 3. PFA films of different sizes, 4. screws of various sizes, 5. glass vials (5a. 1 mL vial vs. 5b. 250 μL vial), 6. stir bars of different sizes.

6.3.4.6 Consumables

Common consumables that need to be replaced regularly are rubber mats, PFA sealing films, and screws (Figure 6.15). Those parts are required to seal the reaction blocks and are available in various sizes fitting different aluminum block layouts from various vendors.[19] In comparison to screws and rubber mats, PFA films that

19 HTE consumables: *Analytical Sales and Services*: (i) **96-well plate** (1 mL, 8 x 30 mm) – Replacement Rubber Mats, SKU: 96965; Replacement PFA Films, SKU: 96967. (ii) **24-well plate** (1 mL, 8 x 30 mm) – Replacement Rubber Mats, SKU: 24248; Replacement PFA Films, SKU: 24247. https://www.analytical-sales.com/

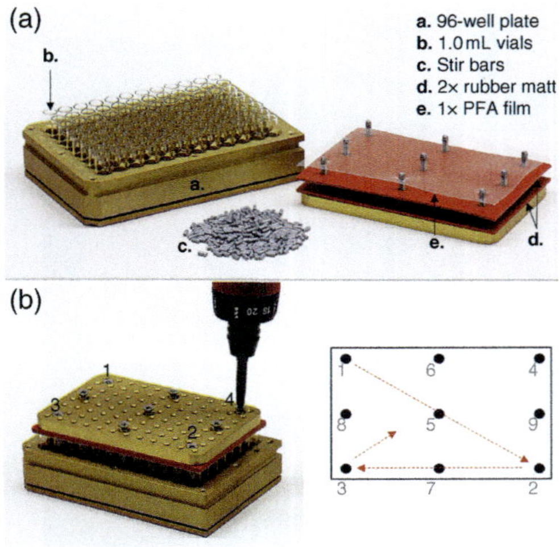

Figure 6.16 **Assembling an HTE plate.** (a) Preparation of HTE plate. (b) Illustration of fastening top plate by screwing and suggested screw pattern.

come in direct contact with the reaction mixtures are never recommended for reuse between high-throughput screens to avoid cross contamination.

The importance of using an *electric screwdriver* for sealing the top sealing plates to the bottom of the reaction plates cannot be underestimated (Figures 6.15 and 6.16). It is both convenient for rapid and even sealing and provides the ability to limit the amount of torque applied. This is a crucial feature, as too much force when tightening screws can lead to a lost experiment, if screw heads get damaged or vials crack. As such, investing in a relatively inexpensive screwdriver and heading the instructions for limited use of torque when sealing (typically included with HTE plates and/or starter kits) can prevent loss of productivity.

6.3.5 Considerations for Experimental Setup

6.3.5.1 Reaction Atmosphere

Setting up HTE plates under inert atmosphere is also part of HTE best practices to avoid any erroneous results caused by the presence of air or moisture. Furthermore, drying HTE glass vials and stir bars in a vacuum oven prior to use is another option to reduce the impact of residual moisture. Both storage and setup are typically performed in commercially available gloveboxes, even though nitrogen-flushed boxes can be used for storage. One important consideration for glovebox usage is the amount and diversity of solvents used during setup. Therefore, many HTE-focused groups have decided to switch off the circulation over a catalyst bed

Figure 6.17 KitAlysis™ inertion box. The box is shown with an open lid (can be further closed by sliding), examples of stock solutions, and a 24-well plate containing 1 mL HTE vials.

and instead constantly purge heavily used gloveboxes with a flow of nitrogen to avoid cross-contamination between HTE experiments. The flow rates of purging can typically be controlled from the glovebox control panel.

In case gloveboxes are not available, an inertion box[20] that allows the creation of an inert atmosphere on the benchtop has been commercialized by Sigma Aldrich. This KitAlysis™ inertion box (Figure 6.17) can be an alternative, particularly for setting up 24-well plates.

The intertion box can be connected to a supply of inert gas and flushed, while the HTE reaction block, vials, and reagent bottles are placed inside. Continuous flushing during pipetting and sealing the plate thus limits the influence of air and moisture on reaction outcomes. The KitAlysis™ system also offers screening kits for common reactions (e.g. Cu catalyzed C–N coupling, olefin metathesis, and Suzuki–Miyaura reaction). These KitAlysis™ kits contain all required reagents, dry solvents, and detailed step-by-step instructions, thus lowering the monetary investment and barrier to testing HTE for commonly used reactions.

6.3.5.2 Reagent Preparation and Dispensing

In traditional singleton or parallel reaction setups, weighing solid reactants is one preferred way to increase the accuracy of reagent ratios. This can be a time-consuming, repetitive, and tedious task, particularly when preparing many reactions at once and in parallel. Therefore, dispensing solid reagents as stock solutions or slurries is preferred when preparing an HTE experiment, which significantly shortens the setup time. Thus, designing an HTE experiment (in Excel, library studio, or

20 Sigma-Aldrich Co. LLC- (i) KitAlysis™ Benchtop Inertion Box, SKU: Z742064-1EA. (ii) KitAlysis™ 24-well Reaction Block and Inertion Box Combination Starter Set, SKU: Z742108-1KT. (iii) KitAlysis™ 24-well Reaction Block and Screwdriver Set, SKU: Z742107-1KT. (iv) Photo KitAlysis™ Starter Kit, SKU: Z742612-1KT.

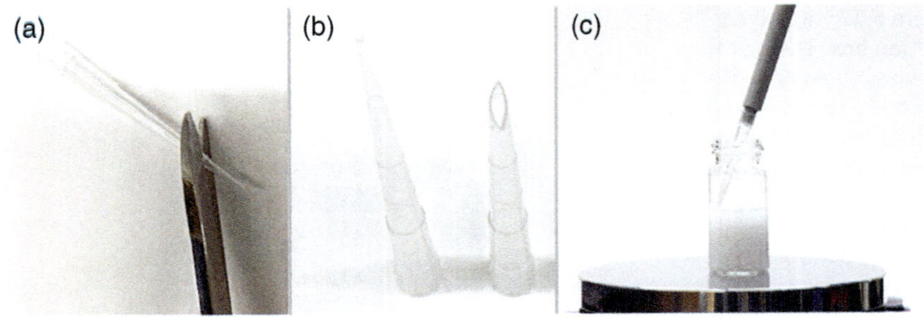

Figure 6.18 Illustration of how to prepare a pipette tip for stock slurry dispensing. (a) Demonstration of enlarging pipette tip by diagonal cut with a pair of scissors. (b). Illustration of normal pipette tip (left) versus pipette tip with enlarged opening (right). (c) Illustration of dispensing a stock slurry: slurry is stirred vigorously on a stir plate while desired amount of slurry is dispensed with a single channel pipette with enlarged pipette tip.

another software) typically requires decisions regarding suitable concentrations of stock solutions and exact orders of addition of these solutions to the reaction wells.

Reagents insoluble in the reaction solvent (e.g. inorganic bases in organic solvents) are most often handled using one of two approaches: (i) As stock slurries in the reaction solvent or (ii) by predispensing in a different solvent, followed by solvent evaporation. Stock slurries are typically dispensed using a pipette tip with an enlarged opening (i.e. a diagonal cut with a pair of scissors often prevents undesired clogging, Figure 6.18). It is critically important to continue to stir the slurry vigorously with a suitably sized stir bar while dispensing to prevent settling of the solid to the bottom of the source vial. Slurries are most accurately dispensed if suspensions with small grain sizes of the solid can be prepared. This can be achieved by vigorously stirring slurries up to several hours before dispensing into an HTE plate; in practice, 30 minutes are often good enough. When preparing slurries, targeting concentrations of less than 0.5 M is often prudent to allow the use of enough solvent to form a dispensable suspension. This is a general guideline, but in case a slurry is too thick, doubling the amount of solvent used can often transform a non-dispensable suspension into a suspension that can be easily handled.

Hand-held solid dispensers are an alternative to slurry dispensing. The Biodot Powder Dispensers (currently discontinued and no longer commercially available), the Mettler Toledo Quantos HPD, and the Drummond digital microdispenser are examples of such solid dispensers.[21] In comparison to slurry dispensing, quantitative dispensing typically requires prior calibration with the solid powder

21 (a) Mettler Toledo Quantos HPD (Handheld Powder Dispenser) https://www.mt.com/dam/P5/labtec/06_Automated_Sample_Preparation/07_HPD_Series/03_Documentation/HPD_Brochure_EN.pdf. (b) Drummond positive displacement digital microdispensers https://www.drummondsci.com/product/microdispensers/positive-displacement-digital-microdispensers/.

in question. As such, these solid dispensing solutions are preferred for dispensing the same solid across a larger number of HTE vials.

Predispensing reagents instead of slurry dispensing can be a good alternative when the reagents in questions are broadly useful (e.g. metal precursors, additives, ligands, and bases) and can be dispensed in a solvent that can be readily evaporated. Preplating reagents that are commonly screened (i.e. the preparation of HTE kits) for later use can lead to considerable time savings and lower barriers to access HTE workflows [3a]. However, compatibility and stability of reagents should be considered when preparing preplated HTE kits. Best practices include dispensing under inert atmosphere, covering the prepared plates with chemically suited mats,[22] and storing plates under inert atmosphere.

With regard to suitable concentrations of slurries and solutions for predispensing reagents, the maximum concentrations depend on the solvent and reagent of choice. Generally, a good guideline is to target dispensing volumes between 10 and 200 μL. Smaller volumes are typically not very accurate, especially when volatile reagents are involved, while larger volumes take longer to evaporate. Dispensing concentrations between 0.2 and 0.01 M are typical, but will vary depending on the exact recipe.

Another consideration related to dispensing is connected to the order of addition of reagents. For example, one common type of HTE experiment is a solvent/base screen. For these screens, solid bases are frequently used in preplated form, followed by plating the substrate. If the substrate is not volatile, solvent evaporation followed by addition of reaction solvents is typically the easiest way to enable diverse screen designs. In a last step, liquid bases are added to the plate, as these are typically soluble and can rapidly interact with substrates in solution - in contrast to the solid bases that typically do not dissolve as readily and thus have limited interactions with the substrates before stirring and heating initiate the desired reaction.

6.3.5.3 Storage of Preplated Reagents

Reactions on small scale with less than 100 μL volume are typically more sensitive to air and moisture than their larger counterparts. For example, inorganic bases such as K_2CO_3 remain a fluffy solid when stored in bulk in an air-tight source bottle, but water accumulation from the atmosphere occurs once dispensed into smaller reaction vials, even when stored in a desiccator. Thus, it is recommended to store predispensed reaction vials under an inert atmosphere and in sealed bags. Some laboratories use zip-lock bags for storage of preplated reagents; Mylar bags are another alternative and are used in commercial settings (e.g. for Sigma-Aldrich's Kitalysis kits). Furthermore, disposing of predispensed inorganic bases after one year of storage is good practice

22 *Analytical Sales and Services*: (i) Silicone/PTFE Cap Mat, SKU: 96726; (ii) Clear Ultra Thin Round Well Cap Mat, SKU: 964007. https://www.analytical-sales.com/.

6.3.5.4 Pipetting

Single channel, multichannel, and repeater pipettes are crucial tools for preparing HTE experiments; all are offered by various vendors[23] and enable high accuracy for dispensing reagents, stock solutions, or stock slurries when calibrated on a regular basis (once or twice per year; Figure 6.19). Typically, pipettes are calibrated using aqueous solution, which introduces a systematic error into experiments. This needs to be considered whenever using volumetric devices of

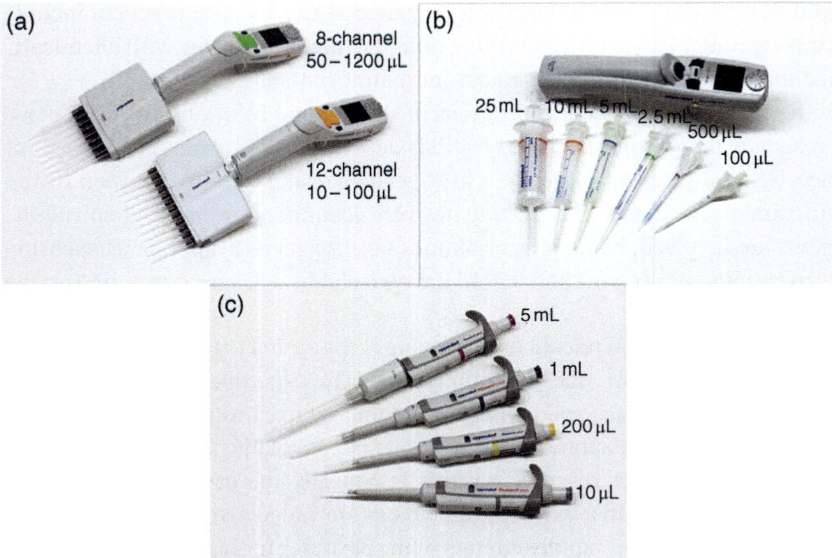

Figure 6.19 Example of available multichannel (a), single channel (c), and repeater pipettes (b).

---

23 Single-, multichannel – and (electronic) repeater pipette vendors are *Eppendorf, Rainin, Fisher, VWR*.
Eppendorf – (i) Eppendorf Research® plus, 3-pack 1-channel pipettes: Volume: **Option 1**: 0.5–10 µL, 10–100 µL, 100 –1000 µL, **Option 2**: 2–20 µL, 20–200 µL, 100–1000 µL, **Option 3**: 100–1000 µL, 0.5–5 mL, 1–10 mL), Catalog No. – 3123000900 (**Option 1**), 123000918 (**Option 2**), 3123000926 (**Option 3**). (ii) Eppendorf Research® plus, multichannel pipettes (8-, 12-, 16- and 24-channel): 8-channel pipette - Volume: 120–1200 µL, Catalog No. 3125000214; 12-channel pipette - Volume: 10–100 µL, Catalog No. 3125000044. (iii) Repeater® E3 Pipette (1 µl–50 mL): Catalog No. 4987000118. For more information regarding pipette selection visit https://www.eppendorf.com
Fisher – (i) Fisherbrand™ Elite Pipette Kit – 1-channel pipettes: Volume: 0.2–2 µL, 2–20 µL, 20–200 µL, 100–1000 µL, Catalog No. 14-388-100. (ii) Fisherbrand™ Elite™ variable volume, multichannel pipettes (8- and 12-channel): 12-channel pipette – Volume: 10–100 µL, Catalog No. FBE1200100. (iii) Thermo Scientific™ E1-ClipTip™ Bluetooth™ Electronic Multichannel Pipettes (8-, 12 and 16-channel): 8-channel pipette – Volume: 15–1250 µL, Catalog No. 14-3879-57BT. For more information regarding pipette selection visit https://www.fishersci.com/us/en/home.html.

any kind, especially upon scale-up, when weighing out reagents increases the accuracy of reagent ratios. Multichannel and repeater pipettes are particularly useful for dispensing liquid reagents or stock solutions for large reaction arrays. In order to dispense slurries accurately, the use of single-channel, manual pipettes are recommended in combination with scissor-cut pipette tips as discussed above.

6.3.5.5 Solvent Evaporation

Solvent evaporation is common practice in synthetic laboratories and often performed using rotary evaporators or vacuum manifolds. However, those tools are not amenable for solvent evaporation from HTE plates. Two options to remove solvents in HTE setups are Genevac evaporators[24] or more cost-efficient blow-down blocks[25] (Figure 6.20). Both are ideally installed in a glovebox under inert

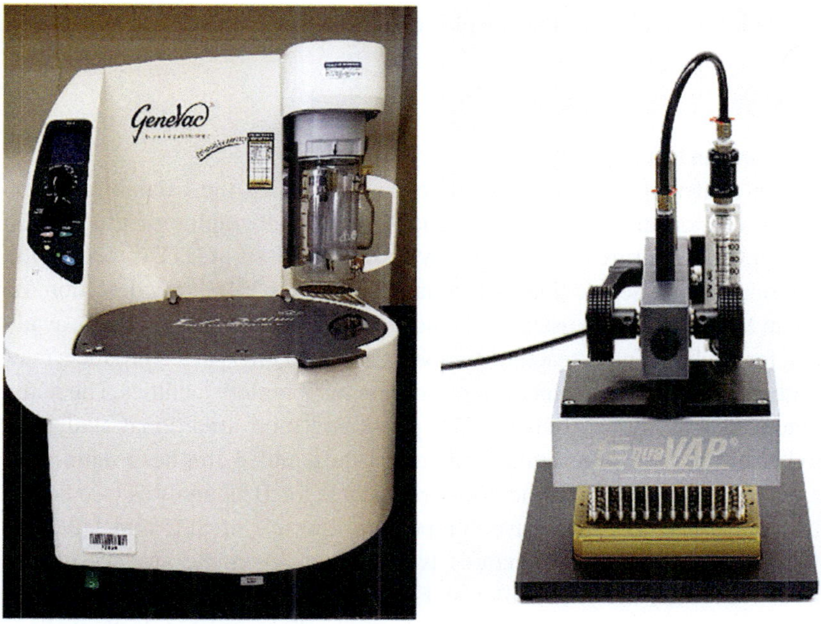

Figure 6.20 Solutions for solvent evaporation from HTE plates: Genevac evaporator (left) (*Source:* SP Industries, Inc.) and blow-down block for 96 well reaction block (right) (*Source:* Analytical Sales and Services, Inc.).

24 GeneVac prices are dependent on vendor contracts and quoting systems in place. Installing Genevacs in a glovebox may cause additional costs due to potentially required modifications (e.g, for plumbing, lower/raising the work surface).

25 Analytical Sales & Services offers EquaVAP® **X**-Well Evaporators (**X** = 24, 54, 96, 384 -Well). https://www.analytica-sales.com/product-category/collection-plates-accessories/devices-accessories/equavap-evaporators-devices-accessories/.

atmosphere to aid with plate preparation, which in turn can cause additional rebuilding costs (e.g. for the installation of vacuum lines and/or nitrogen lines leading into the glovebox).

If the glovebox used for HTE setup is in constant purge mode, another option for evaporation (especially with relatively volatile solvents such as THF) is to plate reagents the day before and leave the plate to evaporate overnight. This approach is very common when preplating is performed using automation platforms: Robotic plating itself is a multi-hour process and often set up overnight, which commonly leads to arriving in the morning with a series of preplated HTE kits ready to use. It is possible to plate several reagents this way into the same vial (e.g. ligands and bases), if desired. This is more rarely done than plating only one reagent per vial (e.g. ligands), since automated plating using a robotic device typically serves to establish reagent libraries to be customized at a later date. A word of caution for this approach: Ideally, all reagents that are plated in this fashion should be nonvolatile.

6.3.6 Analysis of HTE Screens

6.3.6.1 Suitable Instrumentation

The key to analyzing high-throughput screens is finding the balance between managing reaction quantity and data quality. Chromatographic methods (GC, HPLC, UPLC, or SFC) are most commonly used for analysis of HTE screens, with optional mass spectroscopic (MS) detection in addition to FID or UV detection. In particular, MS detectors are useful when no authentic product standards are available and for applications in reaction discovery, in which unexpected products may be of interest. All these instruments are common in laboratory facilities. The main difference between the analysis of HTE plates and more traditional analytical methods is the quantity of reactions that needs to be handled; this has a significant influence on which analytical methods are suitable for HTE analysis (see below for further details). For UV-active compounds, the use of SFC and UPLC is typically preferred, while GC analysis is often favored in the absence of UV chromophores in the analytes of interest. Finally, for ultra-HTE with thousands of reactions to be tested, MS-based methods have also been utilized, as they allow rapid data acquisition [22–24].

6.3.6.2 Autosampler Configurations

One convenient way to perform high-throughput analysis is the use of specialized 96-well plates[26] that fit into instrument autosamplers. These plates are optimized for parallel pipetting with multichannel pipettes for more efficient workup and

26 Analytical Sales & Services: 96-Well – Square Well Format – SKU: 17P687; Autosampler Compatible 1 mL Cap Mat-SKU: 96057. https://www.analytical-sales.com/.

are compatible with Waters and Agilent chromatography systems. However, this option can lead to additional costs for consumables and requires nonstandard instrument additions. As a more cost-effective alternative, common GC/UPLC vials (2 mL volume) can be used for both reaction setup in a 48-well plate[27] and analysis (after quench and dilution).

6.3.6.3 Analytical Methods

HTE analysis using chromatographic methods is particularly sensitive to the length of the used method. With a 2-minute method (the most common method length for UPLC/MS analysis), analyzing a 96-well plate requires 3.2 hours of analysis time; in reality, the time is more typically around 4 hours due to delays between samples (e.g. for column equilibration). Taking this timing into account, analyzing two to three plates per night can be reasonable in a shared analytical facility; more, if the UPLC/MS instrument can run around the clock without interruption by users or significant maintenance (e.g. solvent refill, waste solvent removal, other preventative maintenance). However, in certain cases (e.g. when determining *ee* for asymmetric transformations or purity for late-stage process development applications), longer methods may be necessary, which limits data output.

6.3.6.4 Internal Standards and Assay Yields

The use of internal standards is generally recommended for HTE experiments, as it allows detection of substrate and product disappearance. Considerations for internal standards for chromatographic methods are somewhat different for internal standards used in more common NMR methods.

An ideal internal standard has the following properties: (i) inexpensive; (ii) unreactive under the reactions conditions, so that it can be added at the very beginning of the reaction together with the substrate; (iii) nonvolatile to avoid accidental evaporation and data inaccuracies; (iv) sufficiently soluble in the reaction solvent; (v) fragments readily for MS detection; (vi) has good UV absorbance for UV detection; and (vii) has a retention time that does not overlap with the analytes in a relatively short method suitable for HTE analysis (two to three minutes).

This list is difficult to completely fulfill and therefore, internal standards that adhere to most but not all of these criteria are commonly used. The best example is biphenyl, which fulfills criteria (i), (ii), (iv), and (vi) but shows some volatility and does not fragment readily in ESI-MS detectors. Furthermore, its signal peak typically shows up in the middle of a typical two-minute run on HTE UPLC/MS

---

27 Analytical Sales & Services: 48 x 2 mL Aluminum Multi-well Reaction Block, SKU: 48012; 11 mm, 12 x 32, 2 mL Clear Glass Wide Mouth Crimp/Snap Vials, SKU: 11211-CASE; Assembled 48-Well Vial Tray with Numbered Crimp Vials, SKU: 48221M; 11 mm Silver Aluminum Crimp Caps with PTFE/White Silicone Liners, SKU: 20032-CASE. https://www.analytical-sales.com/

methods (95% H_2O/5% MeCN to 0.1% H_2O/99.9% MeCN; 0.8 mL min^{-1} flow; Acquity BEH C18 1.7 μm 50 mm column; 50 °C) and thus can overlap with solvent and analyte peaks. However, its big advantage is its chemical stability and as such, biphenyl is typically a good choice when accurate relative activity measurements of catalysts are needed.

More expensive internal standards can be added at the end of the reaction with the quench solution to provide an alternative to adding an internal standard into the reaction itself. Standards that work well for this include 4,4′-di-*tert*-butyl-biphenyl and 4,4′-di-*iso*-propyl-biphenyl; however, both compounds could be reactive under highly acidic reaction conditions (retro-S_EAr) and are thus typically not added at the beginning of a reaction. Regardless of the standard chosen, adding 0.1–0.5 M equivalent of a standard to a reaction mixture typically allows for a good readout, as long as the absorption of standard, substrate, and analytes are similar at the chosen wavelength.

In an ideal scenario, standards for reaction products and side products are available (this is common in most late-stage process development situations). If this is the case, chromatographic methods can be used to generate calibration curves (with or without the use of an internal standard) that allow an estimation of crude yields (also called assay yield).

However, product standards are often not readily available. In those cases, LCAP (liquid chromatography area percent) values are frequently used to visualize data and/or discuss reactivity trends. Common UV wavelengths for determination of LCAPs of organic compounds are 210 or 215 nm. In this region of the UV spectrum, aromatic substructures in molecules of interest are the primary absorbing features, which allows practitioners to focus on the most important data in their experiments: the relative change of LCAP values across an HTE plate. Other wavelengths such as 245 nm also detect compounds that are typically of less interest (e.g. solvents) and are thus not used frequently for HTE analysis. However, comparing a complete UV spectrum to an available standard is a useful tool, especially if MS detection is not available; therefore, UV detectors that provide access to the complete UV spectrum (such as DAD or PDA) are useful.

The information obtained from HTE data analysis is frequently shown as ratio of LCAP product and LCAP internal standard (LCAP Pdt/LCAP IS). For reactions that do not suffer from substrate and/or product decomposition, both LCAP Pdt and LCAP Pdt/LCAP IS should show the same patterns when plotted (more on data visualization below). However, comparing patterns for both values is good practice to spot inconsistencies between reactions and prevents data misinterpretations.

6.3.6.5 Data Visualization and Analysis

As mentioned briefly in the last paragraph, the analysis of patterns of LCAP Pdt and for LCAP Pdt/LCAP IS is one key tool for analysis of HTE data. To enable this type of analysis, data need to be exported into a format that can be easily manipulated and visualized using common data analysis tools.

The key to success for this critical step is the availability of suitable software. Most commercially available chromatographic systems come equipped with the respective data analysis tools (e.g. Agilent instruments include ChemStation software) that allow definition of peaks of interest in a set of analytical traces. Alternative solutions exist (e.g. Virscidian or ACD Labs Katalyst) that can read and analyze analytical data files from a variety of sources. Once peaks of interest (e.g. substrate, product, and internal standard peaks) have been identified using data source files, all these software packages allow exporting peak data as tabulated data (e.g. as .xlsx or .csv files). Using these exported tabulated data files, common data visualization tools can be used. The following paragraph will only describe a small number of visualization tools that are considered specifically useful for HTE data analysis, but the designers of visualization software are continuously improving these functions. Thus, the authors would not be surprised if even better visualizations will be broadly available in the near future.

Likely the most straightforward way to visualize HTE data is the use of Excel, the same program that many HTE practitioners use to design screens. One of the advantages of this approach, as mentioned earlier, is that this software is typically widely available, and most scientists and engineers have a basic familiarity with Excel's functions. One function the authors would like to particularly point out as useful for HTE visualization is the "heatmap" function (in Microsoft Office 365: Home tab » Styles Toolbar » Conditional Formatting » Color Scale). This tool adds automatic color gradients (as shown in Figure 6.21, top) and thus clearly points the eye toward data maxima/minima and patterns in the data. Alternatively, bar diagrams (Figure 6.21, bottom) can serve a similar purpose, but often need to be formatted heavily for clarity.

A more advanced and powerful option for data visualization is the use of Spotfire,[28] which allows the use of size and color in the same plot to provide insight into two variables at a time (see Figure 6.22). This approach is not just visually attractive but allows condensing a lot of data into one graphic. This becomes especially important when larger sets of data (>48) are being considered at once and in cases where different reaction outcomes (e.g. ee and conversion) matter for subsequent decision making.

6.3.7 The Role of Automation and Robotics in HTE

As the readers of this chapters may have noticed, we have not yet mentioned automation or robotics as a key part of HTE experimentation, except for discussing

---

28 TIBCO Spotfire® https://www.tibco.com/products/tibco-spotfire/visual-analytics.

ee	1	2	3	4	5	6	7	8	9	10	11	12
A	63	51	53	39	−1	100	57	100	44	−5	n/a	−38
B	n/a	n/a	n/a	n/a	−100	−100	100	n/a	−100	95	58	87
C	24	66	30	38	5	100	64	100	56	−27	−2	n/a
D	24	53	22	38	n/a	100	40	100	54	−37	−29	n/a
E	76	75	75	75	100	76	76	68	74	75	66	100
F	−44	24	−100	n/a	−100	54	−26	−100	21	24	100	n/a
G	66	38	38	48	33	49	51	20	53	50	32	50
H	78	49	42	63	100	49	50	42	58	57	52	52

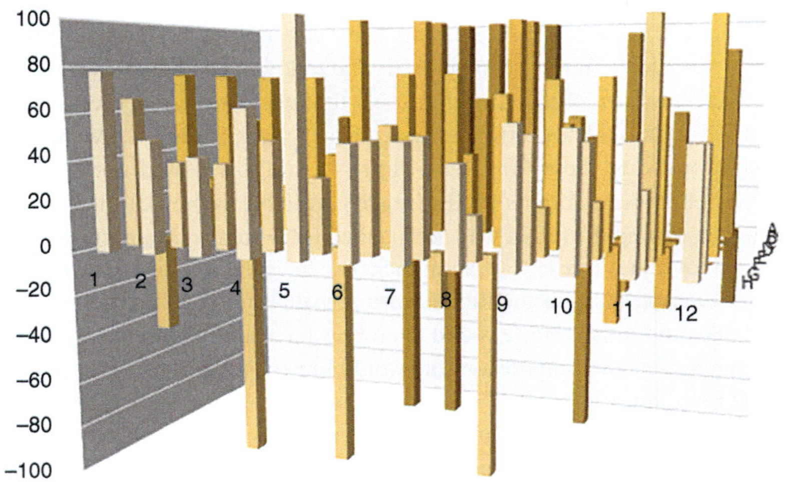

Figure 6.21 Examples of Excel plots visualizing enantiomeric excess (ee) as heatmap (top) and as bar diagram (bottom) for same set of data obtained from a 96-well screen.

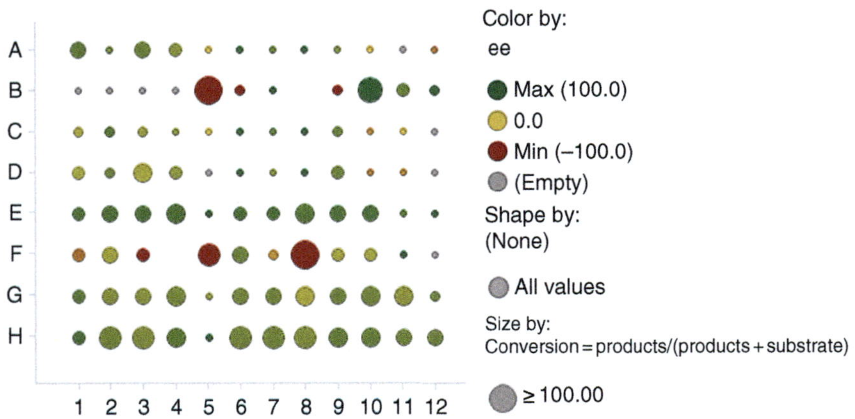

Figure 6.22 Example of Spotfire plot, visualizing ee (by color) and conversion (by circle size) vs. location on a 96-well plate.

autosamplers for UPLCs and GCs, which enable higher data throughput. This may be surprising to many HTE novices, as the current literature showcasing HTE as an enabling technology for reaction discovery, development, or optimization often refers to specialized and expensive automatization tools, which are uncommon outside of specialized facilities. This abundance of automated tools in the HTE literature can create the impression that advanced tools are a requirement for successful HTE experimentation.

We hope that uncovering some of the detailed practical aspects of HTE in the previous sections may have helped to shift this impression. Typical reaction optimization or reaction screening to identify new reactivity can be enabled by a 96-, 48-, or 24-well plate format, which is practical and allows for rapid turnaround and data analysis. With these types of HTE experiments, robotic tools are helpful when it comes to repetitive tasks (e.g. dispensing inorganic bases or ligands into a large number of plates with identical designs). Those scenarios are rather rare in cases that are most common in academic laboratories, which are often focused on reaction discovery. If the need for repetitive plate preparation arises in these and similar surroundings, a new trainee may do a better job than a robot while learning from the experience of executing a routine task.

However, robotics and automation have significant upsides for supporting a highly trained workforce and for executing on tasks that require significant precision. As such, it is not surprising that many industrial laboratories and large, specialized academic facilities heavily rely on such tools, as they increase the throughput of repetitive workflows. It is clearly an exciting and innovative field with extraordinary advances in the last decade. For this chapter, we would like to point out two cases, in which robotics are of paramount importance for the success of HTE workflows.

1) Further miniaturized HTE (e.g. in 384- or 1536- well plates) can enable parallel syntheses or reaction optimizations on very small scales, for example in drug discovery. In those cases, the availability of robotic dispensers and other automated technologies is highly beneficial due to the necessity for accuracy with very small volumes across a large array of reaction wells. These conditions are met by most liquid handling platforms.[29]
2) Automated solid handling platforms have started to emerge onto the market in the last decade.[30] These are particularly useful (as true for many robotic platforms) for highly repetitive tasks (e.g. weighing out solid reagents; see

29 Vendors for **liquid handler robots**: Unchained Laboratories/Freeslate/Symyx, Chemspeed Swing, Swing XL, Tecan Freedom EVO, Thermo Matrix 2 × 3, Andrew Alliance, and Rainin Benchsmart.

30 Vendors for **solid handler robots**: Mettler-Toledo (Quantos QB5 and QX96, and XPE205) Unchained Laboratories (Solid-Junior), Chemspeed (SDU, GDU-P FD), and JKem (Eclipse).

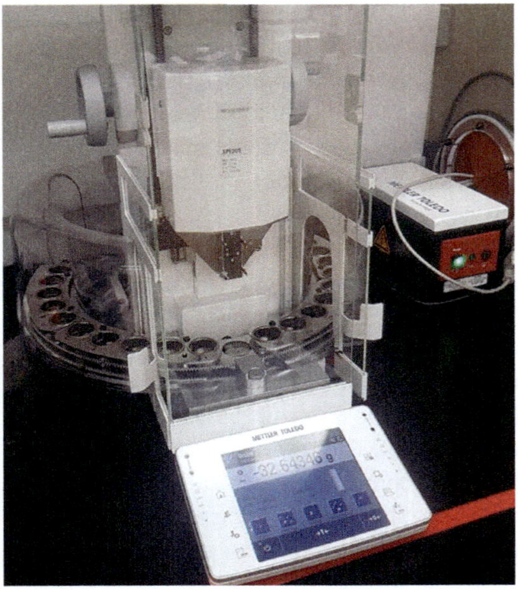

Figure 6.23 Automated balance XPE205 with sample changer (Mettler Toledo).

Figure 6.23 for an example of a Mettler Toledo solid dispensing platform with sample changer). There are specific challenges with solid handling robots, which have been discussed in literature [25–27]. Aside from the cost of such automatization tools, maintenance, operation, and programming require dedicated training to enable efficient operation.

Overall, automated and robotic tools are very useful for repetitive tasks (such as preplating solid reagents for HTE kits with the intent to store those kits). Collaborating with key vendors (similar to the Sigma Aldrich KitAlysis™ collaboration model) to outsource those types of tasks may be one of the most promising options for the democratization of HTE and would impact many prospective HTE users. In several cases, high degrees of automation can achieve extraordinary feats [27]. However, for the typical academic or industrial chemist who may be interested in acquiring first-hand HTE knowledge, robotics, and automation do not need to be the first area of concern.

6.4 Section Summary and Outlook

Overall, this section details a series of practical tips and tricks for HTE users. Miniaturizing reactions has led to adapting slightly different workflows for HTE experiments than for synthetic chemistry experiments set up one at a time. All these small changes allow for more streamlined experimentation, resulting in the impacts detailed in the introduction to this chapter (shorter setup time per reaction, efficient reagent use, multivariable optimizations, understanding reactivity patterns, space for increased serendipity in reaction design).

In the next part of this chapter, two case studies describing HTE platform development and HTE reaction discovery/optimization are discussed in detail. Each example highlights how different HTE workflows contribute to the success of the overall project. We hope that these studies are useful templates for future HTE designs and convey the logic of working in high-throughput formats.

6.5 Case Study 1: Development of an HTE Platform for Nickel-Catalyzed Suzuki–Miyaura Reactions

6.5.1 Motivation

Suzuki–Miyaura reactions are an asset to synthetic chemistry and ae frequently applied across academia and industry [28, 29]. Identifying ideal conditions for efficient Suzuki–Miyaura reactions is an important if somewhat repetitive task and its success depends on the complex interplay of different reaction variables (e.g. substrate structures, base additive, catalyst, solvent). As such, HTE is an efficient approach to address this challenge and can enable rapid screening of all variables. A representative example of such HTE enablement is demonstrated in a recent publication developing a general platform for the Ni-catalyzed Suzuki–Miyaura coupling of heterocycles (Scheme 6.1) [30].

6.5.2 Design of Test Reaction and Initial Ligand Screen

The authors selected (2-methoxyphenyl)boronic acid (**1**) and 2-chloro-5-fluoropyrimidine (**2**) as a first set of model substrates (Scheme 6.2a). This choice

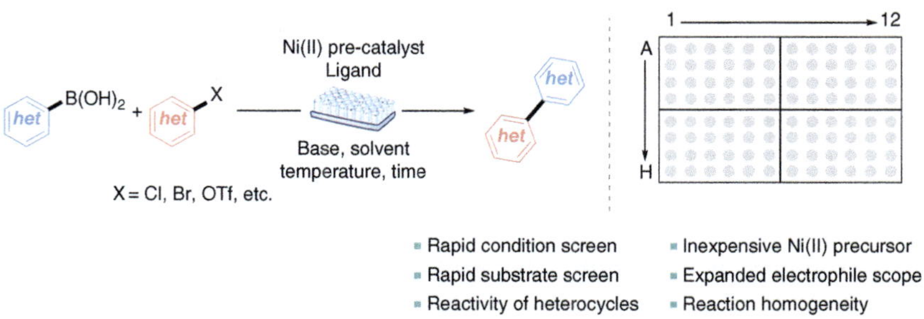

Scheme 6.1 HTE screening platform for the Ni-catalyzed Suzuki–Miyaura coupling of heterocycles. *Source:* Adapted from Goldfogel et al. [30].

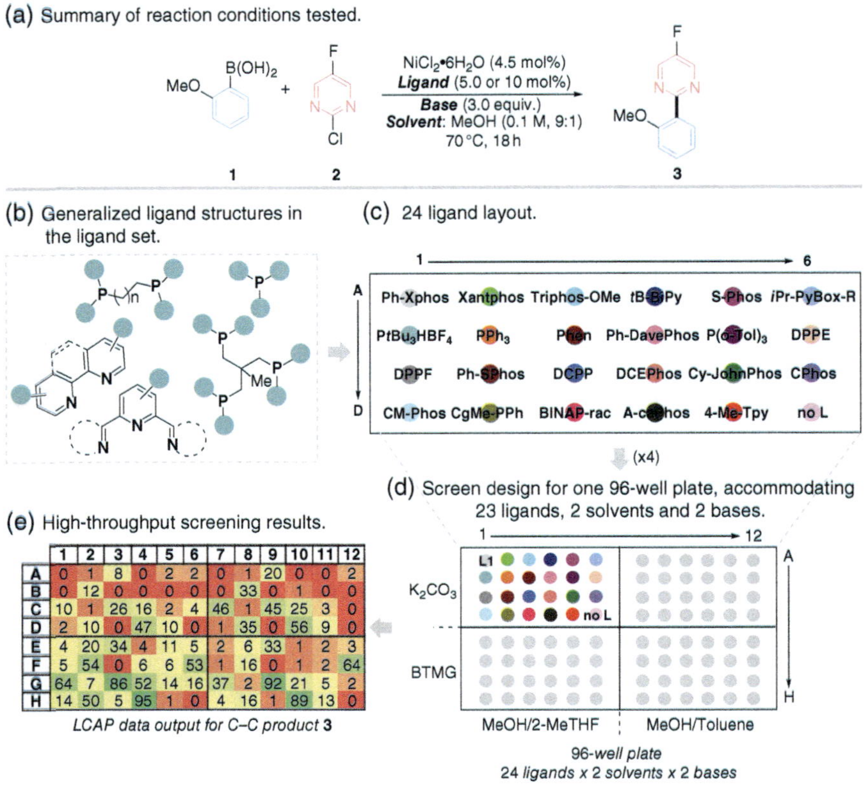

	1	2	3	4	5	6	7	8	9	10	11	12
A	0	1	8	0	2	2	0	1	20	0	0	2
B	0	12	0	0	0	0	0	33	0	1	0	0
C	10	1	26	16	2	4	46	1	45	25	3	0
D	2	10	0	47	10	0	1	35	0	56	9	0
E	4	20	34	4	11	5	2	6	33	1	2	3
F	5	54	0	6	6	53	1	16	0	1	2	64
G	64	7	86	52	14	16	37	2	92	21	5	2
H	14	50	5	95	1	0	4	16	1	89	13	0

Scheme 6.2 Screening round 1: HTE evaluation of ligand, solvent mixtures, and bases for Ni-catalyzed Suzuki–Miyaura coupling of heterocycles. Reaction results are shown in LC area percent (LCAP). BTMG = 2-*tert*-Butyl-1,1,3,3-tetramethylguanidine.
Source: Adapted from Goldfogel et al. [30].

of substrates (*N*-containing heterocycles and heteroaryl chloride electrophiles) is critical, as it incorporates traditionally less-reactive substrates from the very beginning of reaction development. Furthermore, it mirrors "real-world" situations in pharmaceutical development: Heterocycles are frequently encountered drug substructures, but many publications do not include them in their substrate scope. As such, it is often difficult to predict if a novel catalytic system will be impactful for the synthesis of active pharmaceutical ingredients (APIs). Using heterocycles as model substrates addresses this issue from the very beginning, thus enhancing the likelihood that the resulting catalytic system has significant utility for medicinal and process chemistry.

The authors' initial screen is designed to elucidate the effects of ligand scaffold, solvent, and base on product formation. The ligand plate design (Scheme 6.2b, c) includes 23 distinct and diverse ligand scaffolds and one well without ligand to gauge the amount of background reactivity in the absence of ligand. Due to the reaction evaluating a first-row metal catalyst, the choice of ligands includes both phosphine and N-based ligands, as previous reports of Ni-catalyzed C–C bond formations [31] have used both types of ligands. In addition, two solvent mixtures (MeOH/2-MeTHF and MeOH/toluene) as well as one inorganic and one organic base (K_2CO_3, BTMG) were evaluated, filling a 96-well plate design (Scheme 6.2d).

The analysis of the HTE screen by UPLC (Scheme 6.2e)[31] showed that the ligand scaffold had a significant impact on reaction performance: Phosphine ligands were more efficient than N-based ligands (e.g. phenanthroline or 2,2′-bipyridine). The data also demonstrated the importance of homogenous reaction conditions for product formation: Reactions in the presence of BTMG as base in the bottom half of the plate showed generally good reactivity when compared to their counterparts in the presence of K_2CO_3 in the upper half of the plate. As such, the identity of ligand and solvent are crucial reaction variables, while the identity of the base is also impactful. The highest performing ligands DCPP (wells C3/9, G3/9) and A-caPhos (wells D4/10, H4/10) afforded higher reactivity than the other ligands under any of the tested conditions.

6.5.3 Second Round of Ligand/Base/Solvent Screens

Based on this initial screen, 6 ligands were chosen for further optimization, using the same substrate model system (**1** and **2**). These ligands include the two best ligands DCPP and A-caPhos, as well as JohnPhos, PPh_3, DPPF, and DCEPhos. The latter four ligands all show at least some reactivity under the initially tested conditions (2–64 LCAP), but all show detectable product formation regardless of solvent mixture or base tested. This demonstrates one of the major differences in HTE-driven vs. traditional reaction optimization: A range of initial hits can be followed up on due to the relative ease of reaction setup. This allows for the selection of ligands for follow-up across the entire spectrum of reactivity and ligand structure as well as consideration of other factors in that design choice, such as commercial availability, cost, and ligand stability.

These 6 ligands were further evaluated in combination with 15 solvent/alcohol combinations, all in a 96-well plate format (Scheme 6.3b). Visualization of the reaction data (Scheme 6.3b) illustrates that among all the alcohol additives tested, MeOH shows the highest reactivity; one hypothesis that explains this reactivity boost is that MeOH enables homogeneous reaction mixtures that lead to an

31 Product formation quantified by Area Percent (=AP).

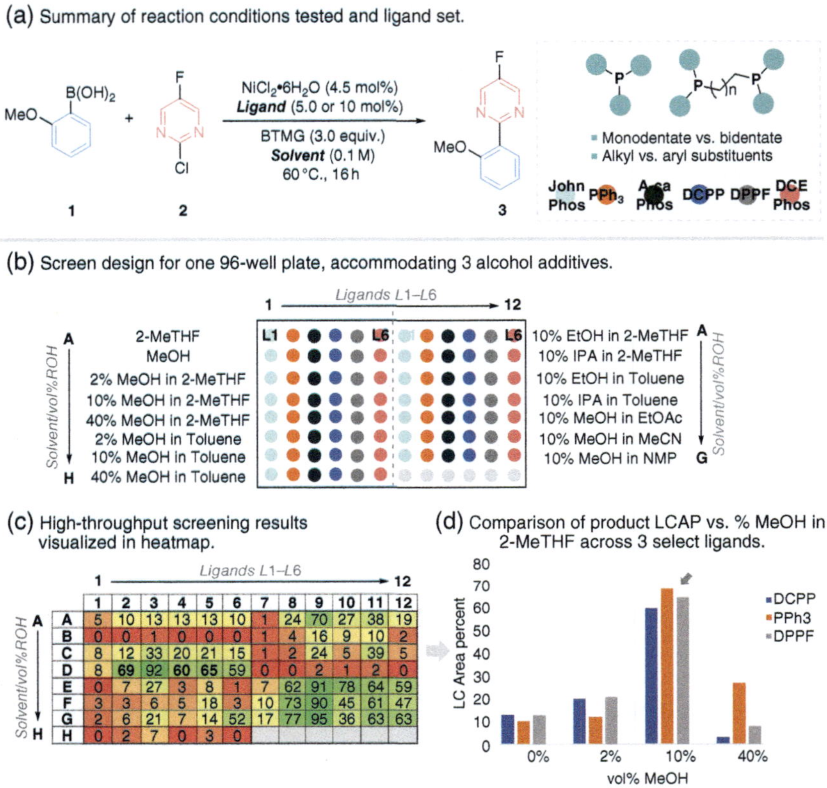

Scheme 6.3 **Screening round 2: Influence of solvent and % MeOH on coupling reactivity.** Reaction results are shown in LCAP. *Source:* Adapted from Goldfogel et al. [30].

increase in product formation. This trend is consistent across different ligand classes (e.g. DCCP vs. PPh_3 vs. DPPF) regardless of the co-solvent chosen. Furthermore, these data illustrate that the stoichiometry of the alcohol additive has a crucial impact on product formation (Scheme 6.3d): 10% MeOH is superior and showed high efficiency across different ligands scaffold and solvents.

After these initial key observations, several more screens were performed to elucidate the importance of base and Ni precatalysts [30]. The data from those high-throughput prescreens enabled the authors to distinguish between relevant reaction variables greatly affecting product formation (e.g. ligand scaffold, solvent) and less influential reaction variables (e.g. Ni(II) precatalyst, alcohol additive, base). The results further suggested that base strength and the ability to form homogeneous reaction conditions are important for efficient product formation.

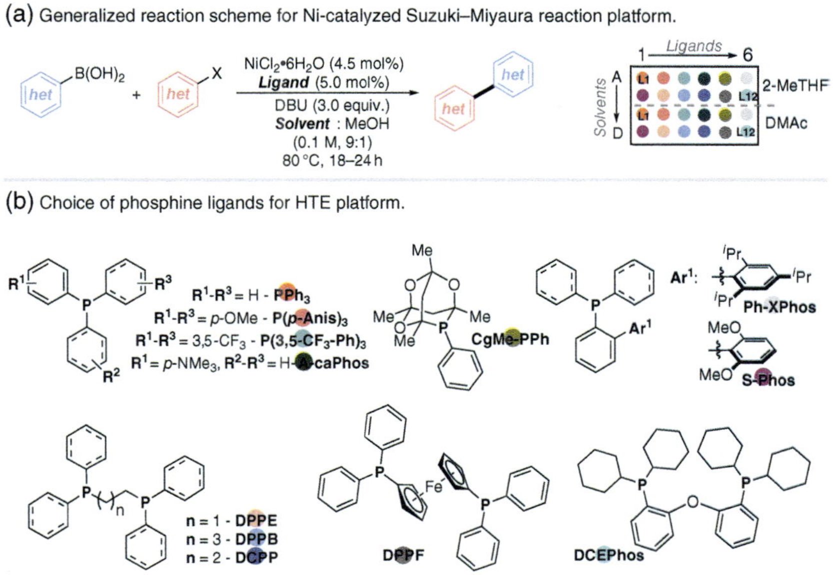

Scheme 6.4 HTE platform for Ni-catalyzed Suzuki–Miyaura reactions.

6.5.4 Final Platform Design

Based on all these results, the general reaction platform for Ni-catalyzed Suzuki–Miyaura reactions was outlined to incorporate 12 phosphine ligands with different stereoelectronic features (Scheme 6.4). For screening unknown substrates, the authors suggest combining this phosphine collection with two solvents of opposite polarity and good solvation characteristics (2-MeTHF and DMAc). Further features of the platform include the use of $NiCl_2 \cdot 6H_2O$ as cost-efficient and active Ni(II) precatalyst; 10% MeOH as non-optional alcohol additive; DBU as a cost-efficient and readily available base; 80 °C as the reaction temperature; and 18 hours as a convenient reaction time.

6.5.5 Validation of Platform Design

The designed platform was further validated to demonstrate that it enables hit identification for Ni-catalyzed Suzuki–Miyaura coupling in a diverse chemical space. To this end, (3-methoxyphenyl) boronic acid (**4**) and 1-benzyl-5-bromo-1*H*-indole (**5**; Scheme 6.5) were chosen as a second set of model substrates. **5** is an electronically rich bromoarene and is thus distinct from the electronically poor heteroaryl chloride chosen in the first set of model substrates. Therefore, the expectation is that successful identification of good reaction conditions for this

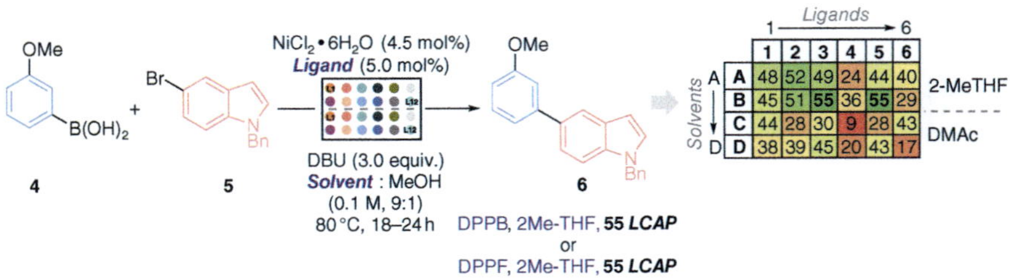

Scheme 6.5 Second model reaction for validation of Ni-catalyzed Suzuki–Miyaura platform. Reaction results are shown in LCAP.

second set of substrates would be a significant step toward demonstrating the generality of the Ni-catalyzed Suzuki–Miyaura platform.

The results of the screen are shown in Scheme 6.5. Interestingly, different ligands are identified as ideal for this substrate combination (wells B3/B5, corresponding to ligands DPPB and DPPF), and 2-MeTHF generally showed better reactivity than DMAc. This suggests that the second set of substrates requires different reaction conditions than the first set of substrates (ideal ligands: DCPP and A-caPhos). This highlights the power of an available screening platform and necessity to evaluate a larger set of reaction conditions to identify optimal reaction conditions for diverse heterocyclic coupling partners.

The generality of the platform was further supported with a broader scope of boronic acid and aryl halide coupling partners (Scheme 6.6). For these studies, the 24-well platform expanded into a 96-well reaction block, enabling the testing of 4 different substrates in one reaction setup in parallel (once for the data shown in Scheme 6.6a, once for the data shown in Scheme 6.6b).

In each case (Scheme 6.6a, 4 boronic acids **7a-d**; Scheme 6.6b, 4 electrophiles **11a-d**), a ligand/solvent combination could be identified that afforded at least 13 LCAP; in many cases, significantly higher LCAPs were observed, up to 93. This demonstrates that the developed HTE platform allows for the rapid identification of Ni-catalyzed Suzuki–Miyaura reaction conditions, regardless of the given substrate combination. This platform establishes several conditions that potentially can be "best" for a given substrate combination.

The approach of platform design is differentiated from typical workflows in organic chemistry that most often establish optimal condition for a given substrate combination, which is subsequently tested for a series of substrates in a substrate scope study. HTE-enabled synthesis most often uses platforms of conditions instead of one "best" condition as the starting point for an investigation. The above discussed study establishes such a platform for Suzuki–Miyaura couplings. The availability of the HTE platform represents significant time savings, since it allows the testing of different catalysts in parallel. In comparison to the traditional

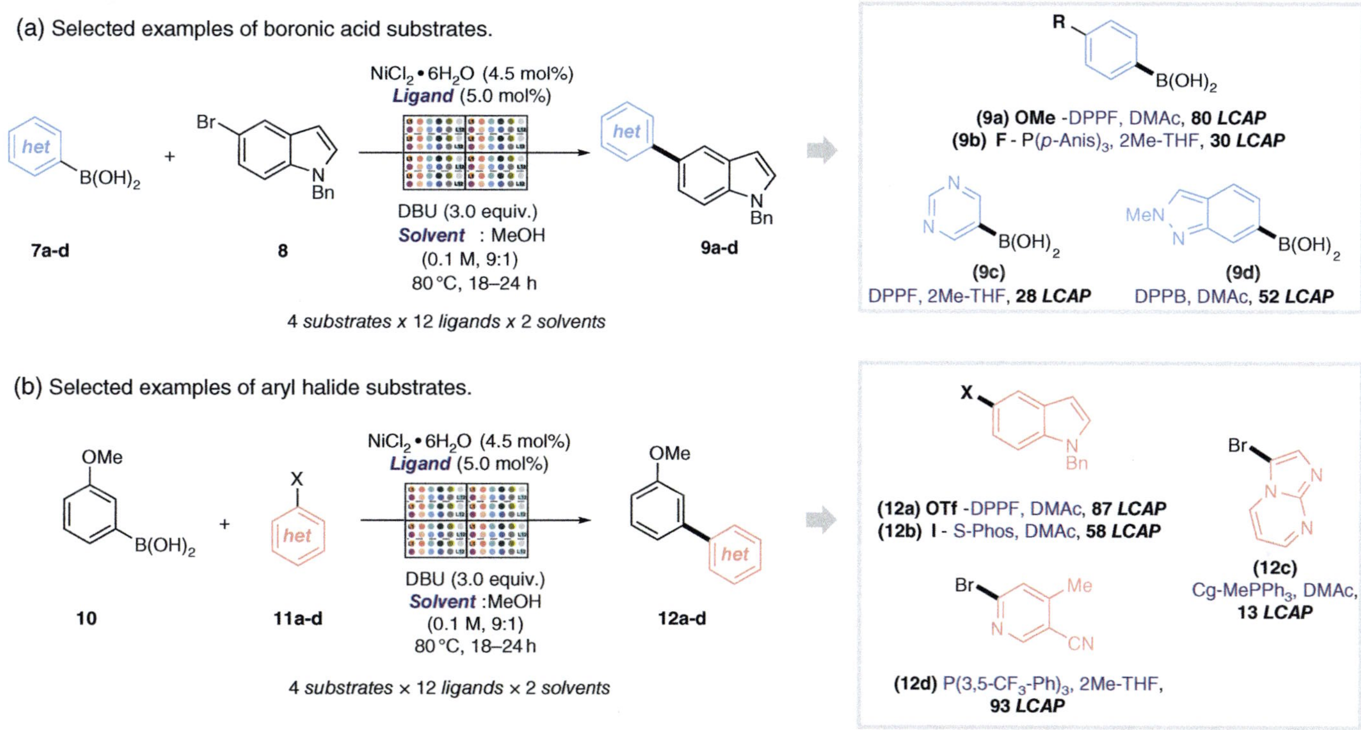

Scheme 6.6 **Demonstration of generality of the developed screening platform by establishing product-forming conditions for broad series of different reactants.** (a) Results for series of diverse (hetero)aryl boronic acids. (b) Results for series of diverse heteroaryl electrophiles. Reaction results are shown In LCAP.

approach of identifying suitable catalysts by testing one ligand at a time, the pre-investment of platform design thus increases the likelihood of identifying the best conditions in a given experiment.

6.6 Case Study 2: HTE Enabled Reaction Discovery and Optimization of Silyl-Triflate-Mediated C–H Aminoalkylation of Azoles

6.6.1 Motivation

Functionalized azole scaffolds are abundant core structures in APIs (see Scheme 6.7a for examples). Therefore, synthetic approaches enabling efficient late-stage diversification of such building blocks are valuable for SAR studies. Most methods to achieve the synthesis of diversified azoles rely on *de-novo* syntheses of the azole core [32]. However, these strategies are often substrate dependent, which can limit the accessibility of desired chemical matter (Scheme 6.7bi). Thus, synthetic strategies toward diversifying C–H functionalization of azoles are of broad interest. The case study reviewed in this section is inspired by the need to access diverse azole building blocks and demonstrates access via a three-component coupling (Scheme 6.7bii) of a nucleophilic azole building block [33–36] (through C–H functionalization) with *in situ* formed iminium electrophiles [37].

As discussed within this chapter, HTE is an efficient approach to invent new chemical reactivity. The case study [38] reviewed in this section includes all three types of HTE experiments: Reaction discovery, reaction optimization, and parallel synthesis of a compound library. Only reaction discovery and optimization will be covered in detail in this section, as these are the most pertinent sections for HTE workflows.

6.6.2 Reaction Discovery Plate Design

To discover the desired reactivity, benzoxazole (**16**), 1-(trimethylsilyl)piperidine (**13**), and benzaldehyde (**14**) were chosen as readily available model substrates (Scheme 6.8a). The silylated amine **13** and benzaldehyde (**14**) react readily to form iminium intermediate **15**, as documented previously [37]. Based on literature precedent, it was initially speculated that the combination of base, azole **16**, and a nickel catalyst precursor would enable the formation of an azole-nickel species *via* C-H deprotonation of the azole reactant. The resulting azole-nickel species was hypothesized to act as a nucleophile with the *in situ* formed iminium intermediate **15**, resulting in the formation of a new C–C bond.

In order to identify conditions that would allow the desired reactivity to proceed, the literature on Ni-catalyzed azole C–H functionalizations was reviewed.

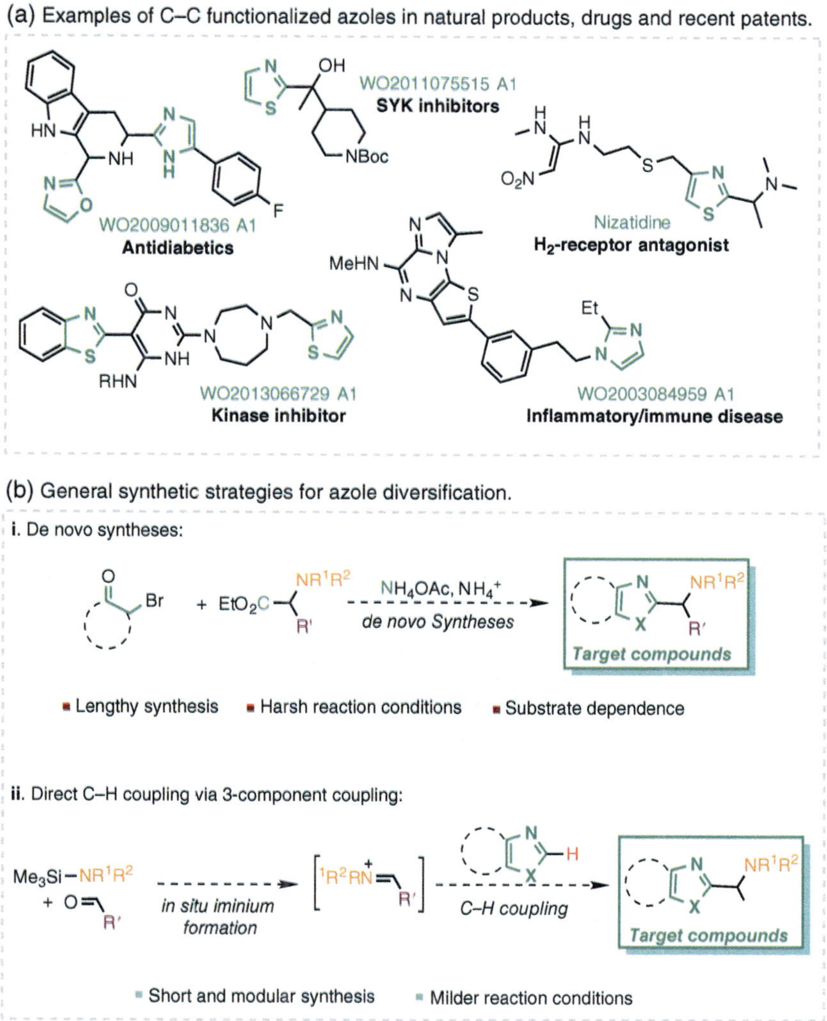

Scheme 6.7 Application and significance of functionalized azole scaffolds.

Based on this review, two solvents (MeCN and dioxane) and three bases (KO^tBu, Cs_2CO_3, and NaOAc) were identified that are frequently used in combination with Ni catalysts. The bases were dispensed onto 96-well plates (Scheme 6.8c), keeping select rows empty for a control reaction (i.e. reactions in the absence of base).

This array was then diversified by choosing 11 ligands (both phosphine and nitrogen-based ligands of different bite angles and denticities; Scheme 6.8b), some of which had previously shown productive reactivity in azole C–H functionalization when combined with Ni(II) catalyst precursors. Column 12 was intentionally

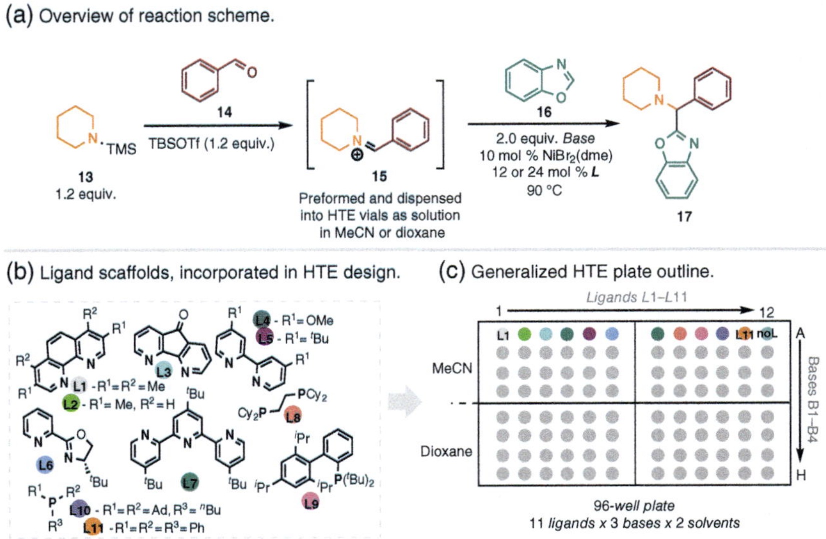

(d) LCAP product/ LCAP internal standard.

Ligands L1–L11 (1 → 12); Bases B1–B3 (A → H)

	Base	1	2	3	4	5	6	7	8	9	10	11	12
MeCN	NaOAc	1.8	1.8	1.8	1.8	3.7	1.9	1.8	1.8	1.9	1.8	1.8	1.8
	Cs_2CO_3	0.0	0.0	0.2	0.0	0.0	0.6	0.0	0.0	0.0	0.0	0.0	0.3
	KO*t*Bu	0.0	1.0	1.1	0.0	0.0	1.1	0.0	0.0	0.0	0.0	0.8	0.0
	no base	4.3	4.3	4.4	4.3	**5.8**	4.3	4.3	4.3	4.2	2.6	3.0	4.3
Dioxane	NaOAc	2.1	1.5	1.7	1.6	3.4	1.6	1.6	1.5	1.5	1.6	1.5	1.6
	Cs_2CO_3	0.0	0.0	0.0	0.0	0.0	0.0	0.0	0.0	0.0	0.0	0.0	0.0
	KO*t*Bu	0.0	0.0	0.0	0.0	0.6	0.0	0.0	0.0	0.9	0.0	0.0	0.7
	no base	3.0	0.0	3.0	2.9	**5.0**	3.1	3.1	3.3	0.0	1.2	1.7	3.0

Scheme 6.8 HTE evaluation of ligand, solvents, and bases for Ni-catalyzed C–H aminoalkylation of 1,3-azoles. Reaction results are shown in LCAP.

performed in the absence of ligand to ensure that a comparison for background reactivity was available in the same plate.

The results of the HTE experiment are shown in Scheme 6.8d, visualized as ratios of LCAP product and LCAP internal standard. The highest value of product was detected in wells D5 and H5, containing bidentate 4,4′-di-*tert*-butyl-2,2′-bipyridine as ligand in the absence of base. Overall, the wells without base performed better, while the wells with the stronger bases (Cs_2CO_3 and KO^{t}Bu) showed only minimal product formation. NaOAc, a mild base, showed intermediate reactivity. MeCN showed generally higher LCAP of product formation (top half of the plate) in contrast to analogous reactions performed in dioxane (bottom half of the plate). These data overall suggest that azole C–H aminoalkylation is a viable reaction under some of the evaluated conditions.

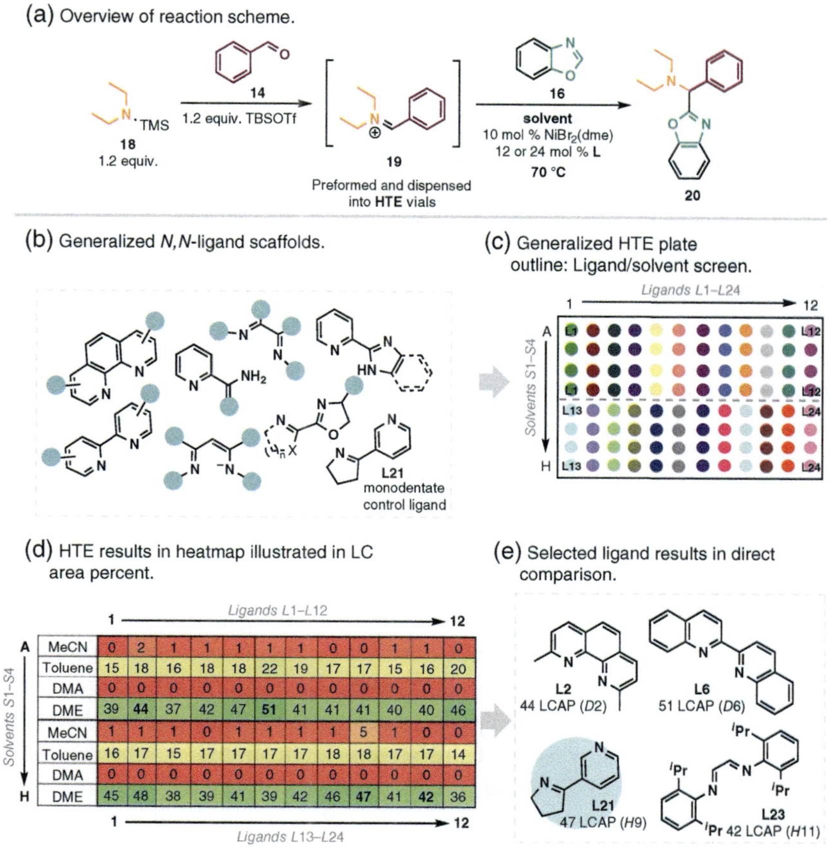

Scheme 6.9 Follow-up HTE experiment, further elucidating the effect of N,N-ligands on C–H functionalization reactivity in different solvents. Reaction results are shown in LCAP.

6.6.3 Ligand Screen

To follow up on these initial hits, the effect of different *N,N*-ligand scaffolds was further evaluated (Scheme 6.9a). The extended ligand screen incorporated 23 bidentate *N,N*-ligands (**L1-20** and **L22-24**) in addition to N,N ligand (**L21**) that can only bind in a monodentate fashion as control (Scheme 6.9b). Additionally, 4 solvents of different polarity and coordinating behavior were evaluated to gain more insight into how solvents affect C–H aminoalkylation reactivity (Scheme 6.9c). Furthermore, the identity of the silylated amine component was changed to $Et_2N\text{-}SiMe_3$ (**18**) due to its better commercial availability as compared to **13**.

The results of the screen are illustrated in Scheme 6.9d. Overall, the screening results illustrate a clear solvent dependence of the reaction, yielding highest product LCAPs for reactions conducted in DME (rows D and H). Polar solvents such as acetonitrile and DMA showed only little product formation (rows A, C, E, and G), while reactions in nonpolar toluene displayed intermediate reactivity. Interestingly, no major difference was observed between reaction outcomes with different ligands in DME: All reactions afford 44 ($\pm$ 8) LCAP of product, including the reaction in the presence of the monodentate control ligand **L21**.

This suggested that the tested additives may not act as ligands – and it also questions the necessity of a metal catalyst for reactivity. Notably, a control experiment *in the absence of catalyst* showed similar levels of product formation (~32 LCAP after 5 minutes at room temperature; 39 LCAP after 18 hours at 70 °C). These results indicate that the reaction is not metal catalyzed and proceeds readily at room temperature.

6.6.4 Parallel Optimization of Three Reagents

After the critical discovery that the reaction proceeds in the absence of a metal catalysis, another reaction optimization was performed to gain insight into which reaction components influence the reactivity. This experiment included optimizing reactant loadings of azole **16**, silylated amine **18**, and TBSOTf in a multidimensional fashion (Scheme 6.10a, b) while keeping the amount of benzaldehyde (**14**) at 1.0 equivalent. The results of this screen, as illustrated in Scheme 6.10c, show that the amounts of all tested reactants have a crucial impact on reaction performance: The highest product assay yields (AY) were observed in wells containing an excess of all three reactants (3.00 equiv. TBSOTf, 1.5 equiv. **16** and 2.0 equiv. **18**). This implied that the reaction is promoted by acidic (TBSOTf) and basic reactants (**16, 18**) alike.

6.6.5 Base Screen

With this key insight, another way to improve the reaction without wasting possibly expensive basic additives **16** and **18** (which are both incorporated into the product) was investigated: The addition of exogenous bases to the reaction while keeping the loadings of **16** and **18** close to 1 equivalent. For this purpose, a 24-well plate HTE array was designed to evaluate five bases (Scheme 6.11b; sixth row without added base) in two different loadings (0.5 or 1.0 equiv.) in the presence of 2.0 and 1.25 equiv. of $TMSNEt_2$ (**18**). As the initial reaction discovery plate had shown that inorganic bases are detrimental to reactivity, only amine bases were evaluated.

The data obtained from this screen are shown in Scheme 6.11c and illustrate that the highest product assay yield (96% AY) is obtained in well **B3** in the

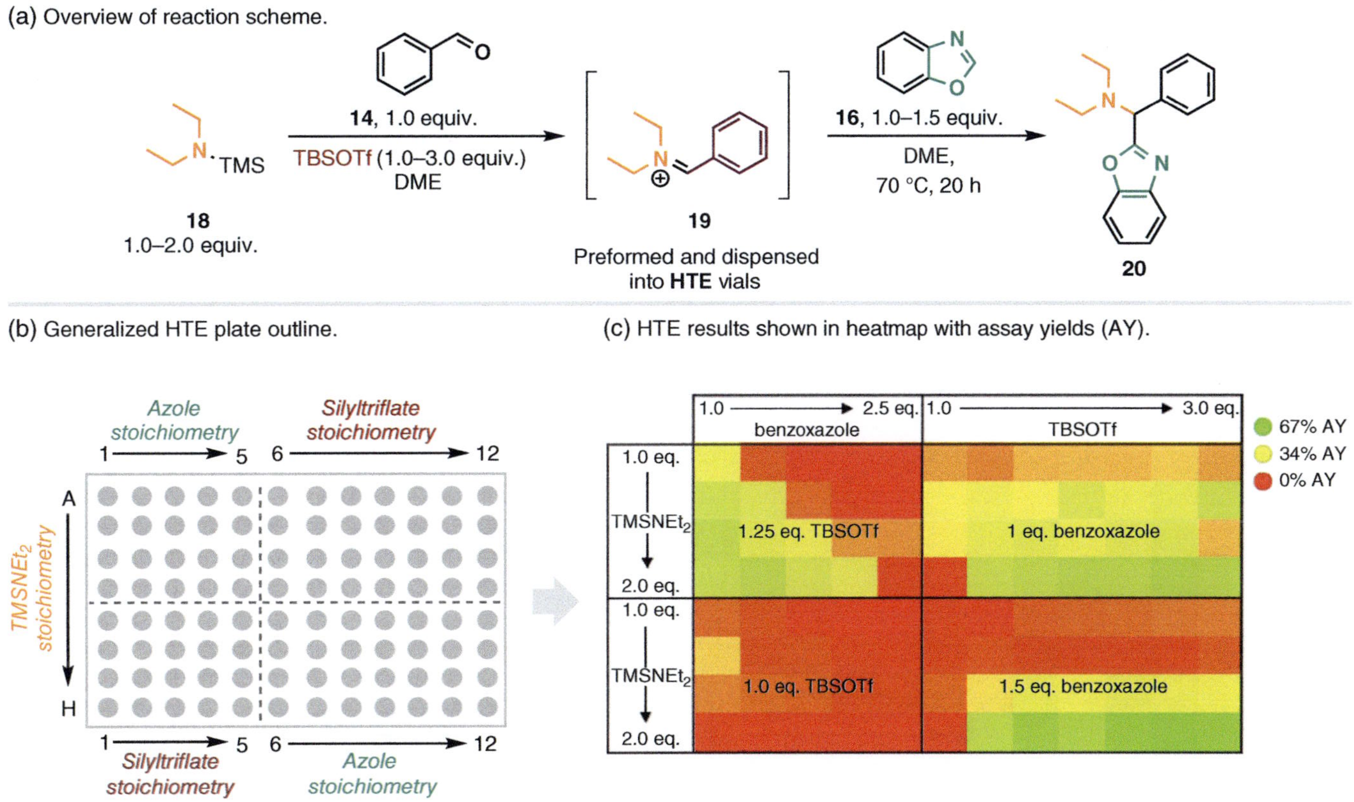

Scheme 6.10 Multidimensional optimization of reactants.

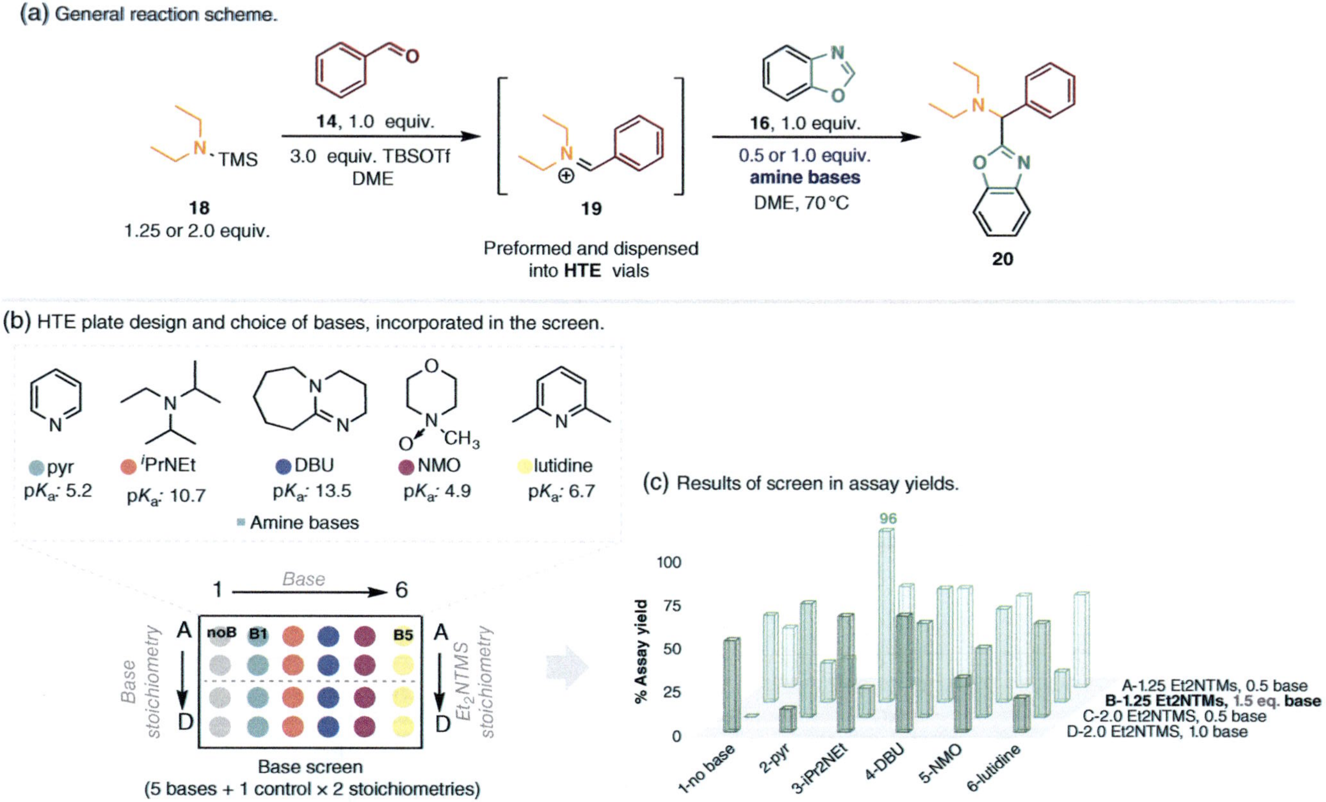

Scheme 6.11 HTE screen with added exogenous bases.

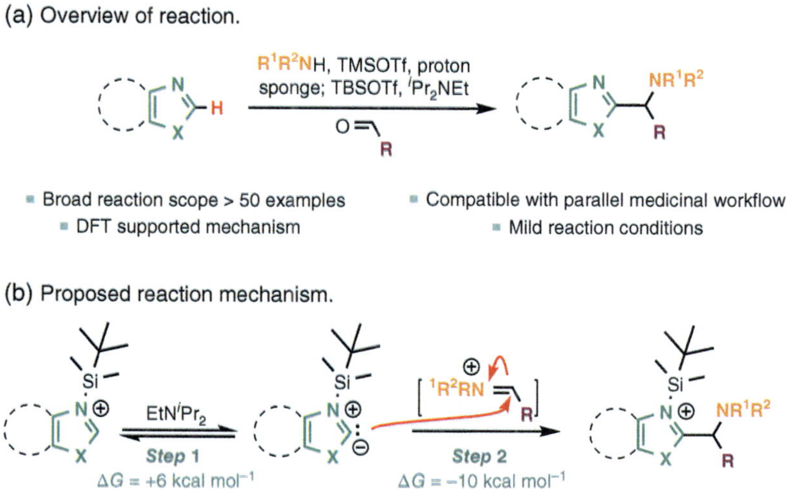

Scheme 6.12 General overview of developed reaction (a) and proposed mechanism (b).

presence of 1.0 equiv. Et_2N^iPr as a base with only a slight excess of $TMSNEt_2$ (1.25 equiv.). These data together with further evaluation of the reaction *via* reaction scope and DFT studies led to the formulation of a mechanistic proposal inspired by the Stetter [39] reaction (Scheme 6.12).

This reactivity was subsequently further elaborated and can employ a broad scope of azoles (including imidazole and benzimidazole), aldehydes, and amines, highlighting the mildness and generality of the developed method [38]. In summary, this study illustrates how HTE experiments can allow for rapid progression of a research program. Initial reaction discovery is followed by studies providing insights into the effects of each reaction component. With optimized reaction conditions established, the chemistry was also applied in library synthesis, highlighting the use of the three-component reaction in medicinal chemistry workflows. Furthermore, many of the shown examples highlight the importance of building controls into optimization plates and illustrate the role of hypothesis-driven design for successful HTE campaigns.

6.7 Current Challenges and the Future of HTE

6.7.1 Summary and Conclusions

We hope that the previous sections of this chapter have sufficiently illustrated how HTE workflows can aid in reaction discovery and optimization (HTE techniques are significantly more established for parallel synthesis applications and were thus not covered in detail here). We expect that the practical tips assembled in this chapter will be helpful for those intending to join the ever-growing field of HTE enthusiasts. HTE misconceptions have been addressed directly, while cost-efficient alternatives to expensive tools have been discussed and clearly indicated where available.

However, the authors would also like to acknowledge that HTE as described in this chapter is certainly not the only pathway to success for reaction discovery and optimization: The long and colorful history of chemistry in the pre-HTE era is proof of that. However, we argue that HTE workflows can provide insights that simple one-factor optimizations often cannot.

6.7.2 Remaining Challenges: The Next Frontiers

Despite their great success in the last decades, HTE experiments come with their innate drawbacks and limits, just like any other laboratory technique. Some of these drawbacks will likely be solved with new technology development in the (not so far) future, while others may remain fundamentally challenging.

6.7.2.1 Biphasic Reaction Mixtures

Biphasic reaction mixtures are frequently encountered in synthetic organic chemistry. HTE workflows generally perform well when testing liquid/solid reaction mixtures, if stirring across the plate is uniform (e.g. by using a tumble stirrer; see the previous sections). However, for biphasic liquid reaction mixtures (e.g. in phase transfer catalysis [40]), common stirring solutions often do not provide sufficient mixing to evaluate catalysts with HTE screens. Solutions to this issue have included miniaturizing reactions in combination with sonication [41] instead of tumble stirring, which allows for efficient phase transfer, improved reactivity, and HTE results that can be predictive for larger scale reactions.

6.7.2.2 Flow Chemistry and HTE

Flow chemistry has significant advantages when scaling up chemistries that require well-controlled reaction times, temperature, and mixing [42]. However, using batch HTE arrays to predict exact reaction conditions for flow can be challenging. Luckily, flow chemistry has its own approaches to high-throughput screening: for example, the use of small-scale flow reactions (often referred to as microfluidics) or droplets that can allow the generation of similar heatmaps [43–45] as those obtained with more traditional, batch-style HTE setups. Additionally, flow reactors are the perfect tool to study the impact of continuous variables, like temperature, reaction time, or stoichiometry – instead of using classical HTE workflows to gain similar insights. To harness the benefits of flow to substitute many HTE experiments, various platforms have been developed to perform rapid reaction optimization in a resource-sparing manner [46–48]. Currently, those techniques are somewhat further removed from the typical laboratory experience of a synthetic chemist, but future developments in software and "off-the-shelf" flow systems may allow routine implementation of microfluidic techniques.

6.7.2.3 Reaction Profiling

Chemists with physical organic training frequently use kinetic traces to gain insight into reactivity and generate hypotheses regarding reaction mechanisms. Currently, HTE plates do not accommodate rapid sampling of all reactions in a timely manner, but the field is in the process of addressing this need. Solutions will likely include the use of robotic sampling, as well as new developments in plate sealing consumables.

6.7.2.4 Building Machine Learning Models to Predict Reactivity

HTE is poised like no other technique to enable acquisition of large quantities of data to fuel developments in machine learning (ML) [49, 50] and its application to reactivity prediction. In theory, HTE and ML are perfect partners: One supplies large data sets that the other one can use to make useful predictions (see Chapter 10). In practice, some read-outs from HTE are easier to use for ML purposes than others. For example, data such as *ee* that are ratios of measured integrals are less prone to errors than yield measurements. LCAP data, the most common HTE readout derived from UPLC/MS analysis, is particularly prone to errors, as the relative areas measured are dependent on the extinction coefficient of each analyte. Another challenge lies in the variability of analytical data: Minor shifts in retention times for chromatographic methods can easily lead to false negatives, when analyzing diverse conditions within or between plates. Such shifts can be caused by the different reaction conditions screened on the plate (e.g. when evaluating different pH ranges). Therefore, more manual modes of analysis often serve as a backup, as long as the analysis can be handled in reasonable time frames. Scaling this workflow to larger data sets is a significant challenge. Finally, one major concern regarding data quality is the question of systematic data capture across a user group containing different individuals. If clear standards for data capture can be implemented, the creation of larger data sets for machine learning purposes will be more successful than in the absence of standards. All these concerns emphasize that obtaining HTE data of "good" quality is one core concern for any reactivity predictions using ML methods, but that data quality can present a substantial hurdle at the current state of the art.

6.7.2.5 Addressing Future Challenges

All challenges discussed in the last paragraphs have one thing in common: It is possible that future technology and software development will address them. Reflecting on the last decades of HTE in synthetic chemistry, new developments and innovative minds have the potential to impact a whole field of chemistry. With change and growth at the heart of chemistry innovation, the future is bright if the field can continue to recruit diverse skillsets and interests. Those backgrounds and motivations will allow us to tackle the most pressing and urgent problems – even those we currently do not even perceive of yet. Automation and technology will certainly have a role to play; exciting current developments in these areas are covered in the other chapters of this book.

Acknowledgments

The authors acknowledge Ellen Y. Aguilera, Cyndi Q. He, Francois L. Levesque, Byron Peters, Rebecca Ruck, Renee Sifri, Neil Strotman, and Cristhian Umana Gamboa (all Merck & Co., Inc., Kenilworth, NJ, USA) for critical feedback on this chapter.

Further Recommended Reading

HTE case studies from academia, drug discovery, and drug development [5]

Design of reaction arrays for reaction optimization [51]

Overview of HTE in pharmaceutical development [4]

HTE evaluation of reaction protocols using informer libraries and nanoscale reaction screening (1536-well plates) [52]

Additive screening to improve reaction generality [53]

High-throughput strategies for the discovery and optimization of catalytic reactions [54]

References

1 Stocker, S., Csányi, G., Reuter, K., and Margraf, J.T. (2020). Machine learning in chemical reaction space. *Nature Communications* 11 (1): 5505. https://www.nature.com/articles/s41467-020-19267-x.

2 Anderson, M.J. and Whitcomb, P.J. Design of experiments. In: *Kirk-Othmer Encyclopedia of Chemical Technology*, 1–22. Wiley Online Library https://onlinelibrary.wiley.com/doi/abs/10.1002/0471238961.0405190908010814.a01.pub3.

3 Parallel Library Synthesis:(a) Dombrowski, A.W., Aguirre, A.L., Shrestha, A. et al. (2021). The chosen few: parallel library reaction methodologies for drug discovery. *Journal of Organic Chemistry* https://pubs.acs.org/doi/abs/10.1021/acs.joc.1c01427; (b) Vasudevan, A., Bogdan, A.R., Koolman, H.F. et al. (2017). Chapter one – enabling chemistry technologies and parallel synthesis—accelerators of drug discovery programmes. In: *Progress in Medicinal Chemistry*, vol. 56 (ed. D.R. Witty and B. Cox), 1–35. Elsevier. https://www.sciencedirect.com/science/article/pii/S0079646816300029. (c) DiRico, K.J., Hua, W., Liu, C. et al. (2020). Ultra-high-throughput acoustic droplet ejection-open port interface-mass spectrometry for parallel medicinal chemistry. *ACS Medicinal Chemistry Letters*. 11 (6): 1101–1110. https://pubs.acs.org/doi/abs/10.1021/acsmedchemlett.0c00066 (d) Hu, Q., Peng, Z., Sutton, S.C. et al. (2012). Pfizer global virtual library (PGVL): a chemistry design tool powered by experimentally validated parallel synthesis information. *ACS Combinatorial Science* 14 (11): 579–589. https://pubs.acs.org/doi/abs/10.1021/co300096q. (e) Wess, G., Urmann, M., and Sickenberger, B. (2001). Medicinal chemistry: challenges and opportunities. *Angewandte Chemie International Edition*. 40 (18): 3341–3350. https://onlinelibrary.wiley.com/doi/full/10.1002/1521-3773%2820010917%2940%3A18%3C3341%3A%3AAID-ANIE3341%3E3.0.CO%3B2-D.

4 Mennen, S.M., Alhambra, C., Allen, C.L. et al. (2019). The evolution of high-throughput experimentation in pharmaceutical development and perspectives on the future. *Organic Process Research & Development* 23 (6): 1213–1242. https://pubs.acs.org/doi/10.1021/acs.oprd.9b00140.

5 (a) Leitch, D., Emmert, M.H., and Jouffroy, M. (ed.) (2022). *Volume 1: The Power of High Throughput Experimentation for Synthesis and Catalysis: General Topics and Enabling Technologies*. ACS ebooks. (b) Leitch, D., Emmert, M.H., and Jouffroy, M. (ed.) (2022). *Volume 2: The Power of High Throughput Experimentation: Case Studies from Drug Discovery, Drug Development, and Catalyst Discovery*. ACS ebooks.

6 Christensen, M., Yunker, L.P.E., Adedeji, F. et al. (2021). Data-science driven autonomous process optimization. *Communications Chemistry*. 4 (1): 112. https://www.nature.com/articles/s42004-021-00550-x.

7 Christensen, M., Adedeji, F., Grosser, S. et al. (2019). Development of an automated kinetic profiling system with online HPLC for reaction optimization. *Reaction Chemistry & Engineering*. 4 (9): 1555–1558. https://pubs.rsc.org/en/content/articlelanding/2019/re/c9re00086k.

8 Lin, B., Hedrick, J.L., Park, N.H., and Waymouth, R.M. (2019). Programmable high-throughput platform for the rapid and scalable synthesis of polyester and polycarbonate libraries. *Journal of the American Chemical Society*. 141 (22): 8921–8927. https://pubs.acs.org/doi/abs/10.1021/jacs.9b02450.

9 Bédard, A.-C., Adamo, A., Aroh, K.C. et al. (2018). Reconfigurable system for automated optimization of diverse chemical reactions. *Science.* 361 (6408): 1220–1225. https://www.science.org/doi/10.1126/science.aat0650.

10 Tabor, D.P., Roch, L.M., Saikin, S.K. et al. (2018). Accelerating the discovery of materials for clean energy in the era of smart automation. *Nature Reviews Materials.* 3 (5): 5–20. https://www.nature.com/articles/s41578-018-0005-z.

11 Allen, C.L., Leitch, D.C., Anson, M.S., and Zajac, M.A. (2019). The power and accessibility of high-throughput methods for catalysis research. *Nature Catalysis.* 2 (1): 2–4. https://www.nature.com/articles/s41929-018-0220-4.

12 Cook, A., Clément, R., and Newman, S.G. (2021). Reaction screening in multiwell plates: high-throughput optimization of a Buchwald–Hartwig amination. *Nature Protocols.* 16 (2): 1152–1169. https://www.nature.com/articles/s41596-020-00452-7.

13 Kassel, V.M., Hanneman, C.M., Delaney, C.P., and Denmark, S.E. (2021). Heteroaryl–heteroaryl, suzuki–miyaura, anhydrous cross-coupling reactions enabled by trimethyl borate. *Journal of the American Chemical Society.* 143 (34): 13845–13853. https://www.nature.com/articles/s41929-018-0220-4.

14 Combinatorial chemistry:(a)Terrett, N.K., Gardner, M., Gordon, D.W. et al. (1997). Drug discovery by combinatorial chemistry—the development of a novel method for the rapid synthesis of single compounds. *Chemistry – A European Journal.* 3 (12): 1917–1920. https://chemistry-europe.onlinelibrary.wiley.com/action/showCitFormats?doi=10.1002%2Fchem.19970031204;(b) Nicolaou, K.C., Hanko, R., and Hartwig, W. (2002). Combinatorial chemistry in perspective. In: *Handbook of Combinatorial Chemistry*, 1–9. https://onlinelibrary.wiley.com/doi/abs/10.1002/3527603034.ch1.

15 Jouyban, A. and Soltanpour, S. (2010). Prediction of dielectric constants of binary solvents at various temperatures. *Journal of Chemical & Engineering Data.* 55 (9): 2951–2963. https://pubs.acs.org/doi/10.1021/je1000632.

16 Legacy, C.J., Hope, T.O., Gagné, Y. et al. (2021). Mechanistic insights into Fe catalyzed α-C—H oxidations of tertiary amines. *ChemCatChem.* 13 (1): 235–246. https://chemistry-europe.onlinelibrary.wiley.com/doi/abs/10.1002/cctc.202001382.

17 Hansch, C., Leo, A., and Taft, R.W. (1991). A survey of Hammett substituent constants and resonance and field parameters. *Chemical Reviews.* 91 (2): 165–195. https://pubs.acs.org/doi/10.1021/cr00002a004.

18 Peker, S.M., Helvacı, Ş.Ş., Yener, H.B. et al. (ed.) (2008). 7 – Mixing in solid–liquid systems. In: *Solid-Liquid Two Phase Flow*, 385–438. Amsterdam: Elsevier https://www.sciencedirect.com/science/article/pii/B978044452237550009 6#cesec11.

19 Schultz, D.M., Levesque, F., DiRocco, D.A. et al. (2017). Oxyfunctionalization of the remote C-H bonds of aliphatic amines by Decatungstate photocatalysis. *Angewandte Chemie Interntional Edition in English* 56 (48): 15274–15278. https://onlinelibrary.wiley.com/doi/full/10.1002/anie.201707537.

20 Rein, J., Annand, J.R., Wismer, M.K. et al. (2021). Unlocking the potential of high-throughput experimentation for electrochemistry with a standardized microscale reactor. *ACS Central Science.* 7 (8): 1347–1355. https://pubs.acs.org/doi/10.1021/acscentsci.1c00328.

21 Shevlin, M., Friedfeld, M.R., Sheng, H. et al. (2016). Nickel-catalyzed asymmetric alkene hydrogenation of α,β-unsaturated esters: high-throughput experimentation-enabled reaction discovery, optimization, and mechanistic elucidation. *Journal of the American Chemical Society.* 138 (10): 3562–3569. https://pubs.acs.org/doi/full/10.1021/jacs.6b00519.

22 Welch, C.J. (2019). High throughput analysis enables high throughput experimentation in pharmaceutical process research. *Reaction Chemistry & Engineering.* 4 (11): 1895–1911. https://pubs.rsc.org/en/content/articlelanding/2019/RE/C9RE00234K.

23 Zhang, H., Liu, C., Hua, W. et al. (2021). Acoustic ejection mass spectrometry for high-throughput analysis. *Analytical Chemistry.* 93 (31): 10850–10861. https://pubs.acs.org/doi/abs/10.1021/acs.analchem.1c01137.

24 Haslam, C., Hellicar, J., Dunn, A. et al. (2015). The Evolution of MALDI-TOF Mass Spectrometry toward Ultra-High-Throughput Screening: 1536-Well Format and Beyond. *Journal of Biomolecular Screening.* 21 (2): 176–186. https://journals.sagepub.com/doi/full/10.1177/1087057115608605.

25 Bahr, M.N., Damon, D.B., Yates, S.D. et al. (2018). Collaborative evaluation of commercially available automated powder dispensing platforms for high-throughput experimentation in pharmaceutical applications. *Organic Process Research & Development.* 22 (11): 1500–1508. https://pubs.acs.org/doi/abs/10.1021/acs.oprd.8b00259.

26 Jorner, K., Tomberg, A., Bauer, C. et al. (2021). Organic reactivity from mechanism to machine learning. *Nature Reviews Chemistry.* 5 (4): 240–255. https://www.nature.com/articles/s41570-021-00260-x.

27 Christensen, M., Yunker, L.P.E., Shiri, P. et al. (2021). Automation isn't automatic. *Chemical Science.* 12 (47): 15473–15490. https://pubs.rsc.org/en/content/articlelanding/2021/SC/D1SC04588A.

28 Taheri Kal Koshvandi, A., Heravi, M.M., and Momeni, T. (2018). Current applications of Suzuki–Miyaura coupling reaction in the total synthesis of natural products: an update. *Applied Organometallic Chemistry.* 32 (3): e4210. https://onlinelibrary.wiley.com/doi/full/10.1002/aoc.4210.

29 Beletskaya, I.P., Alonso, F., and Tyurin, V. (2019). The Suzuki-Miyaura reaction after the Nobel prize. *Coordination Chemistry Reviews.* 385: 137–173. https://www.sciencedirect.com/science/article/pii/S0010854518306581.

30 Goldfogel, M.J., Guo, X., Meléndez Matos, J.L. et al. (2021). Advancing base-metal catalysis: development of a screening method for nickel-catalyzed Suzuki–Miyaura reactions of pharmaceutically relevant heterocycles. *Organic Process Research & Development*. https://doi.org/10.1021/acs.oprd.1c00210.

31 (a) Hansen, E.C., Li, C., Yang, S. et al. (2017). Coupling of challenging heteroaryl halides with alkyl halides via nickel-catalyzed cross-electrophile coupling. *The Journal of Organic Chemistry*. 82 (14): 7085–7092. https://pubs.acs.org/doi/abs/10.1021/acs.joc.7b01334; (b) Kim, S., Goldfogel, M.J., Gilbert, M.M., and Weix, D.J. (2020). Nickel-catalyzed cross-electrophile coupling of aryl chlorides with primary alkyl chlorides. *Journal of the American Chemical Society*. 142 (22): 9902–9907. https://pubs.acs.org/doi/abs/10.1021/jacs.0c02673. (c) Han, F.-S. (2013). Transition-metal-catalyzed Suzuki–Miyaura cross-coupling reactions: a remarkable advance from palladium to nickel catalysts. *Chemical Society Reviews*. 42 (12): 5270–5298. https://pubs.rsc.org/en/content/articlelanding/2013/cs/c3cs35521g. (d) Tasker, S.Z., Standley, E.A., and Jamison, T.F. (2014). Recent advances in homogeneous nickel catalysis. *Nature*. 509 (7500): 299–309. https://www.nature.com/articles/nature13274.

32 De-Novo Synthesis: (a) Attenni B, Ferrigno F, Jones P, Ingenito R, Kinzel O, Llauger Bufi L, et al., inventors; Istituto di Ricerche di Biologia Molecolare P. Angeletti S.p.A., assignee. Preparation of heterocycle derivatives as histone deacetylase inhibitors patent WO2006061638. 2006; (b) Adib, M., Ansari, S., Feizi, S. et al. (2009). A one-pot, four-component synthesis of N-substituted 2,4-diarylimidazoles. *Synlett*. 2009 (20): 3263–3266. https://www.thieme-connect.de/products/ejournals/abstract/10.1055/s-0029-1218381. (c) Sorrell, T.N. and Allen, W.E. (1994). A regiospecific synthesis of 1,4-disubstituted imidazoles. *The Journal of Organic Chemistry*. 59 (6): 1589–1590. https://pubs.acs.org/doi/10.1021/jo00085a056.

33 (a) Lewis, J.C., Bergman, R.G., and Ellman, J.A. (2008). Direct functionalization of nitrogen heterocycles via Rh-catalyzed C—H bond activation. *Accounts of Chemical Research*. 41 (8): 1013–1025. https://pubs.acs.org/doi/10.1021/ar800042p. (b) Chen, S., Ranjan, P., Voskressensky, L.G. et al. (2020). Recent developments in transition-metal catalyzed direct C–H alkenylation, alkylation, and alkynylation of azoles. *Molecules*. 25 (21): 4970. https://www.mdpi.com/1420-3049/25/21/4970.

34 Examples of Cu-catalyzed azole C–H functionalization: (a)Do, H.-Q. and Daugulis, O. (2007). Copper-catalyzed arylation of heterocycle C—H Bonds. *Journal of the American Chemical Society*. 129 (41): 12404–12405. https://pubs.acs.org/doi/10.1021/ja075802%2B; (b) Yoshizumi, T., Tsurugi, H., Satoh, T., and Miura, M. (2008). Copper-mediated direct arylation of benzoazoles with aryl iodides. *Tetrahedron Letters*. 49 (10): 1598–1600. https://www.sciencedirect.com/science/article/pii/S0040403908000804?via%3Dihub.(c) Nishino, M., Hirano, K., Satoh, T., and Miura, M. (2012). Copper-mediated and copper-catalyzed cross-coupling of indoles and 1,3-azoles: double C-H activation. *Angewandte Chemie*

International Edition. 51 (28): 6993–6997. https://onlinelibrary.wiley.com/doi/full/10.1002/anie.201201491. (d) Ohmiya, H., Zhang, H., Shibata, S. et al. (2016). Construction of quaternary stereogenic carbon centers through copper-catalyzed enantioselective allylic alkylation of azoles. *Angewandte Chemie International Edition.* 55 (15): 4777–4780. https://onlinelibrary.wiley.com/doi/10.1002/anie.201600619; (e) Gandeepan, P., Mo, J., and Ackermann, L. (2017). Photo-induced copper-catalyzed C–H chalcogenation of azoles at room temperature. *Chemical Communications.* 53 (43): 5906–5909. https://pubs.rsc.org/en/content/articlelanding/2017/cc/c7cc03107f.

35 Examples of Ni-catalyzed azole C–H functionalization: (a)Amaike, K., Muto, K., Yamaguchi, J., and Itami, K. (2012). Decarbonylative C–H coupling of azoles and aryl esters: unprecedented nickel catalysis and application to the synthesis of Muscoride A. *Journal of the American Chemical Society.* 134 (33): 13573–13576. https://pubs.acs.org/doi/10.1021/ja306062c. (b) Muto, K., Yamaguchi, J., and Itami, K. (2012). Nickel-catalyzed C–H/C–O coupling of azoles with phenol derivatives. *Journal of the American Chemical Society.* 134 (1): 169–172. https://pubs.acs.org/doi/10.1021/ja210249h. (c) Yamamoto, T., Muto, K., Komiyama, M. et al. (2011). Nickel-catalyzed C-H arylation of azoles with haloarenes: scope, mechanism, and applications to the synthesis of bioactive molecules. *Chemistry – A European Journal.* 17 (36): 10113–10122. https://chemistry-europe.onlinelibrary.wiley.com/doi/full/10.1002/chem.201101091. (d) Meng, L., Kamada, Y., Muto, K. et al. (2013). C–H alkenylation of azoles with enols and esters by nickel catalysis. *Angewandte Chemie International Edition.* 52 (38): 10048–10051. https://onlinelibrary.wiley.com/doi/full/10.1002/anie.201304492. (e) Larson, H., Schultz, D., and Kalyani, D. (2019). Ni-catalyzed C–H arylation of oxazoles and benzoxazoles using pharmaceutically relevant aryl chlorides and bromides. *The Journal of Organic Chemistry.* 84 (20): 13092–13103. https://pubs.acs.org/doi/abs/10.1021/acs.joc.9b02094.

36 Examples of metal-free azole C–H functionalization: (a)He, T., Li, H., Li, P., and Wang, L. (2011). Direct amidation of azoles with formamidesvia metal-free C–H activation in the presence of tert-butyl perbenzoate. *Chemical Communications.* 47 (31): 8946–8948. https://pubs.rsc.org/en/content/articlelanding/2011/cc/c1cc13086b; (b) Ahmed, J.P.S., Vijaykumar, G., Jose, A. et al. (2017). A new face of phenalenyl-based radicals in the transition metal-free C–H arylation of heteroarenes at room temperature: trapping the radical initiator via C–C σ-bond formation. *Chemical Science.* 8 (11): 7798–7806. https://pubs.rsc.org/en/content/articlelanding/2017/sc/c7sc02661g;. (c) Li, Z.-l., Jin, L.-k., and Cai, C. (2017). Efficient synthesis of 2-substituted azoles: radical C–H alkylation of azoles with dicumyl peroxide, methylarenes and cycloalkanes under metal-free condition. *Organic Chemistry Frontiers.* 4 (10): 2039–2043. https://pubs.rsc.org/en/content/articlelanding/2017/qo/c7qo00396j. (d) Qiu, Y., Struwe, J., Meyer, T.H. et al. (2018). Catalyst- and reagent-free electrochemical azole C–H amination.

Chemistry – A European Journal. 24 (49): 12784–12789. https://chemistry-europe.onlinelibrary.wiley.com/doi/full/10.1002/chem.201802832. (e) He, T., Yu, L., Zhang, L. et al. (2011). Direct C2-alkylation of azoles with alcohols and ethers through dehydrogenative cross-coupling under metal-free conditions. *Organic Letters* 13 (19): 5016–5019. https://pubs.acs.org/doi/10.1021/ol201779n. (f) Xu, W.-X., Dai, X.-Q., and Weng, J.-Q. (2019). K2S2O8-Mediated Hydroxyalkylation of Benzothiazoles with Alcohols in Aqueous Solution. *ACS Omega.* 4 (6): 11285–11292. https://pubs.acs.org/doi/10.1021/acsomega.9b01695.

37 Heinz, C., Lutz, J.P., Simmons, E.M. et al. (2018). Ni-catalyzed carbon–carbon bond-forming reductive amination. *Journal of the American Chemical Society.* 140 (6): 2292–2300. https://pubs.acs.org/doi/10.1021/jacs.7b12212.

38 Emmert, M.H., He, C.Q., Shah, A.A., and Felten, S. (2021). Lewis acid mediated, mild C–H aminoalkylation of azoles via three component coupling. *Chemical Science.* 12 (11): 3890–3897. https://pubs.rsc.org/en/Content/ArticleLanding/2021/SC/D0Sc6868C.

39 Yetra, S.R., Patra, A., and Biju, A.T. (2015). Recent advances in the N-Heterocyclic Carbene (NHC)-organocatalyzed stetter reaction and related chemistry. *Synthesis.* 47 (10): 1357–1378. https://www.thieme-connect.com/products/ejournals/abstract/10.1055/s-0034-1378692.

40 Maruoka, K. and Ooi, T. (2003). Enantioselective amino acid synthesis by chiral phase-transfer catalysis. *Chemical Reviews.* 103 (8): 3013–3028. https://pubs.acs.org/doi/full/10.1021/cr020020e.

41 Ooi, T., Tayama, E., Doda, K. et al. (2000). Dramatic rate enhancement by ultrasonic irradiation in liquid-liquid phase-transfer catalytic reactions. *Synlett.* 2000 (10): 1500–1502. https://www.thieme-connect.de/products/ejournals/pdf/10.1055/s-2000-7652.pdf.

42 Plutschack, M.B., Pieber, B., Gilmore, K., and Seeberger, P.H. (2017). The Hitchhiker's guide to flow chemistry. *Chemical Reviews.* 117 (18): 11796–11893. https://pubs.acs.org/doi/10.1021/acs.chemrev.7b00183.

43 Jaman, Z., Logsdon, D.L., Szilágyi, B. et al. (2020). High-throughput experimentation and continuous flow evaluation of nucleophilic aromatic substitution reactions. *ACS Combinatorial Science.* 22 (4): 184–196. https://pubs.acs.org/doi/10.1021/acscombsci.9b00212.

44 Sun, A.C., Steyer, D.J., Allen, A.R. et al. (2020). A droplet microfluidic platform for high-throughput photochemical reaction discovery. *Nature Communications.* 11 (1): 6202. https://www.nature.com/articles/s41467-020-19926-z.

45 Perera, D., Tucker, J.W., Brahmbhatt, S. et al. (2018). A platform for automated nanomole-scale reaction screening and micromole-scale synthesis in flow. *Science.* 359 (6374): 429–434. https://www.science.org/doi/full/10.1126/science.aap9112.

46 Wyvratt, B.M., McMullen, J.P., and Grosser, S.T. (2019). Multidimensional dynamic experiments for data-rich process development of reactions in flow.

Reaction Chemistry & Engineering. 4 (9): 1637–1645. https://pubs.rsc.org/en/content/articlelanding/2019/re/c9re00078j/unauth.

47 Florit, F., Nambiar, A.M.K., Breen, C.P. et al. (2021). Design of dynamic trajectories for efficient and data-rich exploration of flow reaction design spaces. *Reaction Chemistry & Engineering.* 6 (12): 2306–2314. https://pubs.rsc.org/en/content/articlepdf/2021/re/d1re00350j.

48 Breen, C.P., Nambiar, A.M.K., Jamison, T.F., and Jensen, K.F. (2021). Ready, set, flow! automated continuous synthesis and optimization. *Trends in Chemistry.* 3 (5): 373–386. https://www.sciencedirect.com/science/article/abs/pii/S2589597421000368.

49 Li, X., Maffettone, P.M., Che, Y. et al. (2021). Combining machine learning and high-throughput experimentation to discover photocatalytically active organic molecules. *Chemical Science.* 12 (32): 10742–10754. https://pubs.rsc.org/en/content/articlelanding/2021/SC/D1SC02150H.

50 Eyke, N.S., Koscher, B.A., and Jensen, K.F. (2021). Toward machine learning-enhanced high-throughput experimentation. *Trends in Chemistry.* 3 (2): 120–132. https://www.sciencedirect.com/science/article/abs/pii/S2589597420303117.

51 Shevlin, M. (2017). Practical high-throughput experimentation for chemists. *ACS Medicinal Chemistry Letters* 8 (6): 601–607. https://pubs.acs.org/doi/10.1021/acsmedchemlett.7b00165.

52 Dreher, S.D. and Krska, S.W. (2021). Chemistry informer libraries: conception, early experience, and role in the future of cheminformatics. *Accounts of Chemical Research* 54 (7): 1586–1596. https://pubs.acs.org/doi/10.1021/acs.accounts.0c00760.

53 Kullmer, C.N.P., Kautzky, J.A., Krska, S.W. et al. (2022). Accelerating reaction generality and mechanistic insight through additive mapping. *Science.* 376 (6592): 532–539. https://www.science.org/doi/10.1126/science.abn1885.

54 Isbrandt, E.S., Sullivan, R.J., and Newman, S.G. (2019). High throughput strategies for the discovery and optimization of catalytic reactions. *Angewandte Chemie International Edition* 58: 7180–7191. https://onlinelibrary.wiley.com/

7

Concepts and Practical Aspects of Computational Chemistry

Martin Breugst

Institut für Chemie, Technische Universität Chemnitz, Chemnitz, Germany

Glossary

In this section, important names, concepts, methods, and more are briefly summarized in a very simplified form to help the reader to better understand the following sections.

Basis function A basis function is the mathematical function that describes an atomic orbital within a compound.

Basis set The basis set is the set of basis functions and is used to construct the wave function.

Basis-set limit Computational methods converge to the basis-set limit, when increasingly large basis sets are used. Sometimes, the basis-set limit is also called the HF limit.

Cartesian coordinates The location of every atom within a structure is described with a set of numerical coordinates, which are the signed distances to a fixed origin located at (0,0,0).

Computational chemistry The field of research that uses theoretical methods to better understand chemical phenomena. The expression is sometimes also used as a differentiation to theoretical chemistry, which focuses more on the development of more advanced theoretical concepts that are then used in computational chemistry.

Enabling Tools and Techniques for Organic Synthesis: A Practical Guide to Experimentation, Automation, and Computation, First Edition. Edited by Stephen G. Newman.

Computational method The computational method refers to the mathematical models that are employed to calculate the behavior of a chemical system. Prominent examples include Hartree–Fock theory (HF), or B3LYP as an example for density functional theory (DFT).

Computing cluster Typically, computational investigations are not routinely performed on normal desktop computers. Instead, a large number of specially designed workstations are used in parallel within a computing cluster. Those clusters can be organized within a research group, a department or university, or on a much larger scale. These machines are also referred to as supercomputers or HPC (high-performance computing) systems.

Correlation energy The correlation energy is the energetic difference between a Hartree–Fock calculation with an infinitely large basis set and the exact solution of the Schrödinger equation. Thus, the correlation energy indicates how much the movement of one electron is influenced by the other electrons.

Force field A force field contains a large series of parameters that allow the calculation of the energy of a system from a given geometry relying on classical, Newtonian mechanics.

Input file Any calculation is controlled via a text-based input file that contains all information for that calculation.

Keyword Keywords are used to control the desired calculation and tell the computer what needs to be done.

Model chemistry Model chemistry is a term that refers to the employed computational models, which typically consists of a computational (or theoretical) method and a basis set. Sometimes, the expression "Level of Theory" is used instead.

Molecular mechanics (MM) Molecular mechanics relies on classical mechanics to model molecular systems as a system of balls and springs.

Optimization In an optimization, the program tries to locate the closest energetic minimum for a given structure. Thus, the program will automatically alter the initial structure until a "minimum structure" is reached.

Quantum mechanics (QM) Quantum mechanics relies on the Schrödinger equation to describe molecular systems.

Single-point calculation This is the default type of calculation in almost every computational program and determines the potential energy for a given structure.

Stationary point In calculus, a stationary point of a function is a point along the graph, where the first derivative is zero. Thus, stationary points are either minima or maxima.

Thermal correction Typically, most calculations take place on the potential (or electronic) energy surface, which does not include any thermal contribution. Thermal corrections are additive corrections that can be computed to convert the potential energy to, e.g. the enthalpy H or free energy G at 298.15 K.

7.1 Introduction

The synthesis of natural products and biologically active compounds has reached an impressive level of complexity over the last decades. To achieve these remarkable results, a large variety of different synthetic pathways has been developed allowing us to synthesize even very complex molecules. However, despite these achievements, we are frequently still far away from the "ideal" synthesis. In contrast, the scope of reactants or the optimal catalyst for a chosen transformation is typically identified by an extensive trial-and-error approach. Screening a large number of molecules, catalysts, solvents, or reaction conditions is costly, wasteful, and inefficient.

To address and eventually overcome these problems, we ultimately need a better understanding of how organic reactions proceed on the molecular level. Currently, we still lack these insights for many reactions because the proposed intermediates are often unstable, short-lived, or present in very small concentrations. Bonney and Schoenebeck concluded in a review in 2014 that, for palladium-catalyzed reactions, experimental or spectroscopic investigations alone are limited due to the fleeting nature of many intermediates in organometallic catalytic cycles [1]. This hinders the rational design of new transformations and more effective catalysts, which depends on a detailed understanding of the factors governing the reactivity and selectivity of organic transformations.

Computational chemistry is the ideal method to overcome these limitations in reaction and catalyst development as it allows the study of intermediates that are too unstable to be observable experimentally [2, 3]. Over the last decades, computational investigation have evolved to become a highly valuable tool to understand reactivities and selectivities alike which is also reflected in awarding the 2013 Nobel prize for chemistry to Karplus, Levitt, and Warshel for *the development of multiscale models for complex chemical systems* [4–6]. The rapid advances in technology as well as the development of efficient algorithms have strongly supported this process. In the 1980s, it took several months to calculate the energy profile of a Diels–Alder reaction using a very simple computational method (Hartree–Fock theory, see below), while today this reaction can be studied within less than one minute [3]. Currently, molecules with approximately 100 atoms can be modelled adequately within a reasonable time-frame. Following Moore's law, the "computational power" will approximately double every two years. Thus, it can be expected that larger organic molecules will be routinely studied computationally in the future.

Within the last decade, computational organic chemistry is frequently used to determine the energies of ground states and transition states and, consequently, activation energies of reactions. Currently, there are still problems (errors up to 20 kJ mol^{-1}) to determine absolute activation energies (i.e. the actual energetic difference between reactants and transition states) accurately. However, the calculation of relative activation energies (i.e. the difference in activation energies of two similar or competing reactions, a value which controls the stereochemical outcome of a reaction) can be achieved at levels close to chemical accuracy (<4 kJ mol^{-1}) [2, 3, 7]. Therefore, one of the most common uses of quantum mechanical techniques is to enable the direct comparison of different mechanistic pathways. This may help to identify improved reaction conditions or catalyst structures. Furthermore, computational approaches are also able to predict the outcome of new transformations or the influence of substituents. Thus, computational investigations will not only explain organic reactivity but can also limit the number of "inconclusive" or unnecessary experiments [1, 8, 9].

However, the application of computational investigations in a black-box approach also bears some risks. In general, there is no single method or procedure that performs best in all possible scenarios. Instead, the researcher has to select a computational (or theoretical) method (see Section 7.3) and a basis set (see Section 7.4). Generally speaking, the term computational method refers to the mathematical models that are employed to calculate the behavior of a chemical system, while basis sets describe the atomic orbitals (i.e. the location of a given electron). The chosen combination of computational method and basis set is sometimes also referred to as a model chemistry and for any computational comparison, exactly the same model chemistry has to be employed. We will later see that the choice of an appropriate model chemistry is crucial to the success of computational investigations. For example, we will discuss below that the popular B3LYP/6-31G(d) method (historically one of the most popular methods in organic chemistry) can no longer be recommended. As a consequence, it is important to choose a computational method that gives the correct result for the right reason. This can be achieved by testing various computational methods and gauging the computational results with experimental findings.

A well-known example for the successful application of computational organic chemistry is the development of a mechanistic model for proline-catalyzed aldol reactions. After the initial application of proline catalysis for the synthesis of steroid precursors by Eder, Sauer, and Wiechert [10] as well as by Hajos and Parrish [11], List, Lerner, and Barbas recognized the potential of this concept for stereoselective organic synthesis [12]. Various mechanistic models had been proposed for these reactions over time [11,13,14]. Computational investigations eventually revealed that carboxylic-acid catalysis (as initially proposed by Jung [13]) is not only the lowest-energy pathway [15] but also explains the experimentally observed stereoselectivities [16–19].

Besides the rationalization of experimental observations, computational investigations can also challenge our traditional understanding of how reactions take place. In 2014, an excellent computational and experimental study demonstrated that chemists do not completely understand the mechanisms for simple organic transformations (Scheme 7.1) [20]. The authors challenged two fundamental tenets of the traditional mechanism of electrophilic aromatic substitution between

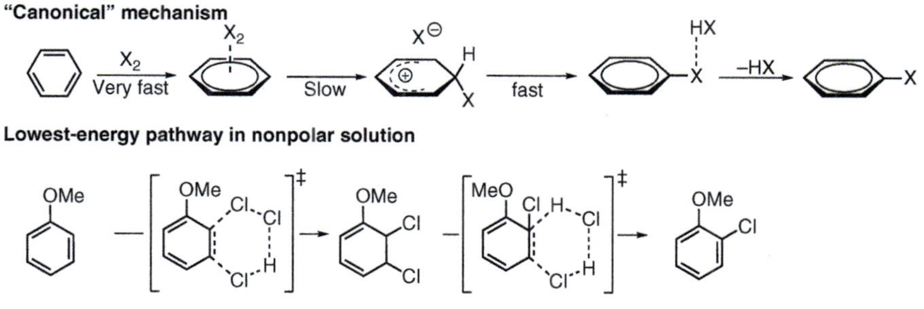

Scheme 7.1 Reaction mechanisms for electrophilic aromatic substitutions.

chlorine and anisole. Their computational results revealed that the direct substitution process – the commonly-accepted mechanism – does not hold for this prototypical example and that an alternative addition–elimination pathway is favored instead (Scheme 7.1) [20]. Subsequent experimental investigations relying on NMR and UV-vis spectroscopy support the computationally proposed mechanism [20].

Today, we can determine the structure of almost all molecules (which also provides information on bond lengths or angles) and their spectral properties. We can further calculate the energies of these structures and derive information on their thermodynamic preferences or how fast a chemical reaction takes place. Computational investigations can probe mechanistic proposals and directly compare energy profiles associated with these pathways. In addition, computational studies will provide deeper insights into the underlying molecular reasons for this behavior and help in developing reactivity models to easily rationalize a given effect. Furthermore, spectral (e.g. IR, NMR, UV) or other physical properties can be determined and help in characterizing unknown structures (see Chapter 8).

In summary, computational studies are vitally important for a modern analysis of reaction mechanisms. Such studies are strongest in fruitful collaborations between experimental and computational researchers [21–23]. While simple analyses can probably also be performed without a strong experience in computational chemistry by following published procedures, it might be beneficial to establish a collaboration with a computational group for more advances questions. This ensures a state-of-the-art computational analysis at a high level and does not "waste" valuable time and resources to familiarize yourself with specific techniques or codes.

In the following sections, we will first briefly look at hardware and software requirements (Section 7.2), before we will present typical computational methods (Section 7.3) and basis sets (Section 7.4) in organic chemistry. We will then focus on frequent applications in organic chemistry (Section 7.5) and their notation in publications (Section 7.6). In Section 7.7, we will use a tutorial reaction to see how computational chemistry can help to understand chemical reactions. Given the tutorial character of this chapter, we will occasionally encounter aspects that will only be discussed at a later stage. In those cases, please also see the Glossary for a very brief explanation of important concepts. Finally, we will provide the reader with some advanced aspects (Section 7.8), a selection of important keywords (Section 7.9), and practical considerations (Section 7.10).

7.2 Hardware and Software Requirements for Computational Investigations

Most beginners in computational chemistry might be interested in the hardware and software required to perform a computational investigation. Unfortunately, this question can often not be answered in an unbiased, general way. This section tries to provide an overview about those requirements, but these suggestions also reflect the personal preferences of the authors and certainly does not include a complete list of alternatives.

Let us first consider the hardware required to perform a computational investigations. Although initial test calculations and the tutorial (Section 7.7) can probably be performed on ordinary desktop computers or notebooks, it is typically not feasible to obtain publishable results only relying on those machines. Calculations would run for several days to weeks and often, those machines are not designed for those tasks. Instead, computational investigations are in most cases performed on dedicated workstations or on special supercomputers (also called clusters or high-performance computing [HPC]). Those computers typically feature a specific hardware like multiple processors. Thus, a substantial speed-up can be obtained when a calculation is performed on several processors rather than on just one ("parallelization"). Frequently, an application has to be filed to get access to a supercomputer, but the specific scenarios and administration varies. In contrast, data handling and analysis can typically be performed on ordinary desktop computers or notebooks.

Besides the suitable hardware, different software packages are also required to perform a computational study. Most importantly, the user needs to have access to a computational program that takes care of the actual calculation. Commercial programs include the programs ADF, Gaussian, MOLPRO, Q-Chem, or Turbomole. Often, research institutions hold site licenses for one of those programs. There are also programs that are freely available (for academic research) and examples include NWChem, ORCA, and PSI4. Each program has advantages and disadvantages and come with a different "content" (implemented methods, basis sets, . . ., see below). In the end, the user has to identify the program of choice based on availability and individual needs.

Besides the computational program, a text editor is required to prepare the input files that are used by the computational program. Input files are basically text files that contain all information to control the subsequent calculation. Any text editor can be employed (plain-text format), but the use of Microsoft Word or related text processing software is not recommended. Furthermore, a graphical program for the visualization of the structures and results of the computational study is very helpful. Again, commercial and free programs are available and selected examples include Avogadro, ChemCraft, GaussView, Molden, TmoleX. The reader might want to test several different programs and then select the most suitable program based on their needs. Eventually, Excel or a related spreadsheet analysis program will be used to analyze the different energies and to calculate reaction energies or activation energies.

7.3 Typical Methods in Computational Organic Chemistry

7.3.1 General Aspects

The general aim of any computational investigation is the use of a computer program to obtain some output (e.g. the energy or the chemical shifts of a molecule) from a given input (typically the molecular structure). For the computational project, the user has to choose a model chemistry consisting of a computational (or theoretical) method and a basis set. While basis sets are discussed in Section 7.4, this section will highlight important computational methods used in organic chemistry.

The mathematical equations required to determine that output are called the computational method. Over the years, many different methods have been developed and some of their names are familiar to synthetic chemists (e.g. Hartree–Fock or Density-Functional Theory). Those methods are usually implemented in computational programs (see Section 7.2), and the user can select the method they want to employ for their investigations.[1] A trained computational chemist will be able to choose the best method for a given problem.

This section describes some of the most commonly encountered methods and provides a simplified explanation about how they work. Knowledge of these underlying principles is important to understand the strengths and limitations of any given method and helps the reader to choose an appropriate method in their own investigations. The reader is referred to specialized text books on theoretical chemistry for a thorough mathematical and physical explanation (selected examples

1 Unfortunately, not all programs include all available computational methods and the user has to check the manual to see if a given method is available in a certain program.

partially with a focus on applications in computational organic chemistry: [2, 24–26] If the reader wishes to first familiarize themselves with applications and practical aspects of computational chemistry, this starts at Section 7.5.

Generally speaking, we will frequently encounter a typical phenomenon in computational chemistry in the following sections: more accurate methods will result in more reliable results but come at the price of (substantially) longer computational time. Thus, crude methods are often referred to as "computationally cheap," whereas more sophisticated methods are considered to be "computationally expensive." This aspect is discussed in slightly more detail in Section 7.3.8.

7.3.2 Molecular Mechanics and Force Fields

Molecular mechanics (MM) is among the simplest computational methods. It relies on classical mechanics (Newtonian mechanics, "ball-and-spring model") to describe a molecule. Each atom is treated as a separate particle ("ball") and bonding interactions are treated as "springs." Electrons and their movements are more or less completely neglected in this approach. The set of mathematical equations and atom-specific parameters that link geometry and energy is called a force field and, therefore, these calculations are sometimes also referred to as force-field calculations. Those parameters are either obtained from experimental and spectroscopic data or from high-level calculations. Most commercial programs include several force fields and typical examples include AMBER (Assisted Model Building and Energy Refinement; [27]), CHARMM (Chemistry at HARvard Molecular Mechanics, [28]), or OPLS (Optimized Potential for Liquid Simulations, [29]). Given the nature of the underlying theory ("ball-and spring model"), molecular mechanics is rather poor at describing bonding events (i.e. chemical reactions).[2]

As molecular mechanics calculations are extremely fast, they are typically applied in the study of very large molecules. A frequent application for this type of calculation lies in molecular dynamics simulations of large biomolecules including enzymes or proteins to study protein folding or the interaction between a protein and a ligand [30, 31]. Within organic chemistry, force-field calculations are often used to assess the conformational flexibility of a system (e.g. to generate starting structures for subsequent calculations, see below) or to quickly visualize organic molecules to get a rough understanding of their three-dimensional shape.

2 Researchers have also developed transition-state force fields (e.g. TSFF, Q2MM, ReaxFF) that account for bond making/breaking, but their applications are still rather limited and will not be discussed here.

7.3.3 Wave-Function Methods I – Hartree–Fock Theory

Wave-function methods rely on the Schrödinger equation. Physical constants are used as the sole input. As such, these methods are also known as *ab initio* methods. The wave function Ψ is a mathematical function that contains all relevant information of a given system and a property (or observable) can be calculated directly from that wave function using the appropriate operator. The most famous of those operators is the Hamiltonian operator ($\hat{H}$), which provides the energy of a system. Unfortunately, the Schrödinger equation cannot be solved exactly for systems containing more than one electron. Therefore, various approximations are used that allow for a solution. The Born–Oppenheimer approximation is probably the best known of these approaches and includes the assumption that electrons and the atomic nuclei can be treated separately. As the electrons are much lighter than protons and neutrons, they move much faster and, thus, motions of electrons do not affect the position of the nuclei.

Hartree–Fock (HF) theory can be considered as the simplest *ab initio* method. The primary assumption within HF theory relies on "non-interacting" electrons. Thus, this theory only considers one electron at a time and assumes that this electron only interacts with an averaged "charge cloud" rather than as pairwise electron–electron repulsions. Following this process, the average repulsive effect is taken into consideration, but not the explicit electron–electron interaction. The wave function is then formed from a linear combination of the atomic orbitals, which are mathematically described by basis functions (see Section 7.4 for more details). The electronic energy of the system is subsequently determined following the variational principle: An initial set of orbital coefficients is selected and the energy of the system is calculated with this trial set. These results are again used as an input and the process continues iteratively until the energy and orbital coefficients do not change anymore ("convergence"). As this procedure is also called a self-consistent field (SCF) approach, HF calculations are sometimes also referred to as SCF calculations.

In contrast to molecular mechanics described above, Hartree–Fock theory allows the description of chemical reactions (i.e. bond making/breaking). Although HF calculations have been used in the past for organic reactions, thus method is seldom used in modern computational chemistry due to the numerous errors.

7.3.4 Wave-Function Methods II – Post-Hartree–Fock Theory

A major limitation of Hartree–Fock calculations is the improper treatment of electrons. As discussed above, only an averaged electron repulsion is considered within HF theory. In other words, the probability for finding an electron at a specific location within an atom within HF theory only depends on the distance from the nucleus and not on the distance from other electrons. This effect is also known as electron correlation and the corresponding correlation energy can be described as the difference between the inaccurate Hartree–Fock energy and the exact solution of the Schrödinger equation. This severe shortcoming within Hartree–Fock theory can result in substantial errors, particularly when the number of electron pairs change in a chemical transformation. To overcome this source of error, various theories have been developed that are summarized as post-Hartree–Fock (post-HF) methods. Examples for those include configuration interaction (CI), coupled-cluster (CC) calculations, or Møller–Plesset perturbation theory (MP), which will be briefly discussed in the following.

In a full configuration interaction (full CI) calculation, all possible distributions of electrons among all available orbitals are considered. Full CI calculations are formally the best calculations that we can do and a full CI calculation with an infinite basis set (see Section 7.4) is an "exact" solution of the Schrödinger equation. However, full CI calculations are computationally extremely demanding. Their demand can be reduced, if only a subset of distributions is considered. For example, CIS calculations only consider single (S) excitations (one electron is excited to an unoccupied orbital), or CISD considers single (S) and double (D) excitations. However, even the simplified CI calculations remain very demanding and are, thus, not routinely used for organic molecules.

Coupled-cluster theory follows a similar concept to CI but is mathematically completely different. Similar to CI, different kinds of excitations can be included, which are termed CCSD, CCSDT, or CCSD(T). Again, the letter S, D, and T stand for single, double, and triple excitations. The very similar acronyms CCSDT and CCSD(T) differ in the way the triple excitation is calculated: While these excitations are calculated exactly for CCSDT, "only" a perturbative treatment is used for CCSD(T). Within organic chemistry, CCSD(T) calculations are frequently considered to be the "gold standard" and are often used as reference values for comparisons of less accurate methods. For standard applications in organic chemistry, CCSD(T) calculations are still too expensive and are typically not performed on a routine level. However, Neese and colleagues recently combined the Domain-based Local Pair Natural Orbital (DLPNO) approach with coupled-cluster calculations and developed the DLPNO-CCSD(T) method [32, 35]. This method is highly accurate and recovers 99.9% of the CCSD(T) correlation energy while having a very modest computational cost comparable to DFT calculations (see Section 7.3.6). Thus, DLPNO-CCSD(T) calculations are currently considered to be among the best affordable methods to routinely calculate the energy of organic structures.

Another approach to include correlation energy was proposed by Møller and Plesset. According to their theory, electron correlation can be treated as an additive perturbation to the Hartree–Fock energy. This method starts from a simple system with a known mathematical solution (HF theory) and adds a "perturbing" Hamiltonian accounting for a (weak) disturbance. This approach is suitable, as long as the perturbation is small compared to the simple system. Different orders of perturbation (second: MP2; third: MP3; fourth: MP4) have been developed, but only MP2 calculations are routinely used in organic chemistry. MP2 calculations have the advantage of being affordable and reasonably accurate (compared to HF calculations). However, Møller–Plesset calculations require rather large basis sets for reliable results, as they show a rather slow basis set convergence (see Section 7.4).

7.3.5 Semiempirical Methods

Semiempirical methods follow the general equations discussed above for Hartree–Fock calculations and have initially been developed to reduce the computational time of HF calculations. Within the philosophy of semiempirical methods, demanding and time-consuming parts of the calculations are not determined directly. To allow for an acceptable accuracy, additional external parameters are included in these calculations to reproduce selected experimental properties or the result of higher level calculations. Furthermore, specially designed basis sets are implemented in the semiempirical method and, thus, are not selected by the user.

One of the oldest examples is the Hückel method to determine the orbitals of conjugated π-systems. Examples that are still occasionally used include AM1 (Austin Model 1, [36]), or the Parametrization Method series (PM3, PM6, PM7, [37, 38]). Semiempirical methods have the advantage that they can be executed on regular personal computers for typical organic structures. However, semiempirical methods are rarely used in organic chemistry anymore, as modern DFT calculations (see Section 7.3.6) can provide more reliable results with comparable times for medium-sized organic molecules. One might consider the use of semiempirical methods for quick pre-optimizations, for illustrative or teaching purposes, or in combination with more accurate methods in QM/MM calculations for very large systems. QM/MM calculations dissect the system into two parts: A more "important" QM part that is described by accurate methods, and a less "important" MM part that is described by force fields or semiempirical methods. For a practical guide on QM/MM, the reader is for example referred to the following review [39].

7.3.6 Density Functional Theory

Density functional theory (DFT) is a common method that relies on a different approach to determine the energy. Here, the energy of a molecule will be calculated from the electron density ρ rather than the wave function Ψ. The mathematical

Table 7.1 Jacob's ladder of different types of DFT functionals.

Step	Type	Description	Selected functionals
4	Double hybrid	+ MP2 contribution	B2PLYP [40] DSD-BLYP [41] DSD-PBEP86 [42]
3	Hybrid	+ HF contribution	B3LYP [43, 44] PBE0 [45] M06 [46] M062X [46] ωB97XD [47]
2b	Meta-GGA	+ Second derivative of ρ(r)	TPSS [48] M06L [49] SCAN [50]
2a	GGA	+ First derivative of ρ(r)	PBE [51] BP86 [52, 53] B97D3[54]
1	LDA	Electron density ρ(r)	VWN [55] PW92 [56]

foundation stems from the Hohenberg–Kohn theorem. This theorem proves that the ground-state properties of a system (e.g. the exact electronic energy) can be uniquely calculated from the electron density with the help of functionals.[3] Unfortunately, the exact nature of this functional is unknown and various approaches have been proposed. In order to differentiate between the different treatments, Perdew used the comparison with "Jacob's Ladder" (Table 7.1, [48]). The ladder stands on the incomplete HF theory "on earth" and ascends to chemical accuracy (the exact functional) "in heaven." With each step of the ladder, the applied theory is getting more sophisticated, and consequently, the obtained results should become more accurate. Functionals belonging to the first step are called *local density approximation* (LDA) functionals. Here, only the local value of the electron density is used for the calculation of the energy. Although it is a very crude approximation to assume a homogeneous electron density, LDA functionals perform surprisingly well due to error compensation. Nonetheless, those functionals tend to predict bonding energies to be too strong. In a first improvement, the *generalized gradient approximation* (GGA) also considers the first derivative of the electron density (meta-GGA functionals even consider the second derivative),

3 Simplified, a functional can be described as a function that acts on another function.

which accounts for the non-homogeneity of the actual electron density. Further improvement can be obtained, when the "pure" DFT contribution is combined with a portion of Hartree–Fock theory (*hybrid functionals*), and additionally of MP2 theory (*double-hybrid functionals*).

On average, it can be expected that functionals from higher steps perform better than those of lower steps. However, exceptions are known and it is highly recommended that users consult recent benchmarks for their properties of interest.[4] In organic chemistry, geometry optimizations are frequently performed with (meta)-GGA or hybrid functionals, while the final electronic energies are obtained from hybrid or double-hybrid functionals.

7.3.7 Dispersion-Corrected Density Functional Theory

Within the last decade, numerous investigations have highlighted the importance of London dispersion and its influence on structures and energies [58–60]. London dispersion is the attractive part of the van-der-Waals interaction, depends on the distance R with R^{-6}, and is consequently particularly important for larger systems [58–60]. Quantum-mechanically, London dispersion is an electron correlation effect, and, therefore, post-HF methods (e.g. MPn, coupled cluster) already account for London dispersion and do not need any additional correction. In contrast, HF theory and standard density functional theory neglect London dispersion and, thus, are likely to be inaccurate in describing molecules and chemical reactions, where dispersion is important.

Extensive investigations by Grimme, Schreiner, Truhlar, and others indicate that standard DFT will, therefore, not result in reliable results. For example, Grimme and others pointed out that B3LYP/6-31G(d) – the standard method frequently employed in computational organic chemistry – can be unsuitable even for very simple tasks and should not be used anymore [62].

As density functional theory is one of the most important computational methods in organic chemistry, different theoretical procedures have been developed to properly account for London dispersion [58–60]. Thus, strictly speaking, London dispersion is not a "computational method" itself. Instead, dispersion-corrected DFT (i.e. functionals with a proper treatment of London dispersion) are now recommended and typically used for modern applications.

Among those, the DFT-D method [54, 62–64] developed by Grimme can be considered as a quasi-standard in computational organic chemistry. In this approach, the dispersion energy is added to the DFT energy based on atom-pairwise interactions over all atom pairs. A Becke–Johnson (BJ) damping function can

4 A first general overview for main group thermochemistry, kinetics, and noncovalent interactions can be found in Goerigk et al. [57].

further be used to improve the description of London dispersion. Several versions of the dispersion correction termed D2, D3, D3BJ, and D4 have been developed over the years and can be requested in most computational programs in addition to the "regular" functionals (like B3LYP or TPSS). Those functionals typically bear the dispersion correction as a suffix (e.g. B3LYP-D3 or TPSS-D4). Truhlar and colleagues followed a different approach in the development of their "Minnesota functionals" (e.g. M06, M06-2X, M11). In these functionals, London dispersion is considered through empirically fitting an (experimental) benchmark set by a large number of parameters. As the inclusion of a dispersion correction does not increase the computational time, only dispersion-corrected functionals or functionals with an additional dispersion correction should be used for any modern application.

7.3.8 Typical Computational Times

In the previous sections, we have encountered different computational methods and have discussed how reliable they are in general. In this section, we will briefly look into the associated computational costs, i.e. the time a calculation takes with such a method. Table 7.2 provides a crude comparison of the computational costs for different methods. The quoted time range refers to the time required to calculate the energy of an organic molecule with approx. 50 atoms for a given geometry on a normal desktop computer.[5] If more than one processor is used for the calculations (this is the default setup, when a computing cluster is used), the computing time will be lower to approx. $1/n$th of the initial time (n stands

Table 7.2 Crude comparison of the time required to calculate the energy of an average organic molecule (ca. 50 atoms) and typical applications in organic chemistry.

Computational method	Time range	Applications in organic chemistry
Molecular mechanics	Seconds	Conformational analysis, crude visualization
Semiempirical	Minutes	Crude optimization, very large molecules
DFT	Hours to days	Many applications
HF	Hours to days	Rarely used anymore
Post-HF	Days to weeks	Very accurate energies

5 In Section 7.4.2, we will have a closer look at those calculations, which are typically called single-point calculations.

for the number of processors). Based on the required computational time, DFT typically results in manageable time scales that allows the study of many different aspects within organic chemistry, while post-HF methods are largely only used for obtaining very accurate energies.

7.4 Basis Sets Used in Computational Organic Chemistry

7.4.1 General Aspects of Basis Sets

As mentioned above, within a model chemistry used to analyze a given problem in organic chemistry, the user has to select both a computational method (Section 7.3) and a basis set (this section). While semiempirical methods rely on predefined basis sets, a basis set needs to be selected for wave-function or DFT methods. A basis set is a series of mathematical functions that describe the atomic orbitals (i.e. the location of a given electron) and are used to construct the wave function (see Section 7.3). Since it is difficult to mathematically describe the "true" shape of an orbital, a simplified model is required. As a minimal requirement, every orbital needs to be described by one mathematical function (termed basis function). For example, a carbon atom consists of five different orbitals (1s, 2s, $2p_x$, $2p_y$, $2p_z$) and, thus, a total of five functions (one per orbital) is at least required to describe all orbitals. Such a basis set is called minimal or single-ζ basis set. If more than one basis function is used, those basis sets are called double-ζ (two functions per orbital), triple-ζ (three functions), quadruple-ζ (four functions), and so on. In general, larger basis sets (more comprehensive models) allow for a more flexible spatial distribution of electrons and do a better job of replicating the true orbital. Thus, the property to be calculated will get more accurate when using larger basis sets (Figure 7.1). As a drawback, larger basis sets are computationally more demanding

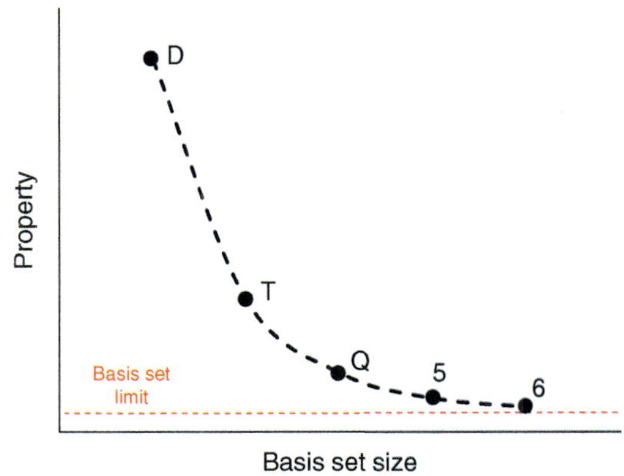

Figure 7.1 Basis-set dependence of a molecular property (D (double), T (triple), Q (quadruple), 5, 6 indicates the size of the basis set).

and require longer calculations. In theory, the best result for a given computational method will be obtained for an infinitely large basis set. This value is typically called the basis set limit (dashed red line in Figure 7.1). While such an infinitely large basis set cannot be used in practice, one can generally say that increasing the number of basis functions per orbital will automatically bring the results closer to the basis set limit. Thus, as a general recommendation, it is advisable to always use the largest basis set that can be handled in a practical manner.

While it is well known that valence orbitals have a substantial effect on the bonding situation, the effect on core orbitals is rather small. Consequently, computational resources can be reduced, when the core and valence electrons are treated differently. The resulting basis sets are called split-valence or valence-multiple-ζ basis sets. In those basis sets, the core electrons are only described with a small number of basis functions (typically a minimal basis set), while the valence electrons are described with a substantially larger number of basis functions.

7.4.2 Introduction to the Mathematical Formalism in Basis Sets

To better understand the nature and quality of different basis sets, we briefly have to look into some aspects of the mathematical formalism. In this section, we will again not provide a thorough explanation of the underlying math but rather focus on important concepts that help in understanding the basic design principles of basis sets.

There are two mathematical functions that are typically used to construct the orbitals: Slater-type orbitals (STOs) and Gaussian-type orbitals (GTOs). A Slater-type orbital uses the exponential function e^{-r}, while a Gaussian-type orbital uses an e^{-r2} function. Both functions are depicted in Figure 7.2. STOs show a much more realistic behavior and closely resemble the atomic orbitals of hydrogen (Figure 7.2).

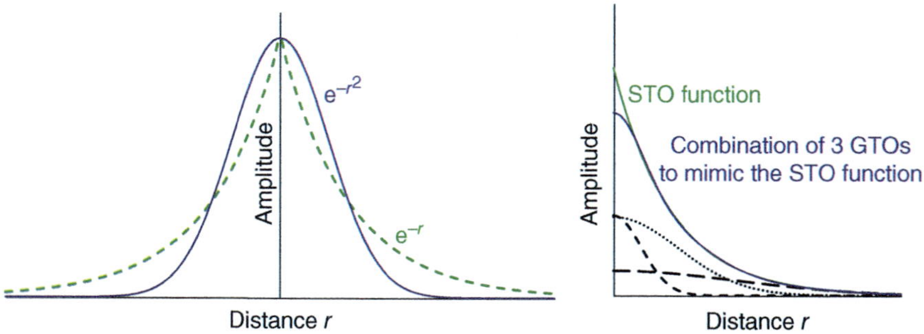

Figure 7.2 Comparison of the exponential functions in Slater-type (green dashed line) and Gaussian-type (blue solid line) orbitals (left) and a combination (blue) of three GTOs (black) to mimic the more accurate STO (green) function (right).

The STOs correctly describe the position at the nucleus (i.e. the cusp at the y-axis), while GTOs have an unrealistic zero slope at that position. Furthermore, GTOs decline substantially faster than STOs for larger distances, which is again not in line with the hydrogen orbitals. In contrast, GTOs can be handled much easier from a mathematical point of view. In the development of basis sets, a compromise has been found that combines the speed of the GTOs and the accuracy of STOs. As shown in Figure 7.2 (right), several (here 3) GTO functions (dashed black lines) can be combined in an additive fashion to mimic the shape of an STO. Thus, most basis sets rely on this procedure and use several GTOs as a linear combination to mimic the shape of a STO.

7.4.3 Polarization and Diffuse Functions

So far, we have considered the typical atomic orbitals (e.g. s-, p-, d-orbitals) and how they can be included in a basis set and technically, a calculation could be performed relying only on those orbitals. However, the computational results will not be very accurate. For example, when we look at the water molecule in Figure 7.3 and only employ the above-mentioned atom-centered s- and p-orbitals, the orbital overlap is rather poor and would not be in line with a bent structure. The reason for this is the fact that those atom-centered orbitals are simply not flexible enough to properly describe the bent structure. This issue can be overcome by including higher angular momentum orbitals (e.g. p-orbital on H, d-orbital on C). Those additional orbitals polarize the existing orbitals and allow for a better orbital overlap. Thus, inclusion of polarization functions is highly recommended to obtain reliable results.

Another shortcoming of the basis sets discussed so far is the treatment of weakly bound electrons that are present, for example, in anions and radicals. These electrons can be localized further away from the core. The basis functions discussed so far do not properly account for this effect. Diffuse functions can be included in the basis set, which typically start with a small amplitude but fall off

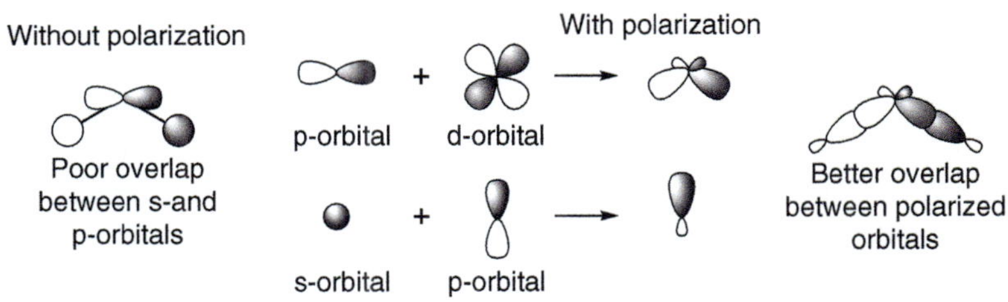

Figure 7.3 Schematic visualization of the influence of polarization functions on the orbital overlap within the water molecule.

very slowly. This provides a reasonable description of weakly bound electrons and diffuse functions should be used whenever structures include these weakly bound electrons.

7.4.4 Basis Set Families

Over the years, different researchers have developed different basis sets independently and each basis set bears a "name" or abbreviation. With a little experience in computational chemistry, the researcher will be able to deduce the quality of the basis set from this abbreviation. Based on the way the individual basis sets have been developed, they are grouped in different "basis-set families." As with the computational methods described above, basis sets have to be specified by the user either by selection from a drop-down menu or by typing the name in the input file (see Section 7.7.2). Within this section, we will briefly discuss three important and frequently employed basis-set families: Pople's a-bcdG basis sets, Ahlrichs' def2-XZP basis sets, and Dunning's cc-pVXZ basis sets. For more information, the reader is referred to thematic reviews [65, 66].

A widely used basis-set family was designed by Pople and colleagues. They include 3-21G, 6-31G, and 6-311G as representative examples and were used as the default basis sets in Gaussian. These basis sets nicely show how basis sets are generally constructed and we will, therefore, use this basis set family to briefly discuss the important elements of a basis set. We will use the 6-31G basis set to discuss this in more detail. The hyphen indicates that this is a split-valence basis set: The number before the hyphen refers to the core orbitals, while the numbers after the hyphen describes the valence orbitals. The number of digits indicates how many basis functions are used per orbital, while the magnitude of each digit specifies how many exponential functions are combined within that basis function. In our example (6-31G), core orbitals will be described by only one function that relies on six exponential functions. Valence orbitals will be described by two functions (digits 3 and 1) that are constructed from three and one exponential function, respectively.

To include polarization functions within Pople basis set, one or two stars (* or **) were initially added after the letter G resulting for example in 6-31G* or 6-311G**. A single star implies that one set of d-orbitals was added to each atom that was not hydrogen or helium, while the second star additionally added one set of p-orbitals to hydrogen and helium atoms. However, subsequent investigations have revealed that it might be useful to include more polarization functions (beyond the single set of d- or p-orbitals). Therefore, the initial Pople nomenclature with */** is no longer recommended and was replaced by parenthesis. The basis set 6-31G* should now be written as 6-31G(d) and 6-311G** as 6-311G(d,p). The entry before the comma (or the only entry in parenthesis) refers to all elements starting at lithium, while the entry after the comma refers to period 1. So the basis set 6-31G(3df,2p) adds three sets of d-orbitals and one set of f-orbitals to "heavy" atoms and two sets of p-orbitals to hydrogen.

Diffuse functions can also be included in the Pople basis-set family to account for weakly bound electrons. The addition of diffuse functions is indicated with a + sign before the letter G as in 6-31+G or 6-31++G. A single + sign implies that one set of s- and p-orbitals is added to each heavy atom, while two + signs indicate that these orbitals are also added to hydrogen and helium.

Another important basis set family was developed by Ahlrichs and coworkers. These basis sets are the default selection within the Turbomole program and are frequently called Ahlrichs bases.[6] This family includes def2-SV, def2-TZV, or def2-QZV as examples for double-, triple-, and quadruple-ζ basis sets. Polarization can be requested by addition of P or PP to the basis set name (e.g. def2-TZVPP or def2-QZVP), while the letter D is used for diffuse functions (e.g. def2-TZVPD).

The Dunning basis sets cc-pVXZ were specifically developed for post-HF methods to properly reflect electron correlation. Within this basis set nomenclature, cc stands for correlation-consistent, p for polarization, V for valence, X for the size (D: double, T: triple, Q: quadruple, . . .), and Z for zeta. Diffuse functions can also be added (or augmented) to the basis set with the prefix aug-cc-pVXZ.

As discussed above in Figure 7.1, larger basis sets will typically result in a more accurate result and are much closer to the basis set limit. However, calculations with large basis sets can require substantially more computational time. For example, the time to complete a DFT calculation depends on the third power of the number of basis functions (N) used to describe the molecule (time ~ N^3). Thus, a basis set that includes twice as many basis functions would result in an eightfold increase in computational time. Table 7.3 provides a direct comparison of the involved basis functions of several basis sets for the benzene molecule. The values of Table 7.3 were derived with the Gaussian program. Note that these values might slightly deviate for other computational programs. This stems from different basis-set implementations in each program, which will not be discussed here any further.[7] Double-ζ basis sets of all families are comparable in size, while significant differences are observed for larger basis sets. For reliable results, it is often required to rely on large Ahlrichs or Dunning basis sets, as Pople basis sets simply do not include a sufficient number of basis functions. Regarding the computational time, changing the basis set from 6-31G(d) to 6-311+G(d,p) will for example

6 The term "def2" stands for default basis set in a second generation. These basis sets should not be confused with the older, first-generation default versions termed "def."

7 The different implementation of basis sets will not change the general outcome of a computational study and relative energies are typically identical regardless of the program used. However, please note that the absolute energy might slightly deviate, when different programs (or even program versions) were used.

Table 7.3 Comparison of basis-set size for selected basis sets within Gaussian: number of basis functions of benzene.

Basis sets	Number of basis functions	Basis sets	Number of basis functions	Basis sets	Number of basis functions
3-21G	66	def2-SV	96	cc-pVDZ	114
6-31G	66	def2-SVP	114	aug-cc-pVDZ	192
6-31G(d)	102	def2-TZV	102	cc-pVTZ	264
6-31G(d,p)	120	def2-TZVP	222	aug-cc-pVTZ	414
6-31+G(d)	126	def2-TZVPP	270	cc-pVQZ	510
6-31+G(d,p)	144	def2-TZVPD	276	aug-cc-pVQZ	756
6-31++G(d,p)	150	def2-TZVPPD	324		
6-311+G(d,p)	168	def2-QZVP	522		
6-311++G(d,p)	174	def2-QZVPP	522		
6-311++G(3df,3pd)	342	def2-QZVPPD	576		

increase the required time by a factor of 3, to def2-TZVPPD by a factor of 9, and to aug-cc-pVQZ by more than a factor of 150.

Most of the basis sets discussed above are already implemented in modern computational methods. The user simply needs to select the correct input (see Section 7.7.2). However, if a different basis set is desired, it can (most likely) be downloaded from a web-based database: https://www.basissetexchange.org.

7.4.5 Effective Core Potentials (Pseudopotentials)

While the basis sets discussed above can easily be used for lighter elements, there are a few challenges that are associated with heavier elements starting somewhere between K and Rb. Thus, this is particularly relevant for studying transition-metal catalyzed reactions. The more electrons there are in a system, the more basis functions are required to describe those electrons. Although most of these electrons are core electrons and can be described with a minimal basis set, the computational cost is substantially increasing.

To reduce the computational complexity, one can use effective core potentials (ECPs, sometimes also called pseudopotentials) rather than explicit basis sets to model those core electrons. These ECPs are analytical functions that describe the core electrons and are much easier to handle computationally. Popular examples include the Los Alamos National Laboratory (LANL) family (e.g. LANL2DZ, [67–69]) or the Stuttgart–Dresden pseudopotential SDD [70].

7.4.6 The Basis Set Superposition Error (BSSE)

While it might be tempting to use smaller basis sets and, thus, faster calculations, those investigations are often less reliable. They suffer from substantial "technical" problem that can result in a substantial error, the so-called basis set superposition error (abbreviated as BSSE). As this abbreviation is sometimes also used in organic chemistry, we will briefly discuss this effect for a bimolecular reaction, where **A** and **B** react to a complex **AB**.

The interaction energy ΔE_{int} is typically calculated as the difference of the energy of the product complex and the energies of free **A** and **B** ($\Delta E_{\text{int}} = E(\mathbf{AB}) - E(\mathbf{A}) - E(\mathbf{B})$). For the calculations of **A**, **B**, and **AB**, we typically rely on finite basis sets (i.e. basis sets with a limited number of basis functions). When we now consider the number of basis functions present in each calculation, the calculation for the complex **AB** will feature substantially more basis functions than the monomers **A** and **B**. This also means that the situation within the complex **AB** is more flexible, one can also say that the substructure **A** "borrows" basis functions from the substructure **B** (or in a very simplified version: electrons of **A** could at least partially also be somewhere within **B**). This effect artificially lowers the energy of the complex **AB** and results in an interaction energy that is too low (i e. the bonding interaction is predicted to be stronger than it is).

There are two options to tackle this problem: A simple solution includes to simply rely on single-point calculations with sufficiently large basis sets. In this case, the energy gain from "borrowing" basis functions is substantially reduced and the BSSE gets small and less important. For many DFT methods, one typically assumes that the use of at least triple-ζ (if possible also quadruple-ζ) basis sets should be sufficient, while post HF-methods could still suffer from BSSE when these basis sets are used. A slightly more complex solution is the so-called counterpoise (CP) correction. This correction includes additional calculations, where the energies of **A** and **B** are determined in the presence of the basis functions of the other species. A drawback of this approach lies in the fact that CP corrections seem to overestimate the BSSE and, thus, introduce another form of error. As a consequence, many researchers prefer the first over the second solution and use the CP correction only for those questions where using larger basis sets is not feasible.

7.5 Typical Computational Tasks in Organic Chemistry

7.5.1 Preliminary Remarks

As this chapter should serve as an introduction for the less experienced reader, the following section will briefly summarize typical computational tasks (colloquially also called "jobs") that are frequently encountered in a computational project.

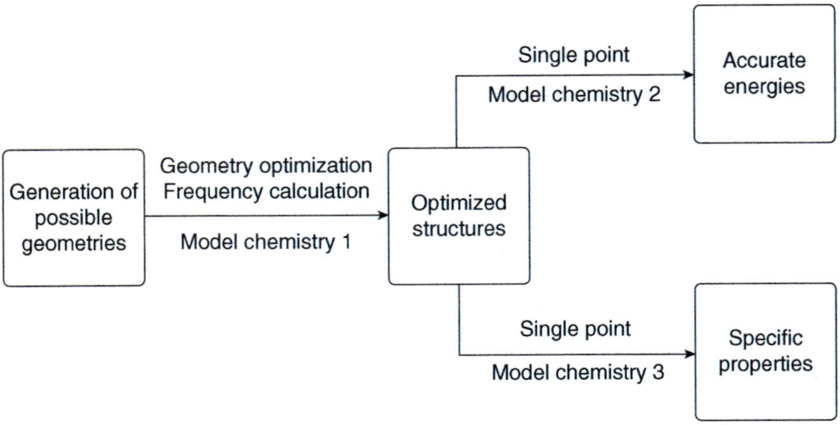

Figure 7.4 Flowchart for a typical computational project.

We will use most of them in the tutorial example discussed in Section 7.7. A typical procedure is summarized in the flowchart in Figure 7.4 and starts with the generation of the initial starting structures (see Section 7.5.6). These structures are then optimized (see Sections 7.5.3 and 7.5.4) using the model chemistry 1. Various additional calculations can follow, which use these optimized structures as input. Those calculations are called single-point calculations (see Section 7.5.2) and will provide properties of interest like energy, chemical shifts, molecular orbitals and many more. In many computational projects, one has to find a compromise between accuracy and computational costs. Ideally, one wants to use a highly accurate method, but in many cases, the geometry optimizations cannot be done within a reasonable time frame with these methods. However, it is generally accepted that the molecular geometry does not change significantly, when going from one method to the other. Therefore, it can be expected that the optimized structures for different model chemistries are very similar. Thus, in many cases, optimizations are performed with a reasonably accurate method to obtain the geometries, and a very accurate method is then used for the energy calculation.

As already mentioned above, all computational tasks are controlled through a text-based input file. These files can either be prepared with a typical text editor from scratch or with the help of a structure editor (see Section 7.7.2 for more details). All aspects of the calculation are defined via specific keywords within that file. To help the reader perform simple calculations by themselves, specific keywords are directly given occasionally in the following section. A list of important and frequently used keywords is also summarized in Section 7.9. While all computational programs should rely on similar keywords, we will highlight commands for two typical programs within this chapter: Gaussian and ORCA. While

Gaussian can be considered to be one of the most frequently used commercial programs, ORCA is currently freely available for academic purposes. The following section is not meant to replace the manual and will only discuss simple approaches. Please consult the respective manuals to learn more details about each keyword. While some parts of the manuals are clearly written for experts in the field, the manual also provides helpful examples and discuss options for beginners.

7.5.2 Single-Point Calculations

Single-point calculations are probably the simplest computational task that are typically encountered. In these calculations, the energy as well as other properties are calculated for a molecule with a given geometry. As the geometry of the molecule does not change in this calculation, reliable structures need to be provided. Frequently, single-point calculations are performed to obtain more accurate energies for previously optimized structures (see Section 7.5.3) or to obtain properties of interest (e.g. charge distributions, molecular orbitals, electrostatic potentials, spectroscopic data, . . .). The keyword[8] for a single-point calculation is "SP" for both Gaussian and ORCA and this is also the default job in both programs. Please note that the energy obtained from a single-point calculation is the potential (sometimes also called electronic) energy, and there is no direct equivalent in the "experimental world."

7.5.3 Geometry Optimizations

Similar to the single-point calculations, a geometry optimization starts from the geometry provided by the user in the input file. The program first calculates the energy for that particular structure and then analyses how changes to that structure alter the energy of the system. The computer then changes the structure in an iterative fashion until a minimum or maximum (i.e. a transition state) of the energy has been reached. These points are called stationary points and from a mathematical point of view, the first derivative of the energy (which is also called the gradient) is zero. Thus, determination of the gradient also indicates in which direction the strongest change in energy occurs. The program then changes the structure to move a little bit into that direction and the entire process starts again until a minimum or maximum is reached. A schematic procedure is depicted in Figure 7.5 for both local minima (i.e. starting materials, products, intermediates; examples A and B) and a maximum (i.e. a transition state; example C).

---

8 Typically, keywords in both Gaussian and ORCA are not case sensitive. Within this chapter, the keywords are written in a way that should help the reader to quickly recognize the requested tasks.

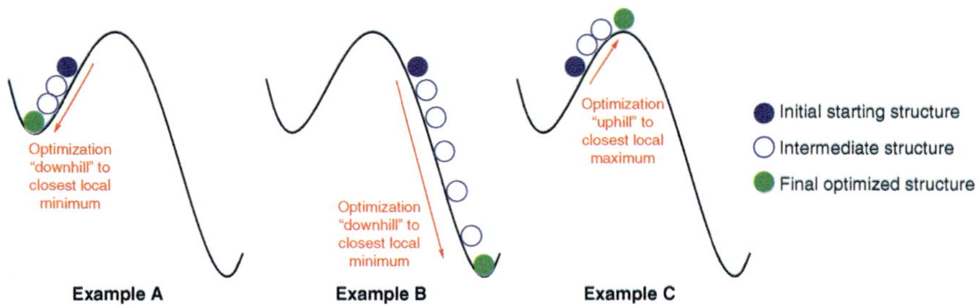

Figure 7.5 Schematic representation of the process of geometry optimization.

To request a geometry optimization in Gaussian, the keywords "Opt" (for a minimum) and "Opt=ts" (for a transition state) have to be specified, while "Opt" and "OptTS" have to be used in ORCA. Reaching a stationary point is indicated by a series of pre-defined "convergence criteria," which analyze if the forces that act on the molecule are sufficiently close to zero.[9] A structure is considered to be optimized when the calculated values of those parameters (e.g. the forces) are below a threshold value. Strictly speaking, this only means that the structure is sufficiently close to the "real" minimum that further computations are not expected to result in a meaningful change in energy anymore. For most applications, the default convergence criteria are probably sufficient, but they can be tightened for more difficult cases or when reaching the "true" minimum is very important (keywords: Opt=tight/verytight in Gaussian; TightOpt/VeryTightOpt in ORCA). During an optimization, several dozens of structure changes are typically required until the stationary point is reached. Thus, with the same model chemistry, an optimization takes considerably longer than a plain single-point calculation.

7.5.4 Frequency Calculations

At the end of a successful geometry optimization, the resulting structure should be a stationary point. Due to the way the optimizations are carried out, it may not be clear if the stationary point found is a minimum or a transition state. However, this information is crucial for any further computational analysis. To distinguish between a minimum and a transition state, a frequency calculation needs to be performed. The keyword for such a calculation in both Gaussian and ORCA is simply "Freq."

One important result of a frequency calculation is the fact that one can distinguish between a minimum and a transition state. During a frequency calculation,

9 Please note that every program relies on slightly different convergence criteria and, thus, minor structural changes can occur for different programs.

all molecular vibrations (or frequencies) are calculated (e.g. C–H stretching vibrations) and most structure editors offer the possibility to look at an animation of those vibration. As a side-note, the calculated molecular vibrations are directly linked to the IR spectrum of that compound. To distinguish between a minimum and a transition state, we need to look if the frequency calculation resulted in any imaginary frequency. Those imaginary frequencies are typically listed as negative entries in the list of frequencies and can also be animated in the structure editor. When looking at the imaginary frequency (which is always recommended), one would expect to "see" the key reaction progress. For example, for an S_N2 reaction between methyl chloride and hydroxide, the imaginary frequency should show that the methyl group vibrates ("moves") between the chloride leaving group and the nucleophile hydroxide. This already provides a strong evidence for having located the correct transition state (see Section 7.5.5 for a more detailed analysis). Alternatively, one could simply open the output file in a text editor and search for the term "imaginary frequency." If one or more imaginary frequencies have been found, the string "****** 1 imaginary frequencies (negative Signs) ******" will be printed within a Gaussian output file, while the ORCA output file indicates the imaginary frequency in the list of frequencies with "***imaginary mode***." For a true minimum, the energy increases in all directions and, therefore, no imaginary frequencies are present in the output file. For a transition state (which is a first-order saddle point mathematically), the energy increases in all but one directions and this direction is associated with exactly one imaginary frequency.

What could go "wrong" in a frequency calculation? It might happen that the visual inspection of the found imaginary frequency does not "fit" to the expected chemical reaction. This means that the optimization algorithm has identified a different transition state. A frequent example are rotations of methyl groups or related bond rotations. It might also occur that the number of imaginary frequencies does not match the expected number (0 for a minimum, 1 for a transition state) or that even more than one imaginary frequency is found for a particular structure. The latter would be called an *n*th order saddle point [71]. In all cases, this means that the program has identified the wrong stationary point and the user has to restart the calculation. One option is to start with a different starting structure, as the chosen one was not close enough to the desired stationary point. Another option could be to look at the animated imaginary frequencies and elongate the molecule along one of the undesired frequencies. By doing so, one is slightly "pushing" the molecule from the mountain top toward the valley and when rerunning the calculation with the modified structure, the program hopefully converges to the desired stationary point.

Besides the identification of minima and transition states, a frequency calculation also provides the so-called thermal corrections, which are important for a more realistic analysis. For almost all computational programs, geometry optimizations (and also single-point calculations) are performed on the potential energy surface, which neglects any contribution from translation, rotation, or vibrations. The potential energy is sometimes also called electronic energy, total energy, or SCF energy. However, for a comparison with experimental data, in most cases we need either enthalpies ΔH or free energies ΔG. The computational programs will use the harmonic-oscillator approximation to determine additive corrections to the potential energy, the thermal corrections.[10] Please note that a frequency calculation is only meaningful for a previously optimized structure (i.e. a stationary point) and cannot be used on a random, non-optimized structure. Within the Gaussian output file, the following thermal corrections will be calculated and these corrections have to be added to the potential energy of the system E_{pot}:

- Zero-point correction: This correction (also called zero-point vibrational energy ZPVE or just zero-point energy ZPE) accounts for the vibrational motion of the system that are present even at 0 K. By adding the zero-point correction to E_{pot}, one obtains the zero-point-corrected potential energy E_0 ($E_0 = E_{pot} + \text{ZPVE}$)
- Thermal correction to energy: This correction includes the internal energy of the system at 298.15 K.[11] This energy is often abbreviated as E_{298}.
- Thermal correction to enthalpy: This correction provides the conversion to the enthalpy of the system at 298.15 K and the final enthalpy is typically termed H_{298}.
- Thermal correction to Gibbs free energy: This correction provides the provides the conversion to the free energy of the system at 298.15 K and, thus, also includes an entropic contribution. Those values for G_{298} are typically used for comparisons with experimental numbers.

The thermal correction of interest is added to the calculated potential energy of the system. Frequently, more accurate potential energies are calculated on previously optimized structures with a better method. In those cases, the thermal corrections are added to the potential energy taken from the single-point calculation. Please note that only one correction is required, as, for instance, the correction to free energies already includes all other corrections.

7.5.5 Intrinsic Reaction Coordinate (IRC) Calculations

As discussed in Section 7.5.4, the animation of the imaginary frequency already provides a reasonable estimate on whether the obtained transition state is the desired one. However, there are cases where this visual inspection does not allow

10 A detailed description on how these thermal corrections are derived within the Gaussian program can be found here: https://gaussian.com/thermo/.
11 The user can generally alter the default temperature (298.15 K or 25 °C) and a default pressure (1 atm) and the reader is referred to the corresponding manuals to check the correct keywords.

a clear answer. In those cases, we can employ a more accurate method – an intrinsic reaction coordinate calculation – that determines the closest minima that are connected through a given transition state. In a very simplified analogy, imagine that you are standing on a mountain top and throw a ball in both directions and check where the ball comes to rest. Similarly, an IRC calculation starts at the transition state and analyzes the pathways leading to the nearest minima in both forward and reverse direction. In the course of an IRC calculation, the geometries are optimized at each point along the reaction coordinate. This requires that the program calculates the force constants (i.e. the forces that act on the transition state) at the beginning of the calculation. In Gaussian, the keyword for a simple IRC calculation is "IRC=calcfc" (calcfc stands for calculate force constants). In ORCA, "IRC" is sufficient to request an IRC calculation. At the end of an IRC calculation, a rough energy profile on the potential energy surface will be obtained. Depending on the nature of the reaction to be studied, the activation energy of the IRC calculation (i.e. the energetic difference between the minimum on the reactant's side and the transition state) can deviate significantly from the calculated activation free energy $\Delta G^{\ddagger}$ that are obtained from the transition state and the reactants (see Section 7.7 for a practical example).

7.5.6 Conformational Analysis

Most compounds can adopt different conformers (e.g. *gauche*- and *anti*-butane) and every conformer contributes to the observed experimental property following a Boltzmann distribution. In computational studies, frequently only the most stable conformers (i.e. the conformer with the lowest energy) of the reactants, transition states, and products are considered to calculate the energy profile of a reaction. This can often be justified by the fact that the most stable conformer dominates the Boltzmann distribution. To identify the "best" (i.e. most favorable) structure for a given compound, all potential conformers need to be considered in an analysis (the conformational analysis). Some compounds (benzene, water, . . .) only adopt one conformation, and the user can probably generate all possible conformers manually in a structure editor for small molecules (acetic acid, cinnamaldehyde, . . .). For larger molecules with dozens of conformers, a manual conformational search is not only very tedious, it is also prone to mistakes. Automated search algorithms can be used for those systems instead. There are various programs available for conformational analyses, and many rely on molecular mechanics (see Section 7.3.2) to allow for fast calculations. Among those programs, the Conformer–Rotamer Ensemble Sampling Tool (CREST) developed by Grimme allows a simple and straightforward workflow [72] with a helpful manual online. However, those programs typically need to be installed separately and are not implemented in the computational programs discussed before.

In simple terms, these programs generate a very large number of structures by random bond rotations and the resulting structures are sorted by energy. All unique structures that fall within a pre-defined energy window (typically around 30–60 kJ mol^{-1}) will be used for further (DFT) investigations.

7.6 Notation of the Model Chemistry

Organic chemists are typically very familiar with summarizing reaction conditions in publications and presentations, and it is probably not surprising that a traditional notation of the model chemistry has been established in computational chemistry. While the computational section of an article typically contains a broader description of methods used in the paper, there also is a short notation that can be used in the text, in captions, or on slides.

In this notation, the computational method and the basis set are typically separated by a slash. MP2/6-31+G(d,p) indicates that the calculations were performed using Møller–Plesset perturbation theory in combination with the 6-31+G(d,p) basis set.

If different computational methods were used for the optimization and the energy calculations (this is typically the case, see Section 7.5.2), the "double-slash" notation is the most common way to denote such a procedure. In this notation, the method for the energy calculation is given before the //, while the method used in the optimization is given after the double slash. For example, B2PLYP/aug-cc-pVTZ// M06-2X/6-311+G(d,p) implies that the initial geometry optimization (and thus also the thermal corrections) have been performed with the M06-2X functional and the basis set 6-311+G(d,p), while more accurate energies have been calculated on these optimized structures with the double-hybrid functional B2PLYP and the larger triple-ζ aug-cc-pVTZ basis set. Further details of the calculations (e.g. solvation, London dispersion, . . .) can also be included in the double-slash notation.

7.7 The Diels–Alder Reaction as a Tutorial Case Study

7.7.1 General Aspects and Requirements

In this section, we will discuss the use of computational chemistry for a specific case study in a tutorial fashion. In this example, the user will first need to create three-dimensional chemical structures. Two common programs are GaussView (a commercial software) and Avogadro (freely available). The user has to build the molecule of interest in three-dimensional space, and the software then converts this representation into Cartesian coordinates that describe the locations of all atoms in a computer-readable format. A so-called *Z*-matrix is an (old-fashioned) alternative to Cartesian coordinates and defines the three-dimensional shape of a molecule by specifying distances, angles, and dihedral angles with respect to other atoms. As most modern programs prefer Cartesian coordinates, we will not discuss the *Z*-matrix in this chapter. Next, one needs a text editor to create an input file, where key features of the job are written, including the Cartesian coordinates from the structure editor (see Section 7.7.2). Those input files are then transferred to the computational programs.

While all computational programs should result in very similar results, we will only highlight typical input files for both Gaussian and ORCA (see Section 7.5.1 on why those programs were chosen). To keep the run time of the computations within a reasonable limit, we will select the M06-2X functional together with the 6-31+G(d,p) basis set for this project. Most of the calculations will be finished within a few minutes (when performed on a HPC cluster with 8 CPUs) or within approx. one hour (when performed on a modern PC) and should, thus, be feasible for beginners. If the local hardware does not allow for those calculations, the user can use the smaller basis set 6-31G(d) for illustrative purposes.[12] The following examples were designed to be run with either Gaussian 16, Revision B.01 or ORCA 5.0.3, but similar results can be expected with other programs as well.

For the case study, we will select the Diels–Alder cycloaddition between 2-methoxybutadiene (**1**) and acrolein (**2**) as shown in Scheme 7.2. The reaction mechanism for this reaction is well known and was extensively studied both computationally and experimentally. For the sake of simplicity, we will only discuss the formation of the favored regioisomer **3** in this example, but the reader could proceed in an analogous fashion to study the regiochemistry of that reaction (not shown in Scheme 7.2).

1 **2** **3**

Scheme 7.2 The Diels–Alder reaction between 2-methoxybutadiene (**1**) and acrolein (**2**) as case study for a computational investigation.

---

12 Calculations with the 6-31G(d) basis set will be substantially faster. However, as this basis set is very small, it is not recommended to use this basis set for a scientific project. As the aim of this tutorial is to provide an easy access to a computational project, these drawbacks are probably acceptable for this task and the reader will be able to use the faster yet less reliable results to make themselves familiar with the tasks.

7.7.2 Preparing Input Files

As discussed above, all information that control the subsequent calculation have to be specified within the input file and a separate input file is required for each calculation. It is highly recommended not to use any spaces or "non-default" characters in the file name, as many queueing systems (i.e. those programs that "handle" the input files on a supercomputer) do not tolerate those within the filename and your job might not run. Although similar things are defined in each input file, the syntax varies from program to program. In the following, we will discuss the input files for simple Gaussian and ORCA calculations. Please note that you might have to adjust the parameters (e.g. number of CPUs, available memory) based on the machine you are using for the calculation.

As shown for a representative example in Scheme 7.3, Gaussian input files (typical file extension .gjf) start with a series of *link 0 commands*. These entries all start with the % sign and are optional (default values are used if not specified). Within this section, several "organizational" things are controlled including the location of important files or the available hardware for this calculation. The first entry (%chk=) defines the location and name of a checkpoint file. This file contains important intermediate results, and information from that checkpoint file can in principle be used in a subsequent calculation. The next entries (%cpu and %mem) are specific to the hardware available for the specific calculation. The %cpu command (older versions of this keyword are %NProcShared or %NProc) defines the number of processors employed for the calculation. While desktop computers often only allow calculations to be performed on one processor, more

Gaussian Input File				Comments
%chk=Acrolein_1				link 0 commands
%cpu=8				
%mem=4000MB				
#P M062X/6-31+G(d,p) opt freq				route section
				blank line
Acrolein_1				title section
				blank line
0 1				charge multiplicity
C	-0.29276	0.87519	0.21186	coordinates
C	-0.43419	-0.43529	-0.02649	
H	0.38962	-1.05199	-0.37263	
H	-1.39558	-0.91913	0.11261	
C	0.96555	1.63406	0.04517	
H	-1.13758	1.47583	0.53108	
O	0.98656	2.85405	0.00814	
H	1.85377	1.11600	-0.04259	
				blank line
				blank line

Scheme 7.3 A typical Gaussian input file (left) with comments (right).

processors (e.g. 8 as in the example below) are used in computational clusters. The %mem command specifies the amount of memory per node to be used (default value: 800 MB, in the example 4000 MB). The reader is referred to the manual for more information on these commands and additional options.

Immediately after those lines, the *route section* specifies the actual computational task to be done. The route section always starts with the # sign and can include various keywords. More detailed information can be printed in the output file with "#P," which could be useful for troubleshooting. The keywords can generally be given in any order and are not case sensitive. Within the route section, the computational method and the desired basis set needs to be specified. This can be done either in the traditional notation method/basisset (here: M062X/6-31+G(d,p)) or as separate entries (here: M062X 6-31+G(d,p)). All keywords that are required for the calculation must be provided in that line.

The route section is followed by a blank line and the title section. The title section can be used to provide more information on the task and is not interpreted by the program. After another blank line, the charge (0 for a neutral molecule) and the multiplicity of the system will be specified. The multiplicity M can be calculated from the total spin S of the molecule ($M = 2S + 1$). For singlets (molecules without any unpaired electrons), the multiplicity is 1, while a value of 2 has to be used for a radical with one unpaired electron. In the next lines, the Cartesian coordinates of the molecule (as obtained from the structure editor or the conformational analysis) are specified and again at least two blank lines are required below the coordinates.

A representative example of an input file (typical file extension .inp) for the analogous calculation within the ORCA program is shown in Scheme 7.4. Here, the job details are given in one or more *keyword lines* that all start with an exclamation mark (!). *Input blocks* can be used to define specific aspects of the calculation. Those input blocks typically start with the % sign. The number of processors

ORCA Input File	Comments
! M062X 6-31+G(d,p) Opt Freq TightSCF	Keyword line
%pal nprocs 8 end	Input blocks
%MaxCore 500	
# Acrolein_1	Comment
*xyz 0 1	charge multiplicity
C -0.29276 0.87519 0.21186	coordinates
C -0.43419 -0.43529 -0.02649	
H 0.38962 -1.05199 -0.37263	
H -1.39558 -0.91913 0.11261	
C 0.96555 1.63406 0.04517	
H -1.13758 1.47583 0.53108	
O 0.98656 2.85405 0.00814	
H 1.85377 1.11600 -0.04259	
*	

Scheme 7.4 A typical ORCA input file (left) with comments (right).

used in the calculations is defined by %npal keyword, while the amount of memory is specified with %MaxCore.

The careful reader might have observed that the values for the amount of memory seem to differ for the Gaussian input file of Scheme 7.3 and the analogous ORCA input file of Scheme 7.4. The reason for that stems from a different notation, the available memory is specified. More advanced computers (i.e. those that are used within clusters or specifically designed work stations) often feature a more advanced hardware setup. In simplified words, those machines consist of several independent computers (called nodes) that each feature several processors (called core or CPU). Within the Gaussian input file, the specified value refers to the total memory per node, while in ORCA, the memory required per core (or CPU) needs to be specified. Thus, for simple desktop computers that only use 1 processor, both numbers will be identical. In our representative example, we assume that our computing node consists of 8 cores/CPUs and that we have a total memory of 4000 MB for the calculation. Hence, in Gaussian the total amount per node (%mem = 4000 MB) is specified, while in ORCA, the amount per core/CPU (%MaxCore 500) is requested in the input file.

Any line that starts with the # sign is considered as a *comment* and will not be interpreted chemically. Finally, the molecular structure is defined between two * symbols. Again, charge and multiplicity need to be defined.

7.7.3 Conformational Sampling – Generation of Initial Geometries

The first step of every computational project typically deals with the generation of the initial input geometries. One of the most important things that has to be considered in this process, is the conformational flexibility of each molecule. Each starting material, product, and transition state can typically exist in a number of conformations, each with their own energy. As discussed in Section 7.5.3, a geometry optimization will only result in the closest local minimum, and it is not guaranteed that a (random) starting structure results in the lowest energy species. Thus, it is required to investigate all plausible conformers (see Section 7.5.6). For the Diels–Alder reaction discussed in Scheme 7.2, only a small number of conformers have to be discussed for each compound (see Scheme 7.5) and the generation of these structures can in this case easily be done manually with a structure editor. When saving those structures, make sure that you can extract the Cartesian coordinates easily. As enantiomers have identical energies, we will only consider one enantiomer of the product. For larger systems, it might be helpful to rely on automated conformational search algorithms (see Section 7.5.6).

Although we will not discuss catalyzed Diels–Alder reactions here, catalyzed reactions are of general interest in organic chemistry. Thus, we will briefly comment on the conformational analysis for the substrate-catalyst interaction

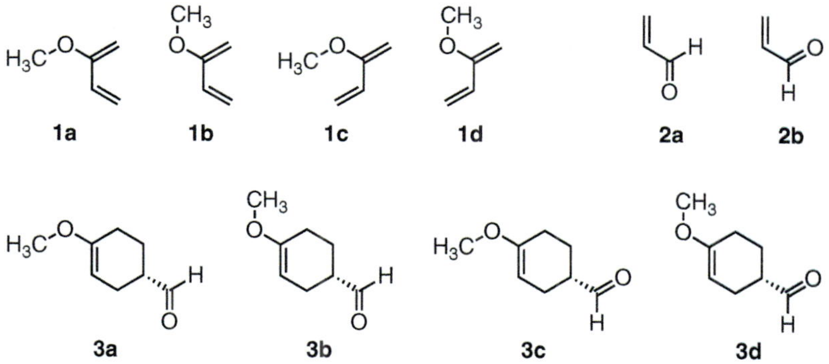

Scheme 7.5 ChemDraw representations for the potential conformers of the starting materials **1** and **2** and the cycloadduct **3**.

in the following: Ideally, a full conformational analysis will be performed for the free catalyst and any complex formed between the catalyst and the reactants, intermediates, and products. The Curtin–Hammett principle states that the most stable structure of the catalyst/reactant might not be the most reactive one, and thus, all potential combinations should be included in the analysis. Furthermore, it is important to consider all potential interactions between the catalyst and the organic molecules. For example, $AlCl_3$, a typical Lewis-acidic catalyst in the Diels–Alder reaction [73], will most likely coordinate to the carbonyl group of the dienophile. However, it might also coordinate to other Lewis-basic sites (oxygen lone pairs, π-bonds). For some reactions, the catalyst-substrate interactions are very clear based on previous investigations (e.g. for the Diels–Alder reaction). The situation might not be as clear for "new" transformations or catalysts, and the insights and creativity of chemists are critical to consider all possibilities.

7.7.4 Geometry Optimizations of Starting Materials and Products

For each geometry identified during the conformation analysis, a separate input file needs to be generated. Representative examples for these input files for an optimization have already been shown above in Schemes 7.3 and 7.4. Next, the user has to "hand-over" those input files to the computational program. The specific routine strongly depends on the available hard- and software and will not be discussed here. The program will then create an output file for each calculation and the content of the output file will be updated during the calculation. Each program prints a typical message within the output file, once the calculation is completed. Gaussian uses the message "Normal termination of Gaussian 16 at" followed by a time stamp, while ORCA includes the line "****ORCA TERMINATED NORMALLY****" at the end of the output file.

Once the calculations are completed, one should first check each file for the occurrence of imaginary frequencies (see Section 7.5.4). As we are currently dealing with the optimizations of the starting materials and products (i.e. minima on the potential energy surface), we expect no imaginary frequencies in the output file. If one or more imaginary frequencies is found, a saddle point has been located rather than a real minimum and there is at least one direction in which the energy is still decreasing. This direction is indicated by the movement within the imaginary frequency and the user can inspect that by visualizing/animating the imaginary frequency within the structure editor (see above). However, we have to identify the closest local minimum for that structure as well and there are different options on how to best achieve that. One way is to use a tool available in most structure editors: when visualizing the imaginary frequency, most programs offer the possibility to "distort" or "elongate" the previously optimized structure along the (undesired) imaginary frequency. The structure is then moved away from the mountain top and brought closer to the next valley. The obtained geometry is then stored and used for a new input file. Additionally, it might be worth to include the keyword "opt=calcfc" rather than only "opt" within Gaussian. When doing so, the force constants at the beginning are explicitly calculated, which should facilitate the optimization. In ORCA, the user needs to specify an input block that includes "%geom Calc_Hess true end" for a similar option.

By now, we should have succeeded in optimizing the geometries for the starting materials and products. The next task of the computational investigation is to construct a free energy profile and to determine the reaction free energy. To do so, we need to determine the lowest energy conformer. As we ideally want to draw comparisons to experimental data, we aim for the calculated Gibbs free energy. As we are not performing any additional single-point calculations in this example, we can directly pick the final value for the free energies from each output file. Within the Gaussian output file, the desired value is printed next to the string "*Sum of electronic and thermal Free Energies*". For ORCA calculations, please look for the string *Final Gibbs free energy*. Those values are given in hartree and a conversion of 2625.4995 (to $kJ\,mol^{-1}$), 627.5094 (to $kcal\,mol^{-1}$), or 27.2113 (to eV) is required. If we were performing additional single-point calculations to obtain more accurate potential (electronic) energies, we would take the potential energy from each single-point calculation (check for string *SCF Done* in Gaussian or *FINAL SINGLE POINT ENERGY* in ORCA) and add the correct thermal correction (*Thermal correction to Gibbs Free Energy=* in Gaussian or *G-E(el)* toward the end of the ORCA output).

We can then identify the most stable conformers (**1d**, **2a**, and **3a**) by the lowest (i.e. most negative) free energy. Ideally, the user copies the relevant information (e.g. free energies or potential energies + thermal corrections) into a spreadsheet. For the calculations described above, these values are summarized in Table 7.4. The reaction free energy for this process can be calculated from those conformers

Table 7.4 Calculated free energies for each conformer of the Diels–Alder reaction between **1** and **2** (in Hartree, rounded to 6 decimal figures).

			Gaussian calculations		
Butadiene_OMe_1 (**1a**)	−270.307465	Acrolein_1 (2a)	**−191.797419**	Cycloadduct_1 (3a)	**−462.157701**
Butadiene_OMe_2 (**1b**)	−270.312256	Acrolein_2 (**2b**)	−191.794754	Cycloadduct_2 (**3b**)	−462.151927
Butadiene_OMe_3 (**1c**)	−270.309641			Cycloadduct_3 (**3c**)	−462.157091
Butadiene_OMe_4 (1d)	**−270.315725**			Cycloadduct_4 (**3d**)	−462.151932
			ORCA calculations		
Butadiene_OMe_1 (**1a**)	−270.304548	**Acrolein_1 (2a)**	**−191.795486**	**Cycloadduct_1 (3a)**	**−462.153057**
Butadiene_OMe_2 (**1b**)	−270.309284	Acrolein_2 (**2b**)	−191.792840	Cycloadduct_2 (**3b**)	−462.146845
Butadiene_OMe_3 (**1c**)	−270.306340			Cycloadduct_3 (**3c**)	−462.152154
Butadiene_OMe_4 (1d)	**−270.312898**			Cycloadduct_4 (**3d**)	−462.147235

The most stable conformer is highlighted in bold.

by subtracting the energies of the starting materials from that of the product. For the Diels–Alder reaction of Scheme 7.2, we end up with a reaction free energy of $-117\,kJ\,mol^{-1}$ or $-28.0\,kcal\,mol^{-1}$. Both programs result in identical results within rounding error and typically energetic values should be given with an accuracy of $1\,kJ\,mol^{-1}$ or $0.1\,kcal\,mol^{-1}$.

7.7.5 Locating the Transition States

As we are interested in a complete free energy profile of the Diels–Alder reaction, we next have to deal with the transition states for the reactions. Locating a transition state is frequently a little bit more tricky than "normal" optimizations of starting materials and products. Different strategies can be used to obtain suitable input structures (also called transition-state guesses) and individual users prefer different strategies for their first attempt. Generally speaking, the closer the transition-state guess is to the "real" transition state, the faster and less problematic the determination of the transition state will be. As the transition state for Diels–Alder reactions is very well established, it is probably easiest to generate the initial input structures in a structure editor. For example, we can start from the cyclohexene product and simply elongate the C–C bonds that form in that reaction to a value between 2.0 and 2.1 Å (typical bond lengths in Diels–Alder transition states). Alternatively, we could also start from the starting materials and add the dienophile part in a suitable distance to the butadiene fragment. The user could also check the literature for published transition-state geometries for similar reactions, extract them (typically from the Supporting Information of the publication), and modify them to the system of interest.

For more complex transition states, or if the manual design described above does not work, it might also be worthwhile to first perform a constrained optimization to obtain a suitable transition-state guess. The idea behind such a constrained optimization is to fix the reactive atoms in a transition-state-like orientation (either from chemical intuition or related computations) and allow the rest of the molecule to adjust to this orientation. Figure 7.6 shows a schematic procedure for the Diels–Alder reaction discussed above. The most pronounced feature of the transition state of the Diels–Alder Reaction are probably the two new C–C bonds formed in that reaction. We know that the transition-state bond lengths for smaller model systems are around 2.1 Å and, thus, we select this value for the preoptimization. These distances are then frozen and will not change during the constrained optimization (see below). All remaining atoms can move freely during the preoptimization and ideally adopt an arrangement very similar to the final transition state. The pre-optimized transition state should then be fairly close to the real transition state and the optimization algorithm should have substantially fewer problems to locate the real transition state later.

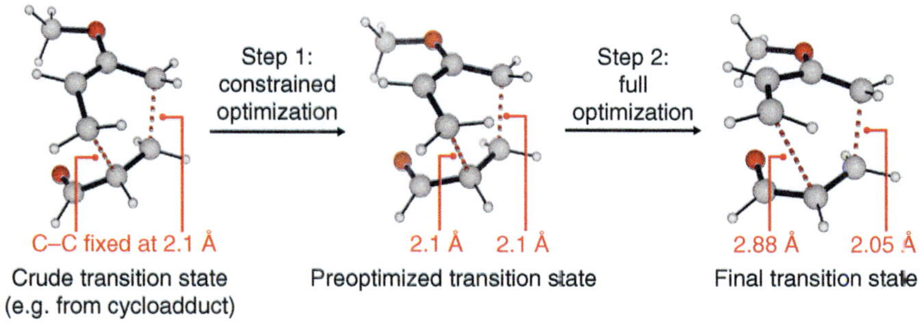

Figure 7.6 Schematic representation of an initial constrained preoptimization to locate the transition state for the Diels–Alder cycloaddition between **1** and **2**.

After the preoptimization is complete, a new input file needs to be prepared that now uses the preoptimized structure as starting structure. Now, all atoms can move freely and the program should be able to locate the closest transition state. For the Diels–Alder reaction, this final transition state is shown on the right of Figure 7.6. A closer look at this structure reveals that the initial assumption of very similar C–C bond lengths was not correct. In line with the polarized nature of diene and dienophile, an asynchronous transition state is obtained with two different C–C bond lengths (2.05 and 2.88 Å). Please note that a constrained optimization is not required to locate the transition state, but it can be very helpful (particularly, when the user has difficulties providing a suitable structure for the initial transition-state guess).

Typical input files for a constrained optimizations are shown in Scheme 7.6. To request such a calculation, the keyword for an optimization needs to be adjusted. In Gaussian, "opt=modredundant" will be required in the route section and the parameters that should be fixed during the optimization are specified after the Cartesian coordinates. To define the constraints, the initial letter indicates which kind of constraint is defined: X stands for a single atom, B for a bond, A for an angle, and D for a dihedral. Next, the atom numbers have to be given that are involved in the constraint. In this context, the atom number is simply the location of that atom within the list of Cartesian coordinates (i.e. as the first oxygen atom is listed as the 13th atom, its atom number is 13). Finally, the letter F indicates that the current value provided within the Cartesian coordinates is fixed and will not change during the optimization. In the example shown below, the first entry "B 2 8 F" indicates that the bond between atoms 2 and 8 will be frozen during the calculation.

Within the ORCA program, the constraints are specified within the input block %geom. Each constraint is then specified in a similar way as discussed before between a pair of curly brackets. Please note that Gaussian and ORCA count atoms in a different way. Gaussian assigns the number 1 to the first atom, while ORCA starts with the number 0 for the first atom. As the same atoms have to be

Gaussian Input File

```
%nosave
%nprocshared=8
%mem=4000MB
#P M062X/6-31+G(d,p) opt=modredundant

TS_Diels_Alder_1_min

0 1
C   -1.47689   0.14381   -0.39766
C   -0.79057   1.39898   -0.70328
C   -0.82456   -1.03585   -0.42553
C   0.60806    -1.14572   -0.70447
C   1.64465    0.20744    0.52494
H   0.90994    -0.74505   -1.67704
H   1.05150    -2.11701   -0.47061
C   0.91352    1.49455    0.52298
H   1.42670    2.35563    0.09593
H   0.34077    1.68198    1.43603
H   -1.36134   2.30666    -0.50359
H   -0.24536   1.40900    -1.64997
O   -2.77266   0.35719    0.00472
C   -3.44877   -0.73035   0.62398
H   -3.61825   -1.54139   -0.09076
H   -4.42319   -0.37084   0.96700
H   -2.89744   -1.08865   1.50051
H   -1.30711   -1.96391   -0.13984
C   3.04390    0.16177    0.01775
H   1.47835    -0.42356   1.40173
O   3.81536    -0.73101   0.34395
H   3.33134    0.97683    -0.66519

B 2 8 F
B 4 5 F
```

ORCA Input File

```
! M062X 6-31+G(d,p) Opt TightSCF
%pal nprocs 8 end
%MaxCore 500
# TS_Diels_Alder_1_min
%geom
Constraints
{B 1 7 C}
{B 3 4 C}
end
end
*xyz 0 1
C   -1.47689   0.14381    -0.39766
C   -0.79057   1.39898    -0.70328
C   -0.82456   -1.03585   -0.42553
C   0.60806    -1.14572   -0.70447
C   1.64465    0.20744    0.52494
H   0.90994    -0.74505   -1.67704
H   1.05150    -2.11701   -0.47061
C   0.91352    1.49455    0.52298
H   1.42670    2.35563    0.09593
H   0.34077    1.68198    1.43603
H   -1.36134   2.30666    -0.50359
H   -0.24536   1.40900    -1.64997
O   -2.77266   0.35719    0.00472
C   -3.44877   -0.73035   0.62398
H   -3.61825   -1.54139   -0.09076
H   -4.42319   -0.37084   0.96700
H   -2.89744   -1.08865   1.50051
H   -1.30711   -1.96391   -0.13984
C   3.04390    0.16177    0.01775
H   1.47835    -0.42356   1.40173
O   3.81536    -0.73101   0.34395
H   3.33134    0.97683    -0.66519
*
```

Scheme 7.6 Representative examples for a constrained optimization in Gaussian (left) and ORCA (right).

constrained in the example, the Gaussian syntax "B 2 8 F" changes to "{B 1 7 C}" in ORCA. The letter C indicates a constrained optimization. The result of these constrained optimizations is an improved input structure for the subsequent transition-state search and the final structure of the constrained optimization will be used as the input structure for the subsequent calculation.

With a reasonable transition-state guess in hand (either by manual design or by constrained optimization), the transition-state optimization can be started. The keyword within the input file for an optimization (Schemes 7.3 and 7.4) has to be changed to reflect the search for a transition state: in Gaussian "opt=(ts,calcfc,noeigen)" needs to be used, while "OptTS" is the keyword for a transition-state optimization in ORCA. For the Diels–Alder reaction discussed here, a total of eight transition-state conformers (and thus eight input files) have to be considered. Both *s-cis*-conformers of the diene (**1a**,**b**) can be combined with both conformers of the dienophile **2** in and *endo*- or *exo*-orientation.

Once these calculations are completed, exactly one imaginary frequency should be present for each structure, as we requested the location of a transition state. As mentioned above, this frequency can be visualized in most computational chemistry programs and provides a first judgement on whether the correct transition state was located. Again, the calculated free energies can be extracted from each file and the lowest transition-state conformer can be identified. From the calculated values summarized in Table 7.5, we can use the following equation $\Delta G^{\ddagger} = G_{298\,\mathrm{K}}(\text{TS\_Diels\_Alder\_8}) - G_{298\,\mathrm{K}}(\text{Butadiene\_OMe\_4}) - G_{298\mathrm{K}}(\text{Acrolein\_1})$ to calculate the activation free energy. After conversion to either $\mathrm{kJ\,mol^{-1}}$ or $\mathrm{kcal\,mol^{-1}}$, we end up with an activation free energy of $114\,\mathrm{kJ\,mol^{-1}}$ or $27.3\,\mathrm{kcal\,mol^{-1}}$ ($113\,\mathrm{kJ\,mol^{-1}}$ or $26.9\,\mathrm{kcal\,mol^{-1}}$ for ORCA) for the Diels–Alder reaction between **1** and **2**.

Although a computational analysis of the Diels–Alder reaction is not the aim of this chapter, we will briefly discuss the calculated activation barrier. As a general rule of thumb, a reaction with a barrier of up to $100\,\mathrm{kJ\,mol^{-1}}$ (or $25\,\mathrm{kcal\,mol^{-1}}$) can occur with a reasonable rate at room temperature. Thus, the calculated barrier indicates that the Diels–Alder reaction between 2-methoxybutadiene (**1**) and acrolein (**2**) will not occur at a measurable rate at 298.15 K. This is qualitatively in line with the literature, as this reaction requires heating the reactants in benzene solution in a sealed tube to 160 °C [74].

Table 7.5 Calculated free energies for each transition-state conformer of the Diels–Alder reaction between **1** and **2** (in Hartree, rounded to 6 decimal figures).

Structure	Gaussian	ORCA
TS_Diels_Alder_1 (**1a**+**2a**, exo)	−462.060853	−462.055467
TS_Diels_Alder_2 (**1b**+**2a**, exo)	−462.064183	−462.058877
TS_Diels_Alder_3 (**1a**+**2b**, exo)	−462.061734	−462.056325
TS_Diels_Alder_4 (**1b**+**2b**, exo)	−462.065974	−462.060735
TS_Diels_Alder_5 (**1a**+**2a**, endo)	−462.062868	−462.057801
TS_Diels_Alder_6 (**1b**+**2a**, endo)	−462.066748	−462.061477
TS_Diels_Alder_7 (**1a**+**2b**, endo)	−462.062170	−462.056925
TS_Diels_Alder_8 (1b+2b, endo)	**−462.069718**	**−462.065471**

The most stable conformer is highlighted in bold.

7.7.6 Verifying the Nature of the Transition State

We have now located a promising transition state for the Diels–Alder reaction. As mentioned above, we can visualize the imaginary frequency (372.66 $i\mathrm{cm}^{-1}$ for Gaussian, 351.47 $i\mathrm{cm}^{-1}$ for ORCA). The movement of atoms resembles the required structural changes of the Diels–Alder reaction and is a good indicator that the transition state is correct. For a more reliable confirmation, IRC calculations can be performed (see Section 7.5.5) and representative input files are shown in Scheme 7.7 using default settings. Please note that the reader's Cartesian

Gaussian Input File

```
%nosave
%chk=TS_Diels_Alder_8_IRC
%nprocshared=8
%mem=4000MB
#P M062X/6-31+G(d,p) IRC=calcfc

TS_Diels_Alder_8_IRC

0 1
C    0.87808   -0.81357    0.05710
C    0.10063   -1.72830   -0.63737
C    0.51909   -0.35294    1.37153
C   -0.62947   -0.71925    1.98181
C   -2.11157    0.13454   -0.33779
H   -0.90616   -0.28089    2.93486
H   -1.27661   -1.49869    1.60070
C   -1.45046   -0.57649   -1.33388
H   -0.78361   -0.00659   -1.97585
H   -1.95330   -1.41971   -1.79961
H   -0.49737   -2.43438   -0.07662
H    0.48833   -2.07721   -1.58892
O    1.91002   -0.28947   -0.62382
C    2.51027    0.91845   -0.16657
H    3.20618    1.20978   -0.95205
H    3.06755    0.74875    0.76080
H    1.74409    1.68748   -0.03157
H    1.12719    0.41960    1.82858
C   -1.56341    1.39865    0.07074
H   -3.03024   -0.22152    0.11691
O   -0.47903    1.82894   -0.33206
H   -2.15297    1.99101    0.79643
```

ORCA Input File

```
! M062X 6-31+G(d,p) IRC TightSCF
%pal nprocs 8 end
%MaxCore 500
# TS_Diels_Alder_8
*xyz 0 1
C    0.88128   -0.81892    0.05812
C    0.10551   -1.73470   -0.63553
C    0.51906   -0.35405    1.37047
C   -0.63186   -0.71395    1.98135
C   -2.11170    0.13284   -0.33838
H   -0.90741   -0.26950    2.93212
H   -1.28321   -1.49169    1.60386
C   -1.44873   -0.57017   -1.33714
H   -0.77586    0.00264   -1.97004
H   -1.94769   -1.41171   -1.80974
H   -0.49874   -2.43609   -0.07556
H    0.49142   -2.08523   -1.58724
O    1.91413   -0.29577   -0.62201
C    2.50878    0.91630   -0.16983
H    3.19957    1.20988   -0.95921
H    3.07137    0.75165    0.75523
H    1.74016    1.68245   -0.03217
H    1.12788    0.41815    1.82747
C   -1.56596    1.39685    0.07727
H   -3.03007   -0.22761    0.11395
O   -0.48322    1.83159   -0.32402
H   -2.15747    1.98523    0.80438
*
```

Scheme 7.7 Representative examples for an IRC calculation in Gaussian (left) and ORCA (right).

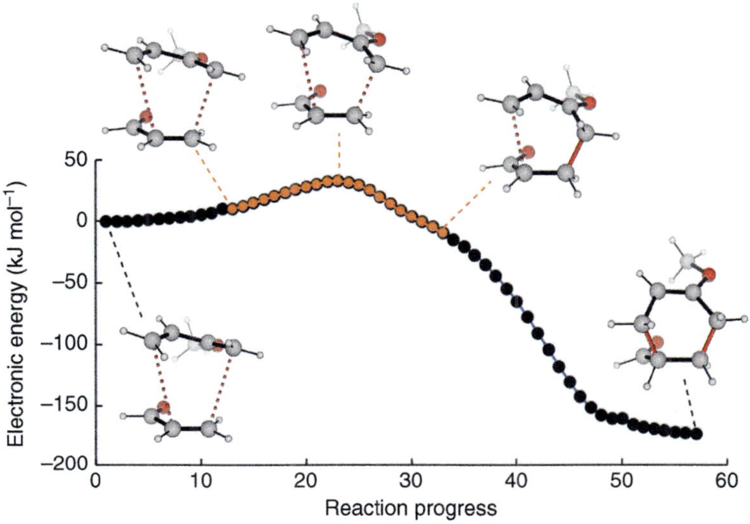

Figure 7.7 Following the Diels–Alder reaction between **1** and **2** with IRC calculations in Gaussian [orange: default step number (10); black: increased step number (up to 100)].

coordinates might look slightly different, depending on how the molecule was oriented in the structure editor.

The result for the Gaussian calculation on the most stable transition state for the Diels–Alder reaction between **1** and **2** with these default settings ("irc=calcfc") is summarized by the orange symbols in Figure 7.7. At a first glance, it seems that the energy profile of this IRC calculation does not fit at all to the previous results. The barrier for this reaction based on the IRC data is only around 20 (rather than 114) kJ mol^{-1} and similarly, the reaction energy lies at around −20 (rather than −117) kJ mol^{-1}. This deviation is the result of the small default number of steps that Gaussian performs in each direction downhill from the transition state. If nothing else is specified, Gaussian performs only 10 steps in each direction. This often is enough to judge, whether the correct transition state was identified. However, for the Diels–Alder cycloadduct, the lengths of both formed C–C bonds at the end of the orange IRC are 1.61 and 2.51 Å, which is a substantial difference to the fully optimized cycloadduct **3** (1.51 and 1.53 Å). As Diels–Alder cycloadditions could also proceed in stepwise reactions via zwitterions, it would be highly desirable to obtain more information on the right part of the orange IRC. One option would be to simply use the final structures of the IRC, perform a regular optimization, and check if the optimized structure fits to the expectation. A better alternative is to increase the number of steps in each direction through the option: "irc=(Calcfc,MaxSteps=value)."

For the Diels–Alder reaction between **1** and **2**, the black IRC of Figure 7.7 was calculated requesting a maximum of 100 steps per direction.[13] For the first 10 steps around the transition state, both IRCs are more or less identical. The complete black IRC now looks much better and confirms that the located transition state belongs to a Diels–Alder reaction and that the reaction proceeds in a concerted fashion. However, there are still some "deviations" that need to be

discussed here: The barrier is still rather small (32 kJ mol^{-1}) and the reaction energy is now very favorable (−170 kJ mol^{-1}). One reason for this difference stems from the different forms of energy used in the calculations. The calculated barrier and reaction energy refer to Gibbs free energies (ΔG and $\Delta G^{\ddagger}$), while the energy associated with an IRC calculation is the potential (or electronic) energy of the system. Thus, the IRC energy neglects all thermal contributions as well as entropy effects. Another reason that causes those deviations comes from the different termination criteria. Those criteria are not as strict for an IRC compared to a "regular" optimization and, therefore, the endpoint of an IRC calculation can deviate slightly from the optimized structure. Furthermore, in this particular case (this holds for every bimolecular reaction), the IRC ends with a complex between both reactants, while the previous calculations referred to separated reactants. In many cases, the reactant complexes are higher in energy than the separated reactants, and, thus, the barrier will be larger when separated reactants are considered. In other words, IRC calculations are typically used to check the pathway of a chemical reaction and are not directly used to determine activation energies.

7.8 More Advanced Aspects

7.8.1 General Comments

While the previous sections and the tutorial covered only the basics of computational chemistry, this will most likely not be sufficient for more realistic problems. Thus, we will discuss a few more advanced aspects of computational organic chemistry in this section. Again, we cannot provide an in-depth discussion of all aspects, a few settings will be highlighted that might be useful very early on in computational studies.

7.8.2 Influence of Solvation

By default, computational programs assume that all chemistry occurs in the gas phase. However, most reactions are not performed in the gas phase, but in solution

---

13 The calculation only "requires" 23 steps in the backwards direction and 34 steps in the forwards direction. Thus, the calculation does not require the maximum number of steps.

and frequently, solvation is particularly important to stabilize certain structures (e.g. zwitterions, highly reactive intermediates). Therefore, one of the most severe limitations of the previous sections is the neglect of solvation. Historically, the influence of solvation could not be considered during a geometry optimization. Thus, one assumed that the optimized structures do not change (significantly), when going from the gas phase to solution. Therefore, solvation effects were only considered through a single-point calculation on gas-phase structures.

However, while this approach is probably justified for simple, neutral systems, the structures of zwitterionic or highly polarized compounds sometimes differ substantially. For more than a decade, it is now also easily possible to perform geometry optimizations "in solution." Although performing a calculation in solution takes significantly longer than in the gas phase, it is generally recommended to perform geometry optimizations in solution, as the influence of solvation is sometimes not obvious at a first glance.

To account for solution, we would ideally place our organic molecule within dozens of solvent molecules and calculate the properties of the entire mixture. This process is called explicit solvation and is considered to be very accurate. Unfortunately, this procedure is computationally too demanding to be performed routinely for typical applications in organic chemistry. In contrast, implicit solvation models employ a continuous medium to simulate the presence of solvent molecules. The computational program then places the solute in a cavity within this solvent medium and the interaction between the molecule and the cavity is calculated. This procedure is computationally much faster but does not account for specific interactions between the solvent molecules and the solute. Among those models, COSMO (COnductor-like Screening MOdel [75]), CPCM (Conductor-like Polarizable Continuum Model [76]), IEFPCM (Integral Equation Formalism Polarizable Continuum Model [77]), and SMD (Solvation Model based on Density [78]) are considered the most popular and are frequently used in organic chemistry. In Gaussian, implicit solvation can be requested with the keyword "scrf=(model,solvent=item)". Solvents can be chosen from a predefined list which includes most frequently occurring organic solvents.[14] In ORCA, CPCM and SMD can be selected as solvation models. To request the CPCM model, the keyword "CPCM(solvent)" has to be added to the keyword line. In contrast, the SMD model has to be requested via the input block %cpcm (smd true, SMD solvent "item") and the solvent has again to be chosen from a predefined list. In both programs, additional parameters (e.g. atomic radii, dielectric constants, . . .) can also be specified.

14 Please refer to the corresponding manuals for the list of supported solvents.

7.8.3 Integration Grid

Besides the suitability and accuracy of the computational method, there is another source of error for DFT calculations that is important for the determination of almost any selectivities, but particularly for enantioselective processes. Here, an energy difference of $4\,kJ\,mol^{-1}$ (or $1\,kcal\,mol^{-1}$) has a significant effect on the observed product distribution. For the determination of some integrals, a numerical integration is required. A very simple form of this procedure allows the calculation of the area under a curve (i.e. the integral) with a series of equally broad rectangles. Those rectangles are either slightly larger ("upper sum") or slightly smaller ("lower sum") than the actual area. The more (or in other words, the thinner) rectangles are used, the more accurate the estimate for the integral.

Similarly, the underlying grid used for the numerical integration within computational programs is another source of error that could lead to substantial energy changes [79, 80]. These atom-centered grids are defined by the number N^r of shells that are comprised of $N\Omega$ points. Generally speaking, the computing time increases with finer grids. While it is known for some time that particularly the Minnesota functionals require finer grids [80], Bootsma and Wheeler have recently shown that this can also apply to other functionals [79]. In line with these findings, many computational programs have now changed their default integration grids. Within the Gaussian program series, the integration grid can be specified with the keyword "int=name". The default integration grid in Gaussian09 (name: "fine") relied on 75 shells and 302 points, while the default in Gaussian16 (name: "ultrafine") already uses 99 shells and 590 points. An even finer grid (name: "superfine") can also be requested (175 shells, 874 points). Within the ORCA program, the grid size is controlled with the "DEFGRID" keyword. DEFGRID2 is probably sufficient for many applications and set as default, while DEFGRID3 is for example recommended for the Minnesota functionals.[15]

Besides a more accurate energy calculation, finer grids are sometimes also beneficial to avoid small imaginary frequencies (around 30–50 $i\,cm^{-1}$) like methyl-group rotations. Thus, if those small imaginary frequencies occur frequently, it might be an option to select a finer grid. However, please be aware that the same grid has to be used within a computational project.

7.8.4 Standard States

Another shortcoming of the default procedures within most computational programs are the default settings for standard states. For most calculations, the

15 The keyword DEFGRID3 was not used during the tutorial to keep the example simple for the beginner. ORCA will, however, print a warning recommending to change the grid to DEFGRID3 in the output file.

programs assume that all species are ideal gases at a standard state of 1 atm at 298.15 K. According to the ideal gas law,[16] this translates to a concentration of approx. 0.04 mol L^{-1}. This value is typically not a very good estimate for concentrations present in organic reactions that are often used in substantially higher concentrations. To correct for this effect, computational chemists frequently assume 1 M solutions.[17] Following Equation (7.1), one can easily calculate the required correction that accounts for different concentration.

$$\Delta\Delta G = RT \ln \frac{[\text{new concentration}]}{[\text{reference concentration}]} = RT \ln \frac{[1\,\text{mol}\,\text{L}^{-1}]}{[0.04\,\text{mol}\,\text{L}^{-1}]} \approx 7.9\,\text{kJ mol}^{-1} \text{ or } 1.89\,\text{kcal mol}^{-1} \tag{7.1}$$

When going from the default value (1 atm) to 1 M solutions, one has to add 7.9 kJ mol^{-1} (or 1.89 kcal mol^{-1}) to the calculated Gibbs free energy of each species. Therefore, this correction will be particularly important, when the number of particles changes during a reaction. For example, the free energy difference for a bimolecular reaction ($A + B \rightarrow C$) will be lowered by 7.9 kJ mol^{-1} (or 1.89 kcal mol^{-1}), if the default concentration is assumed to increase to 1 M (rather than 0.04 M) solutions. In other words, it can be expected that the barrier for a given bimolecular reaction will be 7.9 kJ mol^{-1} (or 1.89 kcal mol^{-1}) smaller when the reactants are used in 1 M solutions compared to the gas phase (or also very diluted solutions in the mM range). Another application of a different standard state could be a reaction with the solvent molecule, which is typically present in large excess (e.g. the concentration of water [55.5 mol L^{-1}] leads to a correction of approx. 18 kJ mol^{-1} or 4.29 kcal mol^{-1}).

7.8.5 Treating Unpaired Electrons

So far, we only considered "normal" organic molecules with paired electrons and those compounds are typically referred to as "closed-shell systems." However, there are many cases, where a molecule features at least one unpaired electron and those are called "open-shell systems." For those systems, the user has to particularly pay attention to the correct definition of charge and multiplicity. For example, nitric oxide (NO) is a neutral molecule with one unpaired electron (i.e. a doublet). Hence, the user has to enter "0 2" rather than "0 1" in the charge/multiplicity line.

---

16 $pV = nRT$ or $n\,V^{-1} = p\,R^{-1}\,T^{-1}$

17 While the reader is free to choose any concentrations they prefer, 1 M solutions are typically used if a different concentration is desired.

Similarly, dioxygen (O_2) can either exist in the more stable triplet form ("0 3") or in the less stable singlet form ("0 1").

As the calculation now cannot assume that all orbitals are doubly occupied (or empty), the user additionally has to choose a slightly different approach. The assumption of only doubly occupied orbitals is also called a "restricted" calculation. In contrast, an "unrestricted" calculation allows orbitals to be singly occupied. While "restricted" calculations are the default and do not have to be specified explicitly, an "unrestricted" calculation has to be requested. To do so, the user simply has to add the letter "U" to the theoretical method within the Gaussian input file (i.e. UB3LYP or UHF requests an unrestricted B3LYP or HF calculation). Furthermore, it is recommended to add the keyword "guess=mix" in the route section for unrestricted calculations.[18] For ORCA, the user needs to add the keyword "UKS" for unrestricted DFT calculations (e.g. "! B3LYP UKS").

7.9 Important and Frequently Used Keywords

To lower the barrier for entry into computational chemistry, this section will briefly summarize the most important keywords discussed in this chapter. Table 7.6 collects this information for the Gaussian and the ORCA program. Please note that for most keywords, further options are available and only a (small) portion of examples discussed above is included in Table 7.6. To discuss all of them here would be beyond the scope of the chapter and the reader is referred to the corresponding manuals.

7.10 Practical Considerations

Given the size of typical organic molecules, density functional theory is frequently been used as the method of choice, as it can handle larger system at a reasonable computational time. Unfortunately, there is no universal method within the alphabet soup of available functionals that can be used for all purposes. Historically, B3LYP/6-31+G(d) was used as the "default" method for decades, but it turned out that the seemingly good agreement with experiment stems from substantial error compensation [61]. Thus, it is generally advisable to consult benchmarking studies for related systems and select methods that perform reliably for the tasks to be performed.[19]

18 Please check the Gaussian manual for a detailed explanation of this keyword.
19 Selected examples of recent benchmarks include: Transition-metal barrier heights (MOBH35): Iron and Janes [81]; General main-group chemistry (GMTKN55): Goerigk et al. [57]; Non-covalent interactions (S66x8): Brauer et al. [82].

Table 7.6 Selected Gaussian and ORCA keywords.

Topic	Gaussian keyword	ORCA keyword	Brief description
Frequency calculation	Freq	Freq	Additionally provides thermal corrections
Integration grid	Int=*item*(*item*: UltraFine, SuperFineGrid)	DEFGRID*item* (*item*: 1,2,3)	Specifies the integration grid
Intrinsic reaction coordinate	IRC=calcfc	IRC	Requests an intrinsic reaction coordinate calculation
London dispersion	EmpiricalDispersion=*item* (*item*: D2,D3,D3BJ)	*item* (*item* = D2,D3, D3BJ, D4)	Requests a Grimme dispersion correction
Optimization	Opt=*item*(*item*: ts, calcfc, tight, modredundant)	Opt	Requests a geometry optimization
Single point	SP	SP	Provides potential energies and other properties
Solvation	scrf=(*model*, solvent=*item*)(*model*: CPCM, IEFPCM, SMD; *item*: water, Acetone, Dichloromethane)	CPCM(*item*) (*item*: water, Acetone, Dichloromethane)	Requests the use of a solvation model

For most computational investigations, "ordinary" desktop computers or laptops will not be suitable, and access to a high-performing computing (HPC) infrastructure is required. Here, calculations are typically performed on more than one CPU and are, thus, reasonably fast. Depending on the available resources, molecular structures with approx. 100 atoms should be feasible to calculate adequately within a reasonable time frame of hours to several days computing time.

In general, the overall geometry of a structure does not change significantly between different computational methods. Therefore, one can typically employ on a faster (yet still reliable) DFT method with dispersion-correction for the optimization (e.g. a GGA or meta-GGA functional) in combination with at least a double-ζ basis set. To obtain reliable electronic energies, a state-of-the-art method (e.g. DLPNO-CCSD(T), double-hybrid, or hybrid functional) in combination with a large triple-or quadruple-ζ basis set should be used. The choice of the computational method also depends on the available programs and the functionals implemented therein.

7.11 Conclusions

Computational investigations can provide very useful insights into organic reactions, such as the prediction of spectral properties as described in Chapter 8. Nowadays, many of these investigations can almost be performed in a black-box manner and many programs can be used without a lot of specific training. This will certainly lower the threshold for synthetic chemists to engage in computational studies themselves, but it also bears the risk that less reliable methods are chosen. This chapter hopefully provides some initial insights for the less experienced reader to (loosely) understand the basic concepts of computational organic chemistry and allows for first preliminary investigations.

References

1 Bonney, K.J. and Schoenebeck, F. (2014). Experiment and computation: a combined approach to study the reactivity of palladium complexes in oxidation states 0 to iv. *Chem. Soc. Rev.* 43: 6609–6638.
2 Bachrach, S.M. (2014). *Computational Organic Chemistry*. Hoboken: John Wiley & Sons.
3 Sperger, T., Sanhueza, I.A., and Schoenebeck, F. (2016). Computation and experiment: a powerful combination to understand and predict reactivities. *Acc. Chem. Res.* 49: 1311–1319.
4 Karplus, M. (2014). Development of multiscale models for complex chemical systems: from H+H_2 to biomolecules (Nobel Lecture). *Angew. Chem. Int. Ed.* 53: 9992–10005.
5 Levitt, M. (2014). Birth and future of multiscale modeling for macromolecular systems (Nobel Lecture). *Angew. Chem. Int. Ed.* 53: 10006–10018.
6 Warshel, A. (2014). Multiscale modeling of biological functions: from enzymes to molecular machines (Nobel Lecture). *Angew. Chem. Int. Ed.* 53: 10020–10031.
7 Houk, K.N. and Cheong, P.H.-Y. (2008). Computational prediction of small-molecule catalysts. *Nature* 455: 309–313.
8 Peng, Q., Duarte, F., and Paton, R.S. (2016). Computing organic stereoselectivity – from concepts to quantitative calculations and predictions. *Chem. Soc. Rev.* 45: 6093–6107.
9 Peng, Q. and Paton, R.S. (2016). Catalytic control in cyclizations: from computational mechanistic understanding to selectivity prediction. *Acc. Chem. Res.* 49: 1042–1051.

10 Eder, U., Sauer, G., and Wiechert, R. (1971). New type of asymmetric cyclization to optically active steroid CD partial structures. *Angew. Chem. Int. Ed. Engl.* 10: 496–497.

11 Hajos, Z.G. and Parrish, D.R. (1974). Asymmetric synthesis of bicyclic intermediates of natural product chemistry. *J. Org. Chem.* 39: 1615–1621.

12 List, B., Lerner, R.A., and Barbas, C.F. (2000). Proline-catalyzed direct asymmetric aldol reactions. *J. Am. Chem. Soc.* 122: 2395–2396.

13 Jung, M.E. (1976). A review of annulation. *Tetrahedron* 32: 3–31.

14 Puchot, C., Samuel, O., Dunach, E. et al. (1986). Nonlinear effects in asymmetric synthesis. Examples in asymmetric oxidations and aldolization reactions. *J. Am. Chem. Soc.* 108: 2353–2357.

15 Clemente, F.R. and Houk, K.N. (2004). Computational evidence for the enamine mechanism of intramolecular aldol reactions catalyzed by proline. *Angew. Chem. Int. Ed.* 43: 5766–5768.

16 Armstrong, A., Boto, R.A., Dingwall, P. et al. (2014). The Houk-List transition states for organocatalytic mechanisms revisited. *Chem. Sci.* 5: 2057–2071.

17 Bahmanyar, S. and Houk, K.N. (2001). Transition states of amine-catalyzed aldol reactions involving enamine intermediates: theoretical studies of mechanism, reactivity, and stereoselectivity. *J. Am. Chem. Soc.* 123: 11273–11283.

18 Bahmanyar, S. and Houk, K.N. (2003). Origins of opposite absolute stereoselectivities in proline-catalyzed direct Mannich and aldol reactions. *Org. Lett.* 5: 1249–1251.

19 Bahmanyar, S., Houk, K.N., Martin, H.J., and List, B. (2003). Quantum mechanical predictions of the stereoselectivities of proline-catalyzed asymmetric intermolecular aldol reactions. *J. Am. Chem. Soc.* 125: 2475–2479.

20 Galabov, B., Koleva, G., Simova, S. et al. (2014). Arenium ions are not obligatory intermediates in electrophilic aromatic substitution. *Proc. Natl. Acad. Sci. U.S.A.* 111: 10067–10072.

21 Grimme, S. and Schreiner, P.R. (2018). Computational chemistry: the fate of current methods and future challenges. *Angew. Chem. Int. Ed.* 57: 4170–4176.

22 Mata, R.A. and Suhm, M.A. (2017). Benchmarking quantum chemical methods: are we heading in the right direction? *Angew. Chem. Int. Ed.* 56: 11011–11018. https://doi.org/10.1002/anie.201611308.

23 Mata, R.A. and Suhm, M.A. (2017). Quantenchemische Methoden im Leistungsvergleich: Stimmt die Richtung noch? *Angew. Chem.* im Druck. doi: https://doi.org/10.1002/ange.201611308.

24 Cramer, C.J. (2004). *Essentials of Computational Chemistry*. Hoboken: John Wiley & Sons.

25 Jensen, F. (2007). *Introduction to Computational Chemistry*. Chichester: John Wiley & Sons.

26 Lewars, E.G. (2016). *Computational Chemistry*. Springer.

27 Weiner, P.K. and Kollman, P.A. (1981). AMBER: assisted model building with energy refinement. A general program for modeling molecules and their interactions. *J. Comput. Chem.* 2: 287–303.

28 MacKerell, A.D., Bashford, D., Bellott, M. et al. (1998). All-atom empirical potential for molecular modeling and dynamics studies of proteins. *J. Phys. Chem. B* 102: 3586–3616.

29 Jorgensen, W.L. and Tirado-Rives, J. (1988). The OPLS [optimized potentials for liquid simulations] potential functions for proteins, energy minimizations for crystals of cyclic peptides and crambin. *J. Am. Chem. Soc.* 110: 1657–1666.

30 Dror, R.O., Dirks, R.M., Grossman, J.P. et al. (2012). Biomolecular simulation: a computational microscope for molecular biology. *Annu. Rev. Biophys.* 41: 429–452.

31 Lindorff-Larsen, K., Piana, S., Dror, R.O., and Shaw, D.E. (2011). How fast-folding proteins fold. *Science* 334: 517.

32 Neese, F., Hansen, A., and Liakos, D.G. (2009). Efficient and accurate approximations to the local coupled cluster singles doubles method using a truncated pair natural orbital basis. *J. Chem. Phys.* 131: 064103.

33 Neese, F., Wennmohs, F., and Hansen, A. (2009). Efficient and accurate local approximations to coupled-electron pair approaches: An attempt to revive the pair natural orbital method. *J. Chem. Phys.* 130: 114108.

34 Riplinger, C. and Neese, F. (2013). An efficient and near linear scaling pair natural orbital based local coupled cluster method. *J. Chem. Phys.* 138: 034106.

35 Riplinger, C., Sandhoefer, B., Hansen, A., and Neese, F. (2013). Natural triple excitations in local coupled cluster calculations with pair natural orbitals. *J. Chem. Phys.* 139: 134101.

36 Dewar, M.J.S., Zoebisch, E.G., Healy, E.F., and Stewart, J.J.P. (1985). Development and use of quantum mechanical molecular models. 76. AM1: a new general purpose quantum mechanical molecular model. *J. Am. Chem. Soc.* 107: 3902–3909.

37 Stewart, J.J.P. (1989). Optimization of parameters for semiempirical methods I. Method. *J. Comput. Chem.* 10: 209–220.

38 Stewart, J.J.P. (2013). Optimization of parameters for semiempirical methods VI: more modifications to the NDDO approximations and re-optimization of parameters. *J. Mol. Model.* 19: 1–32.

39 Lonsdale, R., Harvey, J.N., and Mulholland, A.J. (2012). A practical guide to modelling enzyme-catalysed reactions. *Chem. Soc. Rev.* 41: 3025–3038.

40 Grimme, S. (2006b). Semiempirical hybrid density functional with perturbative second-order correlation. *J. Chem. Phys.* 124: 034108.

41 Kozuch, S., Gruzman, D., and Martin, J.M.L. (2010). DSD-BLYP: a general purpose double hybrid density functional including spin component scaling and dispersion correction. *J. Phys. Chem. C* 114: 20801–20808.

42 Kozuch, S. and Martin, J.M.L. (2011). DSD-PBEP86: in search of the best double-hybrid DFT with spin-component scaled MP2 and dispersion corrections. *Phys. Chem. Chem. Phys.* 13: 20104–20107.

43 Becke, A.D. (1993). Density-functional thermochemistry. III. The role of exact exchange. *J. Chem. Phys.* 98: 5648–5652.

44 Lee, C., Yang, W., and Parr, R.G. (1988). Development of the Colle-Salvetti correlation-energy formula into a functional of the electron density. *Phys. Rev. B* 37: 785–789.

45 Adamo, C. and Barone, V. (1999). Toward reliable density functional methods without adjustable parameters: the PBE0 model. *J. Chem. Phys.* 110: 6158–6170.

46 Zhao, Y. and Truhlar, D.G. (2008). The M06 suite of density functionals for main group thermochemistry, thermochemical kinetics, noncovalent interactions, excited states, and transition elements: two new functionals and systematic testing of four M06-class functionals and 12 other functionals. *Theor. Chem. Acc.* 120: 215–241.

47 Chai, J.-D. and Head-Gordon, M. (2008). Long-range corrected hybrid density functionals with damped atom-atom dispersion corrections. *Phys. Chem. Chem. Phys.* 10: 6615–6620.

48 Tao, J., Perdew, J.P., Staroverov, V.N., and Scuseria, G.E. (2003). Climbing the density functional ladder: nonempirical meta–generalized gradient approximation designed for molecules and solids. *Phys. Rev. Lett.* 91: 146401.

49 Zhao, Y. and Truhlar, D.G. (2006). A new local density functional for main-group thermochemistry, transition metal bonding, thermochemical kinetics, and noncovalent interactions. *J. Chem. Phys.* 125: 194101.

50 Sun, J., Remsing, R.C., Zhang, Y. et al. (2016). Accurate first-principles structures and energies of diversely bonded systems from an efficient density functional. *Nat. Chem.* 8: 831–836.

51 Perdew, J.P., Burke, K., and Ernzerhof, M. (1996). Generalized gradient approximation made simple. *Phys. Rev. Lett.* 77: 3865–3868.

52 Becke, A.D. (1988). Density-functional exchange-energy approximation with correct asymptotic behavior. *Phys. Rev. A.* 38: 3098–3100.

53 Perdew, J.P. (1986). Density-functional approximation for the correlation energy of the inhomogeneous electron gas. *Phys. Rev. B* 33: 8822–8824.

54 Grimme, S. (2006a). Semiempirical GGA-type density functional constructed with a long-range dispersion correction. *J. Comput. Chem.* 27: 1787–1799.

55 Vosko, S.H., Wilk, L., and Nusair, M. (1980). Accurate spin-dependent electron liquid correlation energies for local spin density calculations: a critical analysis. *Can. J. Phys.* 58: 1200–1211.

56 Perdew, J.P. and Wang, Y. (1992). Accurate and simple analytic representation of the electron-gas correlation energy. *Phys. Rev. B* 45: 13244–13249.

57 Goerigk, L., Hansen, A., Bauer, C. et al. (2017). A look at the density functional theory zoo with the advanced GMTKN55 database for general main group thermochemistry, kinetics and noncovalent interactions. *Phys. Chem. Chem. Phys.* 19: 32184–32215.

58 Ehrlich, S., Moellmann, J., and Grimme, S. (2013). Dispersion-corrected density functional theory for aromatic interactions in complex systems. *Acc. Chem. Res.* 46: 916–926.

59 Grimme, S. (2011). Density functional theory with London dispersion corrections. *Wiley Interdiscip. Rev.: Comput. Mol. Sci.* 1: 211–228.

60 Wagner, J.P. and Schreiner, P.R. (2015). London dispersion in molecular chemistry—reconsidering steric effects. *Angew. Chem. Int. Ed.* 54: 12274–12296.

61 Kruse, H., Goerigk, L., and Grimme, S. (2012). Why the standard B3LYP/6-31G* model chemistry should not be used in DFT calculations of molecular thermochemistry: understanding and correcting the problem. *J. Org. Chem.* 77: 10824–10834.

62 Caldeweyher, E., Bannwarth, C., and Grimme, S. (2017). Extension of the D3 dispersion coefficient model. *J. Chem. Phys.* 147: 034112.

63 Grimme, S., Antony, J., Ehrlich, S., and Krieg, H. (2010). A consistent and accurate ab initio parametrization of density functional dispersion correction (DFT-D) for the 94 elements H-Pssu. *J. Chem. Phys.* 132: 154104.

64 Grimme, S., Ehrlich, S., and Goerigk, L. (2011). Effect of the damping function in dispersion corrected density functional theory. *J. Comput. Chem.* 32: 1456–1465.

65 Hill, J.G. (2013). Gaussian basis sets for molecular applications. *Int. J. Quantam Chem.* 113: 21–34.

66 Jensen, F. (2013). Atomic orbital basis sets. *Wiley Interdiscip. Rev. Comput. Mol. Sci.* 3: 273–295.

67 Hay, P.J. and Wadt, W.R. (1985a). Ab initio effective core potentials for molecular calculations. Potentials for K to Au including the outermost core orbitals. *J. Chem. Phys.* 82: 299–310.

68 Hay, P.J. and Wadt, W.R. (1985b). Ab initio effective core potentials for molecular calculations. Potentials for the transition metal atoms Sc to Hg. *J. Chem. Phys.* 82: 270–283.

69 Wadt, W.R. and Hay, P.J. (1985). Ab initio effective core potentials for molecular calculations. Potentials for main group elements Na to Bi. *J. Chem. Phys.* 82: 284–298.

70 Bergner, A., Dolg, M., Küchle, W. et al. (1993). Ab initio energy-adjusted pseudopotentials for elements of groups 13–17. *Mol. Phys.* 80: 1431–1441.

71 Schlegel, H.B. (2011). Geometry optimization. *Wiley Interdiscip. Rev. Comput. Mol. Sci.* 1: 790–809.

72 Grimme, S. (2019). Exploration of chemical compound, conformer, and reaction space with meta-dynamics simulations based on tight-binding quantum chemical calculations. *J. Chem. Theory Comput.* 15: 2847–2862.

73 Yates, P. and Eaton, P. (1960). Acceleration of the diels-alder reaction by aluminum chloride. *J. Am. Chem. Soc.* 82: 4436–4437.

74 Fiesselmann, H. (1942). Diensynthesen mit Oxyprenen, I. Mitteil.: Über 4-Methoxy-, 4-Äthoxy- und 4-Methoxy-1-methyl-Δ3-tetrahydrobenzaldehyd. *Ber. Dtsch. Chem. Ges.* 75: 881–891.

75 Klamt, A. and Schüürmann, G. (1993). COSMO: a new approach to dielectric screening in solvents with explicit expressions for the screening energy and its gradient. *J. Chem. Soc. Perkin Trans. 2*: 799–805.

76 Barone, V. and Cossi, M. (1998). Quantum calculation of molecular energies and energy gradients in solution by a conductor solvent model. *J. Phys. Chem. A* 102: 1995–2001.

77 Cancès, E., Mennucci, B., and Tomasi, J. (1997). A new integral equation formalism for the polarizable continuum model: theoretical background and applications to isotropic and anisotropic dielectrics. *J. Chem. Phys.* 107: 3032–3041.

78 Marenich, A.V., Cramer, C.J., and Truhlar, D.G. (2009). Universal solvation model based on solute electron density and on a continuum model of the solvent defined by the bulk dielectric constant and atomic surface tensions. *J. Phys. Chem. B* 113: 6378–6396.

79 Bootsma, A.N. and Wheeler, S. (2019). Popular integration grids can result in large errors in DFT-computed free energies. *ChemRxiv*. https://doi.org/10.26434/chemrxiv.8864204.v5.

80 Wheeler, S.E. and Houk, K.N. (2010). Integration grid errors for meta-GGA-predicted reaction energies: origin of grid errors for the M06 suite of functionals. *J. Chem. Theory Comput.* 6: 395–404.

81 Iron, M.A. and Janes, T. (2019). Evaluating transition metal barrier heights with the latest density functional theory exchange–correlation functionals: the MOBH35 Benchmark Database. *J. Phys. Chem. A* 123: 3761–3781.

82 Brauer, B., Kesharwani, M.K., Kozuch, S., and Martin, J.M. (2016). The S66x8 benchmark for noncovalent interactions revisited: explicitly correlated ab initio methods and density functional theory. *Phys. Chem. Chem. Phys.* 18: 20905–20925.

8

NMR Prediction with Computational Chemistry

Amy T. Merrill, Wentao Guo, and Dean J. Tantillo

Department of Chemistry, University of California, Davis, CA, USA

Glossary

Computational NMR spectroscopy Using computers to predict chemical shifts and coupling constants.

Computer-Assisted Structure Elucidation (CASE) Approaches that automatically generate multiple constitutions for a given molecular formula along with corresponding chemical shift predictions.

Conformational searching Sampling various conformations of a molecule to find those that are lowest in energy.

Deep learning One type of machine learning involving complex neural networks.

Explicit solvation Modeling solvent in a calculation by including explicit solvent molecules.

Geometry optimization Computing the structure of a molecule corresponding to a local minimum on a potential energy surface by iteratively calculating the energy for possible geometries and then adjusting the geometry to reduce the energy.

Implicit solvation Modeling solvent in a calculation by using a continuum rather than explicit solvent molecules.

Internal scaling factors Scaling factors created using the molecules of interest for a particular study rather than a general set of molecules.

Isotropic shielding constant The quantum mechanical property describing the shielding of nuclei by electrons, which is converted to a chemical shift.

Level of theory The computational method used to perform a calculation.

Enabling Tools and Techniques for Organic Synthesis: A Practical Guide to Experimentation, Automation, and Computation, First Edition. Edited by Stephen G. Newman.

Machine learning (ML) A computational approach in which models are created based on correlations between input and output data (generally large, empirically derived datasets) with the goal of predicting properties from descriptors.

Metadynamics A method that improves sampling efficiency by introducing bias potentials (or forces) that act on some degrees of freedom.

Molecular dynamics (MD) A computational method in which physical motions of atoms in response to forces are simulated over time.

Monte Carlo sampling A random sampling approach used to find low energy conformers.

Multistandard approach Using multiple reference compounds for chemical shifts, e.g. for atoms of different hybridization.

Neural network A computational network (model) composed of multilayer "neurons," each of which makes decisions based on the output of the previous layer; when many such neurons are connected according to a certain hierarchical structure, an artificial neural network is created.

Scaling factors Multiplication factors used to convert isotropic shielding constants to chemical shifts, which are derived from linear correlations between experimental chemical shifts and isotropic shielding constants computed at a particular level of theory for a given set of molecules in a specific solvent.

Systematic dihedral angle sampling A type of conformational searching in which designated dihedral angles are systematically rotated about to create conformers.

8.1 Introduction

You have just finished synthesizing a new natural product for the first time and its NMR spectra do not match those of the reported natural product. What do you do? Or, after failing to get a crystal structure of your complex synthetic product, you want to try computational NMR for help in assigning its structure. But where do you start? How confident can you be in the results? Do you need to know the mathematics of quantum mechanics or statistics? In this chapter, we review the typical procedures available for predicting and assigning structures using NMR computations.

Computational NMR spectroscopy is perhaps the predictive spectroscopic approach most widely used by organic chemists largely because of the relatively low barrier to obtaining reliable predictions of ^{1}H and ^{13}C chemical shifts of organic molecules. This subfield of computational chemistry can be broadly defined as using computers to predict chemical shifts and coupling constants.

Some computational NMR approaches make predictions based on extrapolations from known experimental data, some use quantum-chemistry-based chemical shift predictions [1], and some use a combination of empirical and quantum-chemistry-derived chemical shifts to arrive at predictions. These approaches require different types of computational resources and different levels of experience and training for their implementation (and assessment of the validity of their results).

Common uses for computational NMR include determining the correct diastereomer of a natural or synthetic product and determining which functional groups are present in a compound. These techniques are also used to provide confirmatory evidence for structural or chemical shift assignments that are difficult to make with traditional approaches, such as those for high-barrier-of-rotation conformers and unusual synthetic intermediates lacking extensive precedent.

Calculations for NMR predictions may take seconds or may take weeks to complete, depending on the chemical system under analysis, the method used, and the computer resources available. With respect to molecular structure, large size and conformational flexibility can lead to large time costs.

This chapter will discuss the range of methods available for computing NMR chemical shifts and coupling constants, examples of their application, the advantages and disadvantages of each type of approach, common sources of error, types of systems known to be particularly challenging, and where the field of computational NMR is headed. The prediction of other spectroscopic and spectrometric methods (e.g. IR, MS, CD, UV-vis) is also possible, but not the focus here. After reading this chapter, you should have a basis for understanding the common approaches used for NMR prediction, an idea of how they might be applied in your own research, and the ability to assess their validity when encountered in the work of others.

8.2 Quantum-Chemistry-Based Computational NMR

The first two questions commonly asked by synthetic chemists when approaching NMR calculations are "Is my system a reasonable one for which to calculate chemical shifts?" and "How long is it going to take to get answers?" The answer to the first question is "yes" for most organic molecules, but the answer to the second question is much more variable [2].

8.2.1 Methods

The typical approach for calculating chemical shifts with quantum chemistry begins with building the compound(s) of interest in a molecular modeling program (see Chapter 7). Once the molecule(s) has been built, an initial calculation, called a **geometry optimization**, is performed (generally with a molecular mechanics method rather than a quantum mechanics method) to get to a low-energy conformation (see Section 8.2.1.2.2). This structure is not necessarily the lowest energy conformer or even one of the few lowest, but is rather a reasonable estimation of the best conformer similar in geometry to the structure that was built by the user. Consequently, the next step is usually to perform a conformational search. This search will provide conformers within a specified energy range calculated quickly with a low level of theory to give a "rough approximation" of the energies. From here, each conformation is further optimized using a level of theory that will give a better prediction of free energies. Then, any conformers within ~3 $kcal\,mol^{-1}$ of the lowest energy conformer are typically used for NMR calculations [1]. NMR calculations in a QM program (e.g. using the NMR keyword in Gaussian) are performed for each of these conformers individually. The results of these calculations are isotropic shielding constants, later converted to chemical shifts. A Boltzmann weighted average of all conformers' chemical shifts is then compared to the experimental chemical shifts. If the goal is to revise or assign a chemical structure, this process is repeated for each proposed structure until a structure with chemical shifts similar to the experimental chemical shifts is found. Each step in this brief overview of the typical quantum chemistry-based chemical shift prediction process (Figure 8.1) will be discussed in detail in the following sections.

8.2.1.1 Time/Resources for Calculations

With regard to the question, "How long is it going to take to get answers?" the answer depends on many factors – computer resources, the size and flexibility of the molecule, how many structures you want to calculate, and so on. If you are a synthetic organic chemist, you are most likely interested in predicting chemical shifts to determine which diastereomer of your synthetic product you have isolated or whether or not the relative configuration of your natural product target for total synthesis was correctly assigned. In the first case, you might be able to get away with calculating the *J*-couplings for a single conformation of each of a few diastereomers (as low as minutes of calculating time) [3]. In contrast, determining the correct diastereomer of a natural product may take days of computation time or more. Specific molecular features that lead to differences in required computational time are discussed below, as are differences in **levels of theory**.

A brief note on the relation between time/effort and proficiency: one of the biggest hurdles many chemists face with computational work is learning new software. We refer readers to the following resources for guidance, the third and fourth of which include useful tutorials:

- "Computational Prediction of ^{1}H and ^{13}C Chemical Shifts: A Useful Tool for Natural Product, Mechanistic, and Synthetic Organic Chemistry" [1]

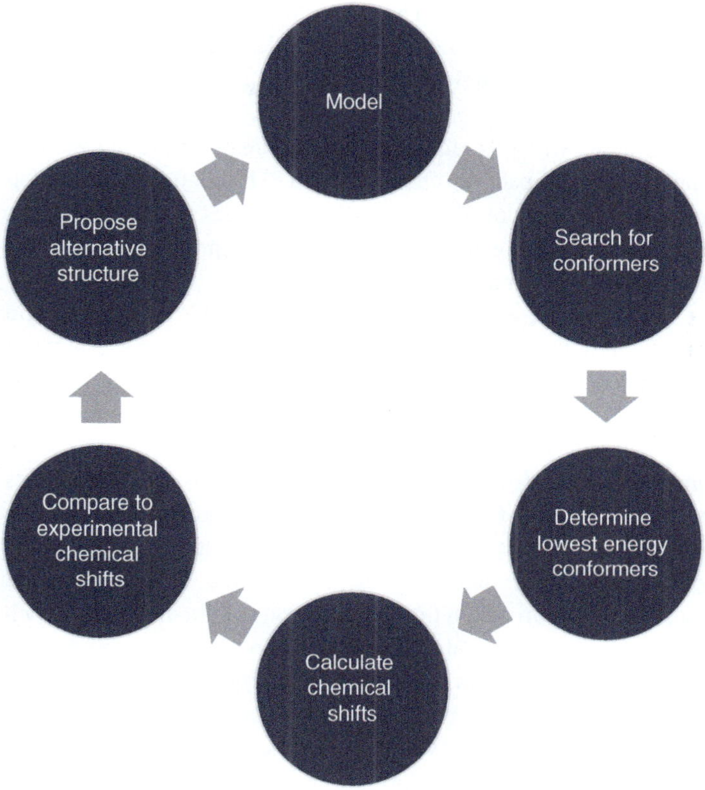

Figure 8.1 Cyclical process of revising structures using computational NMR.

- *CHESHIRE Chemical Shift Repository* (cheshirenmr.info) for scaling factors for ^{1}H and ^{13}C chemical shifts in common organic solvents, directions for calculating coupling constants, and input and excel files for creating custom scaling factors [4]
- "Reassigning the Structures of Natural Products Using NMR Chemical Shifts Computed with Quantum Mechanics: A Laboratory Exercise" (for practice calculating chemical shifts) [5]
- A guide to small-molecule structure assignment through computation of (^{1}H and ^{13}C) NMR chemical shifts [6]

8.2.1.2 Structural Considerations in Modeling

The first decision one must make when undertaking NMR calculation is how to model the structure of a compound of interest. That may sound trivial, but it is not. Depending on the compound, factors such as protonation state and degree of aggregation may need to be considered. The conformational profile relevant to NMR spectroscopy and how solvent is included in the calculations must always be considered and can only rarely be safely neglected.

8.2.1.2.1 Charge Charge, which is related directly to the protonation state of a molecule, can have a large impact on chemical shift. These effects can be felt several atoms away from the site of (de)protonation. Changes to chemical shifts are related to not only changes in electron density upon (de)protonation but also changes to conformational preferences.

Consider, for example, the case of carboxylic acids. In 2017, the neutral and anionic forms of the putative natural product shown below were examined (Figure 8.2) [7]. A difference of >0.3 ppm was predicted for the α-proton chemical shifts (4.05 ppm for the acid vs. 3.66 ppm for the carboxylate).

8.2.1.2.2 Conformational Considerations Since conformational space increases exponentially with the degrees of freedom of a molecule, molecular flexibility drastically increases the number of calculations that must be run and the associated time for arriving at predicted chemical shifts. **Conformational searching** is an automated approach to finding conformers for a molecule, which are then usually filtered by applying an energy criterion. The data observed in laboratory experiments are based on the behavior of macroscopic molecular populations. However, at the molecular level, properties are related to atomic coordinates for particular conformers. When studying solution-phase molecular properties, finding the lowest energy conformation(s) is essential. When only a few conformers are possible one can search manually, but often software is required (e.g. *Spartan* [8], *MacroModel* [9], *CREST* [10]).

If the molecule of interest is large and has a high degree of conformational flexibility, a complete conformational search may not be practical or even possible.

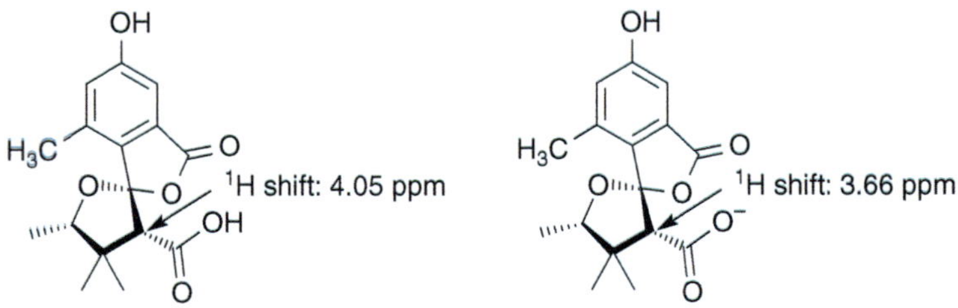

Figure 8.2 ^{1}H and ^{13}C chemical shift prediction differences between the carboxylic acid and carboxylate chemical shifts using the level of theory: SMD(CH_3OH)-mPW1PW91/6-311+G(2d,p)//B3LYP/6-31+G(d,p) with the GIAO method of calculating chemical shifts. *Source:* Saunders and Tantillo [7]/MDPI.

In general, one is limited to molecules comparable in size and flexibility to a small oligopeptide (~100 atoms) for most conformational searching programs.

Conformational sampling approaches have traditionally employed **systematic dihedral angle sampling** or **Monte Carlo sampling** to find low energy conformers. The systematic dihedral angle sampling approach leads to an energy profile like that shown in Figure 8.3, where the interconversion of some conformers (A and C) is associated with relatively large energetic barriers (A→B), and smaller interconversions, such as the rotation of a non-sterically hindered methyl group would have tiny barriers (D→). The latter situation would lead to atoms whose computed shifts would have to be averaged. While this type of conformational searching can take a while to complete, it should capture all relevant conformers if all relevant degrees of freedom are surveyed. In contrast, **Monte Carlo** sampling methods use random sampling to find low-energy conformers that can be changed systematically to find the lowest local free energy minima. This approach can be fast but may miss some structures. In general, applying as many different conformational searches as one can afford and pooling the results provides the best chance of capturing all relevant structures.

Another approach for conformational searching, which has been applied more regularly in recent times, makes use of **molecular dynamics (MD)** simulations. In MD simulations, changes in molecular structure, in response to internal (and external, if desired) forces, *over time* are simulated. Internal forces can be derived from quantum mechanics (QM) calculations (so-called *ab initio* MD or AIMD) or classical mechanics calculations (often called molecular mechanics [MM]). MM approaches use force fields where atoms are generally treated as masses connected by springs. This approach is orders of magnitude faster than quantum chemistry-based approaches but is only effective when the key features of one's molecule(s)

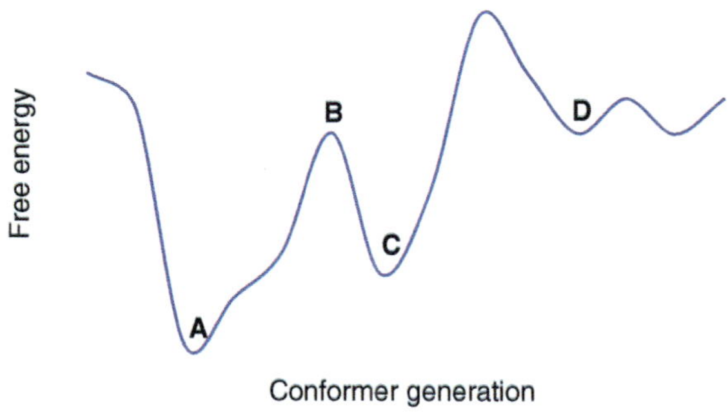

Figure 8.3 Example of a free energy profile from conformational searching. (A) Lowest energy minimum/conformation. (A→B) Barrier for conformational interconversion. (C) Local energy minimum. (D) Minimum preceding small barrier for interconversion of higher energy minima – may not be relevant in NMR spectroscopy.

are captured by the classical physics employed. In order to sample conformations, MD simulations are generally run for a required period of time (typically picoseconds), long enough to explore the energy surface on the scale of atomic movements, and structures are plucked from the resulting trajectories every so often in hopes of obtaining a collection of representative structures that a molecule would sample if left to its own devices. These structures can then be processed in the same manner as those derived from systematic or Monte Carlo searches.

Another strategy to identify all relevant conformers that contribute to a molecule's NMR spectrum employs **metadynamics**. With the aim of exploring a molecule's free energy surface and capturing as many conformations as possible in finite computational time, we need techniques to help said molecule traverse between plausible conformations. In analogy, without boats, an adventurer who is born on an island is very likely to think that the island (a "minimum") is the whole world (of course, there is an infinitesimal probability that he may tunnel to other islands and a small probability that his strong willpower and excellent swimming skills will allow him to cross the sea). But once he invents transportation that floats on water, he could shuttle between different places and get a comprehensive picture of his local world. Metadynamics is a method that facilitates the traveling of a molecule between regions (often minima) of a free energy surface. One first defines collective variables (structural parameters composed of combinations of degrees of freedom) that correspond to particular molecular distortions. For example, if we are interested in rotating a methoxy group, we may pick the dihedral angle corresponding to its rotation as the target variable, or we may examine this degree of freedom coupled to another. By continuously adding potential energy or force to the specified collective variables, the molecules can then be pushed to other regions of a free energy surface – here, to other conformers. This approach is akin to filling (energy) wells with sand, so molecules can reach the top of these wells and escape. When the energy of a molecule exceeds a certain amount, it will leave its local minimum and move to another. In the above case, adding energy helps to overcome the rotational barrier and reach the minimum of another plausible conformer. As we add more energy over time, other minima can be accessed. Perhaps the most complex collective variable for conformational searching is the overall Cartesian RMSD, with which the trajectory gets further from its original minimum and is less likely to go back as shown in Figure 8.4. Different methods for selecting collective variables and strategies for "filling" energy wells have been, and continue to be, explored.

To summarize, given that different conformers of a molecule result in different NMR spectra, computational NMR studies require adequate sampling of conformational space. Different sampling approaches can be used to zoom in on plausible conformations that can then be subjected to optimization with quantum chemical computations.

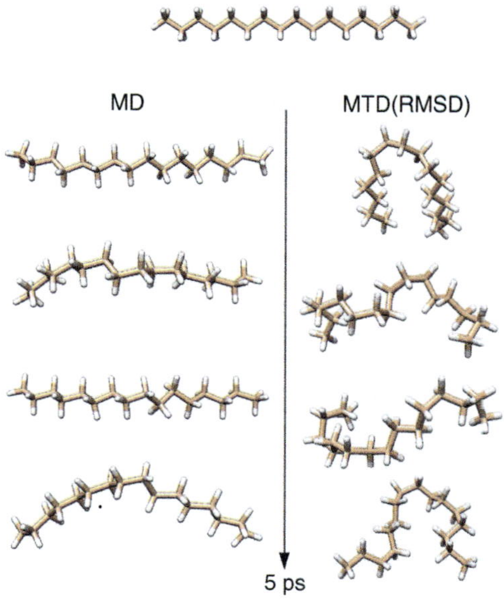

Figure 8.4 Structures of $C_{16}H_{34}$ conformational snapshots in two 5 ps simulations with MD (left) and metadynamics (MTD) (right). *Source:* Grimme [11]/American Chemical Society.

8.2.1.2.3 Solvent While computations are often carried out in the gas phase, it is essential to account for solvent when modeling NMR, since solute–solvent interactions can modulate chemical shifts. Even when nonpolar solvents are used, it is prudent to confirm that they do not have a significant effect on chemical shifts. When it comes to including solvent in calculations, there are two extreme approaches. The most often used approach is called **implicit solvation** (also called continuum solvation). In this approach, the solvent is treated as a continuum with certain bulk properties, including the dielectric constant. There are several commonly used variants of this approach (e.g. the polarized continuum model [PCM] [12] and its relatives and the solvation model based on density [SMD] [13] that differ primarily in how the cavity enclosing the solute is defined and how solute–solvent interactions are computed across this border between solute and continuum).

While different implicit solvent models often provide similar results, sometimes they do not. For example, Zanardi and coworkers found significantly different distributions of conformers when using PCM and SMD energies in an NMR study of bicyclic heterocycles (Figure 8.5) [14].

The second approach is called **explicit solvation** (also known as microsolvation) where one or more molecules of a solvent is modeled explicitly, i.e. one builds solvent molecules along with the compound of interest and includes them in the subsequent modeling steps. The advantage of this method is that it allows for specific solvent–solute interactions to be included in the calculations. When one or

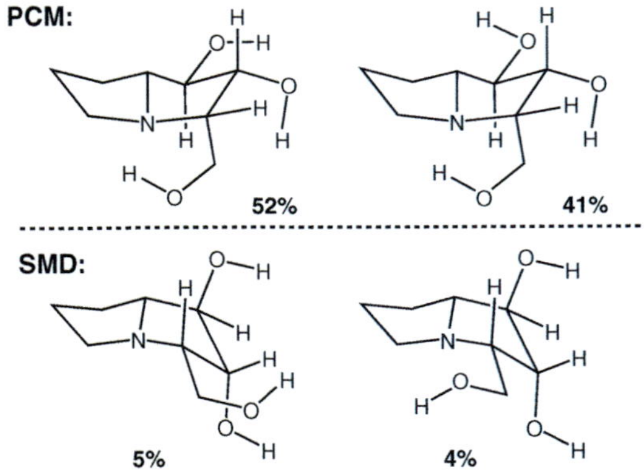

Figure 8.5 Difference in populations of two most populated conformers predicted using two popular implicit solvent methods, PCM and SMD. *Source:* Zanardi et al. [14]/ American Chemical Society.

only a handful of explicit solvent molecules are included, they can be modeled with the same quantum chemical approach as the solute. When large numbers of explicit solvent molecules are included, however, most or all of them must be modeled with a less expensive (e.g. MM or semi-empirical) method for practical reasons.

Including explicit solvent can significantly improve NMR predictions. For example, Da Silva and De Almeida were able to improve the prediction of N-^{1}H and N-C-^{1}H chemical shifts with the inclusion of explicit chloroform molecules in their DFT calculations (Figure 8.6) [15].

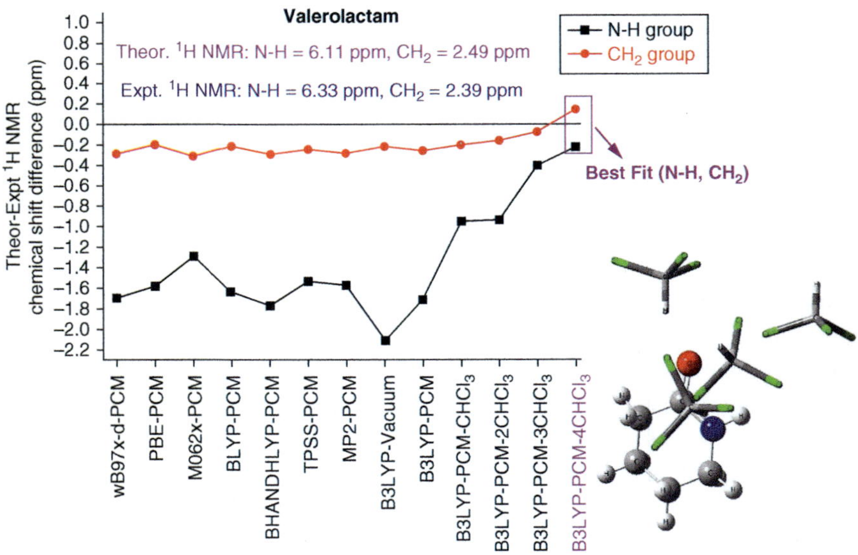

Figure 8.6 Improvement in ^{1}H NMR shift prediction with explicit solvent inclusion in the isotropic shielding constant calculations. *Source:* Da Silva and De Almeida [15]/With permission of Elsevier.

The potential disadvantages of the explicit solvation approach include: (i) A lot of computational time may be needed when including enough solvent molecules to encase your solute to the degree necessary to capture all important interactions. (ii) Strengths of solvent–solute interactions may be overestimated (although this problem can be ameliorated by surrounding a solute•solvent complex with a continuum) [15]. (iii) Adequately sampling different configurations of the solvent molecules around the solute is far from a trivial task, requiring all the considerations associated with conformational searching described above and requiring similar approaches to address.

8.2.1.3 Geometry Optimizations

Once the important (low energy) conformers are obtained from a conformational search, one should refine their structures using a more accurate, i.e. reliable, quantum chemical method. **Geometry optimization** is an iterative process of slightly varying atomic positions and calculating the associated electronic energy until the energy changes less than a specified threshold. Often, multiple structures generated by a conformational search converge to the same optimized structure. It is important that this is verified, however, by examining the geometries and not just the energies of optimized structures.

8.2.1.3.1 Level of Theory Since structures generated from a conformational search can sometimes differ greatly from fully optimized structures and since one wants to use a reliable level of theory for optimization, these calculations can be time-consuming. Consequently, optimizations are often done in two stages, first with a lower (faster) level of theory that is capable of obtaining reliable geometries but less suited for obtaining reliable energies, and then again with a higher (but slower) level of theory good for energies starting from the geometry obtained with the lower level of theory. Not only can this approach save overall computational time, it also allows for a reduction in the number of conformers that must run at the higher level of theory, since, as mentioned above, multiple structures from conformational searches often converge to the same structure when optimized with a QM method. Also, the energies obtained from the lower level of theory are useful for eliminating structures from further consideration; however, a threshold of 5–10 kcal mol^{-1} from the lowest energy conformer is typically used, rather than the ~3 kcal mol^{-1} energy window relevant to experiment, so as not to miss important structures before refining them with a higher level of theory.

There is a wide range of levels of theory that can be used for these calculations and we make no effort to discuss them all here. Practitioners have their own favorites, so we also decline to make specific recommendations. Instead, we point here to Chapter 7 as well as several papers in which various methods were benchmarked. We advise first testing your chosen methods on known compounds similar to your compound(s) of interest. In general, however, most common DFT methods with medium-sized basis sets can provide reasonable results.

- "Extended Benchmark Set of Main-Group Nuclear Shielding Constants and NMR Chemical Shifts and Its Use to Evaluate Modern DFT Methods" [16]
- "Computational ^{1}H NMR: Part 1. Theoretical Background" [17]
- "Benchmarking DFT Calculations of ^{1}H and ^{13}C Chemical Shifts in Monosubstituted Ferrocenes" [18]
- "A Look at the Density Functional Theory Zoo with the Advanced GMTKN55 Database for General Main Group Thermochemistry, Kinetics and Noncovalent Interactions" [19]

Additionally, implicit solvation and empirical dispersion can be included in optimization calculations (and some reviewers demand it), but doing so usually does not change the resulting geometries enough to affect the chemical shifts at the end of the calculations [20].

8.2.1.4 Calculating Isotropic Shielding Constants

After all conformers have been optimized, they are used as input for calculating **isotropic shielding constants** in quantum chemical modeling programs. The isotropic shielding constant is the physical property that is transformed into chemical shift by defining a scale based on a reference compound, e.g. TMS for ^{13}C and ^{1}H. From here, in order to get the final calculated chemical shifts to compare to the experimental chemical shifts, a few steps are taken.

First, the chemical shifts of equivalent atoms within a given optimized conformer should be averaged, e.g. the three hydrogen atoms of a freely rotating methyl group, since these will be indistinguishable in an NMR laboratory experiment. These three hydrogen atoms are computed to have different isotropic shielding constants in a QM calculation (unless they are related by symmetry) because the structure used for the calculation is static and the hydrogens are in different environments.

Second, since the distribution of interconverting conformers in an NMR tube is determined by their relative free energies, conformers with lower energies contribute more to the observed chemical shifts. As mentioned above, the Boltzmann distribution can be used to describe the probability of conformers at a given temperature based on their relative free energies. At room temperature (298 K), a structure that is 2.73 kcal mol^{-1} higher in free energy than another should contribute 1% as much as the lower energy structure to the overall population. Therefore, we typically only consider structures within a ~3 kcal mol^{-1} energy window.

Consider the following representative example of the process just described. 2,6-Pyridinedicarboxaldehyde has three plausible forms (the fourth form is identical to the third but with different labels), which differ in the orientation of their aldehyde groups as shown in Table 8.1. After NMR calculations on optimized versions of these conformers, we collect the isotropic shielding values and convert them to chemical shifts (the conversion process is discussed below). In conf3, we have equivalent pairs of atoms: H14 and H15, H10 and H3, C1 and C2, C8 and C11, C4 and C7, which each share one signal in the experimental NMR spectra. However, as they are distinguishable in this specific geometry, the predicted shifts are different. Thus, we have to average the H14 and H15 shifts to give predictions because they rapidly interchange by rotation on the NMR time scale. The energy-derived weights of the conformations are listed in Table 8.1. Conf2, with the lowest free energy, is designated as the reference. The proportions of conf3 (+9.3 kcal mol^{-1}) and conf1 (+4.0 kcal mol^{-1}) are negligible. In this situation where one specific conformer dominates, the problem is easy to solve and the predicted shifts are in excellent agreement with experimental results (see section 8.2.2 for discussion on confidence analysis).

Below, we show the step-by-step calculation of relative Boltzmann population taking conf2 (+0.0 kcal mol^{-1}) and conf3 (+4.0 kcal mol^{-1}) as an example. Based on the Boltzmann distribution, we know that the population of state i is proportional to an exponential related to its energy:

$$N_i \propto e^{-\varepsilon_i/kT}$$

Thus, the population of N_2 relative to N_1 can be written as:

$$\frac{N_2}{N_1} = \frac{e^{-\varepsilon_2/kT}}{e^{-\varepsilon_1/kT}}$$

We can simplify this equation to:

$$\frac{N_2}{N_1} = e^{-(\varepsilon_2-\varepsilon_1)/kT} = e^{-(\Delta\varepsilon)/kT}$$

Since conf2 has the lowest energy, we take it as the N_1 standard state. Plugging in the +4 kcal mol^{-1} energy difference between conf2 and conf3, we can calculate the relative populations of conf3 and conf2. Note: relative populations also depend on the temperature. The k in the equation is Boltzmann's constant. Typically, NMR experiments are carried out at room temperature (298.15 K) where $kT =$ 0.593 kcal mol^{-1}.

$$\frac{N_{\text{conf3}}}{N_{\text{conf2}}} = e^{-4/0.593} \approx 1.2\times10^{-3}$$

Thus, we get the relative population of conf2 (1.2E-3), as shown in column ***w*** in the table.

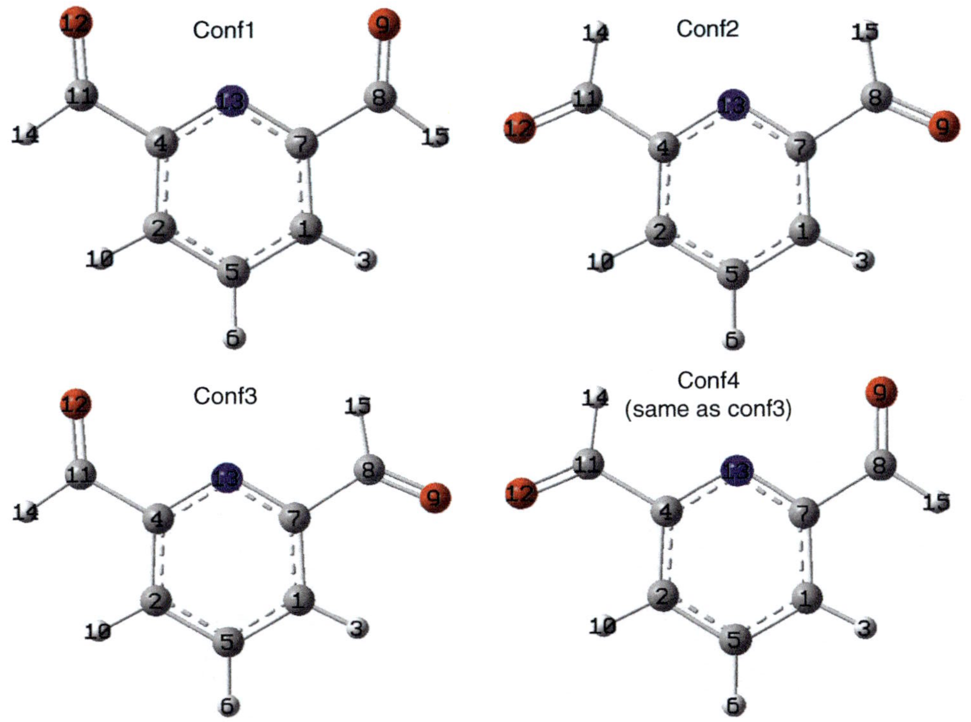
Conf1
Conf2
Conf3
Conf4
(same as conf3)

Table 8.1 Computational and experimental ^{13}C and ^{1}H NMR results.

	ΔG	w	C1	C2	C4	C7	C5	C8	C11	H3	H10	H6	H14	H15
1	9.3	1.5E-07	133.2	133.2	151.6	151.6	139.8	189.5	189.5	7.96	7.96	8.19	9.89	9.89
			133		**152**		**140**	**190**		**7.96**		**8.19**	**9.89**	
2	0.0	1	125.6	125.6	151.7	151.7	138.2	193.1	193.1	8.19	8.19	7.97	9.81	9.81
			126		**152**		**138**	**193**		**8.19**		**7.97**	**9.81**	
3	4.0	1.2E-03	124.5	134.4	151.4	152.1	138.9	193.9	189.3	8.14	8.00	8.07	9.85	9.77
			130		**152**		**140**	**192**		**8.07**		**8.07**	**9.81**	
4	4.0	1.2E-03	134.4	124.5	152.1	151.4	138.9	189.3	193.9	8.00	8.14	8.07	9.77	9.85
			130		**152**		**140**	**192**		**8.07**		**8.07**	**9.81**	
Experiment			**125**		**153**		**138**	**192**		**8.16**		**8.06**	**10.14**	

ΔG column: relative free energies in kcal mol^{-1} of different conformations computed at the M06-2X/6-311+G(2d,p) level of theory. The lowest energy conformation is the second one. **w** column: relative Boltzmann weight derived from **ΔG.** Rows 1, 2, 3 and 4 TOP: chemical shifts (ppm, relative to TMS) for C and H atoms arrived at by scaling isotropic shielding constants as described in the main text. Below: **Bolded shifts** are averaged shifts for equivalent Cs and Hs (ppm, relative to TMS). The bottom row shows experimental ^{1}H and ^{13}C shifts. *Source:* Adapted from Hicks et al. [21].

The Boltzman weighting approach should work well if we have accurate relative free energies of all important conformers. Unfortunately, on rare occasions we are not so lucky. Why? (i) Free energies obtained by affordable quantum chemical methods are not necessarily accurate enough, due to both inherent errors in calculating electronic energies and complications associated with computing entropies. (ii) If the free energy surface is very flat, i.e. conformers can be interconverted exceedingly quickly, using discrete structures does not fully represent the dynamic nature of molecules in solution. (iii) Relative free energies can be affected dramatically by explicit solvent effects and, as described above, these can often be difficult to model accurately.

So, what is one to do? First, one should try Boltzmann averaging and see if a reasonable match with experiment is achieved. Often it is. If it is not, however, we recommend the following approach before abandoning a potential structure. One can simply ignore computed relative free energies and weight the optimized structures so that the best possible match to the experimental data is achieved, i.e. a "fitting" approach. This may sound like cheating and some consider it to be so. We do not. We are well aware of, and openly acknowledge, the limitations of computing relative free energies of interconverting conformers of complex molecules in solution. We believe that insisting on using these numbers and Boltzmann averaging is foolhardy [22].

We also openly acknowledge that the fitting approach does not guarantee that correct relative population of conformers will be found. It will, however, assure that a structure is not discounted prematurely. And if one has reason to believe that the NMR calculations are highly accurate and all contributing conformers have been found, then fitting will indeed tell you the relative populations of those conformers present in the experimental solution.

8.2.1.5 Common Pitfalls and How to Address Them

One of the most common issues in optimizing structures is the failure of the structure to converge on an optimized geometry. This problem can arise for a variety of reasons, but often the quickest solution is to change the geometry of your structure slightly and then try rerunning the calculation.

Another common issue is actually not associated with the computations. Sometimes it is found that certain pairs of computed chemical shifts do match reported shifts as assigned, but do match well if the assignments are switched. This can mean that the original assignments were incorrect (which can easily occur in a complex structure with similar groups in different parts of the molecule), but one should make sure that changing the original assignment does not lead to conflicts with other available experimental data (e.g. coupling constants, 2D NMR results). In general, one should always examine the experimental data carefully when carrying out a computational study [23].

8.2.1.6 Converting to Chemical Shifts

Because chemical shift scales are defined with respect to reference compounds, calculated isotropic shielding constants (the value of which is usually in the 20–50 range for ^{1}H and anywhere from around −5 to 190 for ^{13}C shifts) must be converted to chemical shifts for comparison to experimental data. There are several ways to accomplish this. The seemingly most straightforward approach is to calculate the isotropic shielding constants for the relevant reference compound and take the difference between those and the values computed for your compound of interest. This approach can lead to systematic errors in computed shifts, however, if the method you are using is not up to the task of computing the reference compound accurately. In some cases, multiple reference standards are used to account for different types of atoms, e.g. TMS for sp^3 carbons but benzene for sp^2 carbons, etc. This approach is called the **multistandard approach**.

Another common approach is called linear scaling. In this approach, a regression line is created by plotting experimental chemical shifts for a set of known compounds in a particular solvent on the x-axis and calculated isotropic shielding constants for the same compounds on the y-axis for a particular type of atom. Two **scaling factors** are then obtained: the slope of the regression line and the value for where the line crosses the y-intercept. These scaling factors can then be used to go directly from computed isotropic shielding constants to chemical shifts without worrying about calculating the reference compound [1, 24]. This approach works well, but one must assure that a tight correlation is achieved for the level of theory one wishes to use. The downside of this approach is that different scaling factors are needed for different levels of theory and different solvents. However, many scaling factors (for ^{1}H, ^{13}C and other NMR active nuclei) have already been reported and a large number of these are collected at the *CHEmical SHIft REpository* (cheshirenmr.info) [4]. Most of the time accurate chemical shifts can be predicted using an implicit solvent model, but in cases where strong solvent–solute interactions occur, explicit solvent molecules or a secondary linear regression line for particular types of groups (e.g. Figure 8.7) can be used to improve predictions [20].

A related approach makes use of **internal scaling factors**. In this approach the chemical shifts and isotropic shielding constants used to generate the scaling factors come from the compound(s) for which you are calculating the chemical shifts. This approach is useful when one has a large compound (or family of compounds) with enough atoms to be able to assess the quality of the correlation.

Scaling factors usually do not account for "heavy atom effects," which result from the increasing importance of relativistic effects as one drops down the periodic table. In organic chemistry, this issue is most often encountered in the prediction of shifts for halogenated (most commonly, brominated) carbon atoms. Kutateladze and Reddy were able to successfully create scaling factors for ^{13}C

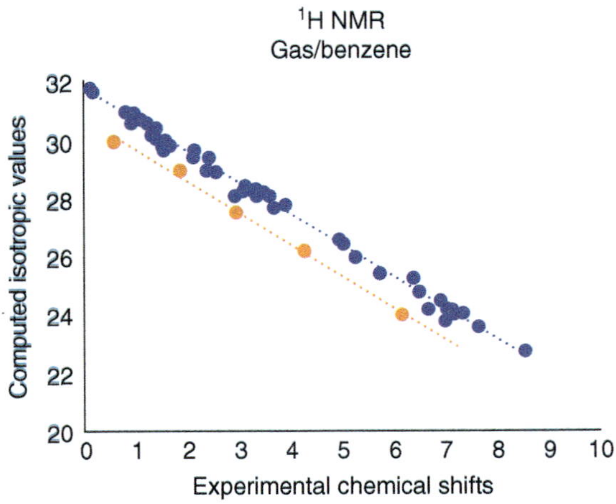

Figure 8.7 C–H–π interaction effect on acidic ^{1}H chemical shifts in benzene necessitating a secondary linear regression line, where the blue line is the standard linear regression line, and the orange line is a secondary linear regression line formed from chemical shifts impacted by C–H–π interactions. *Source:* Merrill and Tantillo [20]/ John Wiley & Sons.

attached to S, Cl, Br, and I and reassigned the structure of several natural products in the process [25].

8.2.1.7 Calculating Coupling Constants

Quantum chemistry also can be used to compute coupling constants [26, 27]. The prediction of coupling constants is undoubtedly valuable for assigning complex structures, although it is remarkable how much can be accomplished with chemical shift predictions alone. The most prevalent method for computing *J*-values is adding up the contribution of Fermi contact (FC), spin–dipole (SD), paramagnetic spin–orbit (PSO), and diamagnetic spin–orbit (DSO) terms. Scaling factors for some coupling constants also are available. ([4]; CHESHIRE Chemical Shift Repository [cheshirenmr.info]). In 2016, Kutateladze reported a new model for geminal, vincinal, 1,3- and long-range coupling constant computations based on four sets of empirical parametric coefficients and natural bond orbital (NBO) hybridization corrections based on QM calculations [29, 30]. Their test set shows good agreement with experimental results (RMSD in ~0.2–0.4 Hz), as shown in Figure 8.8 for the example of nocardiozine A [30].

8.2.2 Confidence Analysis

In the synthetic realm, if your NMR simulation results are in good agreement with your experimental NMR shifts, you can have some confidence that you have

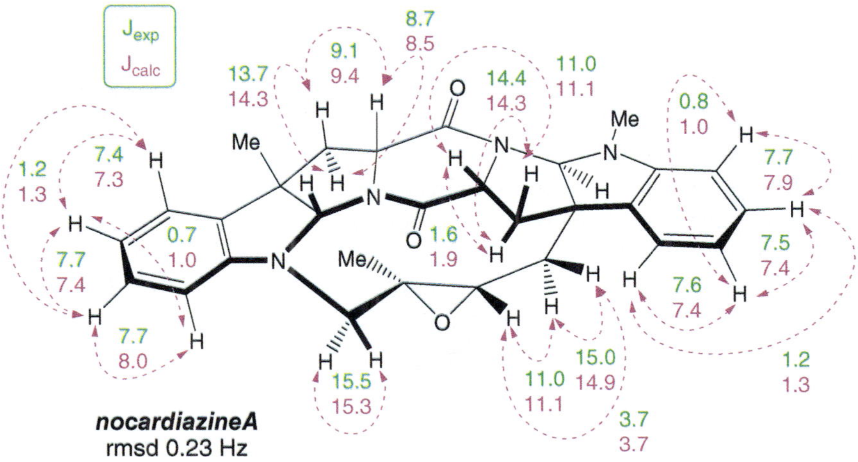

Figure 8.8 Coupling constant predictions. *Source:* Kutateladze et al. [30]/American Chemical Society..

synthesized the correct intermediate or product. In a situation where you do not know the structure of the compound synthesized, NMR calculations can be a powerful means of making structural assignments. However, one must have confidence in the validity of the computed NMR parameters. In this section, we introduce how confidence analysis works and introduce several popular statistical analytic models.

Several factors complicate confidence analysis in quantum chemistry-based computational NMR. First, multiple calculations (generally using multiple levels of theory) are required to predict chemical shifts. Errors in conformer energies and errors in isotropic shielding constants can thus compound themselves. The second complication is inherent to the structure elucidation process using NMR spectroscopy. For example, chemical shifts for natural products and complex synthetic products are not infrequently misassigned. If 2D NMR experiments were not performed (or if misleading correlations were reported), then computationally guided chemical shift reassignments may be suggested but cannot be verified without further experimental evidence. This is where methods like DP4 [31] and DP5 from Goodman and coworkers come in handy. One goal in NMR prediction is to identify how well a given computed NMR spectrum matches an experimental one. Several methods have been developed to provide a score representing how close a match is. The most popular method for assigning probabilities of correct assignments is the DP4 method. This method was designed to statistically rank possible stereoisomeric structures utilizing the best chemical shift assignments from the reported experimental chemical shift list [31]. Since its creation, several updated and specialized versions of DP4 have been reported. An updated DP4+ [34] model based on a higher level of theory was developed by Sarotti and co-workers to address issues dealing with unusually hybridized carbons. DP5 was then developed to facilitate comparisons of predicted chemical shifts of multiple chemical structures (usually diastereomers) in confidence analysis.

An alternative type of confidence analysis is χ^2 analysis. This is a popular test of statistical significance often used in CASE analysis (see Section 8.2.3.1). It is independent of the level of theory used, while DP4-type analyses are not (although statistical tools from them are sometimes used with other levels of theory).

An advantage of DP4-type methods is that both ^{13}C and ^{1}H chemical shifts can readily be used together to determine the probability of a specific structure being correct, reducing the uncertainty created from possible chemical shift misassignments from one nuclei or another. Consequently, the most common practice is to employ a DP4-type method.

8.2.3 Computer-Aided Automated Approaches

Different algorithmic approaches and workflows have been reported to rapidly and systematically simulate NMR spectra and predict reasonable molecular structures. With improvements in the accuracy and speed of QM calculations, such approaches have become more common. Several scripts/programs have been published to help automate the process and enhance computational performance with empirical correction terms. In this section, we briefly introduce automated NMR processing methodology. The history of putting computers in the toolbox of organic synthetic research spans more than 50 years. Beyond DFT calculations, this section aims to show readers the efforts that have been made by computational chemists to integrate procedures for turning an observed NMR spectrum into a molecular structure. These approaches vary in effectiveness and user-friendliness, but both are likely to improve rapidly.

8.2.3.1 CASE

The classic approach for natural product configuration identification based on experimental NMR results is called **Computer-Assisted Structure Elucidation (CASE)**. While specific CASE methods date to the 1960s, the name has become a bit of a catch-all [34]. Typical CASE approaches automatically generate multiple constitutions for a given molecular formula along with corresponding chemical shift predictions. Chemical shifts can be computed empirically using quantum chemical methods, neural networks, or combining the above methods. Structures are then scored based on the match between simulated and experimental NMR data. In many cases, however, multiple candidate structures have similar scores, necessitating further investigation. Now, with sophisticated DFT-based

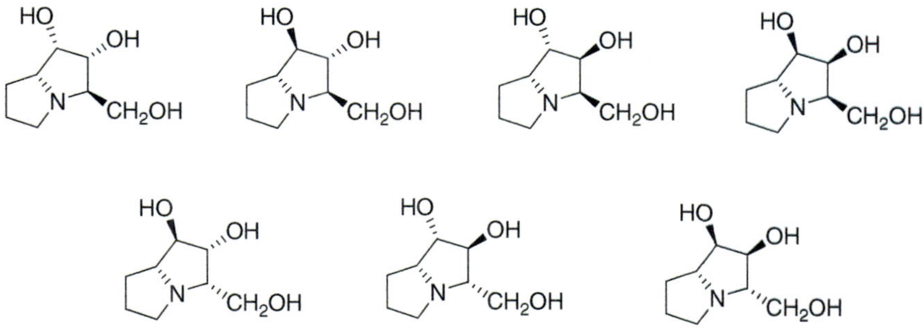

Figure 8.9 Seven diastereoisomeric structures of hyacinthines assigned using CASE-3D. *Source:* Navarro-Vázquez [22]/John Wiley & Sons.

computational NMR techniques and the emergence of machine learning, the term CASE can refer to a total automatic NMR study with computers, including reading and interpreting experimental NMR spectra, generating candidate structures, calculating their NMR parameters, and confidence analysis.

In 2018, the powerful software called CASE-3D was introduced [35]. CASE-3D uses 3D information from Residual Dipolar Couplings (RDCs) and Residual Chemical Shift Anisotropy (RCSA). CASE-3D has been employed to solve many challenging NMR assignment tasks. For example, using CASE-3D software, Navarro-Vázquez et al. successfully determined the configurations for seven reported diastereoisomeric hyacinthines (Figure 8.9) [22].

8.2.3.1.1 CASE with DFT DFT-based NMR computation has been merged into CASE as an important component in the workflow. For example, one recently published program called DP5 (same as the name of the confidence analysis method within the program) automates the procedure illustrated in Figure 8.10 [32]. In this case, DFT-based NMR shifts are generated for candidate structures, processed and assigned, and errors estimated. Computational NMR signals are given by part of another computational engine called DP4-AI packaged with conformational searches, DFT optimization, and NMR calculations based on the GIAO method. Raw NMR data is processed by NMR-AI, including peak picking, modeling, and integration. According to the latest DP5 probability measure algorithms, the probability of candidate structures is computed to reveal the most likely correct structure. Ultimately, a statistical approach is used to rank the candidate structures.

In 2017, a fully automated quantum chemistry-based tool for the computation of spin-spin-coupled NMR spectra was introduced [36]. This workflow comprises conformer/rotamer generation, relative free energy calculations, and NMR

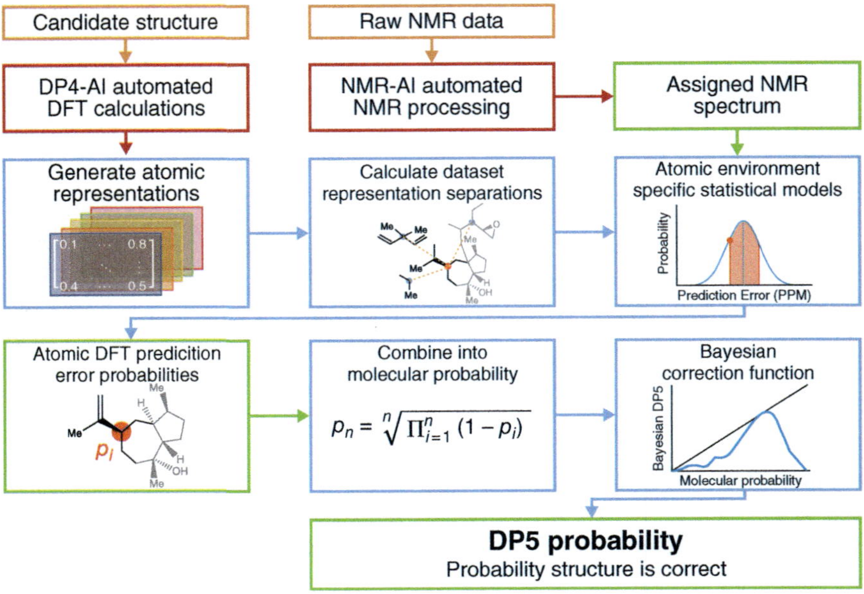

Figure 8.10 Schematic of the DP5 program. *Source:* Adapted from Howarth and Goodman [32].

computations. The conformational energies and NMR results are obtained using reliable DFT levels and Boltzmann weighting. The typical time required for handling structures with ~50 atoms is a few hours. Figure 8.11a shows an example of the good agreement with experiment possible for a sizable and somewhat flexible molecule, vitamin D2. Figure 8.11b shows the utility of this approach in distinguishing isomers. The method is not restricted to typical organic molecules, e.g. Figure 8.11c showcases tricyclopropylgallium.

To reduce the computational cost associated with computing chemical shifts and coupling constants, Kutateladze developed DU8+ [37] and DU8ML [38], the latter involving elements of machine learning. Another machine-learning-based variation, DP4-AI [40], has also been described. In all of these approaches, statistical methods are used to remove human error associated with manual comparison of chemical shifts, and thanks to these computational methods, researchers have corrected many misassigned natural products.

8.2.3.1.2 CASE with Machine Learning A Neural networks is a computational network composed of multilayer "neurons," each of which makes decisions based on the output of the previous layer. Neural network have been applied to NMR simulation since the early 1990s [40–44]. With the continued development of computer hardware, parallel computing based on GPUs has greatly improved

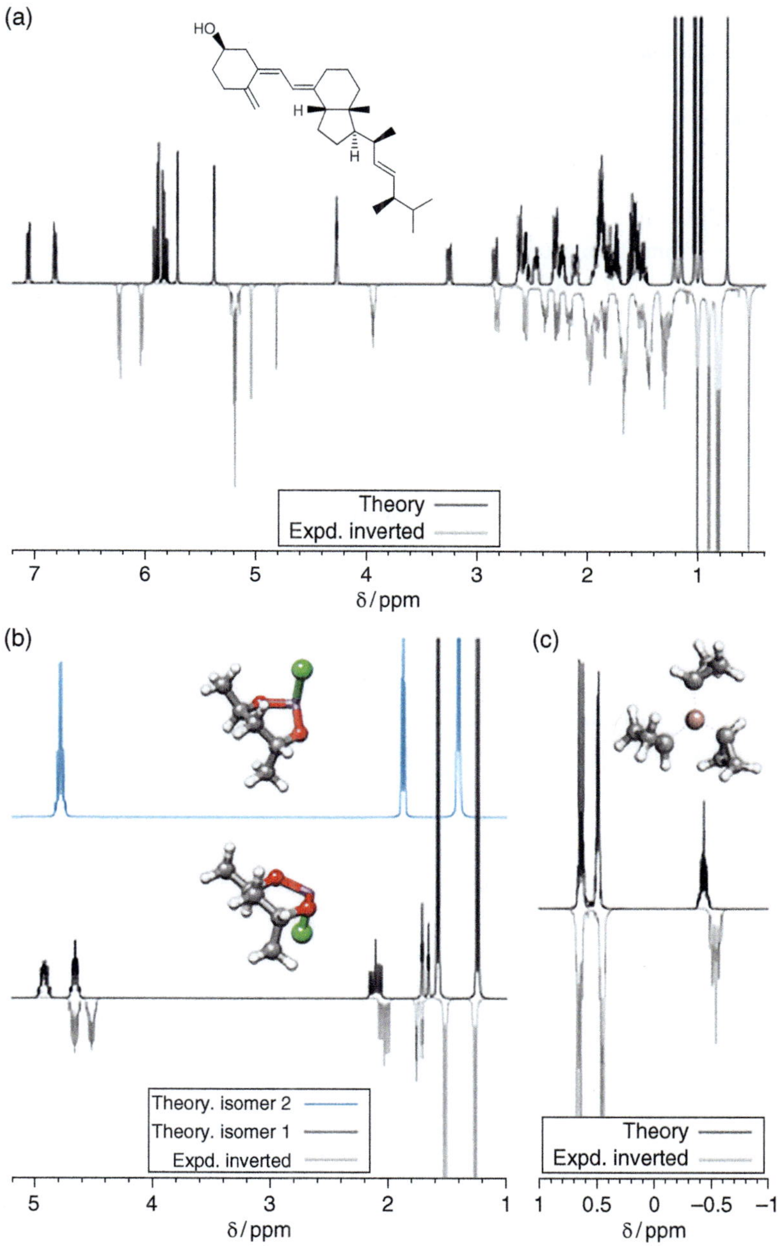

Figure 8.11 ¹H chemical shift prediction compared to experimental ¹H chemical shifts using Grimme's fully automated approach. *Source:* Grimme et al. [36]/John Wiley & Sons.

the feasibility of developing and using complex neural networks based on nonlinear models that can be trained to process huge amounts of data. Recently, the vigorous development of **machine learning (ML)**, a computational approach in which models are created based on correlations between input and output data with the goal of predicting properties from descriptors, has impacted chemistry, including the NMR prediction community [45]. The ultimate goal of ML-facilitated NMR computation is to construct an omnipotent model to predict the NMR spectrum from a molecular structure and to predict a molecular structure from experimental NMR results. These problems have not yet been solved, but progress has been made in using well-trained **deep learning** models (a type of machine learning involving complex neural networks) to automate human or QM-based operations, such as spectra reconstruction [46], peak selection (relevant to denoising of spectra) [47, 48], structure filtration [50], free energy calculation [50–52], NMR parameter prediction [53–55], and structure assignment [56–61]. Although these methods have great potential, the construction of such models requires a large amount of data (either experimental NMR spectra or high-precision QM-calculated NMR parameters). The establishment of a comprehensive and open database of existing NMR spectra will help [23]. In addition, such models can have difficulty making predictions for structures not represented in the datasets used for their training (outliers). And *why* many deep learning models work is still a mystery due to difficulties in deconvoluting layers of algorithms and mathematical issues [62, 63].

In summary, though various (semi-)automatic methods have achieved promising results (some successes have been spectacular), reliable routine identification of correct structures still depends on the knowledge and NMR interpretation skills of scientists. Automated methods can find the correct structure only if it is contained in the pool of structures considered. So one needs either chemists or automated structure generation methods that we can trust. This issue also means that it is much more difficult to predict the molecular structure from NMR data, which is, of course, what most organic chemists need for their research. Nonetheless, these methods can provide useful starting points, especially in cases where the number of seemingly possible structures is dauntingly large. Narrowing down a pool of structures with a CASE method and then having a skilled chemist run a thorough quantum chemical analysis on the best candidates can be an effective approach.

8.2.4 A Case Study

The curcusones (Figure 8.12) [64], a family of diterpenoid natural products, have served as a testing ground for a variety of computational approaches for assigning structures based on NMR. It proved particularly difficult to assign the relative configuration of the stereogenic carbons indicated in Figure 8.12 for curcusones I and J. In mid-2017, Li and Dai reported total syntheses of the originally reported structures of curcusones I and J, revealing that the chemical shifts of these did not match the shifts reported for the isolated natural products [65].

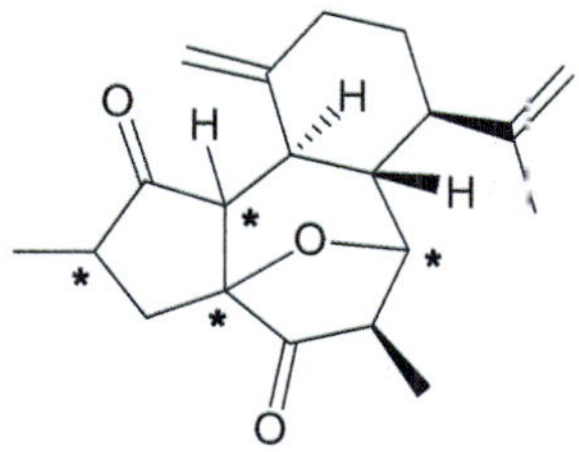

Figure 8.12 The curcusones with stereogenic centers in question marked by asterisks.

Later that year, the relative configurations of these compounds were reassigned by Sarotti based on results of DFT calculations [66]. Chemical shifts for all possible diastereomers were computed at the PCM($CHCl_3$)–mPW1PW91/6-31+G(d,p)//B3LYP/6-31G(d) level (as recommended for DP4+ analysis) [33]. Shifts were obtained by computing TMS at the same level of theory. Conformational flexibility was first considered by carrying out conformational searches using the MMFF force field, retaining all structures within a 10 kcal/mol window of the lowest energy structure for each diastereomer, reoptimizing each with B3LYP/6-31G(d) before carrying out the NMR calculations, which were subsequently Boltzmann averaged. DP4+ analysis was then applied to arrive at corrected structures. Notably, for one diastereomer, computed mean absolute errors and DP4+ analyses for ^{1}H versus ^{13}C chemical shifts did not point to the same structure, but DP4+ analyses using ^{1}H and ^{13}C data together led to clear predictions (an observation validated on other diastereomers whose structures had been confirmed independently). Subsequent calculations of coupling constants (using B3LYP/6-31G(d,p)//B3LYP/6-31G(d) with only the Fermi contact term considered) provided further confidence in the reassignments.

Concurrently, Navarro-Vaźquez et al. reassigned the same structures using a "CASE-3D" approach [35]. In this approach, conformations were generated using the MMFF94 force field, those within ~3 kcal mol^{-1} retained, and chemical shifts computed for each at the B3LYP/6-31G(d) level; these steps were automated using scripts written by the authors. Akaike Information Criterion (AIC) analysis, a method for avoiding overfitting of data [67], was then carried out to rank relative configurations. Although a smaller energy window for conformers was used and a "lower" level of theory was used for the chemical shift computations, the same diastereomers for curcusones I and J were predicted in this study. Karplus-curve-based coupling constant arguments (i.e. based on the expected relationship between coupling constants and dihedral angles) were also used to support the reassignments. Sarotti also noted that preliminary calculations at a low level of theory led to the same predicted diastereomers [66], but one cannot assume that such low levels of theory will be generally reliable.

In 2022, Sarotti and coworkers revisited the curcusone I and J problem, using it (along with many other systems) to test the performance of their new

ML-J-DP4 approach, which integrates machine learning and quantum chemical predictions [68]. In this approach, coupling constants are computed using Karplus-type equations to filter conformations and chemical shifts are computed at a very low but very fast level of theory (e.g. HF/STO-3G) and refined using an ML model developed by the authors, resulting in high-level accuracy at low-level speed, a process that has been automated via a Python script. When applied to curcusone I and J, the same diastereomers as obtained from the traditional quantum chemical approach were obtained, but in minutes rather than days.

8.2.5 Practicing ^{1}H and ^{13}C Chemical Shift Prediction

A good example to practice NMR computations is provided in the *Journal of Chemical Education* article "Reassigning the Structures of Natural Products Using NMR Chemical Shifts Computed with Quantum Mechanics: A Laboratory Exercise" [5]. The steps to follow for this simple NMR exercise are:

1) Model the proposed structure of acremolin (Figure 8.13) using the QM program of your choice.
2) Optimize the structure, here using the very low HF/3-21G level of theory for illustration purposes, but if your computer is fast enough you should jump right into a DFT method like B3LYP/6-31+G(d,p).
3) Calculate the isotropic shielding constants for the compound using the level of theory: here again using HF/3-21G (or B3LYP/6-31+G(d,p), etc.) with the GIAO NMR method, along with the SMD implicit solvent model with DMSO as solvent.
4) Convert the hydrogen and carbon isotropic shielding constants to chemical shifts using the multistandard approach presented by Palazzo et al. [5] to convert each hydrogen and carbon isotropic shielding constant to a chemical shift. Remember, as described above, there are other approaches to converting isotropic shieldings to chemical shifts.
5) Average protons with the same experimental chemical shifts.
6) Compare the ^{1}H and ^{13}C predicted and experimental chemical shifts. Which shifts have the greatest difference between them? How would you change the structure to reduce the inconsistencies between these problem areas? How would your workflow change to consider structures with multiple conformations or diastereomers to consider?

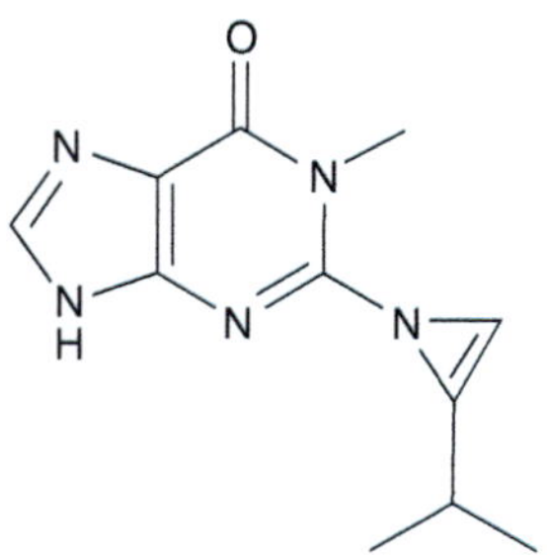

Figure 8.13 The structure first reported for acremolin from the natural product isolation.

This exercise should provide a jumping off point for calculations on more complex systems. Often a simple approach such as this will suffice for simple molecules or can be used as a first pass method in a more elaborate study.

8.3 Summary and Outlook

NMR computations have joined the tool box of routine methods for structure elucidation. Consequently, organic chemists should have some familiarity with their capabilities and limitations. Providing an overview of those was the goal of this chapter. Predicting chemical shifts is generally accomplished using two primary approaches, quantum mechanical or algorithmic, or a hybrid of the two. While the first approach is more reliable when approximations are minimized and is applicable to any system of interest, the second approach is faster. Unfortunately, QM-based approaches still always involve approximations, so users must proceed with caution. Molecular size, flexibility, charge, and potential for strong interaction with solvent affect the difficulty of and length of time for completing a QM-based NMR study. Algorithmic approaches rely on large training sets of empirical or quantum-based data in the development phase, but once that phase is completed, they can be lightning fast. Users of these methods must be cautious, however, if their molecules of interest possess unusual features that are unlikely to have been included in the method training phase. As long as you keep these caveats in mind, we encourage you to dip your toe in the pool of computational NMR!

Key References

For a general overview of ^{1}H and ^{13}C Chemical Shift Prediction, see:

Lodewyk, M. W., Siebert, M. R. and Tantillo, D. J. (2012). Computational prediction of ^{1}H and ^{13}C chemical shifts: a useful tool for natural product, mechanistic, and synthetic organic chemistry. *Chemical Reviews* 112(3): 1839–1862 https://doi.org/10.1021/cr200106v.

For a step-by-step overview of ^{1}H and ^{13}C Chemical Shift Prediction... [6]

Willoughby, P. H., Jansma, M. J. and Hoye, T. R. (2014). A guide to small-molecule structure assignment through computation of (^{1}H and ^{13}C) NMR chemical shifts. *Nature Protocols* 9(3): 643–660. https://doi.org/10.1038/nprot.2014.042.

For an overview of DP4+... [69]

Marcarino, M. O. *et al.* (2021). A critical review on the use of DP4 + in the structural elucidation of natural products: the good, the bad and the ugly. A practical guide. *Natural Product Reports*. https://doi.org/10.1039/d1np00030f.

For an overview of MD for computational NMR. . . [70]

Mazurek, A. H., Szeleszczuk, Ł. and Pisklak, D. M. (2021). A review on combination of ab initio molecular dynamics and nmr parameters calculations. *International Journal of Molecular Sciences* 22(9): 4378. https://doi.org/10.3390/ijms22094378.

For an overview of CASE. . . [34]

Elyashberg, M. and Argyropoulos, D. (2021). Computer Assisted Structure Elucidation (CASE): Current and future perspectives. *Magnetic Resonance in Chemistry* 59(7): 669–690. https://doi.org/10.1002/mrc.5115.

References

1 Lodewyk, M.W., Siebert, M.R., and Tantillo, D.J. (2012). Computational prediction of ^{1}H and ^{13}C chemical shifts: a useful tool for natural product, mechanistic, and synthetic organic chemistry. *Chemical Reviews* 112 (3): 1839–1862. https://doi.org/10.1021/cr200106v.

2 Tantillo, D.J. (2018). Questions in natural products synthesis research that can (and cannot) be answered using computational chemistry. *Chemical Society Reviews* 47: 7845–7850. https://doi.org/10.1039/c8cs00298c.

3 Laws, S.W., Howard, S.Y., Mato, R. et al. (2019). Organocatalytic Mukaiyama Mannich Reactions of 2,5-bis(trimethylsilyloxy)furan. *Organic Letters* 21 (13): 5073–5077. https://doi.org/10.1021/acs.orglett.9b01664.

4 Tantillo, D. J. (2019). *CHESHIRE, Chemical Shift Repository with Coupling Constants Added Too*. http://cheshirenmr.info/ (accessed 28 April 2023).

5 Palazzo, T.A., Truong, T.T., Wong, S.M.T. et al. (2015). Reassigning the structures of natural products using NMR chemical shifts computed with quantum mechanics: a laboratory exercise. *Journal of Chemical Education* 92 (3): 561–566. https://doi.org/10.1021/ed4006584.

6 Willoughby, P.H., Jansma, M.J., and Hoye, T.R. (2014). A guide to small-molecule structure assignment through computation of (^{1}H and ^{13}C) NMR chemical shifts. *Nature Protocols* 9 (3): 643–660. https://doi.org/10.1038/nprot.2014.042.

7 Saunders, C.M. and Tantillo, D.J. (2017). Application of computational chemical shift prediction techniqu es to the cereoanhydride structure problem-carboxylate complications. *Marine Drugs* 15 (6): 171. https://doi.org/10.3390/md15060171.

8 Spartan'10, Wavefunction, Inc.: Irvine, CA, 2010.

9 Guide, Q.S. (2009). *MacroModel 9.7. Quick Start Guide*. Schrödinger Press.

10 Pracht, P., Bohle, F., and Grimme, S. (2020). Automated exploration of the low-energy chemical space with fast quantum chemical methods. *Physical Chemistry Chemical Physics* 22 (14): 7169–7192. https://doi.org/10.1039/c9cp06869d.

11 Grimme (2019). Exploration of chemical compound, conformer, and reaction space with meta-dynamics simulations based on tight-binding quantum chemical calculations. *Journal of Chemical Theory and Computation* 15 (5): 2847–2862.

12 Tomasi, J., Mennucci, B., and Cammi, R. (2005). Quantum mechanical continuum solvation models. *Chemical Reviews* 105 (8): 2999–3093. https://doi.org/10.1021/cr9904009.

13 Marenich, A.V., Cramer, C.J., and Truhlar, D.G. (2009). Universal solvation model based on solute electron density and a continuum model of the solvent defined by the bulk dielectric constant and atomic surface tensions. *The Journal of Physical Chemistry B* 113: 6378–6396. https://doi.org/10.1021/jp810292n.

14 Zanardi, M.M., Marcarino, M.O., and Sarotti, A.M. (2020). Redefining the impact of Boltzmann analysis in the stereochemical assignment of polar and flexible molecules by NMR calculations. *Organic Letters* 22 (1): 52–56. https://doi.org/10.1021/acs.orglett.9b03866.

15 Da Silva, H.C. and De Almeida, W.B. (2020). Theoretical calculations of ^{1}H NMR chemical shifts for nitrogenated compounds in chloroform solution. *Chemical Physics* 528: 110479. https://doi.org/10.1016/j.chemphys.2019.110479.

16 Schattenberg, C.J. and Kaupp, M. (2021). Extended benchmark set of main-group nuclear shielding constants and NMR chemical shifts and its use to evaluate modern DFT methods. *Journal of Chemical Theory and Computation* 17 (12): 7602–7621. https://doi.org/10.1021/acs.jctc.1c00919.

17 Krivdin, L.B. (2019). Computational ^{1}H NMR: part 1. Theoretical background. *Magnetic Resonance in Chemistry* 57 (11): 897–914. https://doi.org/10.1002/mrc.4873.

18 Chamkin, A.A. (2020). Benchmarking DFT calculations of ^{1}H and ^{13}C chemical shifts in monosubstituted ferrocenes. *International Journal of Quantum Chemistry* 12–17. https://doi.org/10.1002/qua.26456.

19 Goerigk, L., Hansen, A., Bauer, C. et al. (2017). A look at the density functional theory zoo with the advanced GMTKN55 database for general main group thermochemistry, kinetics and noncovalent interactions. *Physical Chemistry Chemical Physics* 19 (48): 32184–32215. https://doi.org/10.1039/c7cp04913g.

20 Merrill, A.T. and Tantillo, D.J. (2020). Solvent optimization and conformational flexibility effects on ^{1}H and ^{13}C NMR scaling factors. *Magnetic Resonance in Chemistry* 58: 576–583. https://doi.org/10.1002/mrc.4986.

21 Hicks, R.G., Koivisto, B.D., and Lemaire, M.T. (2004). Synthesis of multitopic verdazyl radical ligands. Paramagnetic supramolecular synthons. *Organic Letters* 6 (12): 1887–1890. https://doi.org/10.1021/ol049746i.

22 Navarro-Vázquez, A. (2019). When not to rely on Boltzmann populations. Automated CASE-3D structure elucidation of hyacinthacines through chemical shift differences. *Magnetic Resonance in Chemistry* 58: 139–144. https://doi.org/10.1002/mrc.4951.

23 McAlpine, J.B., Chen, S.-N., Kutateladze, A. et al. (2019). The value of universally available raw NMR data for transparency, reproducibility, and integrity in natural product research. *Natural Product Reports* 36: 35–107. https://doi.org/10.1039/c7np00064b.

24 Jain, R., Bally, T., and Rablen, P.R. (2009). Calculating accurate proton chemical shifts of organic molecules with density functional methods and modest basis sets. *The Journal of Organic Chemistry* 74 (11): 4017–4023. https://doi.org/10.1021/jo900482q.

25 Kutateladze, A.G. and Reddy, D.S. (2017). High-throughput in silico structure validation and revision of halogenated natural products is enabled by parametric corrections to DFT-computed ^{13}C NMR chemical shifts and spin-spin coupling constants. *Journal of Organic Chemistry* 82 (7): 3368–3381. https://doi.org/10.1021/acs.joc.7b00188.

26 Autschbach, J. and Le Guennic, B. (2007). Analyzing and interpreting NMR spin-spin coupling constants using molecular orbital calculations. *Journal of Chemical Education* 84 (1): 156–171. https://doi.org/10.1021/ed084p156.

27 Fabián, J.S., García De La Vega, J.M., and San Fabián, E. (2014). Improvements in DFT calculations of spin-spin coupling constants. *Journal of Chemical Theory and Computation* 10 (11): 4938–4949. https://doi.org/10.1021/ct500676m.

28 Bally, T. and Rablen, P.R. (2011). Quantum-chemical simulation of ^{1}H NMR spectra. 2. Comparison of DFT-based procedures for computing proton-proton coupling constants in organic molecules. *Journal of Organic Chemistry* 76 (12): 4818–4830. https://doi.org/10.1021/jo200513q.

29 Kutateladze, A.G. and Mukhina, O.A. (2014). Relativistic force field: parametric computations of proton-proton coupling constants in ^{1}H NMR spectra. *Journal of Organic Chemistry* 79 (17): 8397–8406. https://doi.org/10.1021/jo501781b.

30 Kutateladze, A.G. and Mukhina, O.A. (2015). Minimalist relativistic force field: prediction of proton-proton coupling constants in ^{1}H NMR spectra is perfected with NBO hybridization parameters. *Journal of Organic Chemistry* 80 (10): 5218–5225. https://doi.org/10.1021/acs.joc.5b00619.

31 Smith, S.G. and Goodman, J.M. (2010). Assigning stereochemistry to single diastereoisomers by GIAO NMR calculation: the DP4 probability. *Journal of the American Chemical Society* 132 (37): 12946–12959. https://doi.org/10.1021/ja105035r.

32 Howarth, A. and Goodman, J.M. (2022). The DP5 probability, quantification and visualization of structural uncertainty in single molecules. *Chemical Science* 13 (12): 3507–3518. https://doi.org/10.1039/d1sc04406k.

33 Grimblat, N., Zanardi, M.M., and Sarotti, A.M. (2015). Beyond DP4: an improved probability for the stereochemical assignment of isomeric compounds using quantum chemical calculations of NMR shifts. *Journal of Organic Chemistry* 80 (24): 12526–12534. https://doi.org/10.1021/acs.joc.5b02396.

34 Elyashberg, M. and Argyropoulos, D. (2021). Computer assisted structure elucidation (CASE): current and future perspectives. *Magnetic Resonance in Chemistry* 59 (7): 669–690. https://doi.org/10.1002/mrc.5115.

35 Navarro-Vázquez, A., Gil, R.R., and Blinov, K. (2018). Computer-assisted 3D structure elucidation (CASE-3D) of natural products combining isotropic and anisotropic NMR parameters. *Journal of Natural Products* 81 (1): 203–210. https://doi.org/10.1021/acs.jnatprod.7b00926.

36 Grimme, S., Bannwarth, C., Dohm, S. et al. (2017). Fully automated quantum-chemistry-based computation of spin-spin-coupled nuclear magnetic resonance spectra. *Angewandte Chemie International Edition in English* 56 (46): 14763–14769. https://doi.org/10.1002/anie.201708266.

37 Novitskiy, I.M. and Kutateladze, A.G. (2021). DU8+ computations reveal a common challenge in the structure assignment of natural products containing a carboxylic anhydride moiety. *The Journal of Organic Chemistry* 86 (23): 17511–17515. https://doi.org/10.1021/acs.joc.1c02291.

38 Novitskiy, I.M. and Kutateladze, A.G. (2022). DU8ML: machine learning-augmented density functional theory nuclear magnetic resonance computations for high-throughput in silico solution structure validation and revision of complex alkaloids. *Journal of Organic Chemistry* 87 (7): 4818–4828. https://doi.org/10.1021/acs.joc.2c00169.

39 Howarth, A., Ermanis, K., and Goodman, J.M. (2020). DP4-AI automated NMR data analysis: straight from spectrometer to structure. *Chemical Science* 11 (17): 4351–4359. https://doi.org/10.1039/d0sc00442a.

40 Anker, L.S. and Jurs, P.C. (1992). Prediction of carbon-13 nuclear magnetic resonance chemical shifts by artificial neural networks. *Analytical Chemistry* 64 (10): 1157–1164. https://doi.org/10.1021/ac00034a015.

41 Burns, J.A. and Whitesides, G.M. (1993). Feed-forward neural networks in chemistry: mathematical systems for classification and pattern recognition. *Chemical Reviews* 93 (8): 2583–2601. https://doi.org/10.1038/227419a0.

42 Doucet, J.P., Panaye, A., Feuilleaubois, E., and Ladd, P. (1993). Neural networks and ^{13}C NMR shift prediction. *Journal of Chemical Information and Computer Sciences* 33: 320–324.

43 Meiler, J., Maier, W., Will, M., and Meusinger, R. (2002). Using neural networks for ^{13}C NMR chemical shift prediction-comparison with traditional methods. *Journal of Magnetic Resonance* 157 (2): 242–252. https://doi.org/10.1006/jmre.2002.2599.

44 Meiler, J., Meusinger, R., and Will, M. (2000). Fast determination of ^{13}C NMR chemical shifts using artificial neural networks. *Journal of Chemical Information and Computer Sciences* 40 (5): 1169–1176. https://doi.org/10.1021/ci000021c.

45 Chen, D., Wang, Z., Guo, D. et al. (2020). Review and prospect: deep learning in nuclear magnetic resonance spectroscopy. *Chemistry – A European Journal* 26 (46): 10391–10401. https://doi.org/10.1002/chem.202000246.

46 Qu, X., Huang, Y., Lu, H. et al. (2020). Accelerated nuclear magnetic resonance spectroscopy with deep learning. *Angewandte Chemie* 132 (26): 10383–10386. https://doi.org/10.1002/ange.201908162.

47 Klukowski, P., Augoff, M., Zieba, M. et al. (2018). NMRNet: a deep learning approach to automated peak picking of protein NMR spectra. *Bioinformatics* 34 (15): 2590–2597. https://doi.org/10.1093/bioinformatics/bty134.

48 Li, D.W., Hansen, A.L., Yuan, C. et al. (2021). DEEP picker is a deep neural network for accurate deconvolution of complex two-dimensional NMR spectra. *Nature Communications* 12 (1): https://doi.org/10.1038/s41467-021-25496-5.

49 Huang, Z., Chen, M.S., Woroch, C.P. et al. (2021). A framework for automated structure elucidation from routine NMR spectra. *Chemical Science* 12 (46): 15329–15338. https://doi.org/10.1039/d1sc04105c.

50 Lim, H. and Jung, Y.J. (2019). Delfos: deep learning model for prediction of solvation free energies in generic organic solvents. *Chemical Science* 10 (36): 8306–8315. https://doi.org/10.1039/c9sc02452b.

51 Pathak, Y., Mehta, S., and Priyakumar, U.D. (2021). Learning atomic interactions through solvation free energy prediction using graph neural networks. *Journal of Chemical Information and Modeling* 61 (2): 689–698. https://doi.org/10.1021/acs.jcim.0c01413.

52 Riniker, S. (2017). Molecular dynamics fingerprints (MDFP): machine learning from MD data to predict free-energy differences. *Journal of Chemical Information and Modeling* 57 (4): 726–741. https://doi.org/10.1021/acs.jcim.6b00778.

53 Cobas, C. (2020). NMR signal processing, prediction, and structure verification with machine learning techniques. *Magnetic Resonance in Chemistry* 58 (6): 512–519. https://doi.org/10.1002/mrc.4989.

54 Liu, S., Li, J., Bennett, K.C. et al. (2019). Multiresolution 3D-DenseNet for chemical shift prediction in NMR crystallography. *Journal of Physical Chemistry Letters* 10 (16): 4558–4565. https://doi.org/10.1021/acs.jpclett.9b01570.

55 Meiler, J. (2003). PROSHIFT: protein chemical shift prediction using artificial neural networks. *Journal of Biomolecular NMR* 26 (1): 25–37. https://doi.org/10.1023/a:1023060720156.

56 Gerrard, W., Bratholm, L.A., Packer, M.J. et al. (2020). IMPRESSION-prediction of NMR parameters for 3-dimensional chemical structures using machine learning with near quantum chemical accuracy. *Chemical Science* 11 (2): 508–515. https://doi.org/10.1039/c9sc03854j.

57 Jonas, E. (2019). Deep imitation learning for molecular inverse problems. *Advances in Neural Information Processing Systems* 32 (NeurIPS).

58 Shen, Y. and Bax, A. (2015). Protein structural information derived from NMR chemical shift with the neural network program TALOS-N. *Methods in Molecular Biology* 1260: 17–32. https://doi.org/10.1007/978-1-4939-2239-0_2.

59 Shibata, K. and Kaneko, H. (2021). Prediction of spin–spin coupling constants with machine learning in NMR. *Analytical Science Advances* 2 (9–10): 464–469. https://doi.org/10.1002/ansa.202000180.

60 Sridharan, B., Mehta, S., Pathak, Y., and Priyakumar, U.D. (2022). Deep reinforcement learning for molecular inverse problem of nuclear magnetic resonance spectra to molecular structure. *The Journal of Physical Chemistry Letters* 13 (22): 4924–4933. https://doi.org/10.1021/acs.jpclett.2c00624.

61 Zhang, X.X., Deng, T., and Jia, G.Z. (2020). Nuclear spin-spin coupling constants prediction based on XGBoost and LightGBM algorithms. *Molecular Physics* 118 (14): https://doi.org/10.1080/00268976.2019.1696478.

62 Bengio, Y., Deleu, T., Rahaman, N. et al. (2019). A meta-transfer objective for learning to disentangle causal mechanisms. *arXiv* 1–26. https://doi.org/10.48550/arxiv.1901.10912.

63 Amey, J.L., Keeley, J., Choudhury, T., and Kuprov, I. (2021). Neural network interpretation using descrambler groups. *Proceedings of the National Academy of Sciences of the United States of America* 118 (5): https://doi.org/10.1073/pnas.2016917118.

64 Liu, J.-Q., Yang, Y.-F., Li, X.-Y. et al. (2013). Cytotoxicity of naturally occurring rhamnofolane diterpenes from *Jatropha curcas*. *Phytochemistry* 96: 265–272. https://doi.org/10.1016/j.phytochem.2013.09.008.

65 Li, Y. and Dai, M. (2017). Total syntheses of the reported structures of Curcusones I and J through tandem gold catalysis. *Angewandte Chemie (International ed. in English) 56* (38): 11624–11627. https://doi.org/10.1002/anie.201706845.

66 Sarotti, A. M. (2018). Structural revision of two unusual rhamnofolane diterpenes, curcusones I and J, by means of DFT calculations of NMR shifts and coupling constants. *Organic & Biomolecular Chemistry* 16: 944–950. https://doi.org/10.1039/C7OB02916K.

67 Akaike, H. (1974). A new look at the statistical model identification. *IEEE Transactions on Automatic Control* 19 (6): 716–723.

68 Tsai, Y.-H., Amichetti, M., and Zanardi, M.M. (2022). ML-*J*-DP4: an integrated quantum mechanics-machine learning approach for ultrafast NMR structural elucidation. *Organic Letters 24* (41): 7487–7491. https://doi.org/10.1021/acs.orglett.2c01251.

69 Marcarino, M. O. Cicetti, S., Zanardi, M. M., and Sarotti, A. M. (2021). A critical review on the use of DP4 + in the structural elucidation of natural products: the good, the bad and the ugly. A practical guide. *Natural Product Reports*. https://doi.org/10.1039/d1np00030f.

70 Mazurek, A.H., Szeleszczuk, Ł., and Pisklak, D.M. (2021). A review on combination of ab initio molecular dynamics and nmr parameters calculations. *International Journal of Molecular Sciences* 22 (9): 4378. https://doi.org/10.3390/ijms22094378.

9

Introduction to Programming for the Organic Chemist

Jason M. Stevens

Bristol Myers Squibb, Summit, NJ, USA

9.1 Introduction

Organic chemistry has long benefitted from collaboration across subdisciplines. A present example is the advent of photoredox catalysis for organic synthesis [1]. Organic chemists quickly absorbed the volumes of work developed by the photochemistry community over several decades to begin crafting clever ways to make molecules. A similar recent example is the emergence of asymmetric organocatalysis [2], which adopted and exploited knowledge from chemical biology that has long inspired numerous strategies for complex natural product synthesis [3]. While these cross-discipline examples highlight the eagerness of organic chemists to explore external fields of research for application to their work, the underlying techniques remain the same. The catalysts are new, the starting materials may be different, and the transformation might be groundbreaking, but at the end of the day they are still chemical reactions. The application of computer science and programming holds similar promise for inspiring innovation at the intersection of scientific disciplines. However, the productive merger of organic chemistry and computer science may not be as natural an avenue for exploration in organic chemistry because the underlying tasks associated with each discipline are quite different. It can be hard to envision how programming lines of code at your desk will translate into more isolated product. While it may be common for an organic chemist to take several advanced classes in subfields of chemistry or disciplines such as biology or physics, it is less likely that organic chemists receive advanced training in computer science unless they are pursuing research in computational chemistry. This scenario enforces an unfortunate knowledge gap between two

Enabling Tools and Techniques for Organic Synthesis: A Practical Guide to Experimentation, Automation, and Computation, First Edition. Edited by Stephen G. Newman.

scientific fields. However, the growing examples of organic chemistry being impacted by programming and computer science may change this in the near future [4]. This section of the chapter will focus on the current inflection point for the added value that programming can bring to organic chemists and their organizations as well as a brief summary for getting started with programming. The subsequent sections will highlight case studies for programming in organic chemistry to underscore key capabilities, demonstrate value, and provide inspiration. This chapter is written for readers that are curious about the impact that computer science and data science might have on their research and requires no prior background or experience in either discipline.

In fields such as computational chemistry, analytical chemistry, and chemical engineering, where the use of computers to conduct research, execute tasks, and control equipment is more commonplace, programming skills may be more likely to be seen as value added. In contrast, synthetic organic chemistry has historically relied on partnerships across fields to develop the software and applications needed for these tasks. These partnerships include the utilization of software to process and interpret instrument data such as NMR and chromatographic data. Third-party software such as Virscidian [5] and ACD Labs Katalyst [6] automate data import and peak assignment based on observed masses. Similarly, software is available to control reaction equipment, including syringe pumps, autosamplers, and temperature control, like is available on the Mettler-Toledo EasyMax equipment platform. Learning a skill set in a completely different field of science to accomplish a task that could otherwise be done collaboratively may not be viewed as time well spent. This paradigm has sustained organic chemists over time as evidenced by the wealth of software tools at their disposal for a multitude of tasks that continues to expand. Chemists have software to aid experimental design, compute molecular properties, control reaction equipment, and automate the processing of analytical data. Interestingly, it is precisely due to this fact that synthetic organic chemists now find themselves surrounded by computers and software, which makes learning to program a value-added endeavor [7].

A walk-through of an organic chemistry laboratory will likely find computers controlling high-performance liquid chromatography (HPLC) systems, gas chromatography (GC) systems, infrared probes, high-resolution mass spectrometers, nuclear magnetic resonance (NMR) spectrometers, powder dispensing units, high-throughput experimentation (HTE) equipment, reaction sampling equipment, and automated silica gel chromatography stations. Similarly, the equipment that is being purchased for laboratory use is increasingly digital. Items such as heating and stirring plates, vacuum pumps, temperature probes, evaporators, and recirculation pumps are digital and in many cases feature ports to enable external control. Lastly, a tour of an organic chemists' office is now more likely than ever to find an electronic notebook being used to document experiments, electronic files of analytical results from instruments, spreadsheets containing data points, reports prepared in word processing software, and summaries of research prepared in presentation software. In fact, the only hand-drawn structures are likely to be found on whiteboards, hood sashes, and post-it notes. The official documentation of chemical experiments has almost exclusively transitioned to electronic media for most industrial laboratories, and it is this transition that has changed the calculus for organic chemists to learn how to program. They are surrounded by computers, digital equipment, and digital records of experiments (Figure 9.1). Programming has become the means to weave the distinct elements of the digital age of chemistry together [8]. Productivity gains may be realized through storing instrument outputs, such as NMR spectra and HPLC data [9] in a readily accessible location, controlling reaction equipment, applying data science to historical data, and other countless ways. For many individuals and organizations, this has already become apparent, and programming is becoming an increasingly desirable skill for organic chemists.

There are many different avenues to explore for chemists who want to learn programming [10]. Most universities offer courses in data science and introductory programming, which may be avenues for undergraduates and graduate students to explore. For professionals, many universities also offer boot-camp type courses in data science and programming at modest tuition prices that come with a certificate of completion. Companies that are eager to grow their data science capabilities may offer programs through online classes through partnerships with organizations such as Data Camp [11], Code Academy [12], or Coursera [13], to name a few [14]. These sites also offer individual subscriptions for a monthly

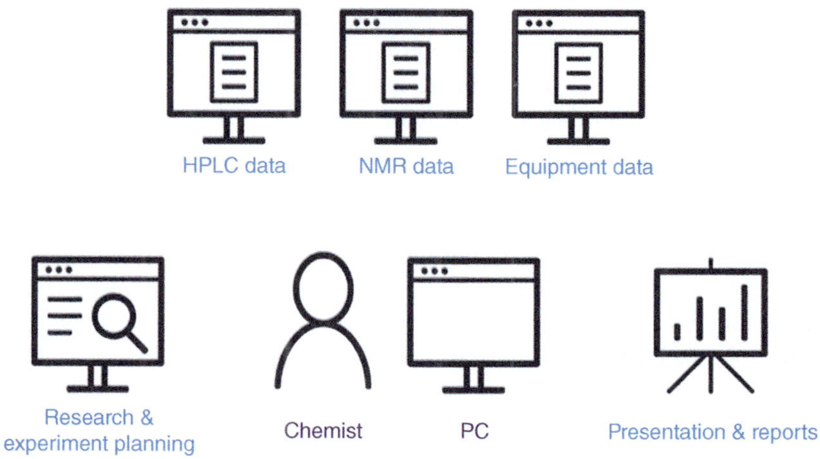

Figure 9.1 Synthetic chemists employ networks of digital equipment and media on a regular basis.

or yearly fee. If options through a university or a company are not available, there are also plenty of free alternatives. Open courseware in programming is available through the Massachusetts Institute of Technology [15], YouTube has troves of content on programming [16], and Stack Overflow [17], a knowledge sharing-forum for troubleshooting code, has lots of training content for specific problems. Particularly rich with open-source training materials is GitHub, a code hosting platform that enables collaboration and the sharing of code either publicly or privately [18]. There are many open-source code repositories on GitHub that contain full books and example code work-throughs in common programming languages [19]. Additionally, developers also provide great resources to learn for free, particularly in the R community [20]. An excellent resource for machine learning is Kaggle [21], that is owned by Google. Kaggle provides free machine learning training and computational resources during the training period. Once users are proficient, they are able to compete in Kaggle competitions for prizes! Recently, Bristol Myers Squibb hosted a Kaggle competition for developing tools to recognize images of chemical structures [22]. In all, there are a wealth of options to learn how to program.

A key ingredient to successfully learning how to program is to have a problem that you are trying to solve. A 10-week boot-camp in machine learning may not make much sense if you do not have a problem that you are actively interested in applying machine learning. This scenario might turn out to be counterproductive if the learned material is not applied within a reasonable timeframe, as it may be forgotten quickly. Potentially worse is a scenario where the learned material is applied to a problem it is not well suited for, just for the sake of having applied it, producing bad results, and generating mistrust of the learned material. Therefore, the first step should be to answer what specific, immediate benefit can be brought to your own work or to your organization by learning how to program. If the answer is clear and invokes a passionate internal response, it is likely to be a worthwhile investment of time and resources.

When learning how to program, one needs to decide where to start and what to learn. Having answered the previous question concerning the underlying problem statement should provide clarity. The first step is to identify an appropriate programming language that is best suited to the problem at hand. A programming language is the computer language and syntax that is used by programmers. There are many different programming languages with the most popular languages according to a Stack Overflow survey in 2022 being JavaScript, HTML/CSS, SQL, Python, and TypeScript [23]. If the answer is in-depth statistics and data visualization, then the R programming language might be a great place to start [24]. If programming robots and executing machine learning pipelines comes to mind, maybe try Python [25]. App development? Try Javascript. State-of-the-art machine learning models? Julia might be for the right choice. There are also other factors to consider when choosing where to start. If most of the people in your organization write code in MatLab, try starting out with MatLab and to get a sense of the core principles of programming and learning collaboratively with your colleagues. Many of the core principles of programming are consistent across languages and starting in one language does not provide significant barriers to learning another language. In fact, it is quite the opposite! Reaching out to community forums on social media is a great way to get advice and can often point you to appropriate resources to get more information.

This chapter provides collection of recent examples highlighting some fundamental aspects of programming applied to organic chemistry: data visualization, text extraction, statistical analysis, machine learning, open-source data, simulation, application development, artificial intelligence, and robotics. The examples were selected to inspire ideas for those curious about programming and the impact it might have on their work. The underlying principles for each of these areas the examples cover will be briefly discussed along with an overview of the work and its impact. User-friendly resources are referenced throughout so that the reader may learn the skills necessary to begin incorporating such tools into their workflows. The underlying principles for each of these areas the examples covered will be briefly discussed along with an overview of the work and its impact.

9.2 Better Visualizations: Communicating Structure–Data Relationships

Data visualization is already common practice for synthetic organic chemists as a means of exploratory data analysis [26] and for the purpose of visually communicating scientific findings to team members and stakeholders. Microsoft Excel is a very common application used for plotting data. Programs such as TIBCO Spotfire [27, 28] and Origin Lab [29] are tailored toward creating detailed graphs and analysis that are quite useful for exploratory data analysis. Also, statistical software such as JMP [30] combines statistical analysis with graphing tools. The advantage that programming offers is that many languages have libraries, pre-packaged bundles of code that can be used in a modular fashion, that are tailored specifically for data visualization and offer immense options for customization. By learning how to utilize the functions within these libraries, blocks of reusable code that perform a single task with variable inputs, a scientist can explore seemingly limitless options for presenting graphical data to stakeholders or collaborators. The increasing amount of data being generated by modern organic chemistry labs makes the flexibility of this approach quite valuable. Additionally, because the plot is generated by specific instructions that are written in code, it can be reproduced by anyone that has access to the same

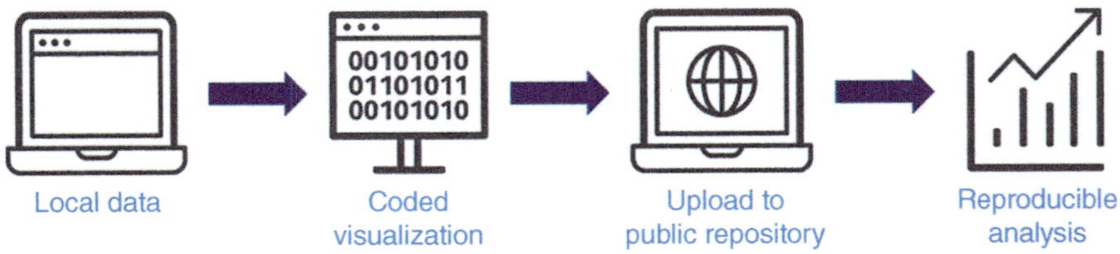

Figure 9.2 Reproducible research and visualization enabled by programming.

underlying data. This is akin to following a recipe. Anyone who has the same ingredients and kitchen appliances can reproduce the dish if the detailed recipe is shared. This enables reproducible research and permits reviewers and potential collaborators to explore the data in the exact same manner as the original author (Figure 9.2) [31]. The use of programming also invites the opportunity to invent new ways of visualizing data for specific applications. This case study explores the development of a visualization method developed specifically for analyzing substrate reactivity trends in chemistry.

Through their efforts toward the development of biocatalytic transformations, the Narayan group made the observation that chemical reaction data has overwhelmingly been presented in table format in publications. The table format is ideal for communicating a small number of results for data where there is minimal substrate change. However, advances in technology have enabled the collection of large volumes of data and accelerated the preparation of increasingly large tables of substrates. When numerous results span across diverse substrate motifs the table format becomes onerous as it requires the reader to refer back and forth between schemes and numerical data presented in the table. This extra effort then makes it difficult to draw conclusions and interpret the presented data in a meaningful way. In 2020, the team sought to develop a new method for conveying structure–data relationships [32].

The Narayan group turned to the python programming language to help address this problem. Python is a popular programming language due to its ease of use and flexibility. Python is developed under an open-source license making it free to use even for commercial use. For data science applications, python is typically accessed using the open-source distribution, Anaconda (Anaconda.com). For software development, the use of a user interface (UI) such as PyCharm is popular option (https://www.jetbrains.com/pycharm/features/). In this case the advantage that python offered the team was that it has numerous visualization libraries (also called packages) such as matplotlib [33] and plotly [34], and robust data wrangling libraries such as pandas [35], would assist with data processing and handling tabular data similar to what may be accomplished in excel but more efficiently through code. Python also features a robust cheminformatics package

RDKit [36] that can calculate descriptors of molecules such as the number of rings, heteroatoms, and hydrogen bond donors/acceptors while also interpreting chemical structures through the passing of string-based identifiers including SMILES and Inchi. Thus, Python contains these numerous code libraries that would be essential for solving this particular problem. In order to utilize these libraries, they simply need to be installed from an online repository and loaded during a python session prior to use. Therefore, all the required elements to build a novel approach for communicating structure–data relationships were present. The next step was to identify an innovative and efficient way to present this data.

The focus of the team's effort was to harness the interpretability of a graph while simultaneously leveraging the organized nature of a classic table. To do this the team came up with a Radial-Scope plot [37]. The Radial-Scope plot fixes the core structure of a molecular scaffold while adding labeled points of variation that contain the target value represented in pie chart format (Figure 9.3). This allows the reader to assess the impact of dozens of structural modifications without having to glance back and forth between a table and graphic, thereby enabling insightful analysis.

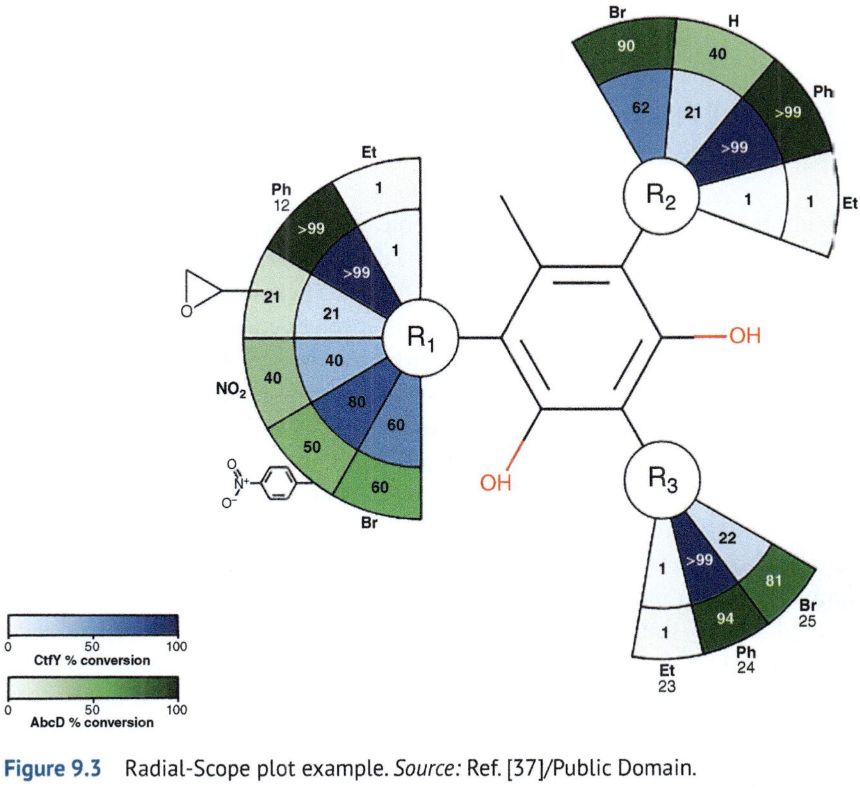

Figure 9.3 Radial-Scope plot example. *Source:* Ref. [37]/Public Domain.

The team created a workflow to enable the use of their developed software through a jupyter notebook. A jupyter notebook is a popular method for sharing research as it blends code blocks and their outputs with free text to allow commentary by the author [38]. A researcher can generate a report style document that has chunks of code that perform specific operations and displays outputs such as tables and graphs. These markdown style notebooks can then be executed by anyone that has the same programming language and supporting libraries installed on their computer. For this example, the user can download the jupyter notebook from the public repository and open the notebook locally on their computer. Once the notebook is open, the user can follow the instructions in the notebook and enter information in the appropriate code chunks. In this case, the user would first enter each SMILES string and corresponding value, such as yield, in the appropriate cell. Next, the user would define the attachment atoms of interest. The code is hosted in a public repository, which enables the broader community to contribute or build on the existing work [39].

In this example, the team was able to expand upon the already vast frameworks available in python for data visualization to create a custom- and domain-specific solution. While the implementation of this solution required experience in python, the end product is accessible to those with beginner level experience.

9.3 Text Extraction: Automating Density Functional Theory Calculations

One of the most mundane tasks for any job is manual data entry. In some cases, the data entry cannot be avoided because the data to be transcribed is handwritten or only available in some other non-digital format. These data entry tasks become particularly annoying when the transcription involves going from one electronic medium to another. Examples of such tasks are transcribing analytical results from an instrument output file to a spreadsheet. In this case the data are usually contained within an exported .pdf or .txt file that needs to be opened, the data copied, and then the data pasted or manually typed into a spreadsheet to complete the desired downstream analysis. These manual tasks of reading files, copying information, transferring the information, and writing a new file in an appropriate format can be accomplished through programming. The advantage of programming is that once the code is written for a particular workflow it can be repeated as many times as necessary without needing to write any additional code (Figure 9.4). The amount of data that chemists generate is enormous. As such, the amount of time allocated to performing mundane data transcription tasks could potentially be alleviated through programmatic means. An initial investment in learning to code and writing a program to execute such tasks may often result in a

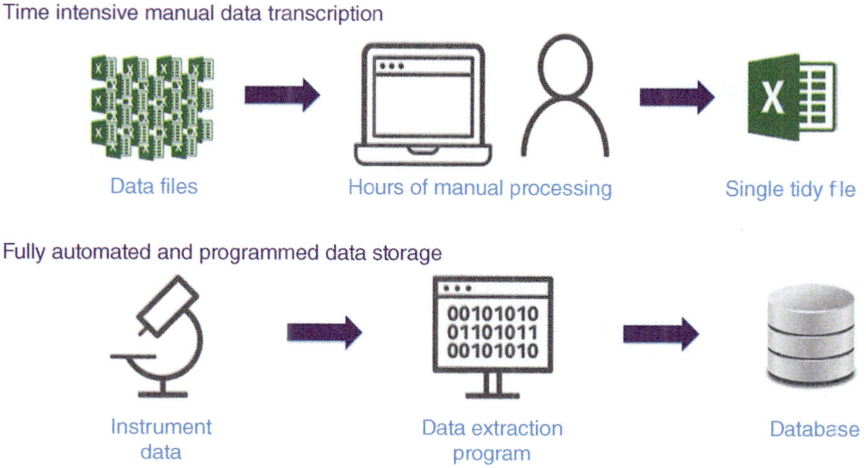

Figure 9.4 Options for data transcription. *Source:* OpenClipart-Vectors/Pixabay.

net gain in overall efficiency, just for the single transcription task alone. When considering the numerous other benefits and efficiencies that programming offers the time investment in learning to code becomes increasingly attractive.

In 2018, the Doyle lab published a landmark machine learning research on the development of predictive models for Pd-catalyzed C–N coupling [40]. The development of the machine learning models requires a great deal of programming expertise and domain knowledge; however, a prerequisite for developing a machine learning model to predict a given reaction yield is that the reaction information needs to be transformed into a format the computer can understand. Thus, the question is how one communicates chemical structure information and properties to a computer.

To overcome this hurdle, the Doyle group sought to utilize an array of properties associated with each reactant or reagent that was obtained through density functional theory (DFT) calculations. For this model, some of the key features from the DFT calculations that the researchers utilized were computed atom NMR shifts, lowest unoccupied molecular orbital (LUMO) energy, atom electrostatic charges, and electronegativity. While the computational software Spartan was able to generate the key features of interest, there was no existing way to export the numerical features in a neat and tabular format that would be required for application with machine learning. This presented a critical problem as the work required DFT calculations for dozens of molecules that each contained potentially hundreds of descriptors. These descriptors would have to be manually transcribed to a spreadsheet one-by-one using existing workflows. Therefore, the group sought to develop code for this specific task to streamline their text extraction process.

The researchers turned to python to develop a program that they named "rxnpredict" to extract the required feature data from the text files generated by Spartan [41]. First, the molecules had to be drawn in Spartan and saved in a specific folder as files. As the team sought to extract features for specific atoms, it was necessary to label the atoms with an asterisk in the Spartan program to alert the code as to which values to extract from the appropriate file (Figure 9.5). Once this initiation step was completed the software could be executed to submit the calculations and extract the properties of interest. An enabling feature of this software is that the data are stored in a tidy tabular format and relieve the scientists of the menial data transcription task that would likely be time intensive and prone to human error.

Based on their initial success for building software for the submission of DFT jobs and subsequent feature extraction, the Doyle group sought a second-generation workflow to allow a public facing implementation that stores the output data in a database that is publicly accessible [42–44]. This solution utilizes programing to develop an open-source repository of computational data that serves to reduce the submission of redundant computational chemistry calculations across institutions.

To implement this workflow, the group first implemented a resource scheduler to manage the submission of jobs such that they can be started and stopped without loss of calculations. In an improvement over the previous workflow, the program takes only a list of smiles strings, text-based representations of molecules, that are entered in the python jupyter notebook. Once the smiles strings are entered, the molecules and conformations are generated. Next, the Gaussian jobs to perform the DFT calculations are created and can be submitted to the server. Once complete, the jobs may be retrieved, resubmitted if necessary, and uploaded to the database.

This work was particularly inspiring as it began as a task-oriented application of programming to solve a specific problem: submitting DFT jobs and extracting the output in a desirable format. In this example, the team utilized python to implement a complex workflow that would be challenging for others to duplicate without extensive experience with python. However, once the initial task was successful the team recognized the value of the work and the impact a streamlined workflow could have for the broader chemistry community. The expansion of the team to include software engineers to manage application development and optimizing the data storage has yielded an incredible open-source asset - a database of DFT

Figure 9.5 Programmatic workflow for extracting values from Spartan job files.

calculations - and also provided a template for how to manage programming projects and collaborations in organic chemistry.

9.4 Statistical Analysis: Deriving Insight from Historical Data

The ability to perform statistical analysis is becoming increasingly necessary for organic chemists due to the increasing prevalence of HTE equipment (see Chapter 6). Robots that can execute a high volume of reactions in a short period of time to furnish troves of data. When the size of the dataset becomes sufficiently large, it may become difficult to assess the results on a case-to-case or reaction-to-reaction basis. Thus, being able to perform statistical tests to identify top performing conditions, and top historically performing conditions, with confidence, become a valuable skill set.

In the development of pharmaceutical chemical processes, the preferred method for isolation and purification of intermediates and active pharmaceutical ingredients is crystallization. The primary source of information to inform crystallization design is the solubility measurements of the target compound in a selection of potential organic solvents and water. Therefore, solvents are selected such that the solubility of the desired compound is low and the solubility of the undesired compounds, such as residual starting materials, reagent by-products, and side products are high. When circumstances arise such that the target compound has high solubility in suitable solvents, the formation of salts with appropriate counter-ions are investigated as salts typically exhibit lower solubility in organic media. At Bristol Myers Squibb, thousands of salt screens are performed annually in multi-well format (see Chapter 6). The standard array generally featured 24 counter-ions across 8 or 12 solvents that were set up and analyzed in a consistent fashion. Each well is analyzed by microscope for solid formation and the yield and purity of the recovered solids is determined. Early on, the team questioned whether meaningful insight could be derived from this rich set of data consisting of 65 acid salt screens and 45 base salt screens that had been collected since 2014 (Figure 9.6).

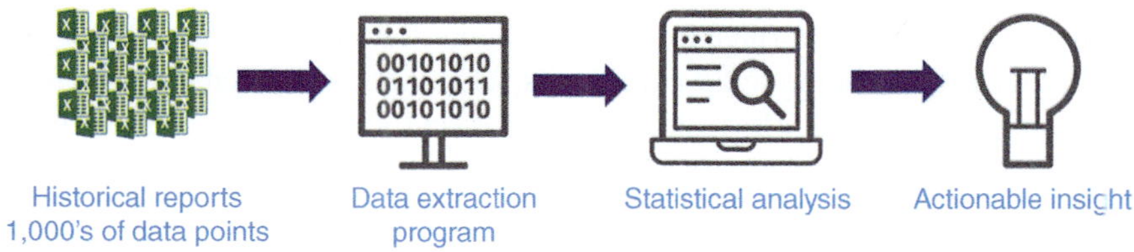

Figure 9.6 Deriving insight from historical data.

The first idea the team considered for analyzing the historical data was to determine hit rates for each of the solvents and the acids for the acid salt screening data. It was thought that the percentage of crystallization hits relative to the number of times the solvent or acid was used would be meaningful. However, this approach does not fully capture the value of a particular combination where the solvent or acid induces crystallization for a substrate where there are only very few hits. Thus, if only a single acid was found to induce salt formation, it would receive the same tally as an acid for a separate screen that induced salt formation along with every other acid in that separate study. An alternative approach would be to standardize the result using the obtained yield as the metric. The standard score would be calculated by subtracting the mean value of each study from the target value then dividing by the standard deviation of the results from that study. The output is the number of standard deviations a given result is from the mean. Using this analysis, the scores are generally in the −2 to 2 range where a score of 0 is the average for the study. Thus, in the previous example the lone acid that provided a hit would receive a very high score while in the scenario where every acid provided a hit, all of the scores would be 0. The team decided the standard score methodology would be more meaningful and applied this basic statistical measure for analysis of their dataset.

To conduct the analysis, the team turned to the R programming language. The R language is widely utilized for its expansive capabilities for performing statistical analysis, although it is a widely versatile language capable that is commonly used for data manipulations, data visualization, and machine learning. The dataset was generated by utilizing R to aggregate hundreds of historical excel reports containing the data for the study. The first insight that the team discovered was an inverse correlation between the average standard score of a solvent across the series and the corresponding dielectric constant that was readily observed through plotting the data in R. Thus, the lower the dielectric constant the better the chances of a high yielding salt. The only outlier in this data was acetonitrile that performs much better than its dielectric constant would suggest. Similarly, an analysis performed in R of the acid standard scores showed that stronger counterions give higher standard scores, which is in line with the pK_a rule. Additionally, it was found that high counterion solubility often leads to low standard scores but low counterion solubility generally does not guarantee a high standard score. A similar analysis was performed for the base counterion series.

The final product of the statistical analysis, which was performed in R, was that the team could confidently assert which acids and which bases outperformed their peers over time (Figure 9.7). The R programming language was utilized to assemble and transform data, conduct exploratory data analysis, and perform statistical analysis to identify trends. This enabled a reconfiguration of the historical salt screen designs to focus more heavily on the proven winners. The net effect was less experiments need to be run to achieve better results. This saves time

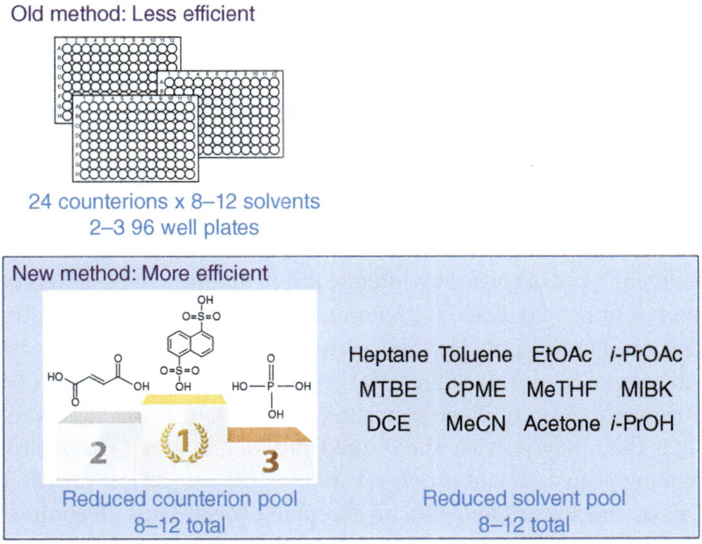

Figure 9.7 Result of statistical analysis of historical crystallization screening data.

through faster screen execution and resources by limiting the size of the experiments and also freeing up valuable time on automation equipment.

9.5 Machine Learning: A Predictive Model for Deoxyfluorination

A subset of programming and data science is machine learning. In a general sense, machine learning is the idea that systems can learn from data. Organic chemistry has long favored linear regression as a means to identify trends and make predictions, where independent features such as Hammett correlation σ-parameters can be correlated with a dependent variable. In this case, if a linear trend is observed, the underlying equation can be used to predict the value of the independent variable if the dependent variable is known. Thus, algorithm has leveraged the data and generated an output function, that when provided an independent parameter such as the σ-parameter, can furnish a dependent value in the form of a prediction such as a chemical yield or reaction rate. Modeling tasks such as linear regression can be executed using common software such as Microsoft Excel and is likely a driving force for many scientists to explore these types of linear regression models. Programming languages, however, offer a multitude of machine learning algorithms, which are a set of rules to be followed by a computer for problem-solving operations. These algorithms are often packaged into convenient libraries, such as caret [45] and tidymodels [46] in the R programming language or SciKitLearn [47] in the python programming language that enable the exploration of numerous algorithms with minimal changes in code. Machine learning for organic chemistry has recently come to the forefront and offers many exciting opportunities. One of the first examples is from the Doyle group, where they sought to develop a predictive model for deoxyfluorination with sulfonyl fluorides [48].

The Doyle group had previously identified the sulfonyl fluoride PyFluor to supplement the existing and often undesirable deoxyfluorination reagents that are commercially available. During their study of deoxyfluorination reagents they had noted that different types of substrates often required unique combinations of deoxyfluorination reagent and base to give optimum yields. Most synthetic methods have been developed to use a single set of conditions that work the best across a large pool of substrates. Under this regime, the generalized conditions are prone to failure when exposed to more diverse substrate pools than those that were reported. Although they may provide the desired product, it may take several rounds of subsequent optimization of the class variables, the categorical variables such as reagent, base, and solvent to obtain an acceptable yield. Thus, in contrast to the development of generalized conditions, the development of a dataset consisting of diverse substrate types and reagent combinations for a given method may enable the development of a model that can predict which reagent combinations are optimal for the target substrate (Figure 9.8). The additional up-front cost of dataset production (i.e. running/analyzing hundreds of reactions) could

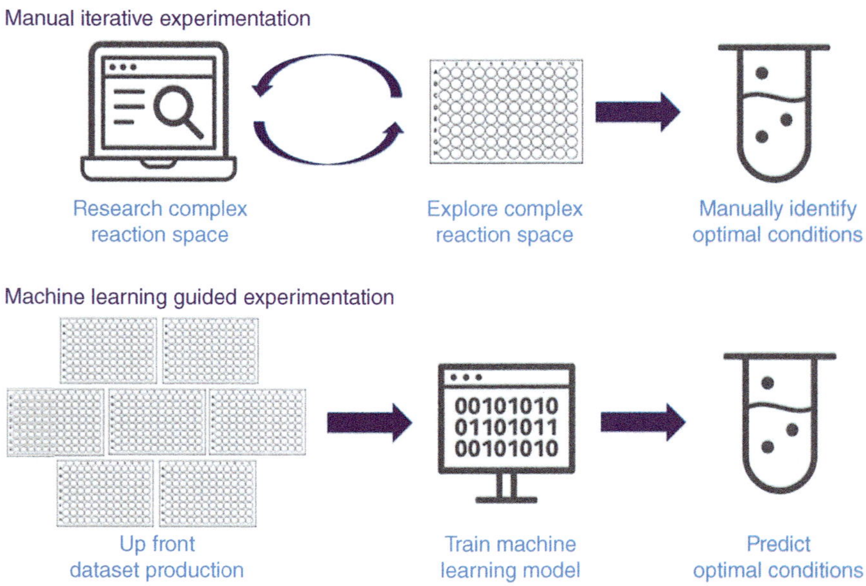

Figure 9.8 Machine learning for reaction prediction.

alleviate downstream optimization of class variables (i.e. base, deoxyfluorination reagent). As it had been noted that distinct pairs of bases and sulfonyl fluorides were required for optimal deoxyfluorination yields, the application of machine learning to the organic chemistry challenge might provide lasting value.

The process for developing a machine learning model first starts with developing a dataset to train the model. Training a model is like fitting a trendline to data points. This step allows the data scientist to expose various machine learning algorithms to the relevant data. The data scientist can adjust several parameters for the algorithms, like one would tune a radio dial, to determine which machine learning algorithms might provide good predictions for a particular dataset. For simple datasets, one might try to train a linear model, or potentially a logarithmic model to fit the data. In more complex cases, decision tree types of models or more advanced mathematical models may perform best. However, the first step in this process is acquiring relevant data.

Toward the first step, the group explored a handful of conditions at the outset to identify suitable operating ranges and typical duration to reaction completion. This step is particularly important as any model trained on data taken before a reaction was complete, or long after a reaction was complete and started to decompose, or at a temperature too low to induce a reaction would be at best uninformative and potentially detrimental to its predictive power. Once the range of reaction conditions were established, an array of five sulfonyl fluorides and four bases were set as the core reaction space (Figure 9.9). These twenty

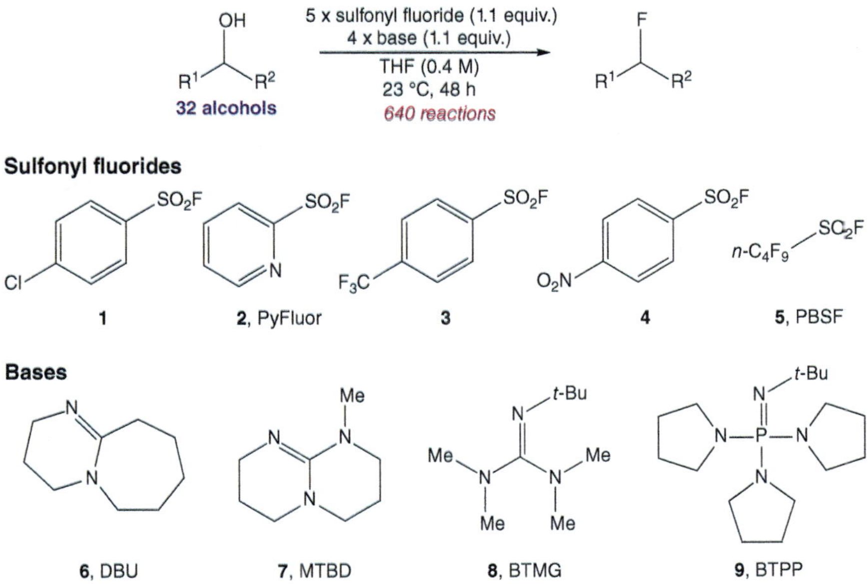

Figure 9.9 Dataset production to support machine learning.

combinations of sulfonyl fluoride and base were examined across a diverse panel of substrates to generate the train/test dataset. This dataset was then used to train a model with the goal of predicting the performance of each of the 20 combinations for new substrates. Toward this end, the team collected data for 32 aliphatic alcohol substrates giving 640 total reactions that would enable the team to pivot their effort to exploring machine learning algorithms to develop a predictive model.

A key decision for the development of machine learning models for reaction prediction is how to represent chemical entities such as PyFluor or DBU in a format that a computer can understand [49]. The Doyle group chose to represent each of these components numerically using atomic and molecular properties obtained through density functional theory (DFT) calculations, as well as numerous binary categorical identifiers [50, 51]. After the categorical variables were represented numerically, the modeling development could commence.

The Doyle group was previously successful using the Random Forest algorithm for predicting C–N coupling reaction performance. The Random Forest algorithm is not specific to chemistry and has been effective for numerous types of problems such as image classification and housing price prediction. Building off their previous success, they turned to the R programming language to build a model using the Random Forest algorithm. The way the random forest algorithm learns is that it builds numerous decision trees containing a subset of features (independent variables) selected at random. This creates a forest of small decision trees that make predictions by committee, an ensemble approach. This model is resistant to overfitting, a scenario where a model may simply memorize the dataset instead of leveraging meaningful relationships in the data. After training their model, they measured performance using root-mean square error (RMSE), the standard deviation of the prediction error, in this case the standard deviation of the error in the yield prediction across all the data points the model was trained on. The RMSE metric is a useful measure to estimate how far off a model might be for a future prediction. Thus, yield is on a scale of 0 to 100%, and the lower the RMSE the better a model would be a predicting chemical yield. An RMSE of 10 would represent that on average, a model would supply predictions that are within 10% yield of the actual result. Of the 640 data points, a subset of 192 data points were withheld to measure how the trained model would perform. This is known as a train-test split. The trained model is evaluated for its predictive performance on the test data that provides the researchers an estimation of how a model might perform on unseen data. The group reported that a 7.4% RMSE was observed, indicating that the test dataset that was withheld from training, provided predictions that on average were off by 7.4% yield from the actual result. Thus, it would be expected that the model provides reasonable predictions for chemical yields when evaluated on new deoxyfluorination substrates.

One risk of machine learning is that the model developed may only be useful for the relatively narrow set of data that it is trained on. To prevent this, the group utilized a diverse substrate pool that was representative of the types of structures that would be of interest in the future. However, the true test for how the model performs in the wild is to use a completely new dataset that the model has never seen. This is known as a validation set. The team collected a new validation dataset using the original 5×4 array across five new substrates (Figure 9.10). They found that the RMSE was 8–23% for the validation set substrates. In the cases where predictions strayed a bit farther from the actual values, it was observed that the top predicted conditions closely followed the top actual conditions, meaning the model correctly identified the appropriate conditions for testing in the lab.

This work was impactful for demonstrating how to construct a machine learning model through data collection, reactant, and reagent featurization, and model optimization. Importantly, the code and data are publicly available and can be implemented by anyone with access to computing hardware and the open-source software. Programming languages such as R, Python, and Julia offer many avenues to explore a wide array of machine learning models for different types of problems. With the explosive growth of machine learning in chemistry, it is likely that it would not be long before algorithms are developed specifically to address problems in chemical research.

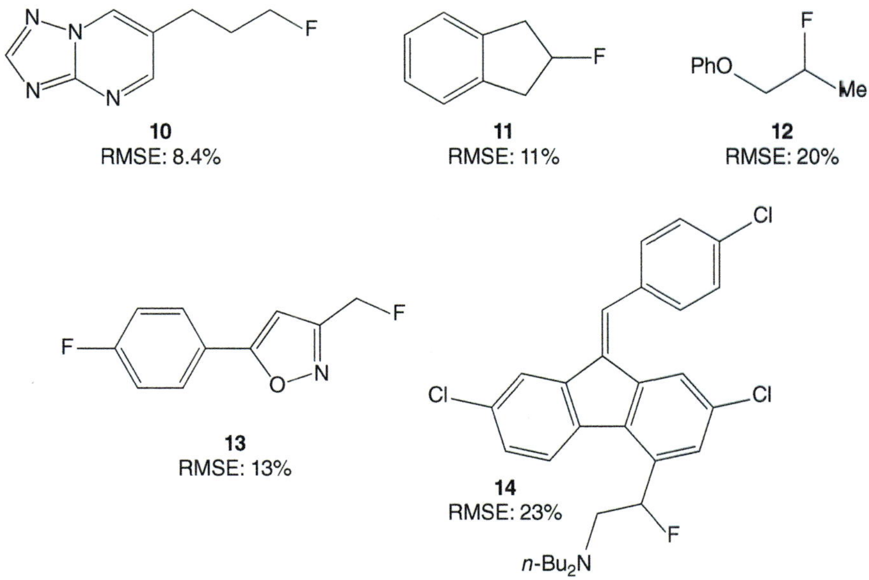

Figure 9.10 Predictive performance of the deoxyfluorination machine learning model.

9.6 Working with Public Datasets: Identifying Reactivity Cliffs

A significant benefit offered by modern programming languages are the code libraries that allow users to pull data from public databases. The ability to pull information from public data repositories can provide scientists with huge troves of information that can be used to develop custom workflows and derive insight into synthetic problems. One of the most common databases in chemistry is PubChem [52]. The PubChem database is hosted by the United States National Institutes of Health and contains a wealth of information for small molecules. PubChem features an application programming interface (API) that allows two applications to communicate with each other. Common everyday activities such as checking the weather through an app, social media use, and instant messaging all utilize an API. In these cases, one requests information through an app or website that forwarded your specific information request to the API that interpreted the request, gathered the information you wanted from the data source, then sent it back to you in a nice and tidy format. Programming languages handle these information requests in the same way. The API for PubChem allows users to request information such as molecular formulas, molecular weights, string identifiers, and molecule images. The significant benefit of using an API through a programming language is when you want lots of data. This circumvents the need to download potentially thousands of discrete files and then have to merge them back into one larger file. Recently, two new databases have been published that enable rapid access to data through an API. The first is the Open-Reaction database [53], an open-access infrastructure for sharing chemical reaction data through a centralized data repository. The second is the Kraken database [54], which contains computed descriptors for thousands of monophosphine ligands. Any user with proficiency in making API calls has the ability to use the millions of datapoints in PubChem, supplement their research with the growing data housed in the open-reaction database, and investigate important reactivity parameters in the Kraken dataset (Figure 9.11). Using this Kraken dataset, the Doyle and Sigman groups sought to identify reactivity trends for Ni- and Pd-catalyzed transformations [55].

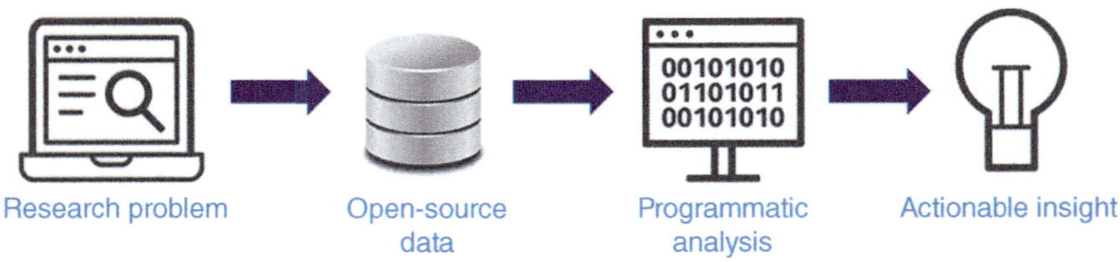

Figure 9.11 Accessing public data to support research. *Source:* OpenClipart-Vectors/Pixabay.

Chemists have relied on numerous tools to describe the various properties of the molecules they work with. In the area of transition metal catalysis, there are parameters that have found utility for the purpose of accounting for observed reactivity trends. Among the most common are the Tolman cone angle [56], sterimol [57], and buried volume parameters [44, 58]. Despite the past success of these parameters, they were unable to correctly classify the outcomes of a new class of monophosphine ligands, the DinoPhos ligands, for Ni-catalyzed cross-coupling of acetals with aryl boroxines (Figure 9.11) [59]. Using the Tolman cone angle as a metric, these ligands would be expected to be similar to bulky bis-aryl ligands that preferentially form monoligated Pd-complexes. In contrast, the %Vbur values for the ligands indicated that they should behave like smaller ligands such as PPh_3, which tend to form bis-ligated species. Based on this divergence observed using known parameters, the team sought to explore the Kraken dataset to identify a parameter that would adequately explain the observed reactivity and that also provided a reasoning for the success of the DinoPhos ligands that could guide future ligand development (Figure 9.12).

To get a clearer understanding of the ligand space, the team expanded on the original work by the Doyle lab by supplementing the dataset with additional ligands to encompass a wider %Vbur range. They also leveraged high-throughput experimentation to collect additional data for Ni-catalyzed Suzuki–Miyaura reactions with aryl chlorides using a large monophosphine library that included only ligands present in the Kraken database. They then evaluated the impact of each of the nearly 200 descriptors provided in the Kraken virtual library against reaction

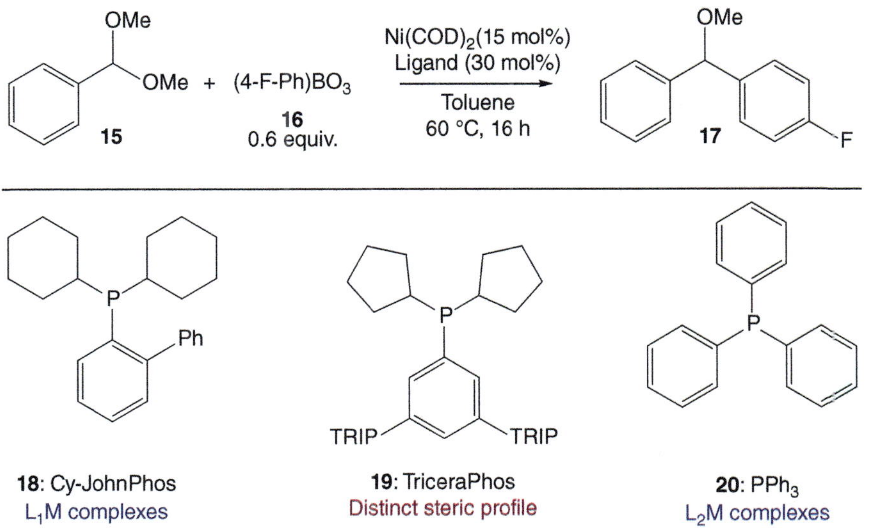

Figure 9.12 Unexpected reactivity for DinoPhos ligands.

performance. Ultimately, the team found that the descriptor, %Vbur (min), the percent buried volume of the conformer with the smallest buried volume, gave a sharp reactivity cliff such that only ligands exhibiting less than 32% Vbur (min) were active. This information was used to form a mechanistic hypothesis that the %Vbur (min) might be representative of the ligation state. The group turned to NMR analysis to confirm this hypothesis and showed that ligands with less than 32% Vbur (min) formed bis-ligated Ni-complexes and were active for catalysis (Figure 9.13). It was observed that not all ligands less than this threshold performed well but the team suggested that the ligands that met the criteria, yet

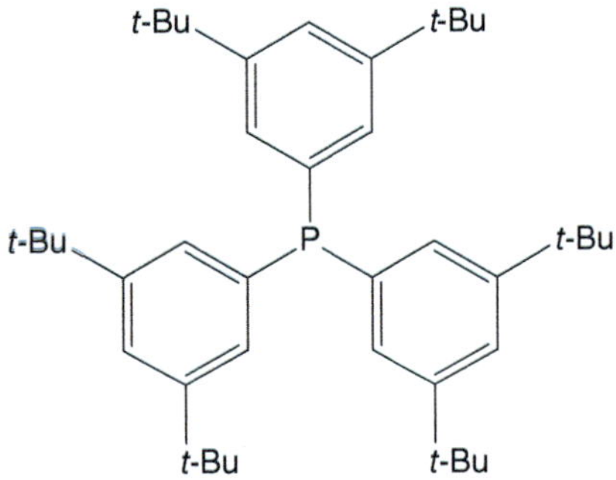

22: DrewPhos
Cone angle = 207°
$\%V_{bur}$ (Boltz) = 27.2%
$\%V_{bur}$ (min) = 27.2%
L_2Ni complex

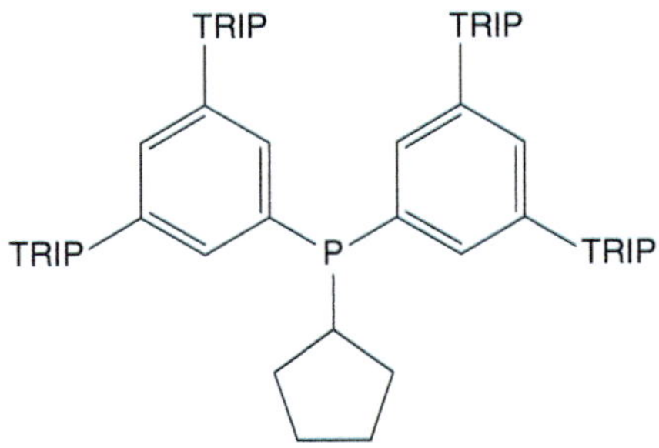

23: PteroPhos
Cone angle = 235°
$\%V_{bur}$ (Boltz) = 29.2%
$\%V_{bur}$(min) = 27.3%
L_2Ni complex

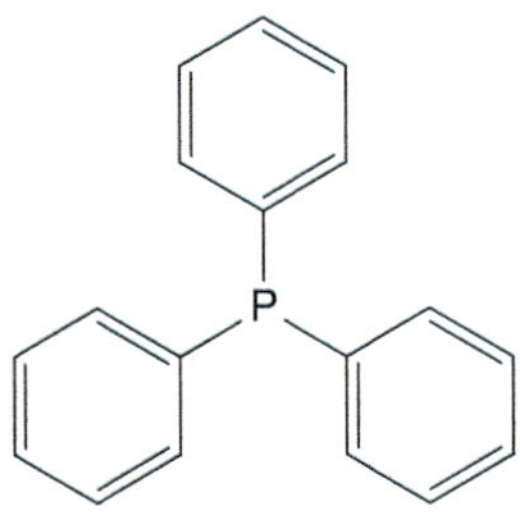

20: PPh_3
Cone angle = 159° %
V_{bur} (Boltz) = 28.2%
$\%V_{bur}$ (min) = 28.2%
L_2Ni complex

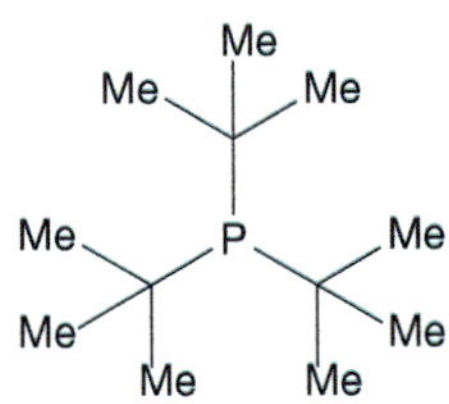

21: P*t*-Bu_3
Cone angle = 176° %
V_{bur} (Boltz) = 36.3%
$\%V_{bur}$ (min) = 36.3%
L_1Ni complex

Figure 9.13 Observed reactivity cliff at 32% Vbur (min).

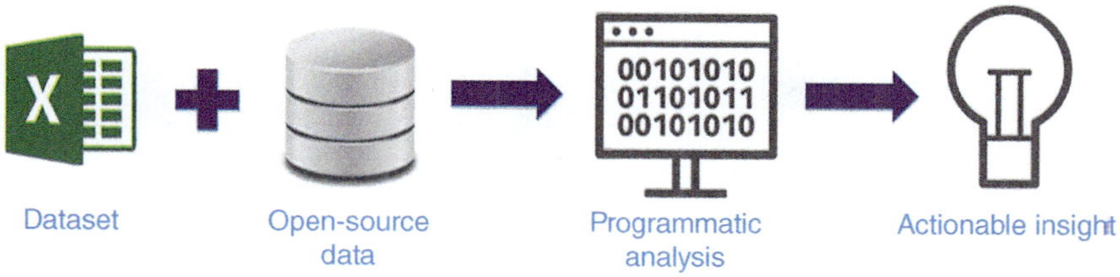

Figure 9.14 Programmatic exploration of data for reactivity cliffs. *Source:* OpenClipart-Vectors/Pixabay.

performed poorly, contained structural features that might poison the metal or be poorly suited to promote catalysis from an electronic perspective. To validate their proposal, the team synthesized a new ligand, pteroPhos, with desirable properties and found that it was catalytically active.

In addition to their incredible work identifying a reactivity threshold using a publicly available dataset, the team used their programming proficiency to formalize and semi-automate the workflow they had developed (Figure 9.14). They provided an algorithm that completes the reactivity cliff analysis across an array of provided descriptors for a given dataset. Thus, new reactivity datasets could be paired with descriptive features, such as ligand properties or similar, and be subjected to this analysis to identify thresholds automatically, which they tested using additional data from the literature. This analysis was similarly able to identify reactivity cliffs for public Pd-catalyzed cross-coupling data.

In summary, the work highlighted here demonstrates the potential impact of open-source data. In this example, a public dataset that is accessible using API calls was coupled with existing data to gain further insight into chemical reactivity, highlighting that the %Vbur min ligand property was critical to determining the ligation state of nickel complexes that explained reactivity trends. The amount and breadth of open-source data for chemistry will continue to grow and the ability to pull this information into an interactive development environment will foster new discoveries and empower data fluent scientists across the globe.

9.7 Running Simulations: Process Greenness

Simulations are an effective way to discern the likelihood of a particular outcome for a series of events that each have a continuous ranges of potential outcomes. The ability to randomly sample an outcome across a range is possible through programming and a simulation can be conducted by repeating the sample process many times. This is like conducting a study of several die rolls. The first die is rolled, the number recorded, the second die rolled, the number recorded, and so

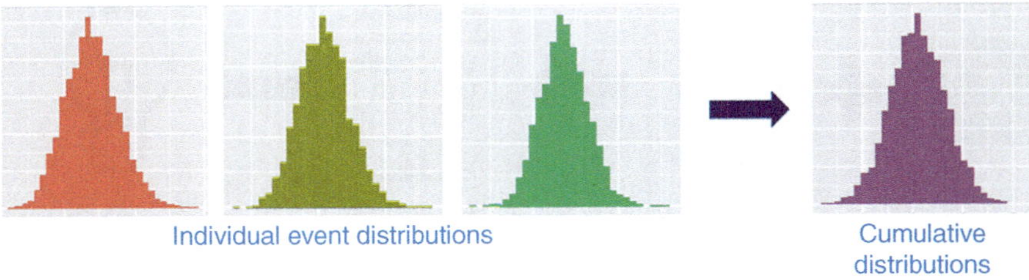

Figure 9.15 Using simulations for determining likely outcomes.

on, until all the recorded numbers are summed at the end. The process can be repeated through programming thousands of times. This simulation then produces ranges for how likely it is to roll a given number at each step or how likely it is that a target number is reached when all the die rolls are summed (Figure 9.15). Thus, the outcome of this type of simulation is itself a distribution of results that can be insightful and put a given problem into context that enables teams to make better decisions. Thus, if a die is rolled three times, how likely is the final sum to be 18? Or 14? In this case study, the simulation will inform teams about whether their chemical reaction process is likely to be efficient.

A team of scientists in the process development group at Bristol Myers Squibb (BMS) sought to develop a methodology that would allow scientists to develop synthetic routes that adhere to "green-by-design" principles [60]. The team recognized that a key problem for designing green synthetic routes is that assessments for greenness are made *after* the reaction or route has been developed and implemented. Namely, during the ideation phase for the reactions targeted for exploration in the lab can be biased toward the individual practitioners' prior experience. Only once proof-of-concept for a given synthetic route has been established does the effort to optimize the chemistry for greenness begin. Thus, as the authors state in their article: "The application of 'green thinking' comes too late in the design phase." Therefore, a method to evaluate prospective syntheses for greenness during the ideation would empower scientists to select chemistry for laboratory evaluation that may produce greener reaction outcomes.

The first step toward achieving their goal was to establish a database to collect information to develop a predictive tool. A critical metric for evaluating process greenness is process mass intensity [61] (PMI, Equation 9.1), a widely used metric to evaluate the efficiency of chemical processes that sums to total mass of all the components of a reaction divided by the total mass of the isolated product (Equation 9.1). The lower the PMI the better. The team set up a custom database to store the critical values of PMI alongside the reaction type (Amidation, Suzuki, etc.), and the program identifier obtained from historical BMS process development campaign's as well as for publicly available data from other institutions that was available in the literature. With a database containing this critical information, the team could then begin to explore how to leverage this knowledge.

$$\text{PMI} = \frac{\text{total mass(kg) used in process}}{\text{total mass(kg) of product}}$$

Equation 9.1. Process mass intensity.

The database that was developed contained historical metrics for chemistry run on large scale at commercial vendors. Thus, individual reaction types could be grouped to find the range of probable outcomes such as PMI and yield for a given reaction type such as amidation and Suzuki coupling. With this knowledge across a large range of reaction types, it then becomes possible to estimate the probable outcomes for a sequence of steps of different reaction types. Thus, comparing outcomes for proposed routes containing different reaction types and sequences of reaction types. The BMS team derived a formula to calculate cumulative PMI for a sequence of steps. This was the critical item that would allow a direct assessment of two different routes and their likely PMIs and would make it possible to assess the likelihood that a proposed process would be green prior to initiating any experiments in the laboratory, much earlier in the design phase than current practice. Thus, two (or more) proposed synthetic routes could be compared side-by-side with a range of likely PMI values prior to running any chemistry.

The workflow developed by the team was to plan the sequence of reaction types necessary for a given synthetic route, retrieve the probable PMI distribution for each reaction in the sequence, sample from each of these distributions, and repeat tens of thousands of times using Monte Carlo simulation to obtain a Cumulative PMI distribution for that route. The methodology was validated against molecules from the BMS portfolio, BMS-911543 [62], Brivanib [63], BMS-741672 [64], and Apixaban [65] (Table 9.1). It was found that two of the four compounds provided actual cumulative PMI within the 95% confidence interval (CI) ranges from the simulation. The other two molecules were outside the 95% CI range and provided an important perspective. Namely, the type of data used to develop models will be good for predicting those type of data. Thus, due to the high barrier to advancing a molecule through the clinic [66], the majority of molecules in the training set were early clinical assets that had not been subjected to the full rigor of process development, whereas the molecules used for validation advanced quite far into the clinic. Nevertheless, with the workflow established the dataset will continue to grow and be more informative over time. This represents an important advance in benchmarking syntheses prior to evaluation in the lab that will enable better decision-making for which synthetic routes to pursue.

Table 9.1 Validation of simulated PMI ranges.

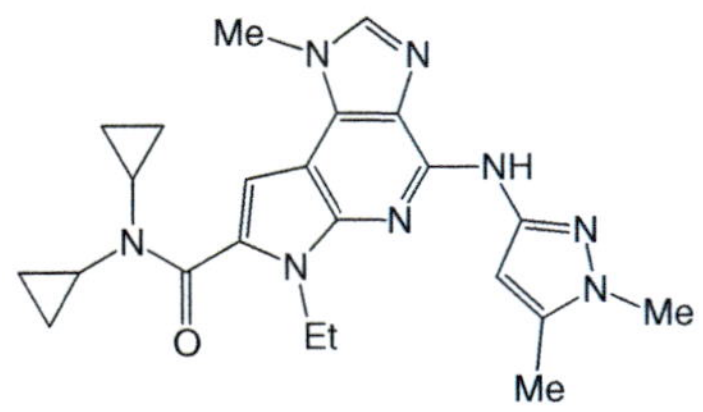

BMS-911543 (**24**)

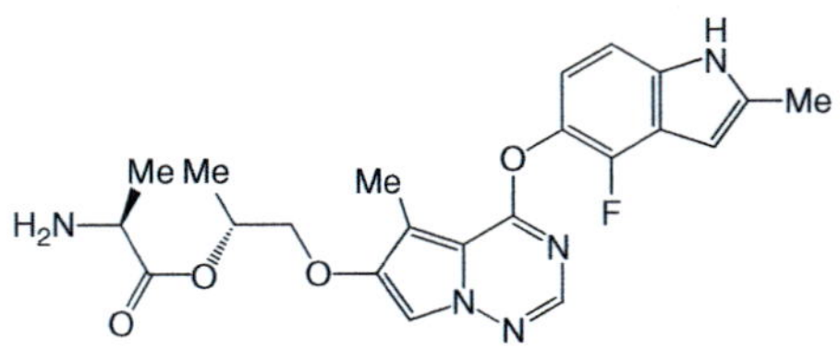

Brivanib (**25**)

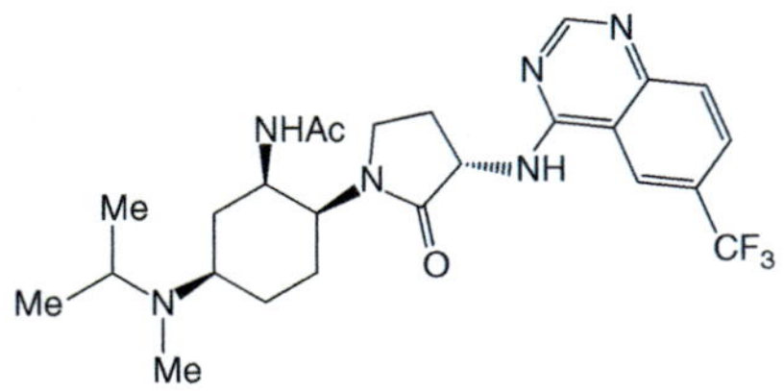

BMS-741672 (**26**)

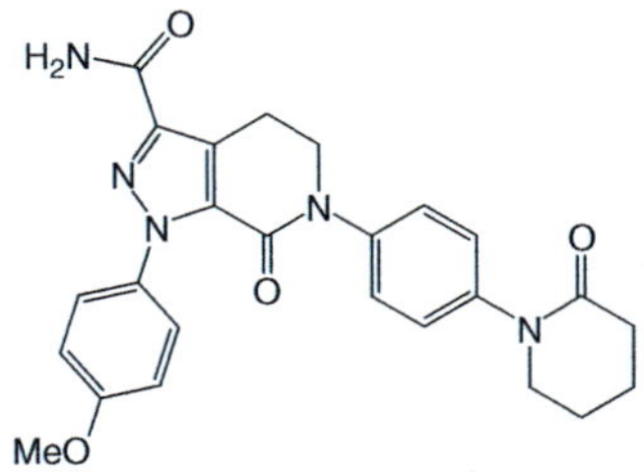

Apixaban (**27**)

Compound	Predicted cumulative PMI range	Actual cumulative PMI
24 (BMS-911543)	1856–4020	3600
25 (Brivanib)	1259–1794	1488
26 (BMS-741672)	1835–2723	1717
27 (Apixaban)	261–480	197

This work also highlights an advantage for the use of programming to advance chemistry and the utilization of digital tools. Namely, once the initial code is written and the underlying data stream is duly supported and hosted in an accessible location, it invites opportunities to improve on previous work without having to rebuild everything from the beginning. This is what the BMS team was able to do as they evolved their predictive algorithm such that it runs more efficiently, handles more complex data, and produces a richer product [67]. The team recognized that the calculations for the original PMI prediction algorithm required very intense calculations that required a workstation with a lot of processing power. The team reevaluated their approach and came up with a much more efficient approach that did not require such intense processing power. Thus, code can be shared and improved upon by a community of people with programming abilities.

In summary, the use of simulations enabled the development of a tool that could inform teams about the likely efficiency of a prospective synthesis based on the knowledge they already had from previous work. Because the workflow was written as program, the team could evaluate many different synthetic routes for a prospective molecule, identifying where the problem steps are, and iteratively proposing new routes with the end goal of a maximally efficient synthesis. The use of programming has given chemists a tool to evaluate synthetic routes for green-by-design principles before initiating any work in the lab, which was a major hurdle in implementing green-by-design principles for chemical process development.

9.8 Application Development: Process Mass Intensity Predictor

Application development is a skill that is associated with computer science and is not likely something an organic chemist might consider learning on the side. However, two of the more popular programming languages, R and Python, have code libraries that makes building applications within these languages quite straightforward. The R language features the Shiny library [68] that enables the creation of dashboards for sharing data. Similar packages available in python are Flask [69], Dash [70], and Streamlit [71]. The advantage of the applications and dashboards that these packages enable is that they allow stakeholders such as team members, project managers, and principal investigators to interact with the data through reactive features such as dropdown menus, sliders, checkboxes, and interactive plots. The ability to program applications such as these has a cascading effect in that they provide greater opportunities for others to interact with and analyze a given set of data than might be expected from simply sharing a spreadsheet or presentation slides. The present case study developed an application to share to the public for the purpose of developing greener chemistry.

The PMI algorithm that was developed by the BMS team was a significant advance in enabling the assessment of Process Greenness and Green-by-Design principles at the synthetic route conception stage. While the results were impressive, a significant drawback is that the code needed to be executed by subject-matter experts proficient in programming. The repackaging of the underlying algorithm within the framework of a web application would enable any user with appropriate domain knowledge to utilize the application. Therefore, the BMS team initiated a collaboration with the American Chemical Society Green Chemistry Institute Pharmaceutical Roundtable to develop such an application [72].

The expanded team set out to develop an interface that would be available free of charge to the public and able to accept the input of a linear or convergent proposed synthetic routes. To be successful, the application needed to be hosted online to alleviate the technical challenges associated with downloading and executing local software. The team chose to develop the application using the Shiny web application framework, part of the R programming language [53, 73, 74].

The PMI application first presents the user with an introduction that defines its purpose, intended use, and finally provides detailed instructions for use (Figure 9.16). In the first tab, the user gets to define their process by entering all the individual steps for their prospective synthesis through a simple set of text inputs and drop-down menus. All that is required of the user is that they have a prepared scheme of their transformation that can be digitally transcribed using the provided tool.

One of the advantages of an application-based approach is that interactive components can be developed. For the PMI prediction tool, the application requests that the user input a molecular weight for each component and select the type of

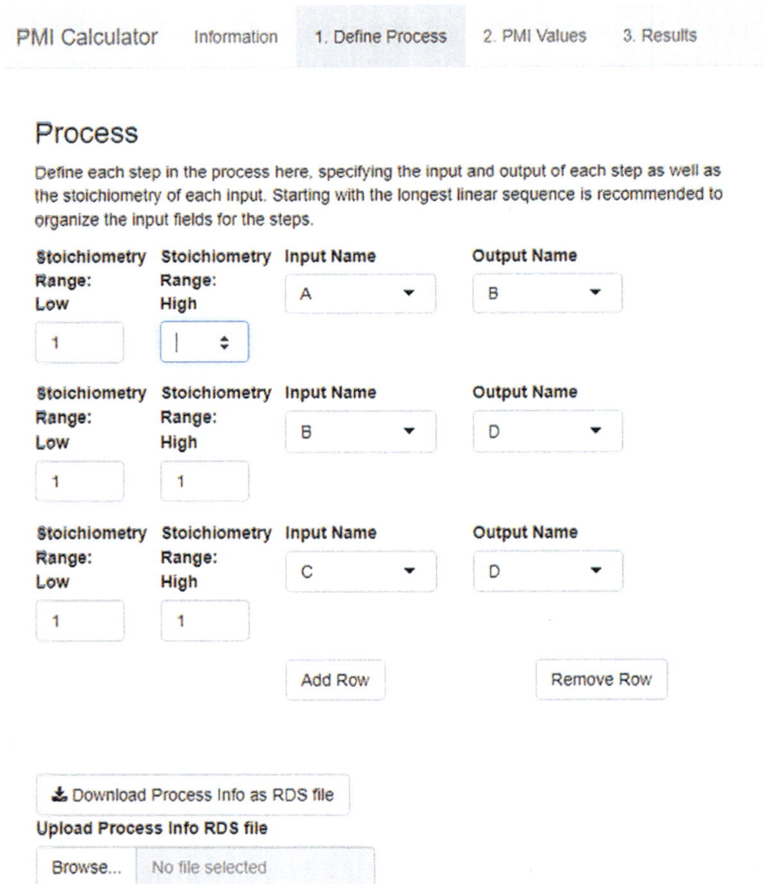

Figure 9.16 Process definition for PMI predictor app.

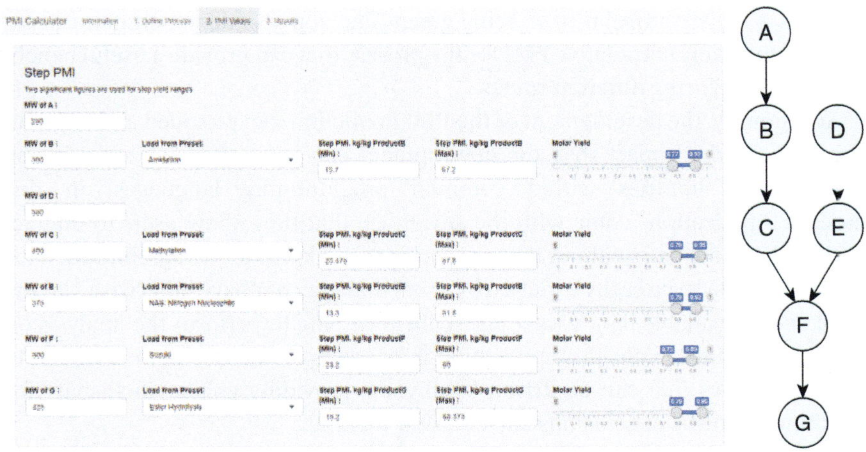

Figure 9.17 Interactive range specification for the PMI predictor app.

reaction. The type of reaction that is selected then pulls the historical yield and PMI ranges from the database to set the bounds of a slider widget (Figure 9.17). However, if the user has experience with the transformation, they can adjust the slider to more accurately represent the knowledge that they have. Similarly, if the user has a specific target for a cumulative PMI, they can use the sliders to benchmark how efficient their processes must be at each step.

The final pane of the app provides visualizations to the user that correspond to the process they have entered. This allows the user to inspect which specific transformations contribute most to the overall PMI for their process using the Step Metrics tab (Figure 9.18). This analysis can provide insight for designing a route

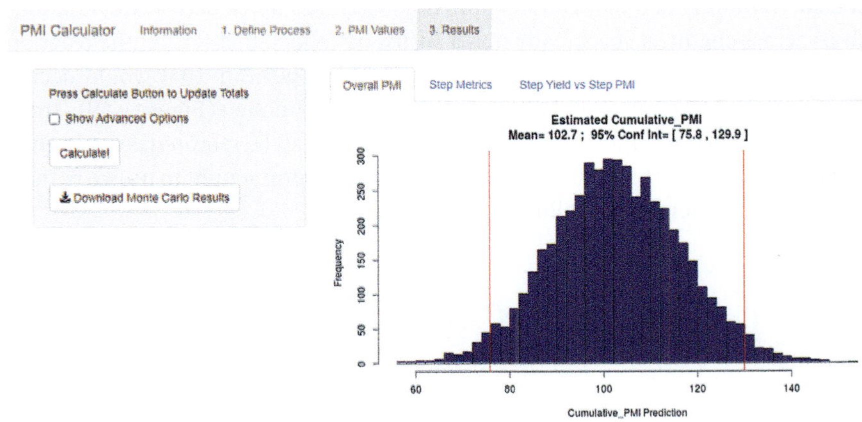

Figure 9.18 Results from PMI predictor app.

that has a higher probability of being green. The app also demonstrates a histogram of the likely cumulative PMI for the process that can provide a useful benchmark for comparing different routes.

In summary, the development of the PMI prediction tool provided an insightful use case for the impact that app development can generate using specific app development libraries within common programming languages, in this case, R. Applications come with the advantage that they allow users to interact with and explore data without themselves having to know any programming. This allows the data to be easily explored by users that may not have otherwise had the time or the available data wrangling skills to be able to perform the analysis on their own. Additionally, this application is an example of a useful mechanism to host workflows that can be driven by anyone, providing greater inclusivity for reaching scientific conclusions and decision-making.

9.9 Machine Learning for Reaction Optimization

This case study features a collaboration from organizations that were early adopters of the use of programming in organic chemistry, including BMS, the research group of Professor Abigail Doyle, and the group of Professor Ryan Adams at the Princeton Computer Science department. The goal of the project was to develop a tool that would select small subsets of reaction conditions from a larger pre-designed list for the purpose of optimizing the reaction as fast as possible. This would enable chemists to augment typical practices, such as setting up small batches of experiments, with artificial intelligence. The programming involved to execute a project of this magnitude was quite extensive as any optimization platform would need to handle tables of data representing the list of reaction conditions to consider, training of a model that approximates the complex details and relationships across chemical space known as a surrogate model, to interpret the reaction space as experimental data are populated, and an algorithm that would decide which reactions to run next known as an acquisition function (Figure 9.19). Thus, data wrangling tasks, modeling, and sampling would all be required components. Given these requirements, it was not surprising the group sought to utilize python as there are numerous capabilities in these areas.

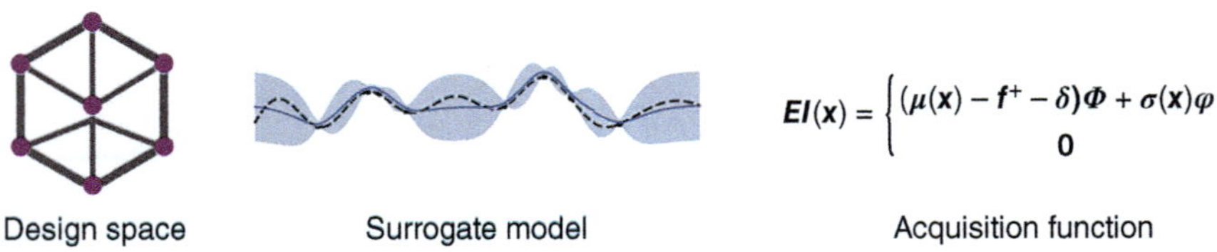

Figure 9.19 Required components for a Bayesian Optimizer.

The initial phase of development focused on utilizing DFT calculated features as numerical representations of chemical reaction design space, and thus, for the surrogate model, a Gaussian process model [75] was investigated. To understand the basics of a Bayesian optimizer, it is easiest to utilize an analogy such as traversing a foggy mountain range with the goal of scaling the highest peak. The design space can be viewed as a mountain range, where the variables for a chemical reaction such as base, solvent, and temperature can be represented as the directions (North, South, East, and West) and the yield is the height of the peak. Because the mountain range is foggy, you know a little bit about the positions nearest to you and where you have previously traveled, but not much else. The actual terrain represents the results of all the possible experiments in the design space, or the true underlying function. In order to know the true underlying function, a scientist would have to run all of the reactions, which would represent the ground-truth, and would be immensely inefficient, much like traversing the entire landscape of the mountain range. The purpose of the surrogate model is to use the available data at hand to emulate the true underlying function that represents the target across the design space. Thus, the surrogate model is used to approximate the mountain range, with all its peaks and valleys, using only limited data points where the traveler has previously investigated. This approach is much more efficient and reasonable for a laboratory scientist running only a few reactions at a time. So, once a few data points are known, the surrogate model is updated. At this stage, the acquisition function comes into play. The job of the acquisition function is to assess the entirety of the approximated mountain range via the surrogate model and select new data points where it thinks the highest peaks might be located, but also data points where the approximated mountain range is the most uncertain and might best inform the surrogate model. Thus, it is exploiting parts of the approximated mountain range where it has the most confidence of a tall peak but also exploring parts of the mountain range where it has the most uncertainty. The process of updating prior beliefs (data you already have) given new information to generate updated (or posterior) beliefs is the essence of Bayesian statistics. Thus, the entire process of iteratively collecting data, updating a surrogate model, and resampling, is known as Bayesian optimization. The group sought to use the two largest public datasets for organic chemistry, namely the Pfizer Suzuki dataset [76] and the Merck C–N coupling dataset [40], to perform this train the optimizer on how to best optimize chemical reactions. After several rounds of optimizer tuning, the team turned to the next task of validating its performance experimentally.

The process of validating an optimizer requires that the optimizer be exposed to an optimization problem that it has not seen before, yet all the answers are known. Under this regime, the performance can be quantified using metrics such as the number or reactions to reach a given threshold or similar. To do this, it would require the generation of a new dataset for a specific reaction. In this case, the

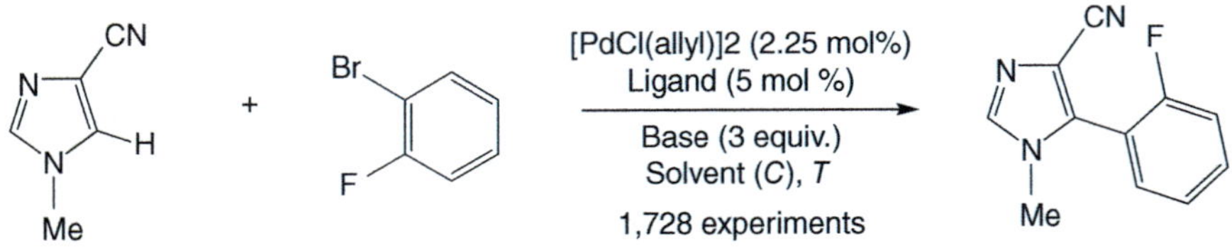

28 **29** **30**

Scheme 9.1 Reaction employed for a ground-truth dataset to test the Bayesian Optimizer.

reaction in question was the palladium-catalyzed direct arylation of 1-methyl-1H-imidazole-4-carbonitrile **28** with 1-bromo-2-fluorobenzene **29**. This ground-truth dataset, meaning a dataset where all the answers (i.e. yield) were known, was composed of 1728 total reactions where there were 12 phosphine ligands, three bases, four solvents, three discrete temperatures, and three different reaction concentrations. All the reaction yields were collected via HTE coupled with HPLC solution yield calculation. Thus, all reaction outcomes were known, and the goal of the optimizer would be to suggest reactions, receive the known results, and iterate this process until the maximum yield was achieved. However, a more interesting question was whether or not the optimizer would perform better than subject-matter experts. Not surprisingly the group turned to programming to answer this specific question (Scheme 9.1).

Since the dataset had already been collected, it would not have been the best use of time and resources to have dozens of synthetic chemists all independently optimizing the same reaction. Therefore, the group sought to develop a digital means for chemists to optimize the reaction. In this case, the group developed a web-based gaming app via the shiny package (R programming language hosted on a private server) where users could virtually optimize the reaction. They designed the app, so users could select five experiments to run at a given time, selecting the variables from drop-down menus, and finally clicking the "execute" button to initiate the experiment (Figure 9.20). At this point, the users would be provided "instantaneous" experimental results and would then be asked to continue the game with another batch of five experiments. This iterative process would continue until the subject-matter expert felt confident that they had reached the optimal result. In total, the game was played by 50 expert chemists and engineers from academia and industry. Concurrently, the Bayesian optimizer was programmed to also play the game 50 times and the results were compared.

The results of the game were quite eye opening! In the case of the human experts, they made better initial choices (i.e. they had a better initial "feeling") for reaction conditions based on their overall advantage of having domain knowledge. In contrast, the optimizer made random initial guesses. However, despite

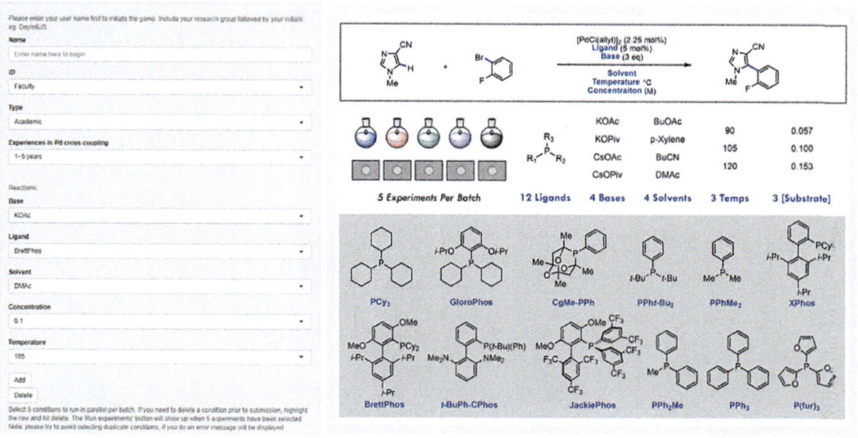

Figure 9.20 Application developed to play the optimization game against the Bayesian Optimizer.

their initial advantage the human experts tended to plateau after only a few rounds of optimization, whereas the Bayesian optimizer kept progressively climbing. To get an estimate of how the optimizer would perform on average, the optimization game was played by the optimizer 50 times. The results of each optimization step were then averaged. The reason this was important is that there is randomness involved with the Bayesian optimization process and thus each time the optimizer is run it may provide a different outcome at each step. The result was that the optimizer, when averaged over the 50 independent runs, exceeded the performance of the top performing human experts.

Following the optimization of the Pd-catalyzed coupling cases, the optimizer was deployed on two new reaction class never before seen by the optimizer (Scheme 9.2). This underscores the point that the optimizer has been tuned to optimize chemical reactions, yet it does not actually know, or need to know, any underlying chemistry principles. It will learn what it needs to know about a given reaction on the fly. In the first example, the team tested a Mitsunobu reaction that gave a 60% yield using standard literature conditions as a performance benchmark. The optimizer started with a random selection of reactions from a broad design space of reactions that varied the Mitsunobu reagent, phosphine, solvent, equivalents of both the Mitsunobu reagent and phosphine, concentration, and temperature. In this first round, the best result achieved a 76% yield. Now, based on its prior knowledge from the first round, the optimizer uses this knowledge to update its model and leverages the acquisition function to select the next set of experiments. After a total of 4 rounds and 40 total experiments, a yield of 99% was obtained despite the optimizer never having been exposed to a Mitsunobu-type

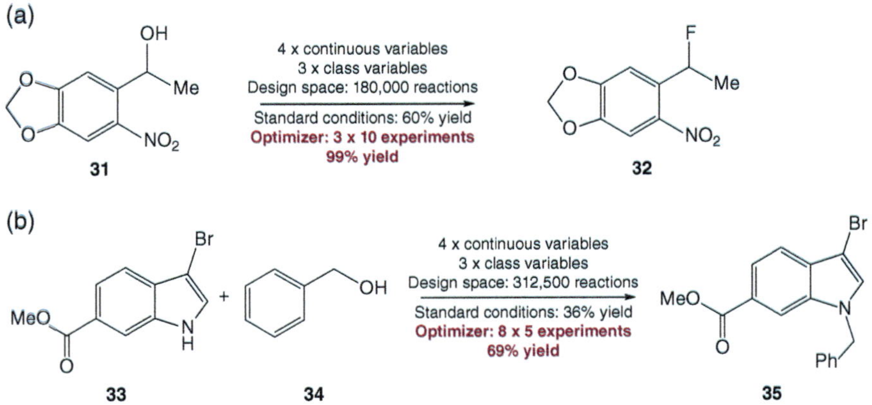

Scheme 9.2 Examples of Bayesian optimization applied to two chemical reactions.

reaction. In this case, the optimizer sampled only 0.02% of the design space to achieve an optimum result. In similar fashion, a deoxyfluorination reaction was also explored (Scheme 9.2). After eight rounds of five experiments, a 69% yield was obtained, which was nearly double that of standard conditions.

This case study has presented an example of what is possible when multiple organizations with significant investment in data science, expertise in programming, and dedicated computational resources can accomplish. In this case, partnership with a dedicated computer science group facilitated the achievement of the end goal. Importantly, the lead investigators from this work were not computer scientists, they were trained synthetic organic chemists with an interest in programming. The end product of the collaboration was an open-source python package that can be installed and used by anyone. A slight drawback is the requirement of basic python knowledge to execute code through a jupyter notebook. In this case, the optimizer has been tuned by the Princeton-BMS team specifically to optimize chemical reactions. Thus, it has already been trained and can be used "off-the-shelf" by scientists comfortable using python and jupyter notebooks. An important and consistent theme for the future is that significant amounts of open-source technology will be developed/released to the general public and those with the ability to harness and deploy these new tools will reap significant benefits.

9.10 Executing Robotic Tasks

Much of the discussion in chemistry regarding the use of programming is centered around data science tasks such as machine learning or software type solutions – application development. On any given day, a synthetic chemist working in a lab interacts with numerous pieces of hardware to do their work (i.e. rotary evaporators, heating and stirring plates, temperature probes, vacuum pumps, syringe pumps), and the application of programming also enables the chemists to remotely control these pieces of hardware, enabling significant productivity gains. An impressive demonstration of this concept was achieved by the Cronin lab for their development of a chemistry computer, or Chemputer [77].

The increasing digitization of laboratory equipment from balances to stir plates and pumps has increased the potential value of developing programs to control them. Additionally, despite the growing inventory of digital lab equipment, there is not a uniform program to supply them with instructions or allow them to communicate. The development of such an architecture could enable the development of new era of chemistry where a standard set of instructions for a chemical reaction could be translated into specific sets of operations for the equipment necessary for its execution. The team recognized that the main unit operations that are commonly executed for organic chemistry are reaction, workup, isolation, and purification. In each case, a specific collection of equipment is typically associated with each, and thus developing a standardized platform to interact with the equipment and schedule their operations could enable and automated execution of a chemical reaction from end-to-end.

To enable the scheduled execution of distinct laboratory, the team developed a "Chempiler" program that produces instructions in the form of commands to control a modular platform, consisting of laboratory equipment that would be required to execute a synthesis such as a reaction flask, heating/cooling, liquid/liquid separation, and evaporation, filtration and chromatography. The Chempiler thus turns a written scheme into code for each of the modules. Once the module receives the instructions from the Chempiler, it can format the instructions for the specific instruments that will be handling the task. This construct represents a means to convert digital code into specific operations for chemical synthesis. The Chempiler provides instructions to the modules that in turn execute the physical tasks, but there needed to be a straightforward mechanism to distribute tasks. Toward this end, the group developed a scripting language called "Chemical Assembly" or ChASM, which provide instructions for operations and supports the definition of functions and variables. The end result is a synthetic sequence translated into a series of operations for execution by the automated apparatus (Figure 9.21).

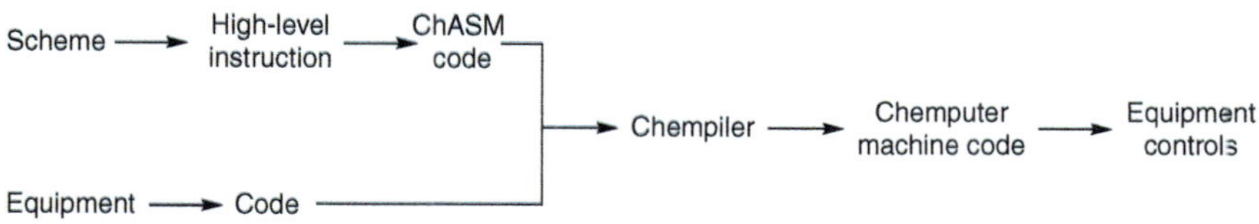

Figure 9.21 Chemputer information flow.

The Chemputer was put to the test for the synthesis of three separate molecules, diphenylhydramine hydrochloride, rufinamide, and sildenafil. These molecules were chosen as they contain published syntheses with a selection of distinct unit operations, such as reagent addition, liquid/liquid separation, and filtrations, and could be manually prepared in the lab to perform benchmarking. The synthesis of diphenylhydramine hydrochloride was a four-step synthesis starting from bromobenzene and benzaldehyde involving Grignard reaction, bromination, etherification, and isolation of the product hydrochloride salt. The process involved automated laboratory operations including solid and liquid reagent additions, stirring, heating, liquid–liquid separation, evaporation, and filtration. The synthesis of rufinamide started from 2-(bromomethyl)-1,3-difluorobenzene and consisted of three chemical steps that utilized liquid reagent additions, mixing, and filtration. Lastly, the synthesis of sildenafil started from ethoxybenzoic acid and required four chemical steps utilizing solid and liquid additions, mixing, liquid–liquid separation, evaporation, and filtration operations. A key takeaway is that the equipment that was used to execute the synthetic tasks are standard pieces of laboratory equipment that were able to be digitally controlled. It was the development of an elaborate suite of programs that enabled them to work in concert to be able to accomplish the feat of fully automating the synthetic task (Table 9.2).

Table 9.2 Results of Chemputer controlled synthesis.

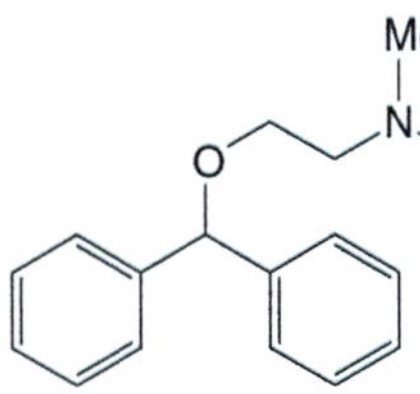

36 (diphenylhydramine HCl)

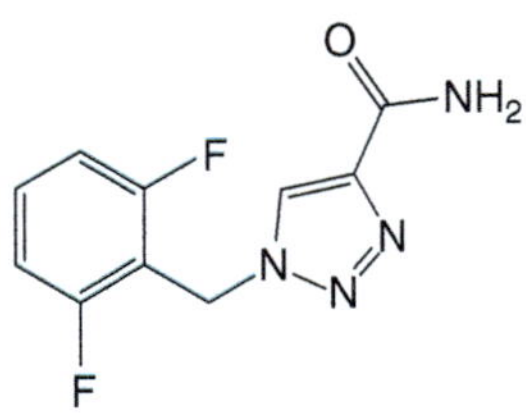

37 (rufinamide)

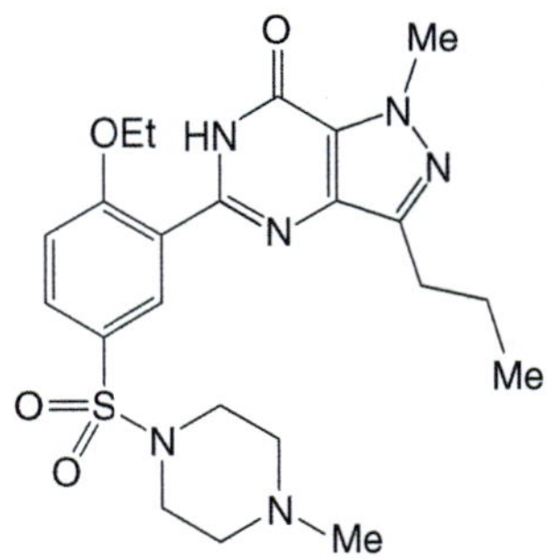

38 (sildenafil)

	36	37	38
Steps	4	3	4
Manual yield(time)	68% (4 days)	38% (n.d.)	n.d. (n.d.)
Chemputer yield (time)	58% (77 h)	46% (38 h)	44% (102h)
Reactor module	Yes	Yes	Yes
Liquid/liquid separator	Yes	No	Yes
Filtration module	Yes	Yes	Yes
Evaporation module	Yes	No	Yes

The outlined work represents a significant advance in synthetic chemistry in that it organized the tasks of synthetic chemistry into programmable modules that could be executed on common digital lab equipment. This is important in the context of creating unambiguous experimental procedures that can be interpreted by humans and machines alike and enables the tasks of organic chemistry to be replicated wherever the appropriate equipment and to be executed day and night. The proportion of this work involving programming was quite high and required significant programming expertise to be able to be accomplished. This is important as it clearly shows how time spent programming lines of code at a desk can lead to more product isolated in the lab and underscores the value programming can offer organic chemists.

9.11 Autonomous Reaction Optimization

The remarkable accomplishments described in the previous sections demonstrate the impact programming can have on augmenting the way synthetic chemists go about their daily work. In the case of the Chemputer, the result was largely a productivity gain through the automation of routine chemical synthesis. This work, if applied more broadly across the field, would free chemists from mundane synthesis tasks and unit operations to provide more time for tackling complex problems. Another challenge that chemists face in their daily work is the optimization of chemical reactions and processes. This includes investigation of different reagent combinations and identifying appropriate operating ranges for continuous variables such as temperature, concentration, and equivalents of reagents. The work previously described in Section 9.9 defined and invented a machine learning augmented workflow for reaction optimization but did not replace the workflow itself. A question asked by the Hein, Sigman, and Aspuru-Guzik labs is whether Bayesian optimization could be coupled with appropriate off-the-shelf robotics equipment to create an autonomous reaction optimizer [78].

The goal for autonomous process optimization is to enable the chemist to decide the bounds of reactivity (e.g. variables and variable identities) but would enable a robotics platform to conduct the experimentation and optimization on its own. Thereby, a chemist would decide what potential experiments would be relevant, assemble the necessary reagents on a robotics platform, and the optimizer would then run reactions independently. An optimizer of this type would need to leverage artificial intelligence to suggest reaction conditions, design and execute these experiments on a robotics platform, analyze the reaction outputs such as yield, and return the results of the experiments to the optimizer to repeat the process until the user-defined goal has been achieved or the experimental budget has been reached. Once the optimizer reaches its target goal or its experimental budget has been reached the chemist can then interpret the results to design subsequent experiments. An advantage of this approach is that the autonomous system can work around the clock and is able to explore multivariate space very efficiently. To accomplish this, the appropriate Bayesian optimization algorithm would have to be coupled with a robotics platform and analytical equipment.

The team chose to use the Chemspeed SWING robotics platform commonly used for HTE. This robotics system is capable of dispensing powders, sampling reactions, heating, and stirring under a controlled atmosphere. For the optimization component, the team selected the Phoenics [79] and Gryffin [80] code libraries developed by the Aspuru-Guzik lab for the Bayesian optimization component. These libraries can accommodate multi-objective optimization, meaning they can simultaneously try to optimize multiple objectives such as higher yield and lower impurity, or higher yield and higher ee, depending on the problem. These programs were modified such that they can suggest batches of experiments that would take advantage of the high-throughput equipment at their disposal. Lastly, the group sought to integrate an Agilent liquid chromatography system for the analysis of reactions. While each of the three components had been selected, the key problem is enabling communication between them. Thus, the team developed a python framework to transmit information between the systems (Figure 9.22). This python framework would operate as a task manager or interpreter that would execute tasks such as collecting and scoring results from the HPLC and supplying them to the Bayesian optimizer, then interpreting results from the optimizer to design the next set of experiments and send protocols and reagent locations to the robotics platform. The versatility of python provided that it was an excellent choice of language to enable communication between all the discrete components of a closed loop reaction optimizer.

The closed-loop autonomous system was fully connected and ready for an optimization task. The team selected a problematic stereoselective Suzuki reaction for initial exploration where the main objective was to optimize the yield of the desired product **41** (Scheme 9.3). For these closed-loop systems, the chemist needs to provide the optimizer with the lists of each of the variables, also decide on the number of reactions per round, and provide the overall experimental budget. To prepare the design space, they selected 12 ligands, varied equivalents of boronic acid (10–20 μmol), varied precatalyst equivalents (0.1–0.5 μmol), varied ligand concentration (0.05–2 μmol), and varied temperature (10–40 °C). An experimental array of eight reactions per round and a total budget of 120 experiments (15 rounds) were set. The reaction had been shown to reach completion within 2 hours, so a total experimental time of 60 hours. With the experimental parameters configured, the autonomous system explored the design space over 15 rounds and found an optimum result at experiment 118 giving a 65% yield (1.6 : 1 E/Z selectivity) of the product that is slightly better than the previously reported 60% yield (1.7 : 1 E/Z selectivity) obtained through traditional HTE. It was interesting that the optimum

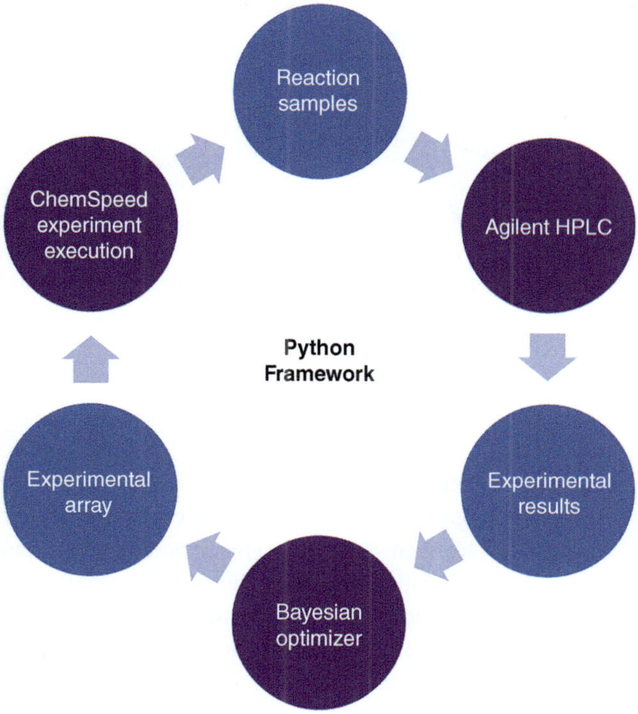

Figure 9.22 Workflow for autonomous reaction optimization.

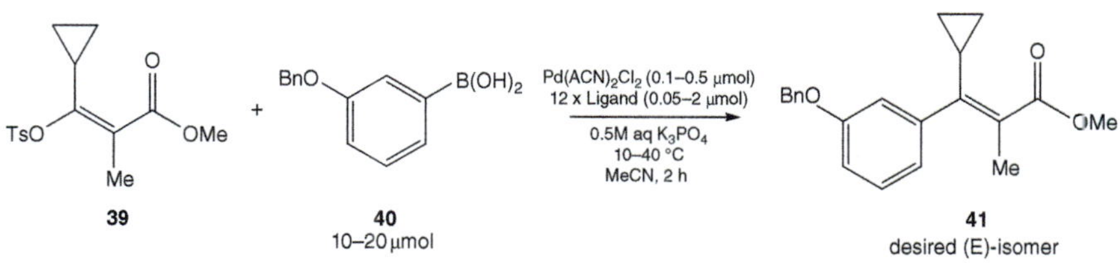

Scheme 9.3 Reaction selected for autonomous optimization.

was achieved through balancing both ligand and metal-to-ligand ratio. The best ligand, diphenylphosphinoferrocene (dppf) **42** did not perform well during early experiments but was continually explored as the optimizer learned the appropriate metal-to-ligand for this system (Figure 9.23).

Following up on these impressive results the team opted for a second optimization effort using a different strategy. The initial set of 12 ligands were selected using expert intuition. The main question is whether the chemist was imparting any bias on the optimization through selection of ligands. To address this, a large

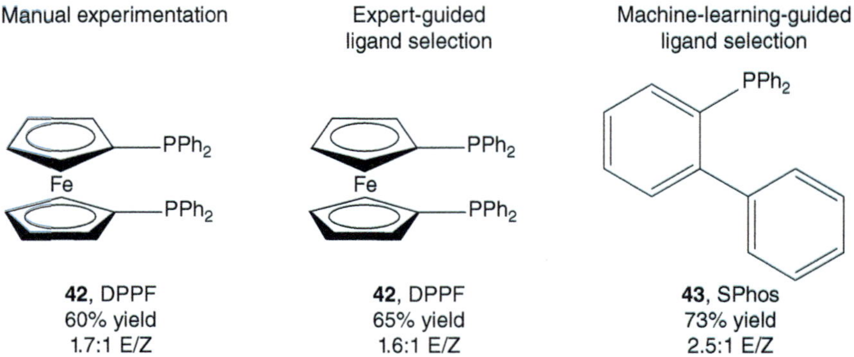

Figure 9.23 Comparison of results using manual expertise, expert-guided autonomous optimization, and machine learning informed autonomous reaction optimization.

pool of ligands was subjected to unsupervised machine learning, using a computer algorithm to group similar entities, to group the ligands into clusters based on their calculated properties. The clusters were then sampled to give a representative and unbiased design space for the ligand component. Since the pool of ligands was larger than the initial set, the team opted to increase the ligand pool to 24 ligands with an increased experimental budget of 192 reactions conducted over 96 hours. Using this approach, the autonomous optimization found a result with an improved 73% yield (2.5 : 1 E/Z) at 161 reactions with Ph-SPhos as the ligand (Figure 9.23).

This work is exciting as the authors were able to program the connectivity between a HTE robotics system, a Bayesian optimizer, and an HPLC system. This autonomous reaction optimization platform utilized expert knowledge from chemist for setting design parameters while relieving the chemist of the mundane task of assembling experiments. Moreover, the system can run day and night until the experimental budget has been exhausted, which provides for additional efficiency gains. While the technical expertise required for programming an experimental setup such as this is quite high, it nevertheless underscores the value that these skills can bring to an organization.

9.12 Conclusion

The past several years has seen a remarkable pivot in the value added by organic chemists supplementing their laboratory skills with programming skills. The shift has seen remarkable advances in performing better data analysis, streamlining workflows, building predictive models, designing user friendly applications, and fully automating robotic tasks. Furthermore, this chapter focused on contributions from individual laboratories, but didn't highlight the amazing work being done in synthetic route planning and prediction, that is being advanced by a number of companies and chemical engineering labs that has an incredibly bright future. Hopefully, the case studies in this chapter sparked ideas, interest, and inspiration for the application of programming to organic chemistry problems you may encounter in your work. Whether or not programming is a skill set for you, the knowledge of what is possible may help you begin your programming journey or seek out those that are able to program to help you advance your research. Whether it be data transcription, file management, reaction prediction, report generation, or autonomous reaction optimization, there is likely a solution using code. In many cases, the solutions have been worked out and are waiting to be found on sites like GitHub or StackOverflow.

Many of the examples in this chapter leveraged python as the programming language. Python is a popular language owing to its relative ease of use and modularity and would be a great place to start for any programmer. There is no specific book or site that is best for learning python or any other language. The best way to learn and to generate impact is to seek out a colleague or collaborator with programming experience and begin working on a problem relevant to your daily work. If there does not appear to be a relevant problem to your work, then seek out collaborators in other departments looking for help on a project. Learning through a community is a great strategy. One recommendation for engaging the coding community more broadly is to follow #TidyTuesday posts on Twitter, and the TidyTuesday GitHub page at https://github.com/rfordatascience/tidytuesday. Tidy Tuesday is a weekly data science problem offered each Tuesday by the R programming community where users then post their solutions and links to their GitHub repositories on Twitter. These datasets can be investigated by anyone using any language and is a great way to learn by trying the problem on your own and then seeing how others solved the problem. No single answer is correct! For those that decide to pursue learning a programming language to advance their research I am confident it will be well worth your effort. Independently learning to code was one of the most rewarding experiences of my career and has opened up areas of research I once thought were never possible and introduced me to collaborators in other fields that have expanded my perspective and grown my capabilities as a scientist. The intersection of programming and organic chemistry has an incredibly bright future and is likely to accelerate research for years to come.

References

1 (a) Prier, C.K., Rankic, D.A., and MacMillan, D.W.C. (2013). Visible light photoredox catalysis with transition metal complexes: applications in organic synthesis. *Chem. Rev.* 113 (7): 5322–5363. https://doi.org/10.1021/cr300503r; (b) Twilton, J., Le, C., Zhang, P. et al. (2017). The merger of transition metal and photocatalysis. *Nat. Rev. Chem.* 1: 0052. https://doi.org/10.1038/s41570-017-0052.

2 (a) Seayad, J. and List, B. (2005). Asymmetric organocatalysis. *Org. Biomol. Chem.* 3: 719–724. https://doi.org/10.1039/B415217B; (b) Xiang, S.-H. and Tan, B. (2020). Advances in asymmetric organocatalysis over the last 10 years. *Nat. Commun.* https://doi.org/10.1038/s41467-020-17580-z.

3 Bao, R., Zhang, H., and Tang, Y. (2021). Biomimetic synthesis of natural products: a journey to learn, to mimic, and to be better. *Acc. Chem. Res.* 54 (19): 3720–3733. https://doi.org/10.1021/acs.accounts.1c00459.

4 (a) Weiss, C.J. (2017). Scientific computing for chemists: an undergraduate course in simulations, data processing, and visualization. *J. Chem. Educ.* 94 (5): 592–597. https://doi.org/10.1021/acs.jchemed.7b00078; (b) Lafuente, D., Cohen, B., Fiorini, G. et al. (2021). A gentle introduction to machine learning for chemists: an undergraduate workshop using python notebooks for visualization, data processing, analysis, and modeling. *J. Chem. Educ.* 98 (9): 2892–2898. https://doi.org/10.1021/acs.jchemed.1c00142.

5 https://www.virscidian.com/ (accessed 5 August 2022).

6 https://www.acdlabs.com/products/spectrus-platform/katalyst-d2d/ (accessed 5 August 2022).

7 (a) www.chemistryworld.com/careers/why-chemists-should-learn-to-code/4013922.article (accessed 5 May 2023); (b) cen.acs.org/articles/95/i35/Perspectives-spreadsheets-programming.html (accessed 5 May 2023).

8 Fante, P., Cinquemani, C., Yaseneva, P. et al. (2021). Transition to sustainable chemistry through digitalization. *Chem.* 7 (11): 2866–2882. https://doi.org/10.1016/j.chempr.2021.09.012.

9 Moser, A., Waked, A.E., and DiMartino, J. (2021). Consolidating and managing data for drug development within a pharmaceutical laboratory: comparing the mapping and reporting tools from software applications. *Org Process Res. Dev.* 25 (10): 2177–2187.

10 Weiss, C.J. (2017). Perspectives: Teaching chemists to code. *Chem. Eng. News.* 95 (35).

11 www.datacamp.com (accessed 5 August 2022).

12 www.codeacademy.com (accessed 5 August 2022).

13 www.coursera.org (accessed 5 August 2022).

14 (a) McLaughlin, M. and Minor, J. (2022). The Best Programs for Learning to Code. https://www.pcmag.com/picks/the-best-programs-for-learning-to-code (accessed 27 March 2022); (b) https://medium.com/javarevisited/7-best-coding-course-to-learn-programming-with-zero-experience-in-2020-52f7d0d9cb80 (accessed 27 March 2022).

15 ocw.mit.edu/courses/intro-programming (accessed 5 August 2022).

16 Gopani, A. (2022). Top YouTube channels to learn programming in 2022. https://analyticsindiamag.com/top-youtube-channels-to-learn-programming-in-2022/ (accessed 27 March 2022).

17 www.stackoverflow.com (accessed 5 August 2022).

18 (a) www.github.com (accessed 4 January 2022); (b) www.github.com/collections/learn-to-code (accessed 4 January 2022).

19 For selected examples: (a) Boehmke, B. and Greenwell, B. (2020). Hands-On Machine Learning with R. https://bradleyboehmke.github.io/HOML/ (accessed 27 March 2022); (b) https://github.com/trekhleb/learn-python (accessed 27 March 2022); (c) https://github.com/jerry-git/learn-python3 (accessed 27 March 2022).

20 (a) Wickham, H., Grolemund, G. (2017). R for Data Science. https://r4ds.had.co.nz/ (accessed 27 March 2022); (b) Peng, R. D. (2020). Exploratory Data Analysis with R. https://bookdown.org/rdpeng/ (accessed 27 March 2022); (c) Mahoney, M. (2019). Introduction to Data Exploration and Analysis with R. https://bookdown.org/mikemahoney218/IDEAR/ (accessed 27 March 2022); (d) Wilke, C. O. (2019). Fundamentals of Data Visualization: A Primer on Making Informative and Compelling Figures. https://clauswilke.com/dataviz/index.html (accessed 27 March 2022); (e) Xie, Y., Allaire, J. J., and Grolemund, G. (2022). R Markdown: The Definitive Guide. https://bookdown.org/yihui/rmarkdown/ (accessed 27 March 2022); (f) Wickham, H. (2019). Advanced R. https://adv-r.hadley.nz/ (accessed 27 March 2022); (g) Wickham, H. (2020). Mastering Shiny. https://mastering-shiny.org/ (accessed 27 March 2022).

21 www.kaggle.com (accessed 5 August 2022).

22 https://www.kaggle.com/c/bms-molecular-translation (accessed 5 August 2022).

23 https://survey.stackoverflow.co/2022/ (accessed 21 July 2022).

24 Tippmann, S. (2015). Programming tools: adventures with R. *Nature* 517: 109–110. https://doi.org/10.1038/517109a.

25 Perkel, J. (2015). Programming: pick up python. *Nature* 518: 125–126. https://doi.org/10.1038/518125a.

26 Tukey, J.W. (1977). *Exploratory Data Analysis*. Reading, MA: Addison-Wesley Pub. Co.

27 Mulrooney, C.A., Lahr, D.L., Quintin, M.J. et al. (2013). An informatic pipeline for managing high-throughput screening experiments and analyzing data from stereochemically diverse libraries. *J. Comput. Aided Mol. Des.* 27 (5): 455–468. https://doi.org/10.1007/s10822-013-9641-y.

28 Bandyopadhyay, D., Kreatsoulas, C., Brady, P.G. et al. (2019). Scaffold-based analytics: enabling hit-to-lead decisions by visualizing chemical series linked across large datasets. *J. Chem. Inform. Modeling* 59 (11): 4880–4892. https://doi.org/10.1021/acs.jcim.9b00243.

29 https://www.originlab.com (accessed 21 July 2022).

30 https://www.jmp.com/en_us/home.html (accessed 5 August 2022).

31 Rule, A., Birmingham, A., Zuniga, C. et al. (2019). Ten simple rules for writing and sharing computational analyses in Jupyter Notebooks. *PLoS Comput. Biol.* 15 (7): e1007007. https://doi.org/10.1371/journal.pcbi.1007007.

32 Benítez, A.R., Dürr, S.L., and Narayan, A.R.H. (2020). Radial scope: a new visualization tool for structure-data relationships. *Trends Chem.* 2 (7): 587–589. https://doi.org/10.1016/j.trechm.2020.04.003.

33 Matplotlib.org (accessed 21 July 2022).

34 Plotly.com/python/ (accessed 21 July 2022).

35 Pandas.pydata.org (accessed 21 July 2022).

36 Rdkit.org (accessed 21 July 2022).

37 github.com/duerrsimon/substrate-scope-plot (accessed 5 August 2022).

38 Perkel, J.M. (2018). Why Jupyter is data scientists' computational notebook of choice. *Nature* 563: 145–146. https://doi.org/10.1038/d41586-018-07196-1.

39 github.com/duerrsimon/substrate-scope-plot (accessed 4 January 2022).

40 Ahneman, D.T., Estrada, J.G., Lin, S. et al. (2018). Predicting reaction performance in C-N cross-coupling using machine learning. *Science* 360 (6385): 186–190. https://doi.org/10.1126/science.aar5169.

41 github.com/doylelab/rxnpredict (accessed 4 January 2022).

42 Żurański, A.M., Wang, J.Y., Shields, B.J., and Doyle, A.G. (2022). Auto-QChem: an automated workflow for the generation and storage of DFT calculations for organic molecules. *React. Chem. Eng.* Advance Article. https://doi.org/10.1039/D2RE00030J.

43 github.com/PrincetonUniversity/auto-qchem (accessed 4 January 2022).

44 Autoqchem.org (accessed 5 August 2022).

45 (a) https://topepo.github.io/caret/ (accessed 22 July 2022); (b) https://cran.r-project.org/web/packages/caret/index.html (accessed 22 July 2022).

46 (a) https://www.tidymodels.org/ (accessed 22 July 2022); (b) https://cran.r-project.org/web/packages/tidymodels/index.html (accessed 22 July 2022).

47 https://scikit-learn.org/stable/ (accessed 22 July 2022).

48 Nielsen, K., Ahneman, D.T., Riera, O., and Doyle, A.G. (2018). Deoxyfluorination with sulfonyl fluorides: navigating reaction space with machine learning. *J. Am. Chem. Soc.* 140 (15): 5004–5008. https://doi.org/10.1021/jacs.8b01523.

49 Pattanaik, L. and Coley, C.W. (2020). Molecular representation: going long on fingerprints. *Chem.* 6 (6): 1203–1221. https://doi.org/10.1016/j.chempr.2020.05.002.

50 Chuang, K.V. and Keiser, M.J. (2018). *Science* 362 (6416). https://doi.org/ 10.1126/science.aat8603.

51 Estrada, J.G., Ahneman, D.T., Sheridan, R.P. et al. (2018). *Science* 362 (6416): https://doi.org/ 10.1126/science.aat8763.

52 Kim, S., Chen, J., Cheng, T. et al. (2019). PubChem in 2021: new data content and improved web interfaces. *Nucleic Acids Res.* 49 (D1): D1388–D1395. https://doi.org/10.1093/nar/gkaa971.

53 Kearnes, S.M., Maser, M.R., Wleklinski, M. et al. (2021). The open reaction database. *J. Am. Chem. Soc.* 143 (45): 18820–18826. https://doi.org/10.1021/jacs.1c9820.

54 Gensch, T., dos Passos Gomes, G., Friederich, P. et al. (2022). A comprehensive discovery platform for organophosphorous ligands for catalysis. *J. Am. Chem. Soc.* 144 (3): 1205–1217. https://doi.org/10.1021/jacs.1c9718.

55 Newman-Stonebraker, S.H., Smith, S.R., Borowski, J.E. et al. (2021). Univariate classification of phosphine ligation state and reactivity in cross-coupling catalysis. *Science* 374 (6565): 301–308. https://doi.org/10.1126/science.abj4213.

56 (a) Tolman, C.A. (1970). Phosphorus ligand exchange equilibriums on zerovalent nickel. Dominant role for steric effects. *J. Am. Chem. Soc.* 92 (10): 2956–2965. https://doi.org/10.1021/ja00713a007; (b) Tolman, C.A. (1977). Steric effects of phosphorous ligands in organometallic chemistry and homogeneous catalysis. *Chem. Rev.* 77 (3): 313–348. https://doi.org/10.1021/cr60307a002.

57 Harper, K.C., Bess, E.N., and Sigman, M.S. (2012). Multidimensional steric parameters in the analysis of asymmetric catalytic reactions. *Nat. Chem.* 4: 366–374. https://doi.org/10.1038/nchem.1297.

58 (a) Falivene, L., Cao, Z., Petta, A. et al. (2019). Towards the online computer-aided design of catalytic pockets. *Nat. Chem.* 11: 872–879. https://doi.org/10.1038/s41557-019-0319-5; (b) Zhao, S., Gensch, T., Murray, B. et al. (2018). Enantiodivergent Pd-catalyzed C–C bond formation enable through ligand parameterization. *Science* 362 (6415): 670–674. https://doi.org/ 10.1126/science.aat2299.

59 Wu, K. and Doyle, A.G. (2017). Parameterizaton of phosphine ligands demonstrates enhancement of nickel catalysis via remote steric effects. *Nat. Chem.* 9: 779–784. https://doi.org/10.1038/nchem.2741.

60 Anastas, P.T. and Zimmerman, J.B. (2003). Design through the twelve principles of green engineering *Env. Sci. Tech.* 37 (5): 94A–101A. https://doi.org/10.1021/es032373g.

61 (a) Curzons, A.D., Constable, D.J.C., Mortimer, D.N., and Cunningham, V.L. (2001). So you think your process is green, how do you know?—Using principles of sustainability to determine what is green–a corporate perspective *Green Chem.* 3: 1–6. https://doi.org/10.1039/B007871I; (b) Jiménez-González, C., Ponder, C.S., Broxterman, Q.B., and Manley, J.B. (2011). Using the Right Green Yardstick: Why Process Mass Intensity Is Used in the Pharmaceutical Industry To Drive More Sustainable Processes. *Org. Process Res. Dev.* 15 (4): 912–917. https://doi.org/10.1021/op200097d.

62 Fitzgerald, M.A., Soltani, O., Wei, C. et al. (2015). Ni-catalyzed C–H functionalization in the formation of a complex heterocycle: synthesis of the potent JAK2 inhibitor BMS-911543. *J. Org. Chem.* 80 (12): 6001–6011. https://doi.org/10.1021/acs.joc.5b00572.

63 (a) Lobben, P.C., Barlow, E., Bergum, J.S. et al. (2015). Control strategy for the manufacture of brivanib alaninate, a novel pyrrolotriazine VEGFR/FGFR inhibitor. *Org. Process Res. Dev.* 19 (8): 900–907. https://doi.org/10.1021/op500125u;

(b) Pesti, J.A., LaPorte, T., Thornton, J.E. et al. (2014). Commercial synthesis of a pyrrolotriazine–fluoroindole intermediate to brivanib alaninate: process development directed toward impurity control. *Org. Process Res. Dev.* 18 (1): 89–102. https://doi.org/10.1021/op400242j; (c) LaPorte, T.L., Spangler, L., Hamedi, M. et al. (2014). Development of a continuous plug flow process for preparation of a key intermediate for brivanib alaninate. *Org. Process Res. Dev.* 18 (11): 1492–1502. https://doi.org/10.1021/op500176z.

64 Deerberg, J., Prasad, S.J., Sfouggatakis, C. et al. (2016). Stereoselective bulk synthesis of CCR2 antagonist BMS-741672: assembly of an all-cis (S,R,R)-1,2,4-triaminocyclohexane (TACH) core via sequential heterogeneous asymmetric hydrogenations. *Org. Process Res. Dev.* 20 (11): 1949–1966. https://doi.org/10.1021/acs.oprd.6b00282.

65 (a) Pinto, D.J.P., Orwat, M.J., Koch, S. et al. (2007). Discovery of 1-(4-methoxyphenyl)-7-oxo-6-(4-(2-oxopiperidin-1-yl)phenyl)-4,5,6,7-tetrahydro-1H-pyrazolo[3,4-c]pyridine-3-carboxamide (Apixaban, BMS-562247), a highly potent, selective, efficacious, and orally bioavailable inhibitor of blood coagulation factor Xa. *J. Med. Chem.* 50 (22): 5339–5356. https://doi.org/10.1021/jm070245n; (b) Shapiro, R., Rossano, L. T., Mudryk, B. M., et al. (2006). Process for preparing 4,5-dihydro-pyrazolo34-cipyrid-2-ones. US Patent 2006/0069258 A1. Filed 28 September 2004 and Issued 30 March 2006; (c) Ramirez, A., Mudryk, B., Rossano, L., and Tummala, S. (2012). A mechanistic study on the amidation of esters mediated by sodium formamide. *J. Org. Chem.* 77 (1): 775–779. https://doi.org/10.1021/jo2021828.

66 Carter, P.H., Berndt, E.R., DiMasi, J.A., and Trusheim, M. (2016). Investigating investment in biopharmaceutical R&D. *Nat. Rev. Drug Discovery.* 15: 673–674. https://doi.org/10.1038/nrd.2016.104.

67 Li, J., Albrecht, J., Borovika, A., and Eastgate, M.D. (2018). Evolving green chemistry metrics into predictive tools for decision making and benchmarking analytics. *ACS Sustainable Chem. Eng.* 6 (1): 1121–1132. https://doi.org/10.1021/acssuschemeng.7b03407.

68 CRAN.R-project.org/package=shiny (accessed 4 January 2022).

69 https://flask.palletsprojects.com/en/2.1.x/ (accessed 27 July 2022).

70 https://dash.plotly.com/ (accessed 27 July 2022).

71 https://streamlit.io/ (accessed 27 July 2022).

72 Borovika, A., Albrecht, J., Li, J. et al. (2019). The PMI predictor – a web app enabling green-by-design chemical synthesis. *Nat. Sustain.* 2: 1034–1040. https://doi.org/10.1038/s41893-019-0400-5.

73 https://acsgcipr-predictpmi.shinyapps.io/pmi_calculator/ (accessed 4 January 2022).

74 https://github.com/chepyle/PMI_Calculator (accessed 4 January 2022).

75 Gardner, J., Pleiss, G., Weinberger, K.Q. et al. (2018). Gpytorch: Blackbox matrix-matrix gaussian process inference with gpu acceleration. *Adv. Neural Inform. Processing Syst.* 31: 7576–7586.

76 Perera, D. et al. (2018). A platform for automated nanomole-scale reaction screening and micromole-scale synthesis in flow. *Science* 359 (6374): 429–434. https://doi.org/ 10.1126/science.aap9112.

77 Steiner, S., Wolf, J., Glatzel, S. et al. (2019). Organic synthesis in a modular robotic system driven by a chemical programming language. *Science* 363 (6423): https://doi.org/10.1126/science.aav2211.

78 Christensen, M., Yunker, L.P.E., Adedeji, F. et al. (2021). Data-Science driven autonomous process optimization. *Commun. Chem.* 112: https://doi.org/10.1038/s42004-021-00550-x.

79 Häse, F., Roch, L.M., Kreisbeck, C., and Aspuru-Guzik, A. (2018). Phoenics: a Bayesian optimizer for chemistry. *ACS Cent. Sci.* 4 (9): 1134–1145. https://doi.org/10.1021/acscentsci.8b00307.

80 Häse, F., Aldeghi, M., Hickman, R.J. et al. (2021). Gryffin: an algorithm for Bayesian optimization of categorical variables informed by expert knowledge. *Appl. Phys. Rev.* 8 (3): 031406. https://doi.org/10.1063/5.0048164.

10

Machine Learning for the Optimization of Chemical Reaction Conditions

A. Filipa de Almeida and Tiago Rodrigues

Instituto de Investigação do Medicamento (iMed), Faculdade de Farmácia, Universidade de Lisboa, Lisbon, Portugal

Glossary

Aleatoric uncertainty Variability is due to random effects
Artefacts Spurious correlations due to experimental design and biases
Bayesian optimization A method for global optimization of mathematical functions
Coefficient of determination A performance metric for regression machine learning models
Epistemic uncertainty Variability due to lack of training data
Gaussian processes A machine learning method commonly applied in optimization problems
Hyperparameters The parameters of a machine learning model
Mean absolute error A performance metric for regression machine learning models
Mean squared error A performance metric for regression machine learning models
Nelder-Mead SIMPLEX A method for optimization
Neural network A type of machine learning algorithm inspired in the biological brain
One hot encoding A method to pre-process categorical variables
Pareto set A collection of optimal solutions according to a trade-off between desirable and undesirable outcomes.

Enabling Tools and Techniques for Organic Synthesis: A Practical Guide to Experimentation, Automation, and Computation, First Edition. Edited by Stephen G. Newman.

Random forest An ensemble machine learning method that aggregates multiple decisions trees.

SNOBFIT A software package for optimization problems with continuous variables.

10.1 Introduction

One can generally consider machine learning as mathematical models that approximate real-world observations through a number of relevant features (or the so-called descriptors) linked to said observations. Those machine learning models are built with datasets containing multiple and ideally diverse observations covering a range of possible outputs. The main goal is to have a mathematical approximation of a given phenomenon and being able to use it to accurately and confidently predict outcomes of yet unseen events. Machine learning models can be categorized into several types according to the required information for learning. While some of them explicitly require information on the output labels or values (supervised learning), others learn by finding structure in the employed descriptors (unsupervised). Other subtle variations can be explored, e.g. self-supervised learning, but the common main goal is to either discriminate or generate outputs, generate research hypotheses in a probabilistically motivated fashion, and minimize the number of experiments conducted in research programs.

Machine learning is transforming research in the chemical sciences and enabling the generation of new knowledge at unprecedented pace [1, 2]. Supported by recent advances in hardware [3], algorithms, data storage capabilities, and its availability, machine learning has seen revolutionizing applications in chemical biology, medicinal chemistry, nanotechnology, materials sciences among others [1]. Several workflows based on statistical learning have been established and prospectively evaluated with the goal of tackling intellectually challenging tasks in chemistry, such as the design of small molecules and materials from scratch [4–8] – the so-called *de novo* design – prediction of protein structure [9, 10], property prediction (e.g. physicochemical [11–14], on- and off-target bioactivity [15–17]), and others. Provided that sufficient quality data is in hand, machine learning tools offer valid assistance to decision-making and promote research avenues that would be discarded otherwise due to biased intuition and established conceptualizations [18, 19]. This does not diminish the importance of expert intuition, since domain knowledge plays an important role in curating predictions but suggests that machine learning has the potential to recommend experiments leading to unexpected and potentially disruptive findings.

It is worth noting that while research in this space is moving toward fully automated systems, including computers for experiment design and data analyses and robots for experiment execution [20–23], one may argue that robots will remain inaccessible to a large portion of the research community in the short term. In that regard, human–machine interfaces [24] may provide an intermediate and necessary step to gradually build trust among end users, which, in the case of chemistry, are invariably bench chemists. One may also argue that remodeling university curricula to provide a more solid background on data science and statistics will also help lowering the entry bar for the future generations and promote more interdisciplinary studies.

An undisputable advantage of fully automated and iterative experimentation cycles is the reproducibility of methods and generation of high-quality data [25]. These are key aspects and without them the quality of any machine learning model may be compromised. As with other fields of chemistry, chemical synthesis has also experienced significant advances in the last five years due to the development of multiple machine learning algorithms [2]. Most prominently, retrosynthetic planning (see Chapter 11) [26–29], prediction of reaction products [30], and even catalyst design for asymmetric syntheses [31, 32] have been explored by multiple research groups with varying degrees of elaboration and success.

The optimization of chemical reactions is among the most experimentally demanding tasks synthetic chemists regularly carry out. Machine learning may play a key role in a chemist's workflow by helping select and interpret experiments. Our lab recently disclosed one such program for achieving this that we named LabMate.ML [18, 19]. This is a software tool that selects the most appropriate experimental conditions according to an estimated reaction yield (under said conditions). In this chapter, the application of LabMate.ML will be used as a case study to demonstrate how machine learning in general and reaction optimization in particular can complement existing tools in the synthetic chemist's toolbox.

We have organized the present chapter into different subsections:

- **Prior art and alternative methods for data-driven reaction optimization**: we provide an overview of methods built by other community members and present the applications of their workflows.
- **Reaction optimization using LabMate.ML**: we break down the concepts underlying the tool, how it is built, advantages/disadvantages, and general advice to its usage from a practical point of view.
- **Primer on evaluation guidelines**: as the field matures further, a general workflow is needed to compare machine learning tools. We recommend a suite of ancillary experiments that may help gauging the utility of machine learning algorithms. In a broader perspective, the guidelines are extensible to any other machine learning applications and are designed to help less-informed colleagues evaluating, interpreting, and selecting the most appropriate workflows for their needs.
- **Outlook**: we provide a forward-looking perspective on unsolved issues and what the future might hold.

10.2 Prior Art and Alternative Methods for Rational Reaction Optimization

The following discussion gives a succinct overview of several recent efforts made toward using machine learning to facilitate reaction optimizations. Here, we also differentiate between machine learning models – the focus of the chapter – and other software alternatives that have been traditionally employed to optimize synthetic processes, although without explicitly learning data patterns. We encourage the reader to read the reference list to gain a better understanding of how the field is evolving. Accordingly, we present a collection of studies that may enrich the readers' knowledge on the topic, but one is welcome to skip ahead to Section 10.3 where the key concepts underlying machine learning and LabMate.ML are described in more detail.

One should bear in mind that the automated optimization of chemical reaction conditions is a complex and multidimensional task. Here, the main challenge for machine learning is creating a model that faithfully approximates a specified endpoint. Thus, the endpoint must be amenable to experimental measurement and verification to allow an iterative refinement of the model (Figure 10.1) [33]. With that in mind, machine learning algorithms can be implemented to predict outputs such as the improvement of reaction yields, purity (as in minimizing side products), or any other criteria of interest.

Understandingly, a machine learning model is built to optimize a specific reaction. Although desirable, the identified conditions are seldom transferrable to other reactants because of the mechanistic dependence toward catalysts, additives, solvents, concentrations, temperatures, and others. The relationships between each of those variables vary according to the reaction type or even substrate scope. To navigate such intricate relationships, chemists draw from previous experience and intuition, and vary one variable at a time to more easily extract knowledge and infer an optimization pathway. While viable, this strategy does not account for potential nonlinear relationships between variables that are more reliably modeled by computer algorithms [18, 19]. Further, there is a well documented intrinsic human inability to process more than four variables concurrently [34]. This limitation can make the optimization of reaction conditions a potential challenge to less experienced researchers. Conversely, while attempting to mimic expert reasoning, computer algorithms can model multiple training examples at a time to identify areas in the reactivity space that lead to desirable experimental outcomes. Another key advantage of computers algorithms is the ability to efficiently process hundreds or thousands of variables, if necessary. Of note, different computational approaches have been utilized to optimize reaction conditions while leveraging distinct concepts and approaches. Their underlying mathematics is out of the scope of the present chapter, but we present select case studies of the

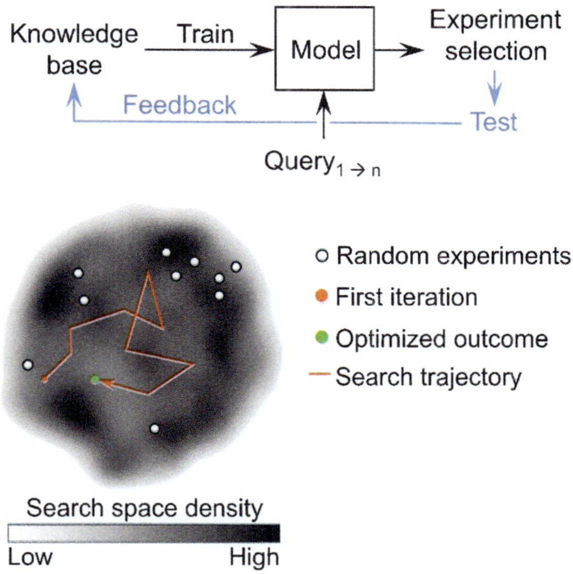

Figure 10.1 The underlying concept of reaction condition optimization with machine learning algorithms. A reaction database is used to train a machine learning model and select experiments for execution. Once the reaction endpoint is measured by an appropriate analytical method (e.g. NMR, LC-UV/Vis-MS), the collected data is fed back to refine the model in iterative fashion. An improved model is used to select the next experiments until a stop criterion or objective is met. The dataset used for training the machine learning algorithm can be assembled with random experiments or existing information.

most popular and/or established methods to exemplify their uses. Those examples do not endorse universal utility but are an important step towards the digitalization of the chemical sciences.

One of the established concepts in process chemistry for reaction optimization is the design of experiments (DoEs, see Chapter 5) [35]. DoE maps all possible reactions in a pre-defined search space by building a grid and conducting fractional/full factorial design strategies to gauge the response landscape, determine relationships between input variables and reaction outputs, and identify a global optimum [36, 37]. As a caveat, the execution of all grid reactions remains highly laborious; one might argue that smarter experiment prioritization strategies can be devised and are in high demand (Figure 10.2).

The Stable Noisy Optimization by Branch and Fit (SNOBFIT) software tool has become a popular software tool for optimizing continuous variables [38] and has been widely applied to chemical reactions. In brief, it computes a response surface to identify local optima and attempts to locate a global optimum. For example, Jensen and colleagues have applied SNOBFIT in together with microfluidic-assisted

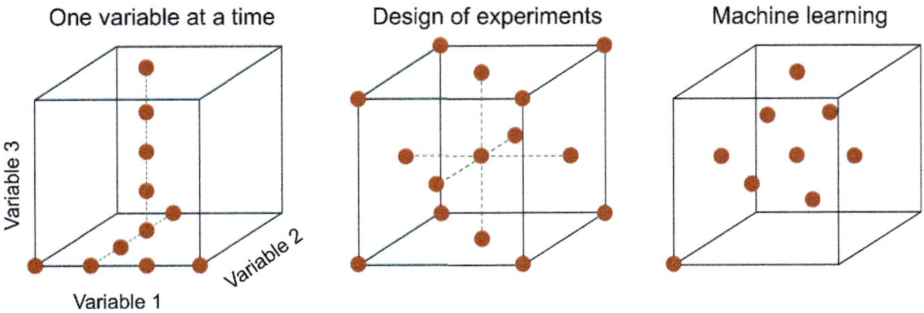

Figure 10.2 Schematics of different concepts and approaches for process optimization. Variation of one variable/factor at a time can be highly inefficient as it neglects interactions between variables and may require a larger number of experiments to identify an optimum in a defined search space (depicted as a cube). In opposition, probing several variables at a time in a defined grid more efficiently leads to an understanding of the response curve and identification of optimal reaction conditions. Machine learning can help identifying optimal reaction conditions without following a predefined set of experiments but by analyzing patterns in data that lead to focus on certain areas of the search space and discard others less promising. Unlike the state-of-the-art tools for design of experiments, the main goal of machine learning is not to fully understand the response surface but minimizing the number of experiments required to achieve optimal endpoint outcomes.

synthesis apparatuses to successfully optimize a Knoevenagel condensation, the oxidation of benzylic alcohol to benzaldehyde [39], C–C/C–N cross-couplings, olefinations, reductive aminations, S_NAr among others [40].

It is worthwhile noting that selecting the most efficient computational method to optimize a chemical reaction can be a difficult step, if possible at all. A specific software tool may perform well in one case but be inefficient in another [39, 41, 42]. Furthermore, each method has limitations that restrict their applicability. For example, it is often desirable to simultaneously optimize real-value (e.g. temperature, time) and categorical variables (e.g. base, solvent). This remains inaccessible or highly impractical to the popular model-based DOE [43–45], SNOBFIT, Nelder-Mead SIMPLEX, or similar methods [46, 47] but has been specifically tackled by Gryffin [48], a software tool that parallelizes the optimization of categorical parameters.

In typical cases, reaction optimizations should also consider conflicting objectives, such as the maximization of product yield and the minimization of side product that may hinder purification routines. In the true sense of multi-objective optimization, algorithms can be exploited to find multiple condition sets or solutions that satisfy both goals. In such real-world scenarios, the optimal solution is not a single point in the search space but rather a set of solutions – the so-called Pareto set – where this trade-off is considered. These optimal solutions can be mathematically represented by a front (for 2 objectives) or surface (for >2 objectives) [49]. Pareto optimal points are defined as a set of solutions where an improvement in one objective necessarily causes a worsening of the other [49]. Lapkin and co-workers [43, 49–51] have demonstrated the utility of Pareto fronts, as well as the efficiency of the Thompson Sampling Efficient Multi-Objective (TSEMO) algorithm at identifying them.

The optimization of reaction conditions offers an ideal scenario to quickly generate new data that may be incorporated into the machine learning model for refinement. Unlike other chemistry tasks with long turn-over times for data generation (e.g. bioactivity data that involves multistep synthesis of a compound of interest followed by a biological assay), the optimization of reaction conditions lends itself to the use of active learning strategies, i.e. where data generated is fed back to the algorithm for retraining and creation of improved models [33]. Aligned with this concept, Zare and coworkers demonstrated how the deep reaction optimizer (DRO) software, which employs a neural network and simulated reaction data, was able to optimize four microdroplet reactions in 30 minutes [52].

Bayesian optimization has also played a role in the identification of optimal reaction conditions. For example, a software named Phoenics [42] employs the Bayesian optimization concept and was designed to work under low experimental and time budgets that are typical in the chemical sciences. The software assists the chemist by proposing experiments to run based on two different concepts. One of those concepts is called "exploration" and its main goal is minimizing model uncertainty and promote more accurate predictive models. The other concept is known as "exploitation," where the goal is focusing on regions of interest to identify high-yielding reactions. Because the machine learning model is iteratively tuned with new data, it is expected that the uncertainty in the predicted outputs decreases over time. Practical examples where Bayesian optimization was successfully employed include the optimization of a stereoselective Suzuki–Miyaura coupling [53], multiple C–H functionalizations and C–N coupling reactions [54]. Specifically, the *edbo* software tool [54] encodes reaction data with density functional theory and uses a machine learning algorithm (called Gaussian Process) to steer the reaction selection routine in batches of predefined sizes. In a benchmarking study, *edbo* performed competitively against expert intuition as it more rapidly identified optimal reaction conditions relative to 50 chemists. One of the drawbacks of the Gaussian processes employed in these tools is how it scales with increasing amounts of data and the assumption of normally distributed noise. Experimentally, input uncertainty can deviate from a normal distribution. Golem [55] was developed with these limitations in mind; it reweights the merit of previous experimental conditions to identify robustness relative to variation and builds a surrogate model that approximates robustness rather than modeling the objective directly. Irrespective of the implemented methodologies, we are currently witnessing a surge of approaches for process optimization that deal with limited amounts of data and harness an active learning strategy with bespoke selection policies for experiment prioritization.

In summary, there are several software tools available, and their development continues an active area of research. Those tools are developed with the clear goal of facilitating the synthetic chemist's daily tasks and assist in decision-making. If employed correctly, said technologies have the potential to alleviate routine work, and save valuable time and brainpower.

10.3 Reaction Optimization Using LabMate.ML

All machine learning workflows have three key pillars that determine their success: data, representations (or the so-called features/descriptors), and algorithms [56]. The models for reaction condition optimization are no exception, and one might argue that data and representations are even more critical in this case. Linked to that, a skilled researcher may recognize that automating the optimization of reaction conditions poses specific grand challenges that do not find parallel in other tasks in the chemical sciences. For example, data is exceptionally scarce because only very rarely has a given reaction been performed and documented hundreds or thousands of times [2].

A universal machine learning algorithm for optimizing chemical reactions should be able to operate with relatively few experimental data points. This is challenging in itself and opposes accepted best practices in machine learning research, which tend to emphasize the importance of using large data sets [18]. Because machine learning models can only be as good as their underlying data [25], choosing the appropriate training data is critical to avoid the exploitation of anthropogenic and experimental design biases [57] in the learning routine. Descriptors ought to be agnostic to reaction mechanisms to allow ample use and generalization. Further, the algorithms should be amenable to iterative refinement and multiple rounds of training, swiftly provide predictions with limited computational resources and, ideally, be interpretable to promote trust among organic chemists (end users) [18].

We have implemented LabMate.ML, which is a machine learning algorithm comprised of two key components. An initializer module that helps the user select a random number of experiments to run and build the initial training data, and an optimizer module, which takes the experimental data and generates a predictive model for what the outcome of a yet-to-be-run experiment may be. LabMate.ML was designed with the abovementioned requirements in mind and has been evaluated by optimizing distinct chemistries for a myriad of goals, such as obtaining optimal yields or diastereomeric ratios (Figure 10.3). Namely, we applied LabMate.ML to multiple small-molecule, glyco-, and protein chemistries

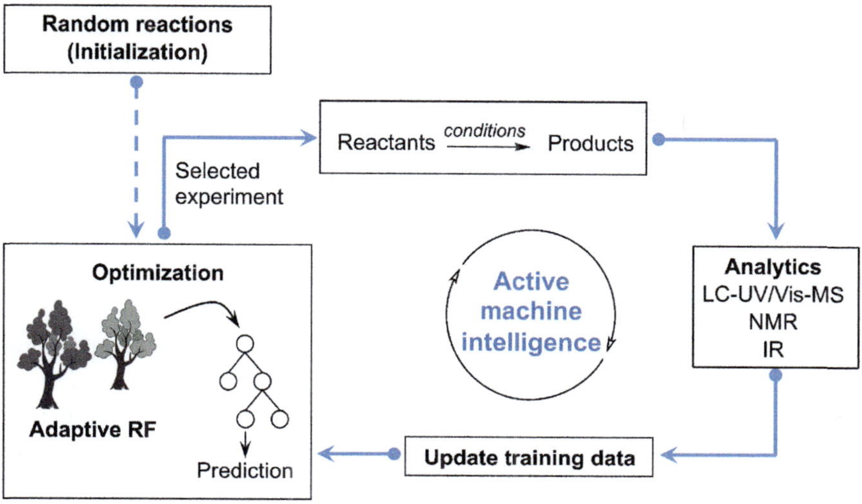

Figure 10.3 Schematics of the LabMate.ML workflow for the optimization of reaction conditions. LabMate.ML includes two Python scripts for initialization and optimization. The training data is iteratively updated as more reaction data becomes available over time.

that are relevant to scaffold generation and decoration as well as late-stage functionalization [18]. LabMate.ML is written in the Python programming language, runs on common laptops, and is meant to democratize the access to machine learning technologies and chemical matter. LabMate.ML uses common and readily available Python "libraries" for basic mathematical analysis (NumPy, Pandas) and ML routines (Scikit-learn), which allows experimentalists to use the program without requiring extensive coding knowledge. In the following sections, we dissect the two modules that serve as workhorses for LabMate.ML, provide a motivation for their implementation, and discuss some examples studied with this reaction optimization routine.

10.3.1 Step One: Accessing the LabMate.ML Code and Installation

The following sections provide a step-by-step walkthrough of each step required to use LabMate.ML to select and interpret experiments with the goal of optimizing them. A potential user can run LabMate.ML with basic knowledge of Python. In the current absence of a graphical user interface, basic knowledge of UNIX commands is required. To install, we direct the reader to https://github.com/tcorodrigues/LabMate.ML while cautioning that LabMate.ML is an active project with continuous updates. The methods that are described herein are updated but may change over time.

1) Install Anaconda (a Python distribution platform) on the local machine;
2) Install the relevant dependencies and files by typing on the terminal:
   a) git clone https://github.com/tcorodrigues/LabMate.ML.git
   b) cd LabMate.ML
   c) conda env create -f environment.yml
3) Activate the created environment in point 2 by typing "conda activate LabMateML."

10.3.2 Step Two: Initializing the Optimization Routine in LabMate.ML

The optimization routine starts by running a script ("initializer.py" using python initializer.py in the terminal) that prepares LabMate.ML for the optimization steps. The initialization script is built with the primary goal of generating all possible combinations of reaction variables (features) that are *a priori* chosen by an expert chemist. These combinations are stored in an output file ("all_combos.txt") that enumerates the so-called search space for which predictions will be made (Figure 10.4). For a more effective use of LabMate.ML, the recommendation is to survey the literature and build a list of conditions that have been employed (both successfully and unsuccessfully) and might be relevant for the reaction under study, such as the range of temperature values, reaction times, catalysts/ligands and others. Further, the search space can comprise both real-valued (e.g. temperature) and categorical features (e.g. solvent type) that are encoded differently.

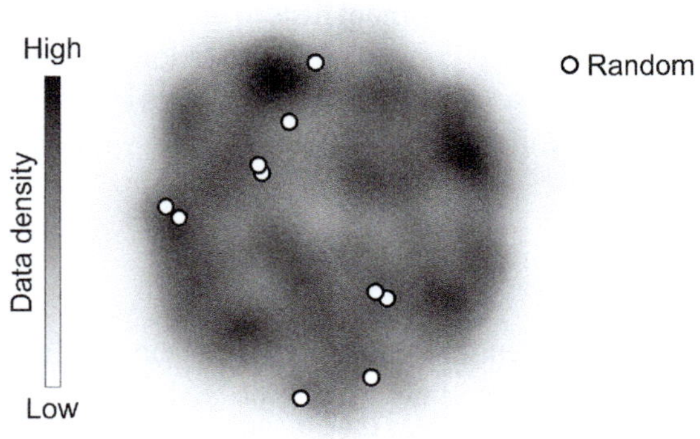

Figure 10.4 The initializer.py script enumerates all possible reaction conditions from a user-defined set of variables and allowed values. The so-called search space (depicted as black surface) comprises all possible reactions for which predictions will be made. A defined number of reactions is then randomly sampled (white dots) from the search space and, upon experimentation, will constitute the initial training dataset.

Those variables are inserted in the "Boundaries.yml" file that works as a reference list for all possible reaction variables and allowed possibilities for each of the variables in the search for optimized conditions. Because LabMate.ML uses random forests (RFs) to build predictive models, no feature normalization is required. Hence real-valued features are used without any further transformation. On the other hand, categorical variables are one-hot encoded to allow mathematical modeling. This encoding method assigns a "1" or "0" value according to the presence or absence of a certain feature. The simplicity of the method comes, however, with the cost of increasing the descriptor vector size and the number of data columns assigned a "0" (absent) value (see Section 10.3.3).

The size of the search space can range from a few hundred to several thousand reactions, wherein features that inherently belong to a continuous space are discretized. This creates a constraint in the search space that is best mitigated by evenly spacing the allowed values for such variables. Equally critical is the balance between the search space size, the initial set of reactions that compose the training data, and the size of the reaction descriptor (i.e. the set of conditions). It is recommended to start with 10 random reactions or, at the very least, with a number of reactions that equals the number of reaction variables being modeled. In general, the more variables being investigated, the greater the challenge it will be for LabMate.ML to build an accurate model. This is sometimes called the "curse of dimensionality," which is likely to occur when there are too many features to parse with few training examples. The fewer variables investigated and the more initial experiments run, the more likely the model will be effective at predicting the yield of yet-to-be-run experiments [58].

Alternatively, the chemist users may have in hand a dataset with information on reactions previously executed and evaluated. While it could in principle bypass the need for performing additional reactions, one should question if those reactions are sufficiently informative and decorrelated from one another. When dealing with low data, it becomes increasingly desirable to ascertain that all training instances are highly informative and do not co-cluster in a specific region of the search space. If the reactions co-cluster in a confined region in the search space, the model will have a small domain of applicability that will ultimately decrease the likelihood of identifying a reaction condition optimum. Further, it has been documented that anthropogenic biases [57] in the dataset construction phase result in models of lower performance relative to those built from random observations. It is worth reemphasizing that the success of machine learning models is highly dependent on the quality of the training data. LabMate.ML is no exception, and the correct identification of an optimum is tightly linked with the diversity of the random reactions, their variables, and the range of allowed values. Once this critical step is executed, the optimization routine can be initiated.

10.3.3 Step Three: Iterative Optimization Routine

The optimization routine in LabMate.ML can be carried out using different reaction experiment prioritization policies, such as exploration, exploitation (or "greedy"), or a mixture of both ("balanced"). While in the exploration mode, the software tool prioritizes reactions that are least understood by the algorithm, in the case of exploitation the routine prioritizes reactions that are confidently predicted as high yielding [18]. In the case of exploration, the main goal is improving the model performance by iteratively including in the training data reactions that are diverse and highly informative. In the case of exploitation, the goal is to swiftly identify the optimal conditions, irrespective of the model performance (Figure 10.5). From our experience, we recommend using exploration when the main intention is understanding the response surface (akin to DoE) or exploitation when the chemist wants to rapidly identify an optimum. Exploitation generally suits the needs in a reaction optimization campaign.

The optimizer.py script runs a series of operations, including loading data that was previously generated ("all_combos.txt" and "train_data.txt" files), training and evaluating a machine learning model through a process called 10-fold cross-validation. In cross-validation, the training data is partitioned into a user-defined number of folds (k-folds), followed by training a model using $k - 1$ folds. The held out fold is then used to assess how good a model was built. In the particular case of LabMate.ML, $k = 10$ (hence 10-fold), which means that a model is trained on

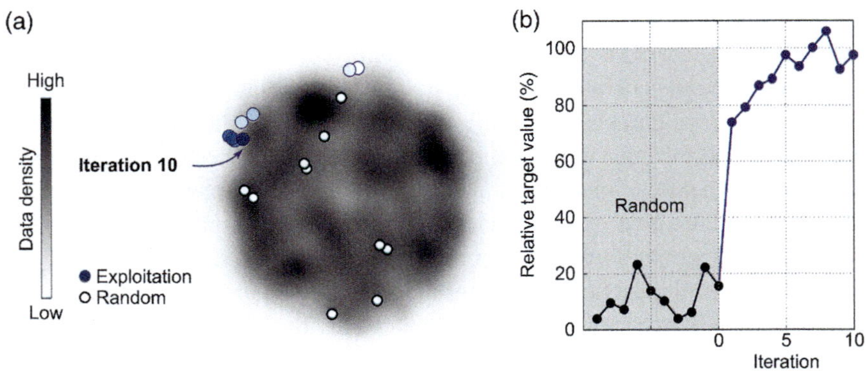

Figure 10.5 LabMate.ML optimizes reaction conditions using random forests (RFs) to mine the search space and an exploration or exploitation approach as experiment selection policy. Generated data is iteratively fed to the algorithm for full retraining (including the selection of hyperparameters). (a) Visualization of the search space and the reactions selected by LabMate.ML using an exploitation approach. The model focuses on a confined region of the search space in an attempt of identifying an optimum. (b) Progress in the optimization of a reaction. LabMate.ML required 7–10 iterations to identify a protocol with quantitative conversion.

9 data partitions and evaluated on a tenth. In fact, the process is repeated 10 times, so that each fold is used once for model evaluation. Once a model is built and retrospectively evaluated through cross-validation, it is then ready to be employed prospectively. The optimizer module in LabMate.ML also predicts outcomes for the reactions in the search space, selects one reaction to execute experimentally, and finalizes by saving all generated information into individual files. Once the reaction is executed and evaluated the user manually updates the "train_data.txt" file and reruns the optimizer.py script to have another experiment suggested.

We have elected RFs as machine learning method of choice for LabMate.ML. RFs harness the wisdom-of-the-crowds concept and consist of multiple decision trees, hence the name forest. Each tree provides a distinct prediction from a certain vantage point on the training data. Ideally, all trees are uncorrelated between themselves, and predictions can (and are desirable to) vary when one analyses trees isolated. Thus, when alone, the trees work as unreliable predictors, but by averaging them a more robust and close-to-reality prediction can be expected. RFs are also suited for the optimization of reaction conditions because they can handle well small amounts of data (unlike complex deep learning methods). Furthermore, RFs are relatively fast to train, allowing predictions on-the-fly and are easier to interpret in comparison to neural networks [18, 19].

Once the required number of experiments are executed, LabMate.ML can build a model to predict reaction outputs, numerically. To achieve this, the LabMate.ML method first queries which are its optimal parameters (the so-called hyperparameters) to achieve a well-performing model. Those parameters are as diverse as the number of trees in the forest (100–1000), the number of descriptors to consider by each tree, and the "depth" of the tree. By depth, the reader should read the number of partitions. Trees with a small number of partitions (e.g. 2) analyses only a small fraction of the data and are expected to provide very crude predictions. Trees without restriction of partitions will be built using the same number of partitions as the number of features being analyses. These result in "pure" leaves that are desirable when one wants to describe the data very accurately (because of overfitting) but are undesirable when the goal is prospective application. This routine for searching model parameters is executed every time that LabMate.ML is fed with new data. In practice, LabMate.ML can choose one out 114 possible combinations. Naturally, selection is made by scoring each model of the 114 possibilities with the mean absolute error (MAE) metric. The MAE provides an average of the sum of distances between predicted and experimental values, over all of the independent models that are built in the cross-validation loop (i.e. training data is partitioned into 10 folds, and each one is iteratively held out for internal evaluation of a model that is built with the remaining 9 folds). Other metrics such as mean squared error could have been employed but using r^2 is in general discouraged (see Section 10.4).

After selecting the best parameter set, the RF model is rebuilt using all available training data. With a model in hand, yields and their variances are calculated for the reactions in the search space that can still be executed. According to the selection policy – which focuses on well-performing reactions and evaluated on multiple instances – the software proposes the reaction with the lowest predictive variance among the top five reactions with the highest predicted yield.

To minimize manual input and increase reproducibility, we have now introduced a series of modifications to the original version of LabMate.ML. We invite the reader to go through more specific documentation and the README file available at https://github.com/DigiChem/LabMate.ML.

10.3.4 Examples

Using LabMate.ML, we have optimized an Ugi four-centred three-component reaction to access a privileged scaffold in drug discovery – imidazopyridines (Figure 10.6a) [18]. By tuning the amount of each of the three building blocks, solvent volume, Lewis acid mol%, temperature, and reaction time, LabMate.ML was able to quickly (over four iterations or <0.05% of all possible experiments) identify reaction conditions leading to quantitative yield. More importantly, the machine learning workflow proved competitive with expert intuition in a limited competition and evolved dynamically, as more data became available for training. In fact, the exploitation approach implemented in LabMate.ML approximates the decision process by expert chemists, providing a viable means to assist in decision-making.

In a more challenging case study, LabMate.ML optimized a C–N coupling that had previously proven challenging to execute by chemists in the pharmaceutical industry (Figure 10.6b) [18, 59, 60]. By optimizing real-value ($\mathbb{R}$) variables and testing different experiment selection policies, LabMate.ML was able to identify conditions favoring the formation of either the C4–N or C2–N products while navigating a common search space. The workflow could also be efficiently extended to other relevant reactions in medicinal chemistry, such as reductive amination, Horner–Wadsworth–Emmons olefination, photocatalytic C–H arylation, *O*-glycosylation, and aza-Michael ligation on proteins, which supports the generalizability of the approach (Figure 10.6c–g) [18]. While optimizing both real-value and categorical variables, LabMate.ML unveiled new chemical insights and discovered methods that improved the synthetic efficiency relative to what had previously been reported in the literature. While it might not be always the case, LabMate.ML was capable of identifying the global optimum in a more efficient fashion (lower amount of experiments and time spent) relative to kinetic modeling in the synthesis of isomacroin through Friedländer chemistry (Figure 10.6h) [19].

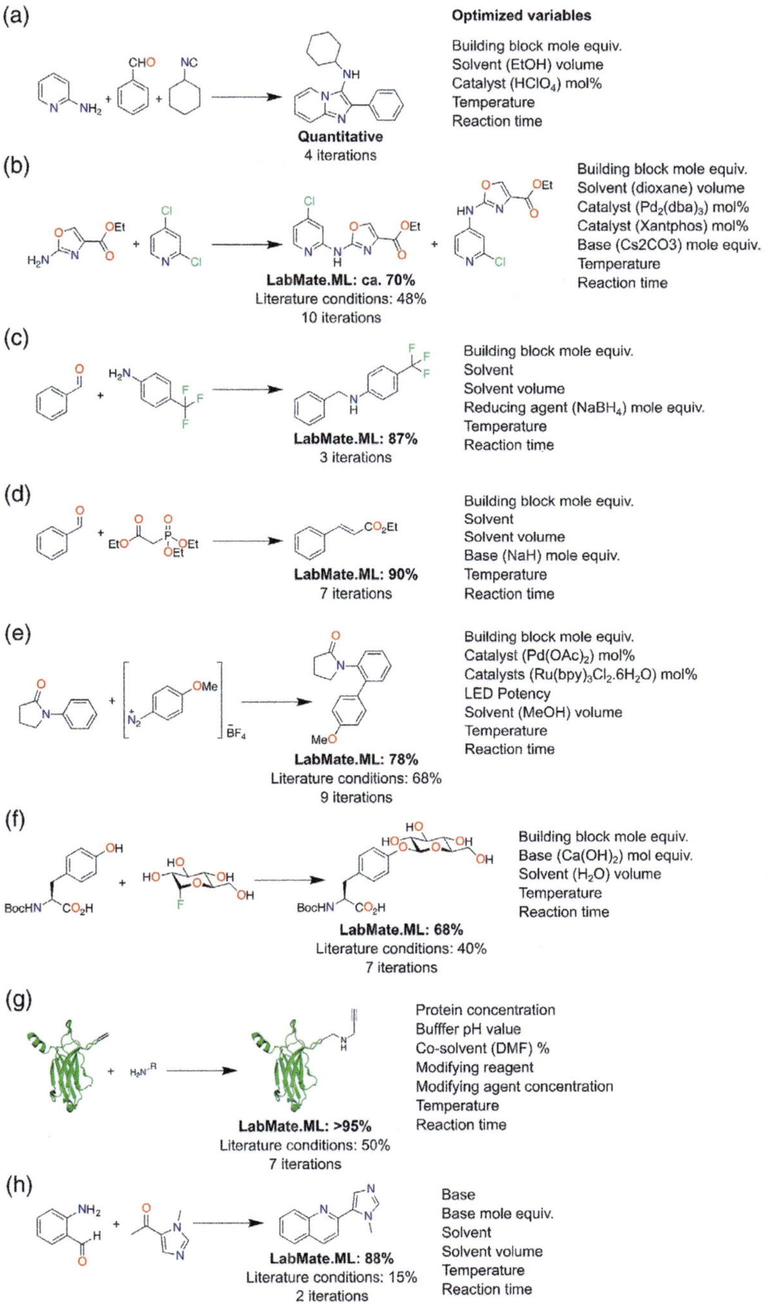

Figure 10.6 Examples of use cases where LabMate.ML was employed to optimize reaction conditions with a small amount of training data. The optimized variables (categorical or real-values) are shown for each example, together with the obtained yield, yield using the literature conditions and the number of iterations used by LabMate.ML. (a) Ugi multi-component reaction. (b) C–N cross-coupling. (c) Reductive amination. (d) Horner–Wadsworth–Emmons olefination. (e) Photocatalytic C–H arylation. (f) *O*-glycosylation. (g) Aza-Michael ligation. (h) Friedländer quinoline synthesis.

10.4 Primer on Evaluation Guidelines

Machine learning algorithms have the potential to impact established research routines and workflows. If employed in an informed way, algorithms can identify patterns, steer experimentation, and support sustainable research. In respect to reaction condition optimization, a growing body of literature showcases that different methods may serve the purpose of assisting bench chemists designing the most informative experiment, according to their needs. A key challenge for the least experienced user and enthusiast is thus selecting the workflow that will be employed to optimize reaction conditions. One of the bottlenecks for an informed decision regarding the selection of an appropriate machine learning tool is extracting the information from the literature that is linked to the algorithm's evaluation experiments. The current heterogeneity in reporting practices can perpetuate the impracticality of comparing tools.

In this section, we provide bonus material on evaluation guidelines with two main purposes. First, with the hope of helping the readers understanding some of the language employed in publications and help them interpreting the information that is provided. We expect this to lower the entry barrier for new researchers. Second, to introduce reporting standards that can help the community to more objectively compare strengths and weaknesses between tools designed for the same purpose, in particular reaction optimization workflows.

On a practical level, we may consider machine learning studies as having one of three main goals: novel concepts, benchmarks, or novel discoveries. According to each of the study types, different evaluation guidelines can be recommended [56]. Those recommendations meant to standardize data reporting, improve quality, transferability, reuse of methods, and, ultimately, help researchers evaluate strengths and weaknesses of machine learning workflows by themselves. Following standardized data reporting procedures will be critical to manage expectations regarding the utility of the tools and assist in the selection of which methods are appropriate to specific use cases. The recommendations for supervised learning algorithms that we summarize here are relevant for the optimization of reaction conditions. Generally, they are focused on reporting practices, evaluation metrics in retrospective studies, comparison to competing methods and baselines, and the design of prospective use cases.

10.4.1 Code and Dataset Availability

It is recommended to curate datasets because machine learning models can only be as good as the underlying data. Is there information available? Are the data reported in a homogeneous way? Which variables were considered? These are some of the questions one must ask before of optimization reaction conditions or, as a matter of fact, any other chemical problem with machine learning. In some cases, it might be needed to engineer features that are meaningful for the problem one is trying to solve or generate additional data. When reporting, it should be very clear if data was excluded from the modeling routine together with a brief explanation. Those datasets ought to be available through repositories (e.g. GitHub) in a machine readable format (e.g. *.txt, *csv) to enable reuse by the community and calculation of data ranges, which are important when assessing the domain of applicability of the machine learning models. Relevant code should also be made accessible to the community.

These aspects influence the selection of appropriate metrics for retrospective evaluation studies, which are critical to evaluate the utility of any machine learning workflow. Ideally, different machine learning pipelines are evaluated using publicly available benchmark datasets with the goal of assessing advantages/disadvantages under identical train/test data [61, 62]. While benchmark datasets are publicly available for several tasks in drug discovery, in the case of chemical synthesis only the USPTO data [63] was accessible until recently. Said dataset has been routinely used to guide the implementation and evaluation of machine learning tools for retrosynthetic planning applications. More recently, the open reaction database was released with the goal of organizing and structuring reaction data from multiple sources [64, 65].

10.4.2 Retrospective Evaluation

Retrospective evaluations are a cheap and accessible means of challenging the generalizability of a machine learning model [66]. They are conducted with data that has not taken part in the training process through one of several possible cross-validation schemes (e.g. 5-/10-fold as described in Section 10.3.3). The main goal of these cross-validations is providing a platform for tuning model-specific parameters and estimating the uncertainty of the model (Figure 10.7) [67]. As such, retrospective evaluations may provide meaningful performance baselines, while mimicking re-discovery if employed on an external evaluation set [68]. Importantly, for a meaningful comparison between tools through cross-validation studies, identical split schemes must be employed.

Depending on the goal of the machine learning workflow, different metrics can be employed to gauge its utility. For regression models, such as LabMate.ML, the MAE and MSE are recommended since they inform on the uniformity of the prediction errors. LabMate.ML automatically computes the MAE without any additional coding. The coefficient of determination (r^2) can also be provided although not all modeled tasks present a linear correlation between features and response [18]. Indeed, much debate on the use of r^2 and q^2 (also known as cross-validated r^2) has been documented [69–71]. For these reasons, the recommended

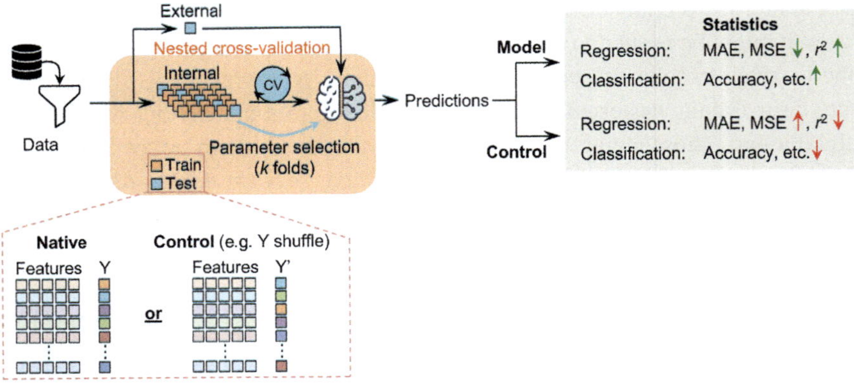

Figure 10.7 Proposed retrospective workflow for machine learning routines. Cross-validation with a defined number of data splits is performed to select the best combination of model hyperparameters. In each iteration, one fold is held out for internal validation of the model that was built with the remaining data. Whenever possible, an external test set should also be employed to evaluate the performance of the model. The model evaluation is performed through metrics that must be selected according to the type of model and data in hand. As a control, the target variables can be shuffled to originate a second dataset of scrambled data. A model that is built with these shuffled data is expected to have lower performance compared to the models built with the native dataset. MAE: mean absolute error. MSE: mean squared error. r^2: coefficient of determination.

practice is evaluating machine learning models according to an ensemble of metrics instead of a single, user-preferred one. An identical principle should be employed in classification tasks, wherein the model predicts "reaction" or "no reaction." Classification models can be assessed by the number of true/false positive and negative predictions and derived metrics. One should however bear in mind that different use cases might require different metrics and a degree of flexibility is recommended.

10.4.3 Baselines and Comparing Tools

The recent widespread interest in machine learning tools for the chemical sciences has resulted in numerous studies with different models and assumptions for myriad applications. Because reports are heterogeneous regarding comparisons between tools, less informed researchers may, arguably, become confused when evaluating and forming an opinion on specific machine learning models. For these reasons, it is important to perform comparisons with null models and other established tools even if they do not employ a statistical learning strategy (Figure 10.8) [72]. Most importantly, claims of superior performance must be accompanied by robust statistical evaluation that is corrected for multiple comparisons and reference to the sample size [73]. Simple comparisons, such as mean

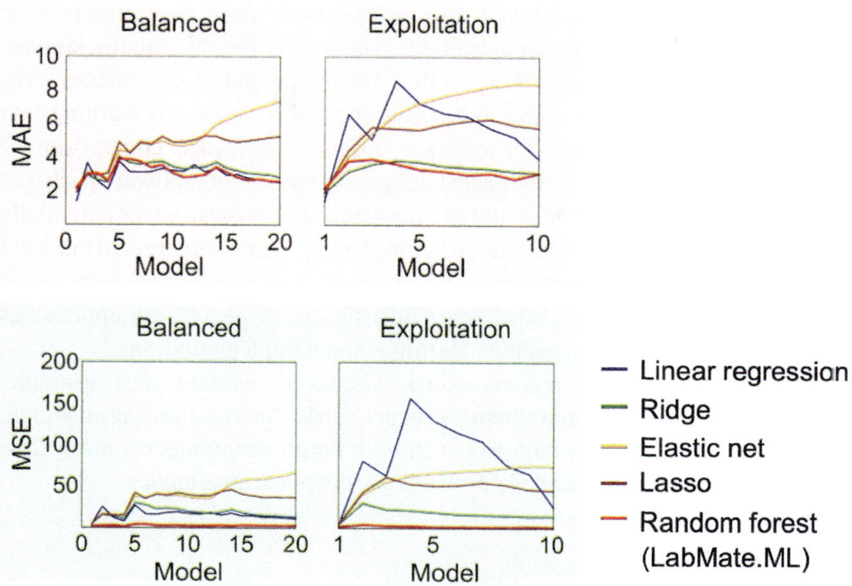

Figure 10.8 Comparison of LabMate.ML with simpler, linear regression methods. The plots show the model performance over several iterations using two different experiment selection strategies. Collectively, low MAE and MSE values are desirable, which attest the lower prediction uncertainty in the cross-validation routine.

values and prior class distributions, are computationally cheap to execute. They provide naïve yet meaningful baselines that are important to motivate the added value of machine learning models with higher complexity. This recommendation is valid for LabMate.ML. Although it was shown that the workflow performed well on a number of different chemistries and was competitive with simpler methods, one cannot guarantee that will be the case for every scenario if the focus is assessing the method's utility.

While assessing the scope of LabMate.ML the effect of artefacts was also studied. Artefacts can be introduced by experimental design and meaningless features have the potential to confound machine learning models [74–76]. An additional baseline and computationally accessible test to rule out the use of confounding variables consists in randomizing target values/classes (the so-called *Y*-randomization/shuffling test) and subsequently building an optimized model from the generated dataset (Figure 10.7) [11]. In doing so, the yield or conversion value will be assigned to a different reaction. The model generated from the scrambled dataset should present lower performance (near random) relative to the counterpart generate with the real dataset. Otherwise, it might suggest the irrelevance of the features used and the poor machine learning model generalizability.

Linear regression models should also be computed since they may provide statistically relevant and more interpretable models [18, 19, 77]. Equally, similarity searches provide relevant baselines that are simple and of easy access; they inform on what a computer is able to discriminate without explicitly learning data patterns. Along these lines, for reaction optimization tasks, comparison of machine learning tools with established design of experiments software packages or kinetic modeling may be relevant [19]. Irrespective of being used for potentially distinct objectives (fast optimization according to an objective function in machine learning *vs.* understanding the response curve in design of experiments) such comparisons may inform on advantages and disadvantages of both approaches and guide developments in machine learning model implementations.

Benchmarking is also recommended whenever possible. For example, Summit [41] contains virtual chemical benchmarks for reaction optimization. Further, Olympus [78] was built to benchmark optimization algorithms against realistic experiments emulated via probabilistic deep-learning models.

10.4.4 Prospective Evaluation

While retrospective studies are an appropriate and inexpensive means of evaluation, not too infrequently they provide an optimistic view on the machine learning model utility. Prospective deployment of machine learning tools in actual discovery programs often leads to lower-than-expected performance, relative to what was retrospectively obtained. The lower performance can result from:

1) Poor model generalizability that was imperceptible in cross-validation routines, due to narrow domain of applicability. This is intimately linked with the second and third points;
2) Dataset limitations in respect to examples and diversity;
3) Inappropriate use of code (e.g. use case outside the domain of applicability or to solve problems for which the model was not designed for).

From a wet-lab, end-user perspective, it is important to conduct an expanded set of experiments to better frame the tool's utility and manage expectations in regards to modeling outcomes and predictability (Figure 10.9). Only then will the wet-lab chemists be able to appreciate the machine learning tool applicability and whether it will serve real-world needs. Here, it is also crucial to emphasize the impossibility of evaluating machine learning tools on their own. Prospective studies do not evaluate algorithms but rather whole processes [80], which include data preparation steps, feature engineering, hyperparameter selection and experiment prioritization policies. Mispredictions can be linked to deficiencies in one or several of these key points. As such, the model cannot be judged by only a few test cases. Further, when deploying a model, one assumes that all variables in the

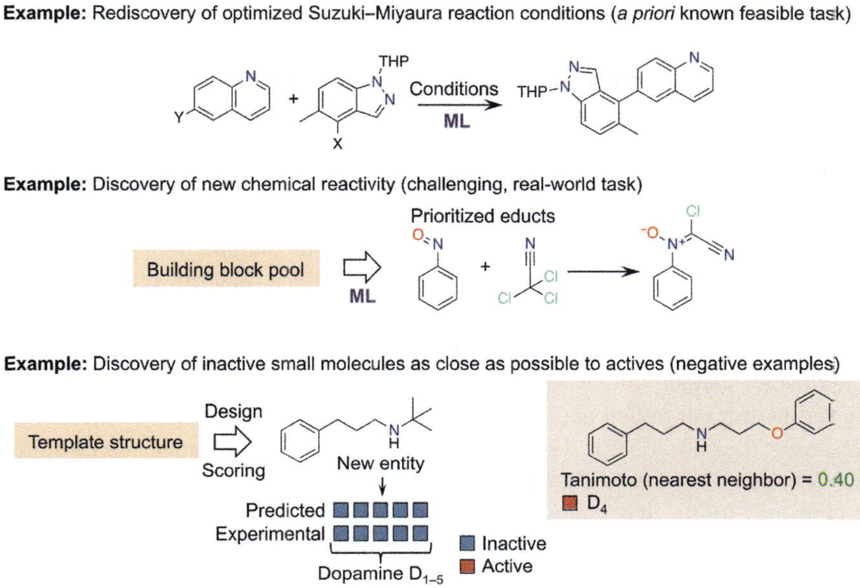

Figure 10.9 Examples of use cases where machine learning was applied for the re-discovery of optimal Suzuki–Miyaura reaction conditions, the discovery of new reactivity (*Source:* Granda et al.[23]/Springer Nature) and design of negative examples (*Source:* Adapted from Reutlinger et al. [79]). THP: tetrahydropyranyl. D_{1-5}: dopamine receptors 1–5.

actual assay are known, controlled, and experimental variability is negligible if not tested in multiple replicates. While errors in experiments testing physical phenomena can be low, one may easily appreciate that biological tests (*in vitro*, *in vivo*, and clinical) are more complex and entail inherent higher variability. This inherent uncertainty adds layers of complexity to the process evaluation routines.

Prospective examples must be relevant and diverse, covering disparate application scenarios and/or probing different aspects of the ML tool and overall process. To meaningfully identify strengths and limitations of the ML tools, three types of experiments should be conducted, following a "divide-and-conquer" evaluation strategy:

1) Rediscovery test;
2) Discovery test;
3) Negative test.

In rediscovery tests, the main goal is prospectively applying the implemented machine learning tool in a semi-controlled, real-world scenario with limited unknown variables. These are configured as examples where a positive outcome can or has been achieved previously but is new to the machine learning routine. For example, to challenge the LabMate.ML routine and identify the most appropriate experiment selection policy, the synthesis of a previously reported imidazopyridine was optimized [18]. While no new knowledge was generated in this case, the re-discovery test provided preliminary proof-of-concept and motivated exploration of the machine learning routine in use cases of actual interest.

In discovery tests, the machine learning tool is employed in a hitherto unsolved research question, with the goal of contributing to the augmentation of domain knowledge. These tests are of utmost importance since they can provide robust support to the transferability of the machine learning model. However, successes should not be treated as definite proof-of-concept. Robust validation is virtually impossible in single studies with a limited number of examples whose selection is biased by human preference. Likewise, inability to reach a positive outcome in discovery tests also does not provide a statistically meaningful measure of utility but can rather point toward gaps of knowledge or new directions of improvement.

One may argue that models that consistently predict positive outcomes are somewhat biased. Since high-quality negative data is as informative and important as positive data [25], a third test is recommended. While the focus of machine learning routines for reaction optimization should be on the identification of positive outcomes (i.e. conditions leading to high yielding reactions), "failure-by-design" tests are necessary. Tests focused on identifying undesired outcomes are underrated, taking into account the difficulty in finding negative outputs or mispredictions in the literature. However, if conducted correctly, they can inform on false-negative/positive rates and clarify the ability of the machine learning model to discriminate experiments that cover the full response spectrum.

In summary, while the development of innovative machine learning tools for reaction optimization is important, it is equally relevant to regularize how models are described, evaluated, and compared with existing codebase. The set of guidelines we provide here (Table 10.1) are specifically designed to compare machine learning tools for reaction optimization tasks but can be readily adapted to other fields of the chemical sciences.

10.5 Outlook

In this chapter, we have presented and discussed how machine learning is impacting reaction condition optimization routines that have traditionally relied on expert intuition and many years of accumulated hands-on experience. Indeed, reaction optimizations can be efficiently configured as a pattern recognition task and are thus amenable to data mining approaches. The integration of intuitive and optimizable features (conditions) as means of describing reactions is

Table 10.1 Summary of required evaluation studies according to the manuscript content.

	Manuscript content		
	Novel concept	**Benchmark**	**Discovery**
Dataset availability	++	+++	++
Code availability	++	+++	++
Retrospective evaluation	+++	+++	n.r.*
Comparison to baseline	+++	+++	n.r.*
Prospective evaluation	n.r.	n.r.	+++

Key: n.r.: not required; +: desirable; ++: important; +++: very important. *if reported elsewhere.

advantageous since they do not rely on quantum chemical computations and detailed reaction mechanisms. We argue this may promote the adoption of LabMate.ML and similar tools as assistants for decision-making.

The field of cheminformatics for synthetic chemistry is now taking important first steps and we expect to witness the implementation of more elaborated routines to explore reactivity spaces in the near future. Such tools ought to focus on key aspects that pose significant challenges to machine learning. While machine learning workflows have gained traction in areas where a significant amount of data is available (e.g. computer vision), reaction optimization problems use cases where the amount of data is typically scarce. This can be countered by emerging high-throughput experimentation (HTE) assisted by robotic systems. While HTE is becoming increasingly accessible to the typical synthetic chemist (see Chapter 6), highly automated systems capable of running thousands of experiments may remain inaccessible to low budget laboratories in the foreseeable future. For this reason, it is imperative to pursue meaningful goals and reaction selection policies that maximize information gains to the model while refining uncertainty estimates from low data. How to gauge said model uncertainty (aleatoric vs. epistemic) is a matter of debate without clear recommendations and standardized methods. Further, because such machine learning models have to operate in low-data scenarios, it is expected that experimental issues that cause high response variance will negatively impact on the quality of the predictions. A means to account for experimental uncertainty is also currently unclear but may boost the applicability of machine learning tools for reaction optimization to a superior level.

The implementation of platforms where humans and machines "collaborate" toward a defined goal will gradually build trust among wet laboratory practitioners. That trust and the realization of the added value that machine learning tools may bring is not only the result of the number of correct predictions but also how much new chemical insight those predictions can afford. To that end, it is important to continue exploring models that directly afford insight and scrutinize black box algorithms to extract new knowledge [81]. A word of caution is warranted in this respect. Model explanations are not holistic as they depend on the employed descriptors and algorithms [82]. Moreover, popular model explanation pipelines, such as SHAP [83] and Lime [84], build model approximations but are not the machine learning model itself. This means that their outputs may not depict the decision process with high fidelity.

Lastly, a perceived entry barrier for the wet-lab chemists is the unfamiliarity with programming languages, data science principles, and/or the underlying math. A solution to this lies in the implementation of web interfaces or applications with graphical interfaces for local use. Overall, much remains to be explored in this space, but the recent success in the field suggests that chemists are gaining viable assistants in the laboratory to conduct research in a more sustainable and probabilistically motivated fashion. Ultimately, the success of machine learning tools will endorse the allocation of expert brainpower to less repetitive tasks, thus promoting productivity.

References

1 Vamathevan, J., Clark, D., Czodrowski, P. et al. (2019). Applications of machine learning in drug discovery and development. *Nat. Rev. Drug Discov.* 18: 463–477.

2 de Almeida, A.F., Moreira, R., and Rodrigues, T. (2019). Synthetic organic chemistry driven by artificial intelligence. *Nat. Rev. Chem.* 3: 589–604.

3 Gawehn, E., Hiss, J.A., Brown, J.B., and Schneider, G. (2018). Advancing drug discovery via GPU-based deep learning. *Exp. Opin. Drug Discov.* 13: 579–582.

4 Zhavoronkov, A., Ivanenkov, Y.A., Aliper, A. et al. (2019). Deep learning enables rapid identification of potent DDR1 kinase inhibitors. *Nat. Biotechnol.* 37: 1038–1040.

5 Segler, M.H.S., Kogej, T., Tyrchan, C., and Waller, M.P. (2018). Generating focused molecule libraries for drug discovery with recurrent neural networks. *ACS Cent. Sci.* 4: 120–131.

6 Yao, Z., Sánchez-Lengeling, B., Bobbitt, N.S. et al. (2021). Inverse design of nanoporous crystalline reticular materials with deep generative models. *Nat. Mach. Intell.* 3: 76–86.

7 Moret, M., Friedrich, L., Grisoni, F. et al. (2020). Generative molecular design in low data regimes. *Nat. Mach. Intell.* 2: 171–180.

8 Gómez-Bombarelli, R., Wei, J.N., Duvenaud, D. et al. (2018). Automatic chemical design using a data-driven continuous representation of molecules. *ACS Cent. Sci.* 4: 268–276.

9 Wang, T., Qiao, Y., Ding, W. et al. (2019). Improved fragment sampling for ab initio protein structure prediction using deep neural networks. *Nat. Mach. Intell.* 1: 347–355.

10 Senior, A.W., Evans, R., Jumper, J. et al. (2020). Improved protein structure prediction using potentials from deep learning. *Nature* 577: 706–710.

11 Lee, K., Yang, A., Lin, Y.-C. et al. (2021). Combating small-molecule aggregation with machine learning. *Cell Rep. Phys. Sci.* 2: 100573.

12 Timmreck, R., Meyer, T., Gilot, J. et al. (2015). Characterization of tandem organic solar cells. *Nat. Photon.* 9: 478–479.

13 Reker, D., Bernardes, G.J.L., and Rodrigues, T. (2019). Computational advances in combating colloidal aggregation in drug discovery. *Nat. Chem.* 11: 402–418.

14 Tian, Y., Yuan, R., Xue, D. et al. (2021). Determining multi-component phase diagrams with desired characteristics using active learning. *Adv. Sci.* 8: 2003165.

15 Rodrigues, T., de Almeida, B.P., Barbosa-Morais, N.L., and Bernardes, G.J.L. (2019). Dissecting celastrol with machine learning to unveil dark pharmacology. *Chem. Commun.* 55: 6369–6372.

16 Conde, J., Pumroy, R.A., Baker, C. et al. (2021). Allosteric antagonist modulation of TRPV2 by Piperlongumine impairs glioblastoma progression. *ACS Cent. Sci.* 7: 868–881.

17 Rodrigues, T. and Bernardes, G.J.L. (2020). Machine learning for target discovery in drug development. *Curr. Opin. Chem. Biol.* 56: 16–22.

18 Reker, D., Hoyt, E.A., Bernardes, G.J.L., and Rodrigues, T. (2020). Adaptive optimization of chemical reactions with minimal experimental information. *Cell Rep. Phys. Sci.* 1: 100247.

19 Almeida, A.F., Ataíde, F.A.P., Loureiro, R.M.S. et al. (2021). Augmenting adaptive machine learning with kinetic modeling for reaction optimization. *J. Org. Chem.* 86: 14192–14198.

20 Burger, B., Maffettone, P.M., Gusev, V.V. et al. (2020). A mobile robotic chemist. *Nature* 583: 237–241.

21 Porwol, L., Kowalski, D.J., Henson, A. et al. (2020). An autonomous chemical robot discovers the rules of inorganic coordination chemistry without prior knowledge. *Angew. Chem. Int. Ed.* 59: 11256–11261.

22 Coley, C.W., Thomas, D.A. III, Lummiss, J.A. et al. (2019). A robotic platform for flow synthesis of organic compounds informed by AI planning. *Science* 365: eaxx1566.

23 Granda, J.M., Donina, L., Dragone, V. et al. (2018). Controlling an organic synthesis robot with machine learning to search for new reactivity. *Nature* 559: 377–381.

24 Tschandl, P., Rinner, C., Apalla, Z. et al. (2020). Human–computer collaboration for skin cancer recognition. *Nat. Med.* 26: 1229–1234.

25 Rodrigues, T. (2019). The good, the bad, and the ugly in chemical and biological data for machine learning. *Drug Discov. Today Technol.* 32–33: 3–8.

26 Segler, M.H.S. and Waller, M.P. (2017). Neural-symbolic machine learning for retrosynthesis and reaction prediction. *Chem. Eur. J.* 23: 5966–5971.

27 Mo, Y., Guan, Y., Verma, P. et al. (2021). Evaluating and clustering retrosynthesis pathways with learned strategy. *Chem. Sci.* 12: 1469–1478.

28 Schwaller, P., Petraglia, R., Zullo, V. et al. (2020). Predicting retrosynthetic pathways using transformer-based models and a hyper-graph exploration strategy. *Chem. Sci.* 11: 3316–3325.

29 Schreck, J.S., Coley, C.W., and Bishop, K.J.M. (2019). Learning retrosynthetic planning through simulated experience. *ACS Cent. Sci.* 5: 970–981.

30 Coley, C.W., Jin, W., Rogers, L. et al. (2019). A graph-convolutional neural network model for the prediction of chemical reactivity. *Chem. Sci.* 10: 370–377.

31 Chen, H., Yamaguchi, S., Morita, Y. et al. (2021). Data-driven catalyst optimization for stereodivergent asymmetric synthesis by iridium/boron hybrid catalysis. *Cell Rep. Phys. Sci.* 2: 100679.

32 Rodrigues, T. (2022). Deriving intuition in catalyst design with machine learning. *Chem* 8: 15–17.

33 Reker, D. and Schneider, G. (2015). Active-learning strategies in computer-assisted drug discovery. *Drug Discovery Today* 20: 458–465.

34 Halford, G.S., Baker, R., McCredden, J.E., and Bain, J.D. (2005). How many variables can humans process? *Psychol. Sci.* 16: 70–76.

35 Reizman, B.J. and Jensen, K.F. (2016). Feedback in flow for accelerated reaction development. *Acc. Chem. Res.* 49: 1786–1796.

36 Leardi, R. (2009). Experimental design in chemistry: a tutorial. *Anal. Chim. Acta* 652: 161–172.

37 Weissman, S.A. and Anderson, N.G. (2015). Design of experiments (DoE) and process optimization. A review of recent publications. *Org. Process Res. Dev.* 19: 1605–1633.

38 Huyer, W. and Neumaier, A. (2008). SNOBFIT – stable noisy optimization by branch and fit. *ACM Trans. Math. Software* 35: 1–25.

39 McMullen, J.P. and Jensen, K.F. (2010). An automated microfluidic system for online optimization in chemical synthesis. *Org. Process Res. Dev.* 14: 1169–1176.

40 Bédard, A.-C., Adamo, A., Aroh, K.C. et al. (2018). Reconfigurable system for automated optimization of diverse chemical reactions. *Science* 361: 1220–1225.

41 Felton, K.C., Rittig, J.G., and Lapkin, A.A. (2021). Summit: benchmarking machine learning methods for reaction optimisation. *Chem. Methods* 1: 116–122.

42 Häse, F., Roch, L.M., Kreisbeck, C., and Aspuru-Guzik, A. (2018). Phoenics: a Bayesian optimizer for chemistry. *ACS Cent. Sci.* 4: 1134–1145.

43 Echtermeyer, A., Amar, Y., Zakrzewski, J., and Lapkin, A. (2017). Self-optimisation and model-based design of experiments for developing a C–H activation flow process. *Beilstein J. Org. Chem.* 13: 150–163.

44 Hsieh, H.-W., Coley, C.W., Baumgartner, L.M. et al. (2018). Photoredox iridium–nickel dual-catalyzed decarboxylative arylation cross-coupling: from batch to continuous flow via self-optimizing segmented flow reactor. *Org. Process Res. Dev.* 22: 542–550.

45 Reizman, B.J., Wang, Y.-M., Buchwald, S.L., and Jensen, K.F. (2016). Suzuki–Miyaura cross-coupling optimization enabled by automated feedback. *React. Chem. Eng.* 1: 658–666.

46 Cortés-Borda, D., Wimmer, E., Gouilleux, B. et al. (2018). An autonomous self-optimizing flow reactor for the synthesis of natural product carpanone. *J. Org. Chem.* 83: 14286–14299.

47 Cortés-Borda, D., Kutonova, K.V., Jamet, C. et al. (2016). Optimizing the Heck–Matsuda reaction in flow with a constraint-adapted direct search algorithm. *Org. Process Res. Dev.* 20: 1979–1987.

48 Häse, F., Aldeghi, M., Hickman, R.J. et al. (2021). Gryffin: an algorithm for Bayesian optimization of categorical variables informed by expert knowledge. *Appl. Phys. Rev.* 8: 031406.

49 Schweidtmann, A.M., Clayton, A.D., Holmes, N. et al. (2018). Machine learning meets continuous flow chemistry: automated optimization towards the Pareto front of multiple objectives. *Chem. Eng. J.* 352: 277–282.

50 Jeraal, M.I., Sung, S., and Lapkin, A.A. (2021). A machine learning-enabled autonomous flow chemistry platform for process optimization of multiple reaction metrics. *Chem. Meth.* 1: 71–77.

51 Amar, Y., Schweidtmann, A.M., Deutsch, P. et al. (2019). Machine learning and molecular descriptors enable rational solvent selection in asymmetric catalysis. *Chem. Sci.* 10: 6697–6706.

52 Zhou, Z., Li, X., and Zare, R.N. (2017). Optimizing chemical reactions with deep reinforcement learning. *ACS Cent. Sci.* 3: 1337–1344.

53 Christensen, M., Yunker, L.P.E., Adedeji, F. et al. (2021). Data-science driven autonomous process optimization. *Commun. Chem.* 4: 112.

54 Shields, B.J., Stevens, J., Li, J. et al. (2021). Bayesian reaction optimization as a tool for chemical synthesis. *Nature* 590: 89–96.

55 Aldeghi, M., Häse, F., Hickman, R.J. et al. (2021). Golem: an algorithm for robust experiment and process optimization. *Chem. Sci.* 12: 14792–14807.

56 Bender, A., Schneider, N., Segler, M. et al. (2022). *Evaluation guidelines for machine learning tools in the chemical sciences. Nat. Rev. Chem.* 6: 428–442.

57 Jia, X., Allyson, A., Huang, Y. et al. (2019). Anthropogenic biases in chemical reaction data hinder exploratory inorganic synthesis. *Nature* 573: 251–255.

58 Chen, L. (2009). Curse of dimensionality. In: *Encyclopedia of Database Systems*, 545–546. Springer US https://doi.org/10.1007/978-0-387-39940-9_133.

59 Noonan, G.M., Dishington, A.P., Pink, J., and Campbell, A.D. (2012). Studies on the coupling of substituted 2-amino-1,3-oxazoles with chloro-heterocycles. *Tetrahedron Lett.* 53: 3038–3043.

60 Blakemore, D.C., Castro, L., Churcher, I. et al. (2018). Organic synthesis provides opportunities to transform drug discovery. *Nat. Chem.* 10: 383–394.

61 Brown, N., Fiscato, M., Segler, M.H.S., and Vaucher, A.C. (2019). GuacaMol: benchmarking models for de Novo molecular design. *J. Chem. Inf. Model.* 59: 1096–1108.

62 Polykovskiy, D., Zhebrak, A., Sanchez-Lengeling, B. et al. (2020). Molecular sets (MOSES): a benchmarking platform for molecular generation models. *Front. Pharmacol.* 11: 565644.

63 Lowe, D. (2012). Chemical reactions from US patents (1976-Sep2016). PhD thesis.

64 Kearnes, S.M., Maser, M.R., Wleklinski, M. et al. (2021). The open reaction database. *J. Am. Chem. Soc.* 143: 18820–18826.

65 McCloskey, K., Sigel, E.A., Kearnes, S. et al. (2020). Machine learning on DNA-encoded libraries: a new paradigm for hit finding. *J. Med. Chem.* 63: 8857–8866.

66 Mathai, N., Chen, Y., and Kirchmair, J. (2020). Validation strategies for target prediction methods. *Brief. Bioinf.* 21: 791–802.

67 Vishwakarma, G., Sonpal, A., and Hachmannm, J. (2021). Metrics for benchmarking and uncertainty quantification: quality, applicability, and a path to best practices for machine learning in chemistry. *arXiv:2010.00110.*

68 del Rosario, Z., Rupp, M., Kim, Y. et al. (2020). Assessing the frontier: active learning, model accuracy, and multi-objective candidate discovery and optimization. *J. Chem. Phys.* 153: 024112.

69 Consonni, V., Ballabio, D., and Todeschini, R. (2009). Comments on the definition of the Q2 parameter for QSAR validation. *J. Chem. Inf. Model.* 49: 1669–1678.

70 Golbraikh, A. and Tropsha, A. (2002). Beware of q2! *J. Mol. Graph. Model.* 20: 269–276.

71 Alexander, D.L.J., Tropsha, A., and Winkler, D.A. (2015). Beware of R2 : simple, unambiguous assessment of the prediction accuracy of QSAR and QSPR models. *J. Chem. Inf. Model.* 55: 1316–1322.

72 Walters, W.P. and Murcko, M. (2020). Assessing the impact of generative AI on medicinal chemistry. *Nat. Biotechnol.* 38: 143–145.

73 Aickin, M. and Gensler, H. (1996). Adjusting for multiple testing when reporting research results: the Bonferroni vs Holm methods. *Am. J. Public Health* 86: 726–728.

74 Chuang, K.v. and Keiser, M.J. (2018). Comment on "Predicting reaction performance in C–N cross-coupling using machine learning.". *Science* 362: eaat8603.

75 Ahneman, D.T., Estrada, J.G., Lin, S. et al. (2018). Predicting reaction performance in C–N cross-coupling using machine learning. *Science* 360: 186–190.

76 Chuang, K.v. and Keiser, M.J. (2018). Adversarial controls for scientific machine learning. *ACS Chem. Biol.* 13: 2819–2821.

77 Reker, D., Rybakova, Y., Kirtane, A.R. et al. (2021). Computationally guided high-throughput design of self-assembling drug nanoparticles. *Nat. Nanotechnol.* 16: 725–733.

78 Häse, F., Aldeghi, M., Hickman, R.J. et al. (2021). Olympus: a benchmarking framework for noisy optimization and experiment planning. *Mach. Learn. Sci. Technol.* 2: 035021.

79 Reutlinger, M., Rodrigues, T., Schneider, P., and Schneider, G. (2014). Multi-objective molecular De Novo design by adaptive fragment prioritization. *Angew. Chem. Int. Ed.* 53: 4244–4248.

80 Kearnes, S. (2021). Pursuing a prospective perspective. *Trends Chem.* 3: 77–79.

81 Jiménez-Luna, J., Grisoni, F., and Schneider, G. (2020). Drug discovery with explainable artificial intelligence. *Nat. Mach. Intell.* 2: 573–584.

82 Sheridan, R.P. (2019). Interpretation of QSAR models by coloring atoms according to changes in predicted activity: how robust is it? *J. Chem. Inf. Model.* 59: 1324–1337.

83 Lundberg, S.M., Nair, B., Vavilala, M.S. et al. (2018). Explainable machine-learning predictions for the prevention of hypoxaemia during surgery. *Nat. Biomed. Eng.* 2: 749–760.

84 Ribeiro, M.T., Singh, S., and Guestrin, C. (2016). "Why should i trust you?": explaining the predictions of any classifier. *arXiv:1602.04938v3.*

11

Computer-Assisted Synthesis Planning

Zhengkai Tu[1,], Itai Levin[2,3,*], and Connor W. Coley[3,4]*

[1] *Computational Science and Engineering, Massachusetts Institute of Technology, Cambridge, MA, USA*
[2] *Synthetic Biology Center, Department of Biological Engineering, Massachusetts Institute of Technology, Cambridge, MA, USA*
[3] *Department of Chemical Engineering, Massachusetts Institute of Technology, Cambridge, MA, USA*
[4] *Department of Electrical Engineering and Computer Science, Massachusetts Institute of Technology, Cambridge, MA, USA*

Glossary

Buyable A colloquial term for a commercially available or otherwise accessible molecule used to define the "stop" criterion for retrosynthetic expansion, i.e., a molecule for which we do not need to continue the search further.

Cheminformatics A broad term used to refer to the digital definition, representation, storage/retrieval, and use of information about molecular structures.

Loss function Also called an objective function, a loss function is the mathematical function that we attempt to minimize when training a machine learning model such as a neural network.

Molecular graph A common abstraction of a molecular structure where nodes are atoms and edges that connect them are bonds.

One-hot A method of describing a categorical (discrete) variable as a list of 1s and 0s (Trues and Falses), where 1 indicates that the variable takes on the value at that position. For example, the elemental identity of an atom could be one-hot encoded by a vector whose entries represent "is this a carbon?," "is this a nitrogen," "is this an oxygen?," etc. There is only one non-zero entry in a one-hot vector.

---

* *Contributed equally.*

Enabling Tools and Techniques for Organic Synthesis: A Practical Guide to Experimentation, Automation, and Computation, First Edition. Edited by Stephen G. Newman.

Reaction template A graph-matching pattern that defines how one or more molecules can be transformed into a different set of one or more molecules. Templates are often encoded as a pair of SMARTS strings describing patterns on both sides of the reaction. Templates can be defined in both the forward and retrosynthetic direction.

SMILES A common language used to describe molecular structures that can be read linearly, with parentheses used to indicate branching. For example, CO is methanol, CCO is ethanol, CCCO is n-propanol, and C(C)CO is isopropanol. Single bonds and hydrogens are generally implicit, and aromatic atoms use lowercase characters.

Token A single element in a sequence representation that may comprise multiple characters. SMILES strings are often tokenized using custom rules to preserve chemical meaning. For example, ethylbromide "CC[Br]" may be treated as three tokens "C," "C," "[Br]" rather than the six individual characters in the string.

11.1 Introduction to Computer-Aided Synthesis Planning

11.1.1 Defining the Tasks and Use Cases

Computer-aided synthesis planning (CASP) is a broad category of computational approaches designed to facilitate the selection of experimental procedures and protocols that produce a molecule of interest. For many, CASP is synonymous with computer-aided *retrosynthesis*, which has been a subject of research for over 50 years. Retrosynthetic analysis is the process by which we typically analyze new molecules and identify simplifying disconnections that connect back to available starting materials. Humans learn to do this task quite successfully as part of an organic chemistry course or through experience reading papers or in the lab. We can similarly teach computers to do the same by either providing them with the "rules of chemistry" (e.g. in the form of reaction templates, which we'll discuss later), or providing them with examples of experimental reaction results and letting them infer the patterns of what retrosynthetic disconnections are possible.

Beyond the task of retrosynthesis are at least two other CASP tasks that relate to follow-up questions one might ask when presented with a retrosynthetic plan: Under what conditions (e.g. what solvent, catalyst, temperature) should I run the reaction? And if I do run the reaction as proposed, what products do I expect to observe in what quantities? Is the desired product the expected product, indicating that a proposed reaction will be a "success" by some definition? The first question has been called the "above the arrow" problem [1] and has elicited only a handful of computational approaches in recent years. The second question is one of reaction outcome prediction, or reaction product prediction, and has seen increasing interest due to the ease of quantitative evaluation by computer scientists. The ability to anticipate what products a chemical reaction will yield would help chemists identify cases of challenging selectivity, or even minor products and impurities that can inform the design of purification strategies. And, importantly, related to the third question, reaction product prediction enables hypothetical retrosynthetic steps to be evaluated in the forward direction for feasibility. In an ideal world, this prediction would be done in a quantitative manner and serve as a fully virtual *in silico* reaction flask. However, tools are still operating at a more qualitative level and instead focus on identifying the identities of major products.

11.1.2 Historical Approaches to Computer-Aided Synthesis Planning

As with many tasks in the field of cheminformatics and reaction informatics, CASP has a long history and has seen many approaches come and go over the years. The field is well-reviewed with an emphasis on retrosynthesis by Ott and Noordik [2], Todd [3], Cook et al. [4], Ravitz [5], and Warr [6]. These articles provide an overview of several early CASP tools and their development and adoption, as well as the beginning of the trend toward data-driven programs. This chapter provides a more updated discussion of available methods and a more colloquial introduction aimed at a chemistry audience, rather than a cheminformatics audience.

11.1.3 The Inflection Point of CASP Methods

Before we elaborate on the techniques that CASP tools employ and the finer details of how they work (and how they fail), we should note that the past few years have seen an inflection point in the utility and promise of these methods. In particular, the expert-program Chematica was sold to Millipore Sigma and rebranded into a commercial software offering known as Synthia. The algorithms and approaches behind data-driven, template-based techniques have now been incorporated into several start-up companies offering commercial retrosynthetic tools, such as Iktos's Spaya and molecule.one. Literature studies have reported that single- or double-blind preference tests show that chemists cannot meaningfully distinguish routes proposed by these tools and routes actually validated and published in the recent literature [7, 8]. And even more evidence for the actual adoption of these tools by practicing chemists comes in the form of industrial interest in these technologies [9, 10] and the participation of pharmaceutical and chemical companies in efforts like the Machine Learning for Pharmaceutical Discovery and Synthesis consortium, which the authors are part of.

Together, these trends all speak to the fact that modern CASP tools are indeed at a useful stage and can facilitate the design of new synthetic routes. Their success has primarily been in the context of retrosynthetic planning. In time, we expect the other tasks of reaction condition recommendation and reaction outcome prediction to have a similar rise in popularity, where a reaction outcome might include details about products and their abundance. In particular, a growing interest in automated chemistry where robotic platforms are capable of executing synthetic steps without human intervention has paralleled (but preceded) a growing interest in the automated *planning* of synthetic steps [11], with the ambitious long-term goal of automating as much of the molecular discovery enterprise as possible [12].

11.1.4 Preliminaries on Molecular Representation and Cheminformatics

Before we go into the details of CASP approaches, it will be helpful to have a high-level understanding of common fundamental concepts in cheminformatics, as recent developments in molecular deep learning build on these ideas. Chemists use a range of methods to describe molecules to each other, including molecular formulas, IUPAC nomenclature, and line structures. The field of cheminformatics requires key pieces of information about chemical structure to be written in a machine-readable format. Digitally, a molecule is typically described by an identifier such as an InChI string or a SMILES (Simplified Molecular Input Line Entry System) string [13], both of which can be converted to and from molecular graph (though there are edge cases and standardization rules that differ, such as how tautomers are handled). The SMILES format is easily extended to represent a reaction using tokens such as "." to separate multiple reactant or multiple product molecules and ">>" to represent the arrow from reactants to products. Often to be input into predictive models, these string-based representations of molecules or reactions must be converted into machine-readable *representations*.

For many machine learning models (e.g. neural networks), a common molecular input representation is a structural fingerprint such as the extended-connectivity fingerprint [14], often implemented as a Morgan fingerprint. Fingerprints represent molecules as vectors of 1s and 0s, where the length of the vector is fixed and does not depend on the size of the molecule or its number of atoms. Each position in the vector corresponds to a molecular substructure, and a 1 or a 0 in that position in the vector indicates the presence or absence of the substructure in the molecule being represented. For models adopted from natural language processing, which take text as input, a SMILES string can be used directly as the input. The SMILES string is tokenized using one-hot encoding. Lastly, for graph-based machine learning models such as message passing neural networks [15] or graph attention networks [16], the input is typically a molecular graph where atoms are nodes and bonds are edges. The graph representation of a molecule has the advantage of being able to incorporate as many atom and bond features as appropriate, which can be easily computed from cheminformatics tools such as RDKit [17].

In addition to representing whole molecules and their reactions, it is useful to be able to represent molecular sub-structures. Incomplete molecular structures can be used to encode a generalization of the chemical transformation in a reaction, also known as the *reaction template* (or *reaction rule*). Reaction templates define the "rules of chemistry" and can be defined in both the retrosynthetic and forward directions. Templates contain the atoms immediately involved in the reaction (the reaction center) and can include additional surrounding atoms. For example, given the Suzuki coupling reaction shown in Figure 11.1, we have depicted a reaction template that captures the atoms of the reaction center and the atoms one bond out (highlighted in pink and green). Any molecule that contains the pattern on the left side of the arrow will be deemed a "match" and can be transformed into two molecules defined by the pattern on the right side of the arrow. One common string-based representation for fragments is the SMILES arbitrary target specification (SMARTS) format, an extension of the SMILES format.

CASP approaches, either data-driven or expert-driven, may operate with or without using reaction templates. Throughout the chapter, we will mostly be introducing data-driven approaches, a terminology that has been recently used interchangeably with *machine learning approaches* (and arguably, *deep learning approaches*). Machine learning, in the narrowest definition, means learning models or patterns from data by using some optimization algorithm with a loss function. Machine learning is a special case of optimization, and deep learning is

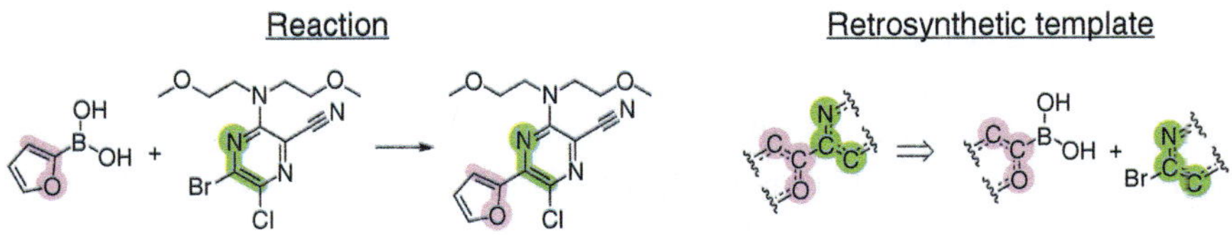

Figure 11.1 An example of a precedent reaction and the corresponding algorithmically extracted retrosynthetic template. This template generalizes the transformation in the precedent reaction into a rule that corresponds to a (retrosynthetic) Suzuki coupling between a 2-furanylboronic acid and a 2-bromopyridine without constraining the substituents on the furan or pyridine otherwise. For this example, the template SMARTS was automatically generated from the reaction SMILES using the Python library RDChiral with default parameters. *Source:* Coley et al. [18]/American Chemical Society.

a special case of machine learning in that the models or patterns are parametrized by *deep neural networks*, which have been theoretically proven to be universal function approximators, meaning that under mild assumptions, any mathematical function can be described to arbitrary precision by a sufficiently large neural network [19]. Deep learning therefore has the potential to train a classifier (predicting a categorical/discrete variable), a regressor (predicting a continuous variable), a generator (proposing new objects/outputs), etc. when given appropriate data. We refer the reader to [20] for a systematic background reading in deep learning and to [21] for an introduction to machine learning in chemistry.

11.1.5 Outline of the Rest of the Chapter

In the remainder of this chapter, we will introduce the wide variety of approaches and algorithms for the core CASP tasks of retrosynthesis, condition recommendation, and forward synthesis prediction, as defined above. The distinction between different strategies will be discussed, select examples of different software tools will be presented, and a few case studies will be shared to provide a snapshot of the current capabilities of CASP tools. We have also previously written a slightly more technical summary of some concepts related to machine learning in synthesis planning [22], as have colleagues at IBM [23]. CASP-related problems also feature heavily in recent reviews on molecular machine learning [24]. Section 11.2 includes an introduction to the inner workings of programs discussed in Section 11.3; we believe it is important to have a basic understanding of how these programs operate in order to have realistic expectations of what they can do well and anticipate when they may fail to provide useful suggestions.

11.2 Approaches and Algorithms for Retrosynthesis

11.2.1 Data-driven v. Expert-Driven Programs

One major axis by which different computer-aided retrosynthetic strategies can be distinguished is by the source of information – data or expert chemists – of potential chemical reactions and allowable transformations. The earliest programs, such as those pursued by Corey [25], were exclusively expert-driven. Chemists would codify all of the "rules of chemistry" as reaction templates, which would define the scope of transformations the program would then be able to propose. Data-driven programs, in contrast, attempt to learn allowable transformations from databases of experimentally validated chemical reactions.

Each approach has distinct strengths and weaknesses from both the development perspective and the user perspective. From a developer's point of view, data-driven programs are straightforward to retrain on different datasets (e.g. company data, or updated literature data) with virtually no additional effort; one program can be continually updated as new reactions are published. However, unless there are explicit implementation decisions made to counteract it, they are inherently biased toward the more popular reaction chemistries and might struggle to learn when, if, or how rare reactions can be applied to a new target product. Expert methods can in principle improve generalization to these rare chemistries, because humans can explicitly define when rare transformations are applicable. This is at the expense of a substantial time investment into writing these rules, as was done during the development of Synthia [26].

The reliance on reaction databases for training also means that data-driven programs may vary in performance depending on what dataset was used for training. Researchers at AstraZeneca analyzed the performance of their retrosynthesis tool, AiZynthFinder [27], on reaction sets extracted from the patent literature, the Reaxys database, and their in-house electronic lab notebook (ELN), and found that there was little overlap in the reaction types covered by each set [28]. The overrepresentation of certain reaction types in medicinal and process chemistry is well documented [29, 30], so one would expect that a data source like Reaxys, which also catalogs reactions from journal articles, might exhibit greater diversity.

11.2.2 Template-Based Approaches

A chemist may be able to envision how these reaction templates (introduced in Section 11.1.4) might be extended to have a greater degree of nuance. The templates used in Synthia, a modern example of an expert template-based program, contain additional details regarding what functional groups are incompatible with a given reaction template, and/or what functional groups may require protecting groups (Figure 11.2). Other expert template sets, including those used in ChemAxon's Reactor, may impose descriptor-based requirements, e.g. requiring that a particular atom has a certain minimum electronegativity, partial charge, or solvent accessible surface area. These descriptors are rapidly calculable and allow for a more sophisticated matching process between one side of a template (pattern) and a molecule than structure alone provides. The reaction SMARTS itself can also be written with varying levels of specificity to ensure that only the "relevant" chemical context – as judged by the chemist composing the rule – is included. When templates are generated algorithmically [18, 31], deciding which atoms outside of the reaction center to include is a major challenge, as these decisions ultimately influence the performance of template-based retrosynthetic models [32]. Template-based methods have intrinsic limitations; there is always a trade-off between how specific and how general we should define the template. In addition, they cannot generalize to unseen templates, i.e. entirely new transformation rules or novel substructures.

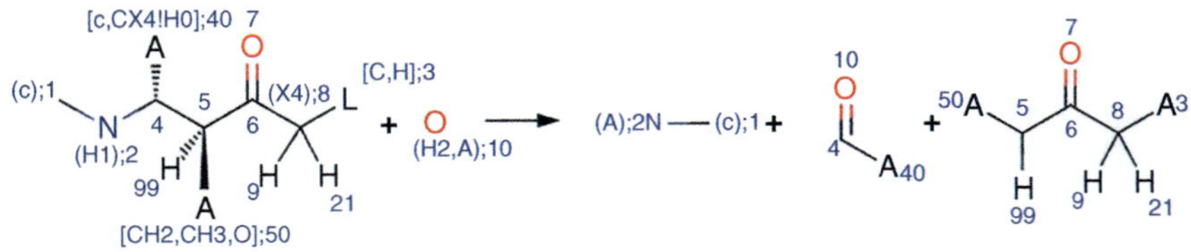

rxn_id: 8382,
name: "Proline-catalyzed Mannich Reaction",
reaction_SMARTS:[c:1][NH:2][C@H:4]([c,CX4!H0:40])[C@:5]([#1:99])([CH2,CH3,O:50])[C:6](=[O:7])[CX4:8]([#1:9])([#1:21])[#6,#1:3].[OH2:10]>>[c:1][N:2].[*:40][C:4]=[O:10].[*:50][C:5]([#1:99])[C:6](=[O:7])[C:8]([#1:9])([#1:21])[*:3]"
products:["[c][NH][C@H]([c,CX4!H0])[C@]([#1])([CH2,CH3,O])[C](=[O])[CX4]([#1])([#1])[#6,#1]", "[OH2]"]
groups to protect: ["[#6][CH]=O", "[CX4,c][NH2]", "[CX4,c][NH][CX4,c]","[#6]C([#6])=O"]
protection_conditions_code: ["NNBl", "EA12"]
incompatible_groups: ["[#6]O[OH]", "c[N+]#[N]", "[NX2]=[NX2]", "[#6]OO(#6]", "[#6]C(=[O])OC(=[O])[#6]", "[#6]N=C=[O,S]", "[#6][N+]#[C-]", "[#6]C(=O)[Cl,Br,I]", "[CX3]=[NX2][*!O]", "[#6]C(=[SX1])[#6]", "[#6][CH]=[SX1]", "[#6][SX3](=O)[OH]", "[CX4]1[O,N][CX4]1", "[#6]=[N+]=[N-]","[CX3]=[NX2][O]"]
typical reaction conditions: "(S)-proline. Solvent, e.g., DMSO",
general references: "DOI: 10.1021/ja001923x or DOI: 10.1021/cr0684016 or DOI: 10.1021/ja0174231 or DOI: 10.1016/S0040-4020(02)00516-1"

Figure 11.2 One of the tens of thousands of retrosynthetic templates in the Synthia program. In addition to defining the reaction SMARTS, it contains a list of incompatible functional groups, groups requiring protecting group strategies, and other metadata. The line drawing representing the template contains "query atoms," such as the left-most "(c);1," which corresponds to the "[c:1]" in the reaction_SMARTS field and denotes that any aromatic carbon should be considered a match, whereas the nitrogen is singly bonded to "[NH:2]" must have a single proton and be aliphatic. *Source:* Szymkuc et al. [26]/John Wiley & Sons.

With a library of retrosynthetic templates, derived through heuristic extraction from reaction databases or encoded by experts, the program developer must then decide how to use them to produce full multistep pathways. We will discuss tree search strategies in a later section, but here note that exhaustive application of every template to every input product or intermediate is typically not pursued. To provide thorough coverage of different reaction types, modern template-based programs typically contain at least tens of thousands of reaction templates; data-driven programs may have some number of closely related or redundant templates, and so hundreds of thousands of templates is not an uncommon size.

To avoid the computational cost of applying every template in their library, data-driven template-based programs generally include an additional prediction step, first proposed by Segler et al. [33] The retrosynthesis problem is formulated as a template prediction task where the goal is to select the most applicable template(s) given an input product molecule – a classification task because the prediction target is a discrete class – after which the top-ranked templates can be applied to the product to generate reactant molecules (using cheminformatics software libraries like RDKit). The practical implications of only applying the top-ranked template are that the model may fail to propose a good disconnection even if the program contains a "good" template because that template might be lower ranked, but it can generate suggestions in just a couple of seconds. This compounds the challenge of proposing rare transformations, which may or may not have a template in the first place and has motivated a line of research aimed at specifically trying to improve the recommendation of rare reaction types [34, 35].

A number of works have used neural networks to model this prediction task as classification with different combinations of molecular representations and model architectures, which this paragraph will describe in more detail than most readers will care for. One example is by feeding Morgan fingerprints into *Deep Highway Networks (DHN)* [33, 36] or by feeding molecular graph representations with atom and bond features into *Graph Convolutional Networks (GCN)* [37]. The representations coming out of these networks are then passed into some output layer(s) in order to assign a probability to each template from a predefined list. There are also retrieval-based approaches that model the task as a template ranking: existing templates that are retrieved are either associated with molecules that are most similar to the given target [38] or deemed as having high relevance to the target with an *affinity model* trained with both molecule and template encodings [35]. The best performing template-based models on the smaller USPTO_50k dataset (a common benchmark dataset) have made use of additional techniques at various stages of the pipeline. Specifically, Dai et al. [39] considers the degree of graph matching between the product/reactant molecules and their corresponding substructures in the template and incorporate this information as score functions in their model. Chen and Jung [40] separates generic templates into atom reaction templates (with no bond breaking), bond reaction templates and multiple change templates, and trains separate classifiers on these more finely defined templates. Performance gain has also been obtained by reranking the list of candidates proposed by template-based methods, using a secondary model trained to increase the rank of the correct prediction in the list [41, 42]. Despite their improved accuracy, most of these improvements have not been integrated into larger multistep pipelines or full CASP tools.

11.2.3 Template-free Approaches with Graphs and Sequences

As a remediation for the limitations of template-based methods, researchers have come up with various *template-free* approaches to perform CASP. These methods are exclusively data-driven and can generally be classified as graph-edit based or translation based. Graph-edit-based methods model retrosynthesis as a sequence (or set) of graph edits that transform the molecular graph of the product into molecular graphs of reactants, for which some neural models can be trained to predict the addition, deletion, or modification of the atoms and bonds as shown in Figure 11.3.

Some variants of graph-edit methods use a two-stage formulation following the "synthon approach" [44–47], modeling retrosynthesis as a single graph edit to first break the product into hypothetical reaction intermediates, or synthons. Thereafter, a second model is trained to recover the full reactants, e.g. by attaching the right leaving groups. Figure 11.4 shows the schematic for these two stages for one such method.

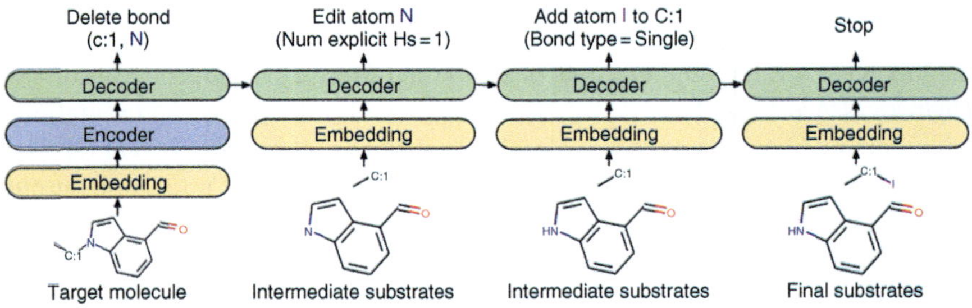

Figure 11.3 Sequential graph edits of the target molecule into reactants. At each step, the target molecule or the intermediate substrates are turned into some molecular embedding, based on which the Decoder predicts the graph edit action to be performed next. *Source:* Sacha et al. [43]/American Chemical Society.

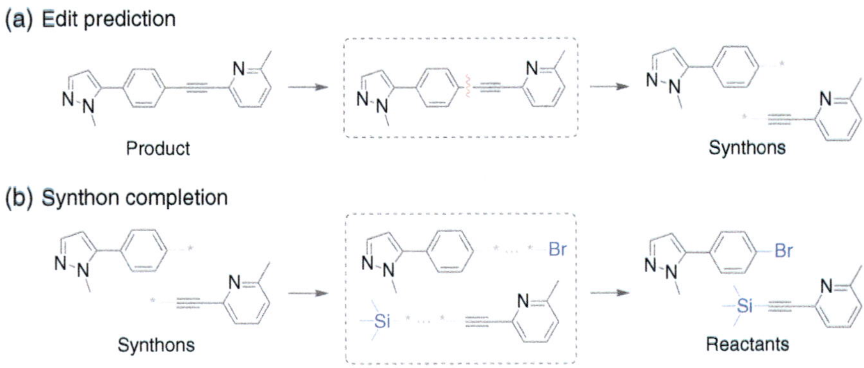

Figure 11.4 Schematic of two-stage template-free approach. At the first stage, a model is trained to predict a single edit (in red) to break the product into two synthons. At the second stage, another model learns to pick the right leaving groups (in blue) from a discrete vocabulary, which are attached to the synthons to recover the reactants. *Source:* Somnath et al. [45]/arxiv.

The other category of template-free methods formulates retrosynthesis as a translation from products to reactants. Since both the target and the precursors can be represented as SMILES strings (i.e. text-based representations of molecular structures), SMILES-to-SMILES translation is an intuitive way of modeling. As an example, the source (input) and the target (output) of a training sample may be in the form of

```
input: O c 1 c c c ( O C c 2 c c c c c 2 ) c ( Br ) c 1
output: Br C c 1 c c c c c 1 . O c 1 c c c ( O ) c ( Br ) c 1
```

where both the source and the target are sequences of SMILES tokens. As an additional benefit, SMILES-to-SMILES models can leverage the well-optimized models and techniques from natural language processing (NLP), although at the expense of neglecting chemical domain expertise. Translation-based approaches typically make use of sequence models such as recurrent neural networks (RNN) or the Transformer architecture [48–52]. Variants of these approaches design additional stages such as pretraining and reranking [53–55], use information about graph topology for enhancement [56–59], or employ performance engineering techniques such as variational inference [60, 61] and data augmentation [62].

There are a few practical comments to be made regarding template-based versus template-free models, now that we have introduced both types. Reaction templates can carry with them metadata regarding additional details of reaction precedents, e.g. the specific reactions from which a template was extracted. Each recommendation from a template-based model can therefore be traced back to the original literature or database entry, allowing users to look for more information and supporting evidence. A chemist can then judge whether there is sufficient reason to believe the reaction will be successful, or whether the template might be missing some key piece of chemical context that would render the proposed substrates incompatible. Template-free models do not allow one to easily reference supporting publications, as they operate in a more opaque end-to-end manner. The mistakes made by template-free models can be more dire than recommending a template that is not actually feasible; nothing constrains the recommendations of translation-based models, so in principle, they may suggest reactants that have no chance of forming the correct products (e.g. missing necessary atoms). In practice, all types of methods have performed well on quantitative benchmarks. Note that template-free approaches are less prevalent among CASP programs than template-based approaches; within template-free approaches, translation methods are more commonly used than graph-based methods.

11.2.4 Multistep Planning Algorithms

The goal of computer-aided retrosynthesis is of course not just to recommend a single-step disconnection, but to identify a full pathway that connects simple building blocks to the product of interest. To extend one-step retrosynthesis models to the multistep setting, an intuitive strategy is to recursively build a synthesis tree (or more generally a graph) while employing a search algorithm to selectively navigate the possible options until one or more pathways is found. An essential consideration, which we will not spend much time discussing, is how one defines a viable "starting material." Identifying pathways that start from commodity chemicals with a price under $10 kg^{-1}$ will be much more challenging than identifying pathways that start from any commercially available compound irrespective of price. Some programs have also defined their expansion stop criteria to include molecules considered sufficiently small or molecules that have known preparations (regardless of step count).

Segler et al. [7] was one of the first studies to combine a neural one-step model with Monte Carlo Tree Search (MCTS) [63]. As illustrated in Figure 11.5, the root node in the tree represents the target product, and any leaf represents a set of precursors obtained by applying the single-step model on the parent node. A search step starts by selecting the best unexpanded node using some score (i.e. the red node A under (1) Selection) and expanding it with the pretrained one-step

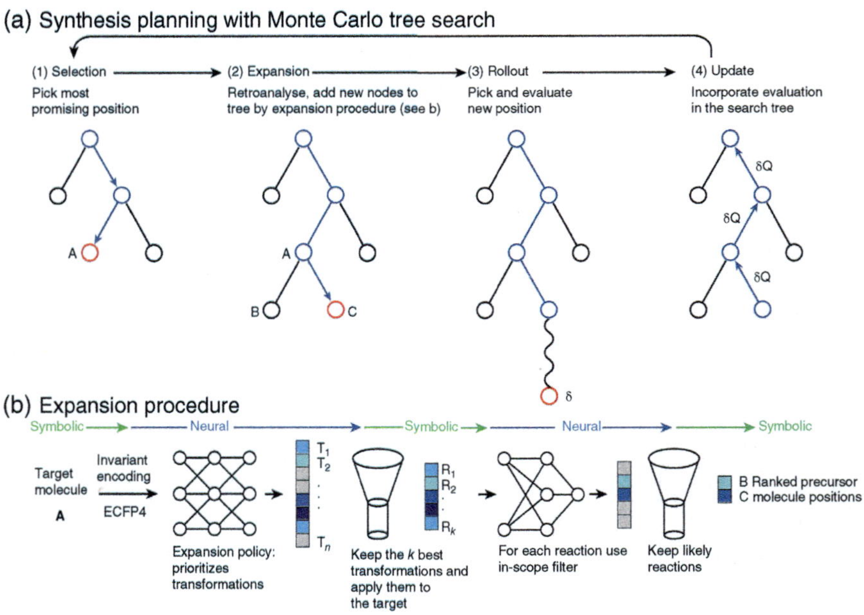

Figure 11.5 An illustration of the four phases of Monte Carlo Tree Search: selection, expansion, rollout, and update. The bottom panel demonstrates how the approach can be seen as *neural-symbolic*, in which heuristic-based symbolic operations such as fingerprinting and template application are combined with neural model training. *Source:* Reproduced from Segler et al. [7]/Springer Nature.

model to get possible sets of precursors, which are filtered with a separately trained model (i.e. the in-scope filter) to retain only the probable ones (i.e. nodes B and C under (2) Expansion). Each of the new precursor sets (i.e. the bottom blue node under (3) Rollout) is then evaluated on whether it is synthesizable from available building blocks, by iteratively applying a separate, lightweight one-step model. Thereafter, the evaluation result is used to update the scores all the way up the tree (i.e. the blue upward arrows under (4) Update). The process repeats until a pathway with some predefined criteria is found (e.g. all bottom leaves are commercially available or maximum search depth is reached).

There have been a few follow-ups, though most can be thought of as variants of, or extensions to this tree search process. Again, these details may not be important for most readers. Different one-step models that have been experimented include the SMILES-to-SMILES Transformer [50, 64], as well as different search structures and/or search algorithms. Alternative variations of Segler et al.'s approach, such as ASKCOS [65, 66] and AiZynthFinder [27], mostly rely on the same core modules of template-based one-step predictor and MCTS, with modifications and extensions for engineering considerations. Specifically, ASKCOS attempts to validate one-step proposals with a condition recommendation and a forward prediction model and, in its original release, used a parallelized UCT search. AiZynthFinder uses the same one-step model for expansion and rollout, but excludes the in-scope filter. As a step to go beyond the default search tree in which a node may correspond to a set of molecules, Schwaller et al. [64] experiments with a hyper-graph search structure where each node only corresponds to a single molecule. In a similar way, Retro* [67] uses an AND-OR tree with two types of nodes to represent a single molecule and single reaction differently, building on work from Heifets and applying Proof-Number Search [68]. Additionally, Retro* also explores a different search algorithm known as "best-first search" and trains a separate surrogate model to give a more accurate estimation of the cost in the update phase. Qualitatively, the overall idea and goal is the same: guide the search in promising directions to avoid the computational cost of a depth-first or breadth-first search and replace the rollout phase with scoring mechanisms that are computationally cheaper.

Some related works also focus on the problem of evaluating proposed pathways [69, 70], the methodology of which itself has yet to reach a consensus among the community. Objectively evaluating and comparing multistep tools is complicated and, in our opinions, nearly impossible to do fairly. Merely evaluating whether they are able to find a pathway, or multiple pathways, toward a variety of target molecules within a certain time or computational budget is insufficient. Programs differ in their training data, set of starting materials, and definitions of what transformations are possible. It is not possible to automatically evaluate with certainty whether a proposed retrosynthetic pathway is realistic, and hence some studies have instead sought to measure the impression of expert chemists instead [7, 8]. Whether a proposed retrosynthetic step is actually feasible requires consideration of *how* it should be carried out (i.e. under what conditions), and whether the product will be made with sufficient yield and/or selectivity. This then motivates the other two primary tasks in CASP, which we will discuss in the next section.

11.3 Approaches and Algorithms for Condition Recommendation and Forward Synthesis

11.3.1 Condition Recommendation Approaches

The task of condition recommendation attempts to fill in procedural details of how a particular reaction step should be performed, perhaps with implicit goals of maximizing the yield. When performing a synthesis experimentally, it is obviously insufficient to just have knowledge of what reactants to use as might be suggested from a retrosynthetic planner. Decisions must be made regarding the choice of solvent (if not neat), any catalysts or reagents, the temperature, concentrations of all species, reaction time, reaction vessel, orders and timing of addition, etc. Expert programs cannot answer all of these questions at present, although one could imagine defining high-specificity templates that could suggest typical conditions for a family of transformations. Data-driven methods can in principle learn substrate-specific conditions, but there are practical limits to the precision of existing reaction datasets. At present, for example, they do not even contain information regarding concentrations. Two recent studies have nevertheless developed data-driven models capable of proposing certain aspects of conditions.

One data-driven approach to condition recommendation is to take the entirety of a reaction database like Reaxys and train a model to generate the recorded conditions when given the reactants and product as input. Because the breadth of chemical species considered reagents, catalysts, or solvents are fewer in number than "reactants" (based on typical guidelines for role assignment [71]), these choices can be treated as discrete options in a classification setting. That is, it does not take a large number of distinct reagents, catalysts, and solvents to describe the majority of the literature. Gao et al. [72] developed a model as part of the ASKCOS software suite to select candidate catalysts, solvents, reagents, as well as the temperature. Because it is trained on virtually all reactions in Reaxys, the model is intended to work with "any" reaction type indiscriminately; as a data-driven model, there is still some bias to be expected, where reaction types with fewer precedents are predicted less reliably.

An alternate approach restricts the domain of applicability of the model in order to improve a cheminformatician's ability to curate a clean dataset and, ideally, learn a more nuanced view of how substrate choice influences suitable conditions. Maser et al. demonstrate this paradigm using a selection of Suzuki, Negishi, C–N coupling, and Pauson–Khand reactions [73]. Each reaction type lends itself to a categorization scheme with more granularity than "reagent," "catalyst," and "solvent," allowing the model to differentiate species more easily (e.g. ligands from bases). This model demonstrated the ability to propose substrate-specific conditions with the same level of detail as what is reported in reaction databases and with an accuracy that clearly exceeds a baseline model that always proposed the most popular conditions.

While neither type of model is yet at a stage where suggestions are immediately actionable, they do provide reasonable first guesses. There are certain reaction types (e.g. metal-catalyzed cross-couplings) where performance may be highly sensitive to the conditions used (e.g. a suitable ligand). These choices can be harder for chemists to predict *a priori* unless they are very familiar with the specific transformation. When ample data exists, we should expect data-driven models to be able to help choose between closely related catalysts [74]. At this stage, we lack strong evidence that condition recommendation models surpass expert chemistry intuition, especially if experts have access to the literature precedent for a proposal and can see what conditions were used there. A different line of research that should be of interest to chemists curious about computational assistance is iterative reaction optimization, where experimental conditions are iteratively improved through multiple rounds of trial-and-error [75, 76].

11.3.2 Forward Synthesis Approaches

In the context of CASP, forward synthesis predictors aim to predict the major product of a chemical reaction given reactants and reagents. Ideally, this would be done with quantitative precision, but we again face a practical data limitation that challenges the utility of these tools. Forward predictors mostly serve as a plausibility check for proposals generated by retrosynthesis models. While forward prediction can be seen as the reverse problem of one-step retrosynthesis, it may be arguably easier since all the heavy atoms are present in the input, unlike in retrosynthesis where heavy atoms in a leaving group have to be inferred or generated *ex nihilo*. Although there is no particular reason why any retrosynthesis approach, either template-based or template-free, cannot be applied for forward synthesis, the progress for forward prediction has been somewhat independent from that for retrosynthesis. In particular, template-free approaches have recently been dominant for this task.

A few earlier works use neural networks to model forward prediction with the help of forward reaction templates somewhere along the pipeline [33, 77, 78], in a similar manner to their counterpart for retrosynthesis. As a variation to the template classification formulation, Coley et al. [78] applies a pre-extracted set of around 1700 templates to any pool of reactants to generate a list of potential products, after which a reaction likelihood classifier (similar to the in-scope filter later used in Segler et al. [7]) is used to rank these candidate products.

Graph-edit-based template-free approaches also exist [43, 79, 80], along with variants that use Markov decision processes [81] or Integer Linear Programming [82]. A slightly different approach that models reaction transformations as electron flow has also been explored [83, 84]. Here, octet rules and valence constraints are considered, and a non-autoregressive (i.e. non-iterative) decoding is employed in order to speed up predictions. These methods formulate the prediction of reaction products as the prediction of graph edits (sequentially, or as a set) that adjust the bonds in a molecular graph. In contrast to retrosynthesis, it is typically assumed that no individual atoms are added or removed from the set of input molecules, which in some cases reduces the problem into the prediction of changes in the atom adjacency matrix.

Lastly, analogous to the retrosynthetic task, machine translation models can be trained to transform reactant SMILES into product SMILES in an end-to-end manner without any regard for chemical rules. Although earlier works appear to be inferior [49, 85], the later adoption of the Transformer model has shown impressive results [86]. The performance on the benchmark USPTO datasets, in terms of top-n accuracy, has been further improved in translation-based variants through various strategies such as data augmentation and pretraining [53, 62], as well as making use of molecular graph representations in the input [56, 59], some of which have already been discussed in Section 11.2.3.

To reiterate: these models have been trained to predict only the major product of chemical reactions as reported in the literature. They can accomplish this task with impressive accuracy, often exceeding 80 or 90% on benchmark datasets from the patent literature. But for the same reason that condition recommendation cannot predict concentrations in their output, forward prediction models have not been designed to take concentration information as input. As a result, there is legitimate ambiguity in what product(s) will be formed if only the identity of reactants, reagents, catalysts, and solvents are provided as input. In fact, anecdotally, these models are not terribly sensitive to the choice of catalysts/ligands and solvents, and so a great deal of chemical nuance is missing. Further, while they are all designed to propose a ranked list of possible product outcomes, lower-ranked products are not necessarily the side products or impurities one might observe experimentally. Additional work in this area is certainly justified, but these tools do exist and on paper perform very well, at least to filter or validate retrosynthetic model suggestions.

11.4 Select Examples of Software Tools for CASP

11.4.1 Open-Source Tools

11.4.1.1 ASKCOS

ASKCOS [87] is a collection of CASP tools developed at MIT. The core modules consist of the step-by-step *Interactive Path Planner* for human-in-the-loop multi-step retrosynthesis, and a more automatic synthesis *Tree Builder* with stepwise evaluation. Additional functionalities include reaction condition recommendation, reaction outcome prediction, atom mapping using graph neural networks, synthetic complexity evaluation based on SCScore [88], and buyable chemical look-up using a database comprising compounds from eMolecules and Sigma Aldrich available for under $100 g^{-1}. The web interface is free to use, but backed by only modest computational resources. For users interested in local deployment and/or scale-up, the code is open-sourced on Github [65].

The Interactive Path Planner (IPP) allows users to build their own reaction route one step at a time. This module is designed for chemists who would prefer to guide synthesis in specific directions, evaluating each reaction step along the way and choosing which options to expand further. The single-step proposer is based on a template prioritizer, as discussed in Section 11.2.2, with template sets from the Reaxys and NextMove Software's Pistachio databases.

Figure 11.6 shows a screenshot of the IPP panel. An interactive tutorial for first-time users can be triggered by clicking the "?" button. In general, the planning process starts by inputting the target product by drawing or as SMILES (easily copied from drawing tools like Chemdraw). Any molecule is represented as a *node* (in box) and can be independently expanded. The user can choose the node to be expanded in a recursive manner (over and over again), until a synthesis path is found in which all precursors are buyable (in green box). More details on precursor ranking can be viewed by clicking on expanded nodes, and the corresponding reaction template information can be accessed by clicking on the indices (e.g. #1) in the circles along the paths. Links to the pages for condition recommendation and reaction product prediction are provided when viewing proposed reactions.

The Tree Builder makes use of the same single-step model as does the IPP but enables the user to run a time-limited tree expansion in an automated multistep search. The expansion runs until either all building blocks are buyable chemicals or the maximum depth (longest linear sequence) has been reached for all molecules in the growing synthetic tree. In practice, the fully automated tree builder can help chemists generate novel synthesis strategies for certain applications. For other chemists, the interpretability and increased control offered by the interactive path planner will prove more useful, even if it is more manual.

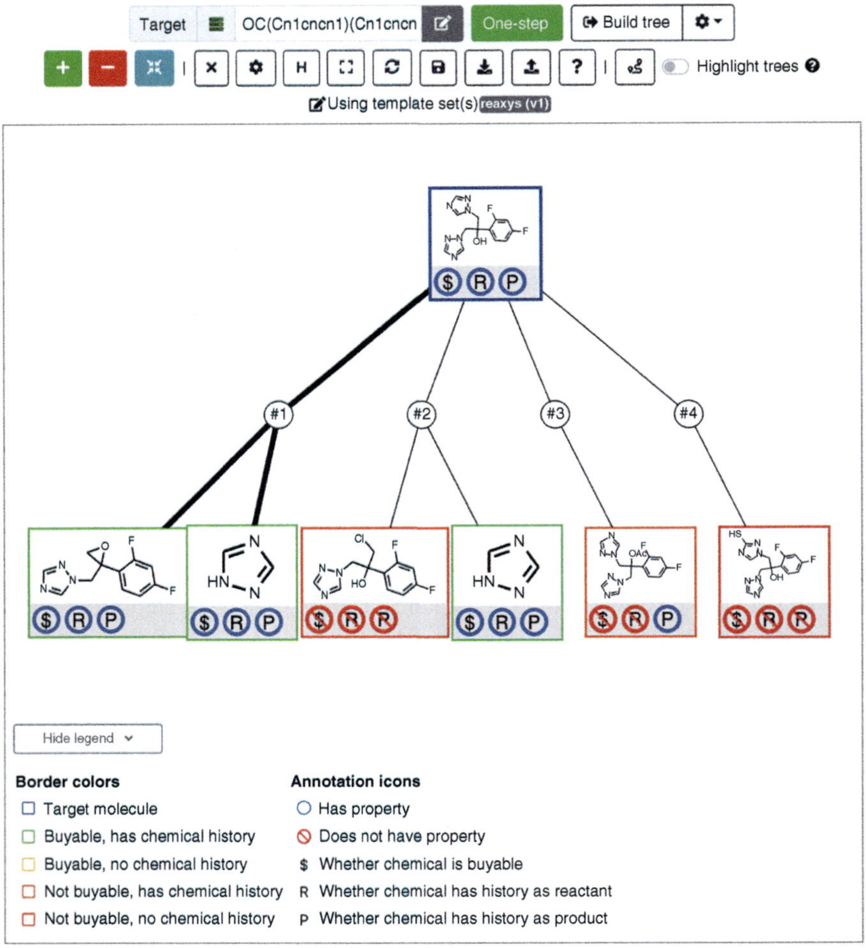

Figure 11.6 A screenshot of the Interactive Path Planner panel for a sample molecule.

Figure 11.7 shows the screenshots of the Tree Builder settings (left), and the sample synthesis tree generated with default settings (right) from the public release of ASKCOS (v. 2021.01). The user can customize settings such as maximum depth and expansion time, as well as stop criteria. More details on the usage of the Interactive Path Planner and the Synthesis Tree Builder, as well as for other applications can be found in the ASKCOS tutorial [89].

11.4.1.2 AiZynthFinder

AiZynthFinder [27] is an open-source tool solely for retrosynthetic planning that is based on a similar template-based single-step model and Monte Carlo Tree

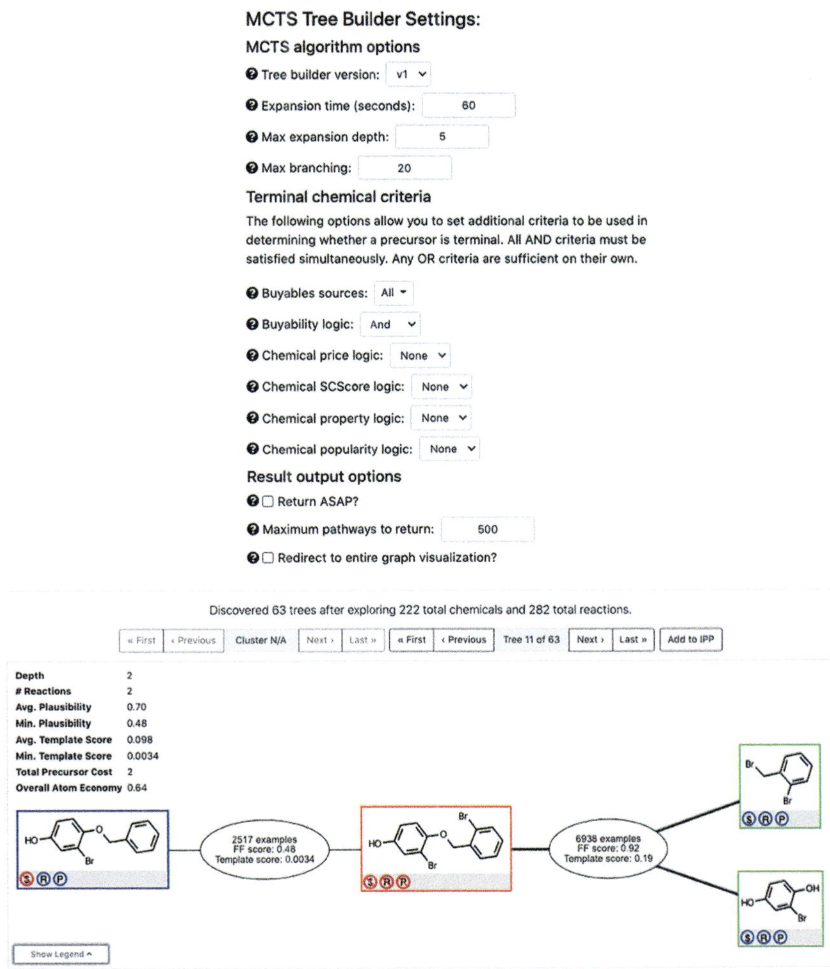

Figure 11.7 Screenshots of the Tree Builder setting page and the generated synthesis tree for a sample molecule.

Search to ASKCOS. It uses a template set from the US Patent and Trademark Office (USPTO) dataset [90] and uses a subset of compounds in the ZINC database [91] as the buyable database. The code is available on Github [92], and basic familiarity with conda [93] and Jupyter notebook [94] is required. By following the instructions and typing in a few commands, the user can create and activate the conda environment, download pretrained models, and open up the graphic user interface (GUI) in a Jupyter environment as shown in Figure 11.8. Similar to other retrosynthetic planning tools, the process starts with a SMILES input, after which the search can be run and results displayed with a few clicks.

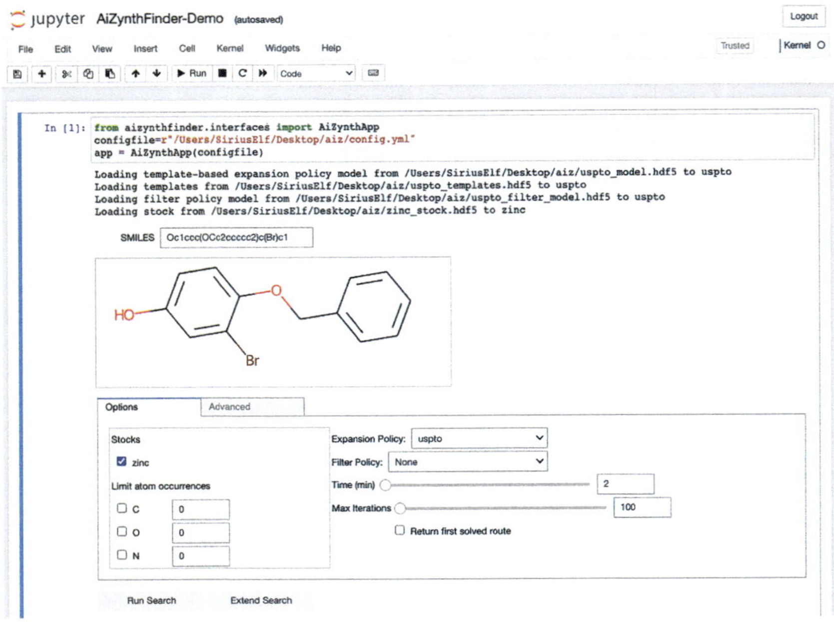

Figure 11.8 The GUI of AiZynthFinder in the Jupyter environment, whose panel in the cell has similar functionalities to ASKCOS with configurable settings.

Figure 11.9 shows the sample result for the same molecule as the sample in the ASKCOS Tree Builder. The results for AiZynthFinder and ASKCOS differ, due to multiple possible reasons as discussed by Genheden et al. [27], which also contains a non-rigorous comparison between these tools.

11.4.1.3 Retro*

The Retro* algorithm discussed in Section 11.2.4 provides an open-source implementation, available on Github [95]. Basic familiarity with Python and conda is required to set up the environment, and the package only provides an importable Python API (application programming interface) without any user interface. A model trained on processed USPTO data and a database of commercially available building blocks based on eMolecules [96] are provided, which are to be downloaded separately by the user.

Figure 11.10 shows a section of the printout from running the sample script in Retro*. The results are arguably less intuitive without visualization modules as in ASKCOS or AiZynthFinder. Retro* is therefore only recommended for advanced users who can export these results with ease for subsequent processing.

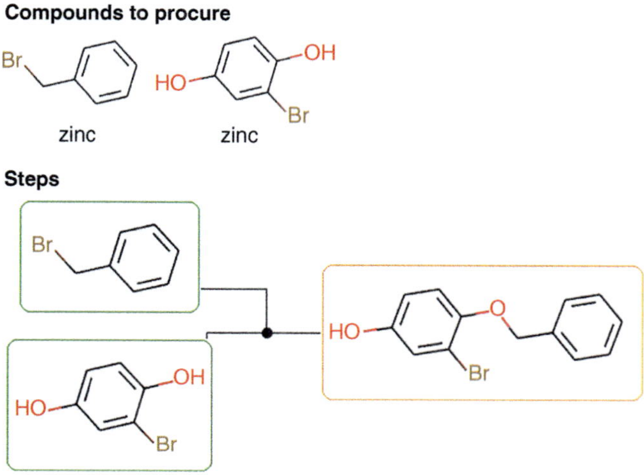

Figure 11.9 Sample result of AiZynthFinder in the Jupyter notebook cell when running with the same molecule as in Figure 11.7.

```
(retro_star_env) % python example.py
...
root: INFO Synthesis route found!
root: INFO Final search status | success value | iter: 7.009890672790091 | 6.905859674323364 | 7
{'succ': True, 'time': 8.103998899459839, 'iter': 7, 'routes':
'CCOC(=O)c1nc(N2CC[C@H](NC(=O)c3nc(C(F)(F)F)c(CC)[nH]3)[C@H](OC)C2)sc1C>0.0010>CCOC(=O)c1nc(Br)sc1C.CCc1[nH]c(C(=
O)N[C@H]2CCN(C(=O)OC(C)(C)C)C[C@H]2OC)nc1C(F)(F)F|CCc1[nH]c(C(=O)N[C@H]2CCN(C(=O)OC(C)(C)C)C[C@H]2OC)nc1C(F
)(F)F>0.9998>CO[C@@H]1CN(C(=O)OC(C)(C)C)CC[C@@H]1N.CCc1[nH]c(C(=O)O)nc1C(F)(F)F|CCc1[nH]c(C(=O)O)nc1C(F)(F)F>0.
9916>CCc1[nH]c(C=O)nc1C(F)(F)F.[O-][Cl+][O-
]|CCc1[nH]c(C=O)nc1C(F)(F)F>0.9994>CCc1[nH]c(C(OC)OC)nc1C(F)(F)F|CCc1[nH]c(C(OC)OC)nc1C(F)(F)F>1.0000>N.O=C1CCC(=O)
N1Cl.COC(C=O)OC.CCC1(C(=O)C(F)(F)F)SCCCS1|CCC1(C(=O)C(F)(F)F)SCCCS1>1.0000>CCC1SCCCS1.CCOC(=O)C(F)(F)F',
'route_cost': 6.905859674323364, 'route_len': 6}
```

Figure 11.10 A sample printout by running the example.py script under Retro* repo. The routes are produced in alternating sequence of molecule sets and confidence scores associated with the single steps.

11.4.2 Closed-Source Tools

There are a growing number of commercial CASP tools that are not open-sourced and typically require a paid license to use. Because the authors of this chapter do not have access to all of these software packages (and because license agreements for scientific software in general may preclude direct evaluations or comparisons), we are limited in the amount of detail we can include in this section. Additionally, each platform listed is a complex piece of software with a unique user interface and set of features that are difficult to capture in any quantitative manner.

Table 11.1 A list of closed-source CASP programs under active commercial development. Literature references to the methods employed in each program are included where available.

Tool	Company	Refs.	Demo availability
Synthia	Merck KGaA	[8, 26, 97]	Upon request
RXN	IBM	[64, 98]	Free
Predictive Retrosynthesis Planner	Reaxys	[7]	Free (limited capabilities)
SciFinder^{n}	CAS		Upon request
ICSynth	DeepMatter	[99]	Upon request
Molecule.one	Molecule.one	[43, 100]	Upon request
Spaya	Iktos		Free
Manifold	PostEra	[51]	Free

We attempt to provide a brief summary of the different commercial CASP tools listed in Table 11.1 to provide readers with a sense of the diversity of available software tools. Where we can, we have attempted to highlight the datasets that each tool is developed from and links back to as well as the methods used to generate retrosynthetic plans. We do not discuss the details of each program's interface as this would be best assessed by a user directly.

Synthia (formerly known as Chematica) differentiates itself from other CASP tools by its database of expert-curated reaction rules. An example reaction template from Synthia is shown in Figure 11.2 in Section 11.2.2, illustrating how the templates explicitly encode various aspects of reaction selectivity, incompatibility, and potential cross-reactivity. Synthia now includes over 100 000 such retrosynthetic reaction templates. For multistep synthesis planning, Synthia applies these templates iteratively to search for precursors that maximize a user-customizable scoring function. The Synthia multistep search algorithm allows the search to sometimes step to precursors that have a worse score if they allow the algorithm to reach better precursors further along the search [8]. Synthia includes a building block catalog from Sigma Aldrich and over 80 other vendors with over 4.2 billion compounds and their corresponding catalog price. Synthia has been shown to find promising pathways to complex organic molecules including pharmaceuticals and natural products, discussed in Section 11.5.3.

The chemistry database Reaxys offers an integrated Predictive Retrosynthesis Planner. The planning tool can piece together known reactions from the Reaxys database, as well as generalize known reaction chemistry to novel products. The synthesis planner applies the "neural-symbolic" [7] approach from Segler et al. to single-step retrosynthesis. This means that a neural network is trained to prioritize a set of retrosynthetic templates for an input product molecule. The templates are algorithmically extracted from the Reaxys reaction database, which of course will continue to grow. The planner incorporates additional information for reaction steps from the Reaxys database such as yield (when known), and reaction steps in returned routes link directly to entries in Reaxys. The buyable stopping points for the search are defined from multiple vendor catalogs and can be customized by the user.

CAS similarly offers Scifinder^{n}, a retrosynthesis tool integrated with the CAS content collection. Scifinder^{n} also enables users to include a mix of exact known reaction steps as well as known templates applied to new products. Evidence is provided for each proposed reaction step from the extensive CAS reaction database. Multistep retrosynthesis searches are stopped once buyable starting materials are reached with a user-defined price limit. In addition to common search parameters like search depth, SciFinder^{n} allows users to specify which bonds to break or to maintain in the retrosynthetic search. The tool also allows users to select whether to include reaction types with few precedents when considering predicted retrosynthetic steps or whether to consider only well-supported chemical transformations.

IBM's RXN is a free-to-use but closed-source CASP platform that uses template-free Transformer-based retrosynthesis models. These models – adapted from the field of natural language processing – are trained to predict precursor molecules directly from an input product molecular structure, with no intermediate prediction of a reaction template. RXN's chemical reaction transformer models are trained on a dataset that was generated from patents in the United States Patent and Trademark Office (USPTO) as well as the Pistachio database from NextMove Software, which is also derived from the patent literature but includes European patents. As a translation-based tool, IBM RXN is particularly sensitive to the data source on which it is trained, as there is no chemical intuition built into the models. IBM recently announced a partnership with Science of Synthesis/Thieme Chemistry, who curate and maintain databases of chemical reactions.

ICSynth, acquired by DeepMatter, uses a set of retrosynthetic "transforms"; templates algorithmically extracted from reaction databases that include InfoChem's own SPRESI, Thieme's Science of Synthesis, and the enzymatic reaction database Rhea. Additionally, users can input custom sets of reactions from which transforms are automatically generated. ICSynth links to the eMolecules and Chemspace databases of buyable building blocks. The reported algorithm for ICSynth [99] first assesses which bond transformations from the target molecule would lead to a simpler precursor and then searches for retrosynthetic transforms to achieve those bond changes, although details are sparse.

Many new commercial CASP tools have been developed in recent years, leveraging new machine learning models, search algorithms, or prioritization techniques for discovered synthesis plans. Beyond these handful of examples are several more commercial programs, including molecule.one (graph-based), Iktos's Spaya (template-based), PostEra (translation-based), and Galixir's Pyxir (founders affiliated with GLN and Retro*). Internal efforts are underway, anecdotally, at dozens of companies who are building their own internal pipelines for synthesis planning even if not selling software directly. We mention these tools briefly but do not mean to understate their utility and importance to the field; chemists interested in finding the CASP system that works best for them should try out a number of programs and find which workflows are most comfortable and suitable for their use case. Ultimately, the best way to assess these different tools is to directly test out their demo versions to see which tool offers the right capabilities for one's project goals. Some considerations we believe to be important are summarized in Section 11.4.4.

11.4.3 CASP Tools for Enzymatic Catalysis

While CASP has a long history for conventional organic synthesis, its application to "retrobiosynthesis" [101] is more recent. However, as manipulating enzymatic chemistry in cells and *in vitro* has become easier (see Chapter 3), interest in tools to automate the design of novel metabolic pathways and biocatalytic cascades has surged. The underlying algorithms and model architectures are similar to organic chemistry CASP tools, but retrobiosynthesis CASP tools are trained on datasets of enzyme-catalyzed reactions, often use metabolite databases instead of vendor catalogs to define building blocks, and can include additional considerations such as estimated catalytic promiscuity for a given enzyme sequence. Examples of user-accessible retrobiosynthesis tools include RetroPathRL [102], BNICE.ch [103], and RetroBioCat [104]. RetroPathRL and BNICE.ch are tuned toward metabolic engineering projects in living cells; they employ reaction templates extracted from MetaNetX and the Kyoto Encyclopedia of Genes and Genomes (KEGG), respectively. Alternatively, RetroBioCat is tuned toward planning of industrial biocatalytic cascades; the platform uses a set of manually encoded reaction templates that correspond to well-explored biocatalytic transformations. In addition to these dedicated retrobiosynthetic tools, organic chemistry CASP tools like RXN, ICSynth and ASKCOS have also incorporated reaction data from various enzymatic reaction datasets [105].

11.4.4 Practical Considerations for CASP Programs

We believe that having a basic understanding of how CASP tools work under the hood helps navigate their settings and proposals more effectively. For example, understanding that a data-driven template-based program may fail to generate an expected idea could be a result of missing the necessary template, having that template be ranked poorly, or having an insufficiently exhaustive tree search that cannot find the right starting materials, which may or may not be in the buyable database. If no results are found during a multistep expansion, examining one-step recommendations using more permissive settings can help diagnose the root cause.

There are a few additional factors we feel warrant more discussion: whether one can connect suggestions to literature precedent, whether one can access the program through a graphical interface, whether using a program requires transmitting structural information to the software provider, and whether one can customize the tool to fit the specific needs of a user or organization.

11.4.4.1 Traceability to Literature Precedent

In our experience, an important feature of a retrosynthetic planning tool for a chemist is the ability to check a suggested reaction or retrosynthetic step against precedents from a database and, ultimately, from published literature. Different tools vary in which databases they link to and how they link to them. For example, in the template-based retrosynthesis tools, such as Reaxys Predictive Retrosynthesis or ASKCOS, each suggested retrosynthetic step corresponds to a retrosynthetic reaction template, which is linked to a set of precedent reactions. On the other hand, for tools that rely on template-free methods, such as IBM's RXN, there is no direct link between a suggested retrosynthetic step and database precedents. Instead, in RXN, precedents from the USPTO or Pistachio database are linked to a suggestion based on the structural similarity of the database example and the proposed reaction; for most use cases, simply searching a reaction database for entries similar to the proposed reaction may be a sufficient way to cross-reference the literature.

11.4.4.2 How to Use CASP: Command Line Versus Graphical User Interface

Synthetic chemists may or may not be comfortable with programming concepts, package installation/management, and the use of the command line to run these programs without an interface. Most open-source tools use Python, fortunately, so the barrier to entry is perhaps lower than it might otherwise be for other languages. Regardless, some free tools and all/most commercial tools are accessed through a GUI and therefore require no knowledge of coding whatsoever. GUIs help users navigate the vast number of ideas CASP tools may generate and make it easier to visualize and sift through pathway options. For chemists who wish to plan routes to only one or a few molecules, this is an ideal workflow. For other use cases in medicinal chemistry, one may wish to examine thousands of potential molecules and triage them based on perceived synthesizability; in these cases, programmatic access through the command line or an application programming interface (API) is necessary, i.e. writing code. It should not be surprising that commercial CASP tools invest a great deal into their interfaces, as the user experience is an important aspect of gaining traction and adoption. For users who are interested in testing a wide variety of approaches including newly published methods from academia, familiarity with Python and the command line will be necessary.

11.4.4.3 Data Privacy

Depending on one's use case, maintaining the confidentiality of target molecules might be essential. If a CASP tool is only available through a web application, or if the computation occurs on the software provider's servers, information security should be a concern unless one does not intend to file intellectual property claims on that molecule. Certain programs, including but not necessarily limited to open-source ones, can be deployed behind an organization's firewall without any communication with external servers. This introduces the need to provide the necessary computational resources oneself, but that tradeoff may be well worth it to most pharmaceutical companies.

11.4.4.4 Customization Ability

Related to the notion of data privacy is the question of whether one can retrain or adapt CASP model(s) to one's own data, with or without sharing that information with a third party. Data-driven CASP models can in principle be retrained or fine-tuned on new reaction datasets to tailor their predictions to the distribution of reaction types a user is most interested in; whether there are convenient scripts and extract-load-transform pipelines to facilitate retraining is a separate matter. The most common customization users desire, in our experience, is the definition of a "solved" molecule during retrosynthetic analysis. Programs for multistep planning will be shipped with some default database of buyable (or otherwise appropriate) starting materials. Users may wish to supplement this database with the internal chemical inventory at their organization, or with a larger collection corresponding to a more inclusive price threshold. If customization would benefit one's use case, we recommend evaluating candidate CASP tools along this dimension. Open-source tools provide the greatest flexibility, at least in principle.

11.5 Case Studies

As mentioned in the Introduction, retrosynthetic planning tools have reached an inflection point where they are starting to be used with greater success and greater popularity than ever before. Three case studies are given in the following sections to highlight the rapid progression of this field in recent years.

11.5.1 Segler et al.'s Data-driven Program and A/B Testing Success

In 2018, Segler, Preuss, and Waller published their landmark paper describing the successful development of a data-driven retrosynthetic planner with the most compelling performance evaluation seen up to that point [7]. This program used a Monte Carlo tree search (MCTS) strategy to balance exploration and exploitation, following a neural network policy trained on templates algorithmically extracted from the Reaxys database. Among their evaluations was a medium-scale analysis of roughly 500 products reported in the literature, which showed that the MCTS algorithm found hypothetical routes to a greater proportion (92%) of those compounds given a 60-second computational budget as compared to a neural best-first search (71%) or heuristic best-first search (4%).

Perhaps the more impressive evaluation, however, was a double-blind preference test conducted with 45 graduate-level organic chemists. Nine synthetic pathways to nine products were extracted from the literature and presented alongside nine model-predicted pathways to the same targets. Each chemist participant was asked to select the preferred route based on their qualitative understanding of feasibility and personal preference. The authors found there was a slight, albeit not significant, preference toward the program's generated route compared to the literature route on average. This was the first notable attempt at demonstrating the quality of routes proposed by a data-driven retrosynthetic planner, taking advantage of machine learning advances from just a few years prior. While the software reported in the paper was never open sourced or made available, it has since been adapted and integrated into the commercial Reaxys Predictive Retrosynthesis model offered by Elsevier.

11.5.2 MIT's ASKCOS Program and Robotic Synthesis Demonstration

A similar template-based retrosynthetic planner features in the ASKCOS software suite, which encompasses the additional CASP tasks of condition recommendation and reaction outcome prediction. While components had been published in earlier work, ASKCOS was first formally disclosed in a 2019 publication describing its integration with laboratory automation [66]. This paper illustrated an approach to target-oriented synthesis wherein synthetic routes could be planned by a data-driven program and executed on a robotic flow chemistry platform. While there are several limitations to be aware of, the integration of AI planning and robotic synthesis brought together two research threads that had to-date been advancing in parallel. A total of 15 medicinally relevant small molecules were synthesized in continuous flow, each requiring 1–3 synthetic steps, including two molecules with defined stereochemistry.

The limitations of this demonstration connect back to the challenge of having incomplete data on chemical reactions with which to train predictive models. In Segler et al.'s evaluation, routes were examined by chemists on paper and only included the identities of reactants and reagents that contribute heavy atoms to the product. For experimental synthesis, one additionally requires all of the additional procedural details to carry out the synthesis. As mentioned earlier, these considerations include far more than what can be predicted by the current generation of condition recommendation tools, which lack the ability to predict quantitative concentration information, reaction times, or purification steps and lack an understanding of flow compatibility, among other factors. Nevertheless, data-driven tools like ASKCOS will almost certainly transition toward a more quantitative mode of operation as data availability improves. This transition would further facilitate the ability to accelerate target-oriented chemical synthesis. For now, the open-source ASKCOS tool provides a starting point for further route development by expert chemists.

11.5.3 Grzybowski's Chematica/Synthia Program's Experimental Validations and Acquisition

Grzybowski and coworkers have made a number of invaluable contributions to the field of computer-aided synthesis planning over the past decade. They have championed the expert approach to retrosynthesis wherein human chemists curate detailed transformation rules (and exceptions) that cover as many known reaction types as possible. Their set of expert templates form the basis of the retrosynthetic engine within the Chematica program, which is summarized and contextualized in a review by Szymkuc et al. [26]. A major milestone in its validation came in 2018 when a team of chemists from Grzybowski's team and MilliporeSigma reported a joint study where Chematica-planned routes to eight challenging targets were validated experimentally [97]. These targets were considered challenging as each had been synthesized either in low yield, with poor scalability, or entirely unsuccessfully. The experimental execution of these routes required a reasonable amount of condition screening and optimization, which is to be expected given that only coarse template-level conditions are suggested, rather than substrate-specific conditions. The success of this demonstration led to the purchase and rebranding of Chematica as Synthia by MilliporeSigma, which is now a commercial offering.

Further validation of Synthia was reported using a suite of complex natural products, rather than smaller medchem-like compounds [8]. The reaction types used for natural product synthesis do not necessarily resemble the transformations that appear most often in reaction databases, so these structures are quite challenging for data-driven techniques. Expert-encoded templates have the potential to generalize more effectively as humans can describe directly when/if the rule should be applied. Of the many total syntheses proposed by Synthia, three were experimentally validated, though again requiring condition screening and optimization. A preference test was performed in a manner almost identical to Segler et al. to gauge whether program-proposed routes were identifiable as more likely to be human-designed or program-designed when displayed side by side by literature (validated) pathways. The result was likewise the same: experts could not tell when a pathway came from the program or the literature. This represents a major accomplishment for Synthia and the field, as these natural product targets were rather complex; the full test set and list of pathways are available in their supporting information. An additional case study involving the use of Synthia for route planning and validation was reported by Lin et al., who applied the retrosynthetic tool to find routes to several antiviral compounds [106]; this paper provides a useful discussion of the successes and failures of computer assistance from a practitioner's perspective.

11.6 Conclusion

The wealth of options when considering using CASP tools to aid in target-oriented synthesis (particularly, for retrosynthetic analysis) is a direct reflection of the field's rapid growth and maturation. While these tools have existed in some form for decades, we do believe that the field hit an inflection point around 2017–2018 that coincided with the demonstration and/or experimental validation of both data-driven and expert-driven methods (Figure 11.11). There are many ways in which data science and machine learning will continue to influence research in the chemical sciences, and CASP is an area ripe for further growth.

Even if a chemist is not interested in paying for a commercial system, free-to-use platforms are highly competitive in the data-driven category. What open-source tools might lack in certain usability features (or customer support) that can improve the user experience, they might make up for in ease of customization and internalization into molecular design workflows. There are plenty of new methods for retrosynthesis emerging seemingly every few months at least, although many are not integrated into a larger multistep path planning program with an interface. These tools are challenging or impossible to compare in terms of raw "performance" due to the complexity of evaluating retrosynthetic pathways quantitatively, so we have instead focused on the relationship between their approaches and the few examples of validation that prove their collective utility.

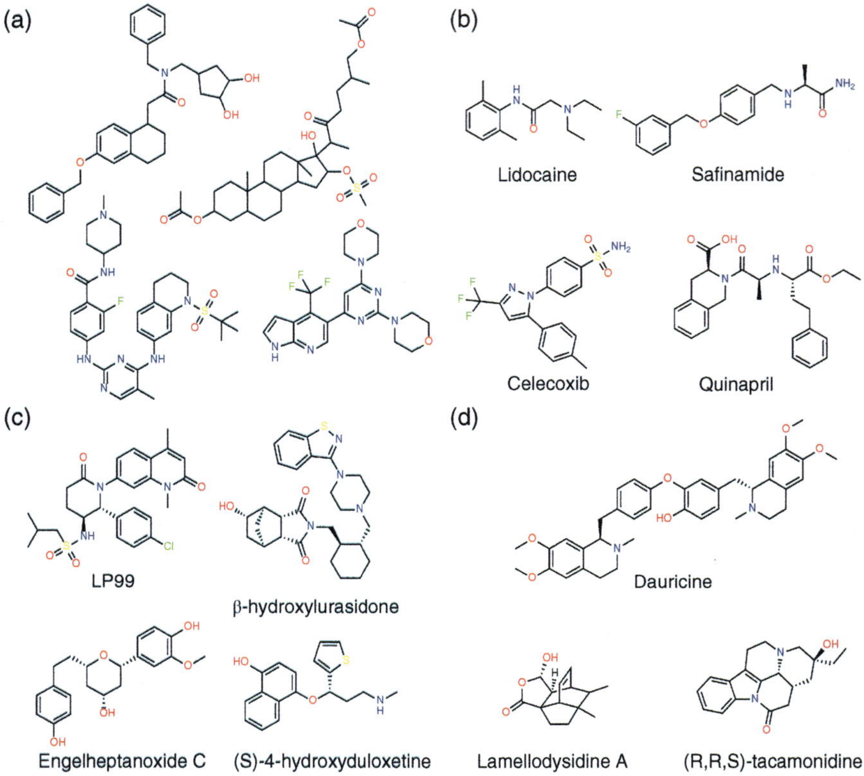

Figure 11.11 Examples of molecular targets from the case studies described in Section 11.5. (a) Targets from [7] whose predicted synthesis plan was rated higher than the corresponding literature synthesis route. *Source:* Segler et al. [7]/Springer Nature. (b) Targets from [66] whose synthesis plan was automatically generated with ASKCOS and executed successfully by a robotic synthesis platform. *Source:* Coley [66]/American Association for the Advancement of Science. (c) Drug molecules from [97] and (d) natural product molecules from [8] whose synthesis plans were automatically generated with Chematica and successfully executed by chemists. Sources: (c) Klucznik et al. [97]/Elsevier; (d) Mikulak-Klucznik et al. [8]/Springer Nature.

Admittedly, much is missing from the current generation of CASP programs. Predicting the stereoselectivity of reactions can be difficult, as can be predicting the full substrate tolerance of a synthetic transformation with little experimental precedent. Data-driven methods work best on compounds that resemble compounds they have been trained on, which are biased toward flat, sp^2-rich med chem-like compounds; programs trained on data from journal articles may have a greater diversity of reactions in their knowledge base than programs trained only on patent information, although medicinal chemistry use cases are well addressed by both. Complex natural products are better left to expert programs like Synthia, at least for now, as the models behind data-driven methods have yet to be fine-tuned for this use case.

Beyond retrosynthesis, there is also plenty of scope for future work in condition recommendation and reaction outcome prediction. These models are likewise already at a useful state but will need to transition to be more quantitative than qualitative in order to drive adoption. Condition recommendation in particular can provide good initial guesses for what species to use, but does not address the challenge of finding appropriate concentrations or equivalence ratios. Reaction outcome prediction does not inform yields (except for specific machine learning models trained to predict yields from narrow high-throughput experimentation campaigns), although it can predict relative selectivities in certain cases. Here, we have focused on models with the broadest possible domain of applicability (i.e. "any" organic transformation), but there are many examples of predictive models built on smaller, more focused datasets that represent a complementary part of the data chemist's toolbox.

Different types of models will prove useful in different settings, so we hope this chapter has left the reader with a sufficient understanding and interest in these techniques to at least give them a try.

Key References

Research articles cited in Section 11.5 Case Studies will provide the interested reader with a strong understanding of the most significant themes in the field.

References

1 Davies, I.W. (2019). The digitization of organic synthesis. *Nature* 570 (7760): 175.

2 Ott, M.A. and Noordik, J.H. (1992). Computer tools for reaction retrieval and synthesis planning in organic chemistry. A brief review of their history, methods, and programs. *Recl. Trav. Chissm. Pays-Bas* 111 (6): 239–246.

3 Todd, M.H. (2005). Computer-aided organic synthesis. *Chem. Soc. Rev.* 34 (3): 247–266.

4 Cook, A., Johnson, A.P., Law, J. et al. (2012). Computer-aided synthesis design: 40 years on. *Wiley Interdiscip. Rev. Comput. Mol. Sci.* 2 (1): 79–107.

5 Ravitz, O. (2013). Data-driven computer aided synthesis design. *Drug Discov. Today Technol.* 10 (3): e443–e449.

6 Warr, W.A. (2014). A short review of chemical reaction database systems, computer-aided synthesis design, reaction prediction and synthetic feasibility. *Mol. Inform.* 33 (6–7): 469–476.

7 Segler, M.H.S., Preuss, M., and Waller, M.P. (2018). Planning chemical syntheses with deep neural networks and symbolic AI. *Nature* 555 (7698): 604–610.

8 Mikulak-Klucznik, B., Gołębiowska, P., Bayly, A.A. et al. (2020). Computational planning of the synthesis of complex natural products. *Nature* 588: 83–88.

9 Engkvist, O., Norrby, P.-O., Selmi, N. et al. (2018). Computational prediction of chemical reactions: current status and outlook. *Drug Discov. Today* 23 (6): 1203–1218.

10 Johansson, S., Thakkar, A., Kogej, T. et al. (2020). AI-assisted synthesis prediction. *Drug Discov. Today Technol* 32–33: 65–72.

11 Gao, W., Raghavan, P., and Coley, C.W. (2022). Autonomous platforms for data-driven organic synthesis. *Nat. Commun.* 13 (1): 1075.

12 Schneider, G. (2018). Automating drug discovery. *Nat. Rev. Drug Discov.* 17 (2): 97–113.

13 Weininger, D. (1988). SMILES, a chemical language and information system. 1. Introduction to methodology and encoding rules. *J. Chem. Inf. Comput. Sci.* 28 (1): 31–36.

14 Rogers, D. and Hahn, M. (2010). Extended-connectivity fingerprints. *J. Chem. Inf. Model.* 50 (5): 742–754.

15 Gilmer, J., Schoenholz, S.S., Riley, P.F. et al. (2017). Neural message passing for Quantum chemistry. *Proc. 34th Int. Conf. Mach. Learn* 70: 1263–1272.

16 Veličković, P., Cucurull, G., Casanova, A. et al. (2018). Graph attention networks. *ArXiv171010903 Cs Stat* 1710.10903v3: 1–12.

17 Landrum, G. (2016). RDKit: Open-Source Cheminformatics Software. https://github.com/rdkit/rdkit/releases/tag/Release_2016_09_4

18 Coley, C.W., Green, W.H., and Jensen, K.F. (2019). RDChiral: an RDKit wrapper for handling stereochemistry in retrosynthetic template extraction and application. *J. Chem. Inf. Model.* 59 (6): 2529–2537.

19 Hornik, K., Stinchcombe, M., and White, H. (1989). Multilayer feedforward networks are universal approximators. *Neural Netw.* 2 (5): 359–366.

20 Goodfellow, I., Bengio, Y., and Courville, A. (2016). *Deep Learning*. MIT Press.

21 Janet, J.P. and Kulik, H.J. (2020). *Machine Learning in Chemistry*. American Chemical Society.

22 Coley, C.W., Green, W.H., and Jensen, K.F. (2018). Machine learning in computer-aided synthesis planning. *Acc. Chem. Res.* 51 (5): 1281–1289.

23 Schwaller, P., Vaucher, A.C., Laplaza, R. et al. Machine intelligence for chemical reaction space. *WIREs Comput. Mol. Sci.* **n/a** (n/a): e1604.

24 Strieth-Kalthoff, F., Sandfort, F., Segler, M.H.S., and Glorius, F. (2020). Machine learning the ropes: principles, applications and directions in synthetic chemistry. *Chem. Soc. Rev* 49 (17): 6154–6168.

25 Corey, E., Long, A., and Rubenstein, S. (1985). Computer-assisted analysis in organic synthesis. *Science* 228 (4698): 408–418.

26 Szymkuc, S., Gajewska, E.P., Klucznik, T. et al. (2016). Computer-assisted synthetic planning: the end of the beginning. *Angew Chem Int Ed* 55 (20): 5904–5937.

27 Genheden, S., Thakkar, A., Chadimová, V. et al. (2020). AiZynthFinder: a fast, robust and flexible open-source software for retrosynthetic planning. *J. Cheminformatics* 12 (1): 70.

28 Thakkar, A., Kogej, T., Reymond, J.-L. et al. (2020). Datasets and their influence on the development of computer assisted synthesis planning tools in the pharmaceutical domain. *Chem. Sci.* 11 (1): 154–168.

29 Roughley, S.D. and Jordan, A.M. (2011). The Medicinal Chemist's Toolbox: an analysis of reactions used in the pursuit of drug candidates. *J. Med. Chem.* 54 (10): 3451–3479.

30 Schneider, N., Lowe, D.M., Sayle, R.A. et al. (2016). Big data from pharmaceutical patents: a computational analysis of medicinal chemists' bread and butter. *J. Med. Chem.* 59 (9): 4385–4402.

31 Law, J., Zsoldos, Z., Simon, A. et al. (2009). Route designer: a retrosynthetic analysis tool utilizing automated retrosynthetic rule generation. *J. Chem. Inf. Model.* 49 (3): 593–602.

32 Heid, E., Liu, J., Aude, A., and Green, W.H. (2022). Influence of template size, canonicalization, and exclusivity for retrosynthesis and reaction prediction applications. *J. Chem. Inf. Model.* 62 (1): 16–26.

33 Segler, M.H.S. and Waller, M.P. (2017). Neural-symbolic machine learning for retrosynthesis and reaction prediction. *Chem. – Eur. J.* 23 (25): 5966–5971.

34 Fortunato, M.E., Coley, C.W., Barnes, B.C., and Jensen, K.F. (2020). Data augmentation and pretraining for template-based retrosynthetic prediction in computer-aided synthesis planning. *J. Chem. Inf. Model.* 60 (7): 3398–3407.

35 Seidl, P., Renz, P., Dyubankova, N. et al. (2022). Improving few- and zero-shot reaction template prediction using modern Hopfield networks. *J. Chem. Inf. Model* 62 (9): 2111–2120.

36 Baylon, J.L., Cilfone, N.A., Gulcher, J.R., and Chittenden, T.W. (2019). Enhancing retrosynthetic reaction prediction with deep learning using multiscale reaction classification. *J. Chem. Inf. Model.* 59 (2): 673–688.

37 Ishida, S., Terayama, K., Kojima, R. et al. (2019). Prediction and interpretable visualization of retrosynthetic reactions using graph convolutional networks. *J. Chem. Inf. Model.* 59 (12): 5026–5033.

38 Coley, C.W., Rogers, L., Green, W.H., and Jensen, K.F. (2017). Computer-assisted retrosynthesis based on molecular similarity. *ACS Cent. Sci.* 3 (12): 1237–1245.

39 Dai, H., Li, C., Coley, C. et al. (2019). Retrosynthesis prediction with conditional graph logic network. *Adv. Neural Inf. Process. Syst.* 32.

40 Chen, S. and Jung, Y. Deep retrosynthetic reaction prediction using local reactivity and global attention. *JACS Au* 1: 1612–1620.

41 Sun, R., Dai, H., Li, L. et al. (2021). Towards understanding retrosynthesis by energy-based models. *Adv. Neural Inform. Process. Syst.* 34.

42 Lin, M.H., Tu, Z., and Coley, C.W. (2022). Improving the performance of models for one-step retrosynthesis through re-ranking. *J. Cheminformatics* 14 (1): 15.

43 Sacha, M., Błaż, M., Byrski, P. et al. (2021). Molecule edit graph attention network: modeling chemical reactions as sequences of graph edits. *J. Chem. Inf. Model.* 61 (7): 3273–3284.

44 Shi, C., Xu, M., Guo, H. et al. (2020). A graph to graphs framework for retrosynthesis prediction. *Proc. 37th Int. Conf. Mach. Learn.* 119: 8818–8827.

45 Somnath, V.R., Bunne, C., Coley, C.W. et al. (2021). Learning graph models for retrosynthesis prediction. *arXiv* 2006.07038v2: 1–15.

46 Yan, C., Ding, Q., Zhao, P. et al. (2020). RetroXpert: decompose retrosynthesis prediction like a chemist. *Adv. Neural Inf. Process. Syst.* 33: 11248–11258.

47 Wang, X., Li, Y., Qiu, J. et al. (2021). RetroPrime: a diverse, plausible and transformer-based method for single-step retrosynthesis predictions. *Chem. Eng. J.* 420: 129845.

48 Liu, B., Ramsundar, B., Kawthekar, P. et al. (2017). Retrosynthetic reaction prediction using neural sequence-to-sequence models. *ACS Cent. Sci.* 3 (10): 1103–1113.

49 Schwaller, P., Gaudin, T., Lányi, D. et al. (2018). “Found in Translation”: predicting outcomes of complex organic chemistry reactions using neural sequence-to-sequence models. *Chem. Sci.* 9 (28): 6091–6098.

50 Lin, K., Xu, Y., Pei, J., and Lai, L. (2020). Automatic retrosynthetic route planning using template-free models. *Chem. Sci.* 11 (12): 3355–3364.

51 Lee, A.A., Yang, Q., Sresht, V. et al. (2019). Molecular transformer unifies reaction prediction and retrosynthesis across pharma chemical space. *Chem. Commun.* **55** (81): 12152–12155.

52 Duan, H., Wang, L., Zhang, C. et al. (2020). Retrosynthesis with attention-based NMT model and chemical analysis of “wrong” predictions. *RSC Adv.* 10 (3): 1371–1378.

53 Irwin, R., Dimitriadis, S., He, J., and Bjerrum, E.J. (2022). Chemformer: a pre-trained transformer for computational chemistry. *Mach. Learn. Sci. Technol.* 3 (1): 015022.

54 Zhu, J., Xia, Y., Qin, T. et al. (2021). Dual-view molecule pre-training. *ArXiv210610234 Cs* arXiv:2106.10234v2: 1–15.

55 Zheng, S., Rao, J., Zhang, Z. et al. (2020). Predicting retrosynthetic reactions using self-corrected transformer neural networks. *J. Chem. Inf. Model.* 60 (1): 47–55.

56 Yoo, S., Kim, Y.-S., Lee, K.H. et al. (2020). Graph-aware transformer: is attention all graphs need? *arXiv* arXiv:2006.05213v1: 1–12.

57 Seo, S.-W., Song, Y.Y., Yang, J.Y. et al. (2021). GTA: graph truncated attention for retrosynthesis. *Proc. AAAI Conf. Artif. Intell.* 35 (1): 531–539.

58 Mao, K., Xiao, X., Xu, T. et al. (2021). Molecular graph enhanced transformer for retrosynthesis prediction. *Neurocomputing* 457: 193–202.

59 Tu, Z. and Coley, C.W. (2022). Permutation invariant graph-to-sequence model for template-free retrosynthesis and reaction prediction. *J. Chem. Inform. Modeling* 62 (15): 3503–3513.

60 Chen, B., Shen, T., Jaakkola, T.S., and Barzilay, R. (2020). Learning to make generalizable and diverse predictions for retrosynthesis. *arXiv* arXiv:1910.09688v1: 1–11.

61 Kim, E., Lee, D., Kwon, Y. et al. (2021). Valid, plausible, and diverse retrosynthesis using tied two-way transformers with latent variables. *J. Chem. Inf. Model.* 61 (1): 123–133.

62 Tetko, I.V., Karpov, P., and Van Deursen, R. (2020). State-of-the-art augmented NLP transformer models for direct and single-step retro synthesis. *Nat. Commun.* 11 (5575): 1–11.

63 Coulom, R. (2007). Efficient selectivity and backup operators in Monte-Carlo tree search. *Comput. Games* 4630: 72–83.

64 Schwaller, P., Petraglia, R., Zullo, V. et al. (2020). Predicting retrosynthetic pathways using transformer-based models and a hyper-graph exploration strategy. *Chem. Sci.* 11 (12): 3316–3325.

65 Coley, C.W. (2019). ASKCOS Software Repository. https://github.com/connorcoley/ASKCOS (accessed 1 June 2019).

66 Coley, C.W., Thomas, D.A., Lummiss, J.A.M. et al. (2019). A robotic platform for flow synthesis of organic compounds informed by AI planning. *Science* 365 (6453): eaax1566.

67 Chen, B., Li, C., Dai, H., and Song, L. (2020). Retro*: learning retrosynthetic planning with neural guided A* search. *Proc. 37th Int. Conf. Mach. Learn.* 1608–1616.

68 Heifets, A. (2014). Automated synthetic feasibility assessment: a data-driven derivation of computational tools for medicinal chemistry. Thesis, University of Toronto.

69 Mo, Y., Guan, Y., Verma, P. et al. (2021). Evaluating and clustering retrosynthesis pathways with learned strategy. *Chem. Sci.* 12 (4): 1469–1478.

70 Genheden, S. and Bjerrum, E. (2022). PaRoutes: a framework for benchmarking retrosynthesis route predictions. *Digital Discov.* 1: 527–539.

71 Schneider, N., Stiefl, N., and Landrum, G.A. (2016). What́s what: the (nearly) definitive guide to reaction role assignment. *J. Chem. Inf. Model.* 56 (12): 2336–2346.

72 Gao, H., Struble, T.J., Coley, C.W. et al. (2018). Using machine learning to predict suitable conditions for organic reactions. *ACS Cent. Sci.* 4 (11): 1465–1476.

73 Maser, M.R., Cui, A.Y., Ryou, S. et al. (2021). Multilabel classification models for the prediction of cross-coupling reaction conditions. *J. Chem. Inf. Model.* 61 (1): 156–166.

74 Fitzner, M., Wuitschik, G., Koller, R. et al. (2020). What can reaction databases teach us about buchwald-hartwig cross-couplings? *Chem. Sci.* 11: 13085–13093.

75 Reizman, B.J. and Jensen, K.F. (2016). Feedback in flow for accelerated reaction development. *Acc. Chem. Res.* 49 (9): 1786–1796.

76 Shields, B.J., Stevens, J., Li, J. et al. (2021). Bayesian reaction optimization as a tool for chemical synthesis. *Nature* 590 (7844): 89–96.

77 Wei, J.N., Duvenaud, D., and Aspuru-Guzik, A. (2016). Neural networks for the prediction of organic chemistry reactions. *ACS Cent. Sci.* 2 (10): 725–732.

78 Coley, C.W., Barzilay, R., Jaakkola, T.S. et al. (2017). Prediction of organic reaction outcomes using machine learning. *ACS Cent. Sci.* 3 (5): 434–443.

79 Jin, W., Coley, C., Barzilay, R., and Jaakkola, T. (2017). Predicting organic reaction outcomes with Weisfeiler-Lehman network. *Adv. Neural Inf. Process. Syst.* 30: 1–10.

80 Coley, C.W., Jin, W., Rogers, L. et al. (2019). A graph-convolutional neural network model for the prediction of chemical reactivity. *Chem. Sci.* 10 (2): 370–377.

81 Do, K., Tran, T., and Venkatesh, S. (2019). Graph transformation policy network for chemical reaction prediction. *Proceedings of the 25th ACM SIGKDD International Conference on Knowledge Discovery and Data Mining*, pp. 750–760.

82 Qian, W.W., Russell, N.T., Simons, C.L.W. et al. (2020). Integrating deep neural networks and symbolic inference for organic reactivity prediction. *ChemRxiv* chemrxiv.11659563.v1: 1–24.

83 Bradshaw, J., Kusner, M.J., Paige, B., et al. (2019). A generative model for electron paths. *International Conference on Learning Representations.*

84 Bi, H., Wang, H., Shi, C. et al. (2021). Non-autoregressive electron redistribution modeling for reaction prediction. *Proc. 38th Int. Conf. Mach. Learn.* 139: 904–913.

85 Nam, J. and Kim, J. (2016). Linking the neural machine translation and the prediction of organic chemistry reactions. *ArXiv161209529 Cs* arXiv: 1612.09529v1: 1–19.

86 Schwaller, P., Laino, T., Gaudin, T. et al. (2019). Molecular transformer: a model for uncertainty-calibrated chemical reaction prediction. *ACS Cent. Sci.* 5 (9): 1572–1583.

87 ASKCOS. http://askcos.mit.edu/ (accessed 8 February 2019).

88 Coley, C.W., Rogers, L., Green, W.H., and Jensen, K.F. (2018). SCScore: Synthetic Complexity Learned from a Reaction Corpus. *J. Chem. Inf. Model.* 58 (2): 251–261. https://pubs.acs.org/doi/10.1021/acs.jcim.7b00622.

89 ASKCOS. tutorial. https://askcos.mit.edu/help/tutorial (accessed 31 March 2022).

90 Lowe, D. (2017). Chemical reactions from US patents (1976-Sep2016).

91 Sterling, T. and Irwin, J.J. (2015). ZINC 15 – ligand discovery for everyone. *J. Chem. Inf. Model.* 55 (11): 2324–2337.

92 Genheden, S., Thakkar, A., Chadimová, V. et al. (2020). AiZynthFinder: a fast, robust and flexible open-source software for retrosynthetic planning. *ChemRxiv* 12: 1–9.

93 Conda — Conda documentation. https://docs.conda.io/en/latest/ (accessed 31 March 2022).

94 Project Jupyter. https://jupyter.org (accessed 31 March 2022).

95 Chen, B. (2022). Retrosynthetic Planning with Retro*. https://github.com/binghong-ml/retro_star (accessed 31 March 2022).

96 eMolecules Free Downloads. https://downloads.emolecules.com/free/ (accessed 31 March 2022).

97 Klucznik, T., Mikulak-Klucznik, B., McCormack, M.P. et al. (2018). Efficient syntheses of diverse, medicinally relevant targets planned by computer and executed in the laboratory. *Chem* 4 (3): 522–532.

98 Probst, D., Manica, M., Nana Teukam, Y.G. et al. (2022). Biocatalysed synthesis planning using data-driven learning. *Nat. Commun.* 13 (1): 964.

99 Bøgevig, A., Federsel, H.-J., Huerta, F. et al. (2015). Route design in the 21st century: The IC *SYNTH* Software Tool as an idea generator for synthesis prediction. *Org. Process Res. Dev.* 19 (2): 357–368.

100 Liu, C.-H., Korablyov, M., Jastrzębski, S. et al. (2020). RetroGNN: approximating retrosynthesis by graph neural networks for de novo drug design. *arXiv* arXiv:2011.13042v1: 1–11.

101 Hadadi, N. and Hatzimanikatis, V. (2015). Design of computational retrobiosynthesis tools for the design of de novo synthetic pathways. *Curr. Opin. Chem. Biol.* 28: 99–104.

102 Koch, M., Duigou, T., and Faulon, J.-L. (2020). Reinforcement learning for bioretrosynthesis. *ACS Synth. Biol.* 9 (1): 157–168.

103 Finley, S.D., Broadbelt, L.J., and Hatzimanikatis, V. (2009). Computational framework for predictive biodegradation. *Biotechnol. Bioeng.* 104 (6): 1086–1097.

104 Finnigan, W., Hepworth, L.J., Flitsch, S.L., and Turner, N.J. (2021). RetroBioCat as a computer-aided synthesis planning tool for biocatalytic reactions and cascades. *Nat. Catal.* 4 (2): 98–104.

105 Levin, I., Liu, M., Voigt, C.A., and Coley, C.W. (2022). Merging enzymatic and synthetic chemistry with computational synthesis planning. *Nat. Commun.* 13: 7747. https://doi.org/10.1038/s41467-022-35422-y.

106 Lin, Y., Zhang, Z., Mahjour, B. et al. (2021). Reinforcing the supply chain of umifenovir and other antiviral drugs with retrosynthetic software. *Nat. Commun.* 12 (1): 7327.

Index

Enabling Tools and Techniques for Organic Synthesis: A Practical Guide to Experimentation, Automation, and Computation, First Edition. Edited by Stephen G. Newman.

c

d

m

n

o

p

Printed and bound by CPI Group (UK) Ltd, Croydon, CR0 4YY

17/09/2026

14964253-0006